Radio Technology Handbook

Radio Technology Handbook

Editor: Francis Schmidt

NY RESEARCH PRESS
New York

Published by NY Research Press
118-35 Queens Blvd., Suite 400,
Forest Hills, NY 11375, USA
www.nyresearchpress.com

Radio Technology Handbook
Edited by Francis Schmidt

International Standard Book Number: 978-1-63238-665-6 (Hardback)

Cataloging-in-Publication Data

Radio technology handbook / edited by Francis Schmidt.
p. cm.
Includes bibliographical references and index.
ISBN 978-1-63238-665-6
1. Radio. 2. Radio--Technological innovations. I. Schmidt, Francis.
TK6550 .R34 2019
621.384--dc23

Contents

Preface

Every book is a source of knowledge and this one is no exception. The idea that led to the conceptualization of this book was the fact that the world is advancing rapidly; which makes it crucial to document the progress in every field. I am aware that a lot of data is already available, yet, there is a lot more to learn. Hence, I accepted the responsibility of editing this book and contributing my knowledge to the community.

Radio technology refers to the technology that is concerned with the use of radio waves to carry information. Radio systems operate by using a combination of a transmitter, an antenna and a receiver. Such systems work within the frequency range between 3 kHz and 300 GHz. The transmitter modulates some aspect of the energy such as frequency, amplitude or phase to impress a signal on it. The purpose of the antenna is the conversion of electric currents into radio waves, and vice versa. The electromagnetic energy generated propagates in space, and is subject to reflection, refraction or diffraction. It is intercepted by a receiving antenna and then demodulated at the receiver. The ever-growing need of advanced technology is the reason that has fueled the research in this field in recent times. This book attempts to understand the multiple branches that fall under the discipline of radiotechnology and how such concepts have practical applications. This book is a resource guide for experts as well as students.

While editing this book, I had multiple visions for it. Then I finally narrowed down to make every chapter a sole standing text explaining a particular topic, so that they can be used independently. However, the umbrella subject sinews them into a common theme. This makes the book a unique platform of knowledge.

I would like to give the major credit of this book to the experts from every corner of the world, who took the time to share their expertise with us. Also, I owe the completion of this book to the never-ending support of my family, who supported me throughout the project.

Editor

1

Electromagnetic Scattering from a PEC Wedge Capped with Cylindrical Layers with Dielectric and Conductive Properties

Hulya OZTURK [1], *Korkut YEGIN* [2]

[1] Dept. of Mathematics, Gebze Technical University, Gebze, 41400 Kocaeli, Turkey
[2] Dept. of Electrical and Electronics Engineering, Ege University, Bornova, 35100 Izmir, Turkey

h.ozturk@gtu.edu.tr

Abstract. *Electromagnetic scattering from a layered capped wedge is studied. The wedge is assumed infinite in z-direction (longitudinal) and capped with arbitrary layers of dielectric with varying thicknesses and dielectric properties including conductive loss. Scalar Helmholtz equation in two dimensions is formulated for each solution region and a matrix of unknown coefficients are arrived at for electric field representation. Closed form expressions are derived for 2- and 3-layer geometries. Numerical simulations are performed for different wedge shapes and dielectric layer properties and compared to PEC-only case. It has been shown that significant reduction in scattered electric field can be obtained with 2- and 3-layered cap geometries. Total electric field in the far field normalized to incident field is also computed as a precursor to RCS analysis. Analytical results can be useful in radar cross section analysis for aerial vehicles.*

Keywords

PEC wedge, dielectric capped wedge, electromagnetic wedge scattering, radar cross section

1. Introduction

Electromagnetic scattering from perfectly electrical conductor (PEC) wedge plays an important role in many applications ranging from radar cross section computation to antenna analysis. Especially, in airborne applications many parts of an airplane body can be represented by a wedge for radar cross section analysis. Since wedge angle can be varied to represent many structures, scattering from wedge is considered as a major canonical problem and has been extensively studied in the past [1–4]. Although many forms of wedge from dielectric covered [5], [6] to impedance boundary treated surfaces [7–10] have been reported, dielectric capped wedge for arbitrary layers of dielectric has not been studied. In this work, we present derivations for N-layer lossy dielectric capped wedge and closed form derivations for 2-layered and 3-layered dielectric capped wedge.

Scattering from PEC wedge due to infinite line source aligned with the longitudinal axis of the wedge is well known and scattering due to transverse magnetic (TM) plane wave excitation can be easily derived [3]. Although less studied, transverse electric (TE) case can also be derived using $\rho-$directed infinite line source. When the wedge is capped with a circular dielectric layer, analytical expressions for this case can be derived after some manipulations [5]. However, when more than one layer is assumed, the derivations get intricate and the effects of multilayer cap on resulting scattered field become less predictable. Especially if conductive losses in dielectric layers are assumed as in practical applications of such configuration, the effect of conductive loss may play a critical role in near- and far-field analysis. Thus, the contributions of present study are rigorous mathematical derivations for the general N-layer capped wedge problem with each layer having different electrical properties and analysis of electromagnetic scattering both in the near- and far-field of the structure, where near-field may represent antenna to antenna coupling and far-field for RCS computation.

Although the geometry of the problem assumes infinite direction on the wedge axis, its solution provides insightful information on electrically large finite objects. Despite scattering from electrically large finite structures can be computed using fast multipole methods (FMM) [11], [12] analytical closed form solutions for two dimensional problems such as the one stated here are still invaluable for researchers as they provide quick and reliable means for the scattered field.

We first present the system of equations for N-layer dielectric capped wedge where the dielectric layers are represented by simple constitutive parameters. In Sec. 3, we derive analytical expressions for 2-layered and 3-layered capped wedge. In Sec. 4, numerical results for 2-layer and 3-layer capped wedge with various thicknesses and constitutive parameters are obtained and compared to standalone

PEC wedge case for two different wedge angles, which are chosen to resemble airplane wing geometries. Conclusions are given in Sec. 5.

2. Arbitrarily Layered Dielectric Capped Wedge

A perfect electric conductor wedge which is capped with multiple (N-layers) dielectric cylinders, is presented in Fig. 1. The electric line current of amplitude I_e is assumed at (ρ_0, φ_0). Throughout the analysis, $e^{j\omega t}$ harmonic time dependence is assumed and suppressed. TM incident electric field is represented by

$$E_z^i = -I_e \frac{\omega \mu_0}{4} \mathrm{H}_0^{(2)} (k_0 \mid \rho - \rho_0 \mid) \tag{1}$$

where $\mathrm{H}_0^{(2)}(.)$ is the Hankel function of the second kind of order zero and $k_0 = \omega\sqrt{\varepsilon_0 \mu_0}$. At wedge surfaces $\varphi = 0$ and $\varphi = 2\pi - \alpha$, $\mathbf{E}$ satisfies

$$\hat{\mathbf{n}} \times \mathbf{E} = 0 \tag{2}$$

where $\hat{\mathbf{n}}$ is the unit vector normal to the wedge surface. Taking into account the contribution on the φ dependence of (2), the z-components of the scattered electric field in their relevant regions can be determined as follows:

$$E_z^I(\rho, \varphi) = \sum_{n=1}^{\infty} a_{1n} \mathrm{J}_\nu (k_1 \rho) \sin \nu\varphi \sin \nu\varphi_0,$$

$$E_z^m(\rho, \varphi) = \sum_{n=1}^{\infty} \left[a_{mn} \mathrm{J}_\nu (k_m \rho) \sin \nu\varphi \sin \nu\varphi_0 + b_{mn} \mathrm{H}_\nu^{(2)} (k_m \rho) \sin \nu\varphi \sin \nu\varphi_0\right], \quad m = 2, \ldots, N$$

$$E_z^{(N+1)}(\rho, \varphi) = \sum_{n=1}^{\infty} \left[a_{(N+1)n} \mathrm{J}_\nu (k_0 \rho) \sin \nu\varphi \sin \nu\varphi_0 + b_{(N+1)n} \mathrm{H}_\nu^{(2)} (k_0 \rho) \sin \nu\varphi \sin \nu\varphi_0\right],$$

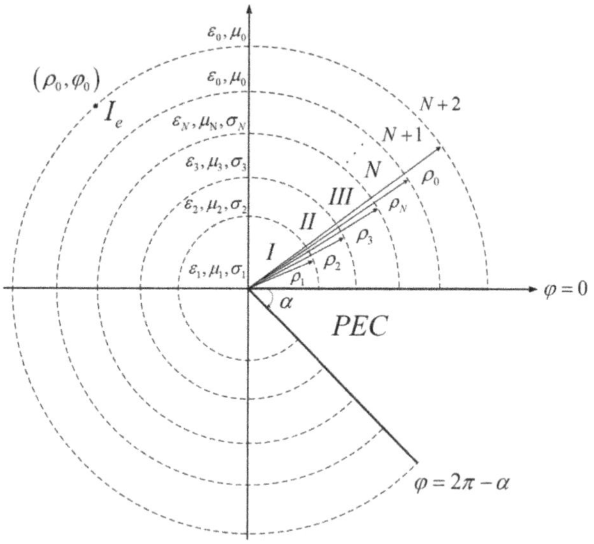

Fig. 1. The geometry of the problem.

$$E_z^{(N+2)}(\rho, \varphi) = \sum_{n=1}^{\infty} b_{(N+2)n} \mathrm{H}_\nu^{(2)} (k_0 \rho) \sin \nu\varphi \sin \nu\varphi_0 \tag{3}$$

where $\mathrm{J}_\nu(.)$ is the νth-order Bessel function of the first kind, while $\mathrm{H}_\nu^{(2)}(.)$ is the νth-order Hankel function of the second kind, $k_m^2 = \omega^2 \varepsilon_m \mu_m - j\omega\mu_m\sigma \; (m = 2, \ldots, N)$ and $\nu = \frac{n\pi}{2\pi-\alpha}$. From Maxwell's equations, the field components can be expressed in terms of $E_z(\rho, \varphi)$ as

$$H_\varphi(\rho, \varphi) = \frac{1}{j\omega\mu} \frac{\partial E_z(\rho, \varphi)}{\partial \rho} \tag{4}$$

and

$$H_\rho(\rho, \varphi) = -\frac{1}{j\omega\mu} \frac{1}{\rho} \frac{\partial E_z(\rho, \varphi)}{\partial \varphi}. \tag{5}$$

Thus, the magnetic field components inside dielectric cylinders can be written as

$$\begin{aligned}
H_\varphi^I(\rho, \varphi) &= \frac{k_1}{j\omega\mu_1} \sum_{n=1}^{\infty} a_{1n} \mathrm{J}'_\nu (k_1 \rho) \sin \nu\varphi \sin \nu\varphi_0, \\
H_\varphi^m(\rho, \varphi) &= \frac{k_m}{j\omega\mu_m} \sum_{n=1}^{\infty} \left[a_{mn} \mathrm{J}'_\nu (k_m \rho) \sin \nu\varphi \sin \nu\varphi_0 + b_{mn} \mathrm{H}_\nu^{\prime(2)} (k_m \rho) \sin \nu\varphi \sin \nu\varphi_0\right], \quad m = 2, \ldots, N \\
H_\varphi^{(N+1)}(\rho, \varphi) &= \frac{k_0}{j\omega\mu_0} \sum_{n=1}^{\infty} \left[a_{(N+1)n} \mathrm{J}'_\nu (k_0 \rho) \sin \nu\varphi \sin \nu\varphi_0 + b_{(N+1)n} \mathrm{H}_\nu^{\prime(2)} (k_0 \rho) \sin \nu\varphi \sin \nu\varphi_0\right], \\
H_\varphi^{(N+2)}(\rho, \varphi) &= \frac{k_0}{j\omega\mu_0} \sum_{n=1}^{\infty} b_{(N+2)n} \mathrm{H}_\nu^{\prime(2)} (k_0 \rho) \sin \nu\varphi \sin \nu\varphi_0
\end{aligned} \tag{6}$$

and

$$H_\rho^I(\rho, \varphi) = -\frac{1}{j\omega\mu_1 \rho} \sum_{n=1}^{\infty} a_{1n} \mathrm{J}_\nu (k_1 \rho) \nu \cos \nu\varphi \sin \nu\varphi_0,$$

$$H_\rho^m(\rho,\varphi) = -\frac{1}{j\omega\mu_m\rho}\sum_{n=1}^{\infty}\left[a_{mn}J_\nu(k_m\rho)\,\nu\cos\nu\varphi\sin\nu\varphi_0 + b_{mn}H_\nu^{(2)}(k_m\rho)\,\nu\cos\nu\varphi\sin\nu\varphi_0\right],\quad m=2,\ldots,N$$

$$H_\rho^{(N+1)}(\rho,\varphi) = -\frac{1}{j\omega\mu_0\rho}\sum_{n=1}^{\infty}\left[a_{(N+1)n}J_\nu(k_0\rho)\,\nu\cos\nu\varphi\sin\nu\varphi_0 + b_{(N+1)n}H_\nu^{(2)}(k_0\rho)\,\nu\cos\nu\varphi\sin\nu\varphi_0\right], \tag{7}$$

$$H_\rho^{(N+2)}(\rho,\varphi) = -\frac{1}{j\omega\mu_0\rho}\sum_{n=1}^{\infty} b_{(N+2)n}H_\nu^{(2)}(k_0\rho)\,\nu\cos\nu\varphi\sin\nu\varphi_0.$$

In (6), prime denotes the first-degree derivative with respect to ρ. The unknown coefficients a_{mn} and b_{mn} $(m = 1,\ldots,N+2)$ can be determined with the help of the following continuity conditions at $\rho = \rho_m$ $(m = 1,\ldots,N)$:

$$\begin{cases} E_z^m(\rho_m,\varphi) = E_z^{m+1}(\rho_m,\varphi) \\ H_\varphi^m(\rho_m,\varphi) = H_\varphi^{m+1}(\rho_m,\varphi) \end{cases},\quad m=1,\ldots,N \tag{8}$$

$$\begin{cases} E_z^{(N+1)}(\rho_0,\varphi) = E_z^{(N+2)}(\rho_0,\varphi) \\ H_\varphi^{(N+1)}(\rho_0,\varphi) = H_\varphi^{(N+2)}(\rho_0,\varphi) + J_e \end{cases}, \tag{9}$$

where J_e denotes the current density. J_e can be expressed as

$$J_e = \frac{I_e}{\rho_0}\frac{2}{2\pi-\alpha}\sum_{n=1}^{\infty}\sin\nu\varphi\sin\nu\varphi_0, \tag{10}$$

Application of (8) and (9) at $\rho = \rho_m$ $(m = 1,\ldots,N)$ yields:

$$a_{1n}J_\nu(k_1\rho_1) - a_{2n}J_\nu(k_2\rho_1) - b_{2n}H_\nu^{(2)}(k_2\rho_1) = 0, \tag{11}$$

$$\frac{k_1}{\mu_1}a_{1n}J'_\nu(k_1\rho_1) - \frac{k_2}{\mu_2}\left\{a_{2n}J'_\nu(k_2\rho_1) + b_{2n}H_\nu'^{(2)}(k_2\rho_1)\right\} = 0, \tag{12}$$

$$a_{mn}J_\nu(k_m\rho_m) - a_{(m+1)n}J_\nu(k_{(m+1)}\rho_m) + b_{mn}H_\nu^{(2)}(k_m\rho_m) - b_{(m+1)n}H_\nu^{(2)}(k_{(m+1)}\rho_m) = 0,\quad m=2,\ldots,N-1 \tag{13}$$

$$\frac{k_m}{\mu_m}\left\{a_{mn}J'_\nu(k_m\rho_m) + b_{mn}H_\nu'^{(2)}(k_m\rho_m)\right\} - \frac{k_{(m+1)}}{\mu_{(m+1)}}\left\{a_{(m+1)n}J'_\nu(k_{(m+1)}\rho_m) + b_{(m+1)n}H_\nu'^{(2)}(k_{(m+1)}\rho_m)\right\} = 0,\quad m=2,\ldots,N-1 \tag{14}$$

$$a_{Nn}J_\nu(k_N\rho_N) - a_{(N+1)n}J_\nu(k_0\rho_N) + b_{Nn}H_\nu^{(2)}(k_N\rho_N) - b_{(N+1)n}H_\nu^{(2)}(k_0\rho_N) = 0, \tag{15}$$

$$\frac{k_N}{\mu_N}\left\{a_{Nn}J'_\nu(k_N\rho_N) + b_{Nn}H_\nu'^{(2)}(k_N\rho_N)\right\} - \frac{k_0}{\mu_0}\left\{a_{(N+1)n}J'_\nu(k_0\rho_N) + b_{(N+1)n}H_\nu'^{(2)}(k_0\rho_N)\right\} = 0, \tag{16}$$

$$a_{(N+1)n}J_\nu(k_0\rho_0) + b_{(N+1)n}H_\nu^{(2)}(k_0\rho_0) - b_{(N+2)n}H_\nu^{(2)}(k_0\rho_0) = 0, \tag{17}$$

$$a_{(N+1)n}J'_\nu(k_0\rho_0) + b_{(N+1)n}H_\nu'^{(2)}(k_0\rho_0) - b_{(N+2)n}H_\nu'^{(2)}(k_0\rho_0) = -\frac{I_e}{\rho_0}\frac{2}{2\pi-\alpha}\frac{j\omega\mu_0}{k_0}. \tag{18}$$

These (2N+2) x(2N+2) system of algebraic equations can be solved numerically and the expansion coefficients a_{mn} and b_{mn} $(m = 1,\ldots,N+2)$ can be fully determined.

3. 2-Layer and 3-Layer Capped Wedge

Explicit forms of the unknown coefficients for $N = 2$ and $N = 3$ can be derived as follows. For $N = 2$ case,

$$a_{1n} = \frac{k_2\left\{H_\nu^{(2)}(k_0\rho_2)J'_\nu(k_0\rho_2) - H_\nu'^{(2)}(k_0\rho_2)J_\nu(k_0\rho_2)\right\}}{\mu_2 D(k_0,k_1,k_2,\rho_0,\rho_1,\rho_2)}\left\{H_\nu^{(2)}(k_2\rho_1)J'_\nu(k_2\rho_1) - H_\nu'^{(2)}(k_2\rho_1)J_\nu(k_2\rho_1)\right\}, \tag{19}$$

$$a_{2n} = \frac{\left\{\frac{k_1}{\mu_1}H_\nu^{(2)}(k_2\rho_1)J'_\nu(k_1\rho_1) - \frac{k_2}{\mu_2}H_\nu'^{(2)}(k_2\rho_1)J_\nu(k_1\rho_1)\right\}}{D(k_0,k_1,k_2,\rho_0,\rho_1,\rho_2)}\left\{H_\nu^{(2)}(k_0\rho_2)J'_\nu(k_0\rho_2) - H_\nu'^{(2)}(k_0\rho_2)J_\nu(k_0\rho_2)\right\}, \tag{20}$$

$$a_{3n} = \frac{\mu_0}{k_0 D(k_0,k_1,k_2,\rho_0,\rho_1,\rho_2)}\left[\left\{\frac{k_2}{\mu_2}H_\nu'^{(2)}(k_2\rho_1)J_\nu(k_1\rho_1) - \frac{k_1}{\mu_1}H_\nu^{(2)}(k_2\rho_1)J'_\nu(k_1\rho_1)\right\}\left\{\frac{k_0}{\mu_0}H_\nu'^{(2)}(k_0\rho_2)J_\nu(k_2\rho_2) - \frac{k_2}{\mu_2}H_\nu^{(2)}(k_0\rho_2)J'_\nu(k_2\rho_2)\right\} + \left\{\frac{k_2}{\mu_2}H_\nu^{(2)}(k_0\rho_2)H_\nu'^{(2)}(k_2\rho_2) - \frac{k_0}{\mu_0}H_\nu'^{(2)}(k_0\rho_2)H_\nu^{(2)}(k_2\rho_2)\right\}\left\{\frac{k_2}{\mu_2}J'_\nu(k_2\rho_1)J_\nu(k_1\rho_1) - \frac{k_1}{\mu_1}J'_\nu(k_1\rho_1)J_\nu(k_2\rho_1)\right\}\right], \tag{21}$$

$$b_{2n} = \frac{\left\{ \mathrm{H}_{\nu}^{(2)}(k_0\rho_2)\,\mathrm{J}_{\nu}'(k_0\rho_2) - \mathrm{H}_{\nu}'^{(2)}(k_0\rho_2)\,\mathrm{J}_{\nu}(k_0\rho_2)\right\}}{D(k_0,k_1,k_2,\rho_0,\rho_1,\rho_2)} \left\{ \frac{k_2}{\mu_2}\mathrm{J}_{\nu}'(k_2\rho_1)\,\mathrm{J}_{\nu}(k_1\rho_1) - \frac{k_1}{\mu_1}\mathrm{J}_{\nu}'(k_1\rho_1)\,\mathrm{J}_{\nu}(k_2\rho_1)\right\}, \tag{22}$$

$$\begin{aligned} b_{3n} = {} & \frac{\mu_0}{k_0}\frac{1}{D(k_0,k_1,k_2,\rho_0,\rho_1,\rho_2)} \\ & \left[\left\{ \frac{k_2}{\mu_2}\mathrm{H}_{\nu}'^{(2)}(k_2\rho_1)\,\mathrm{J}_{\nu}(k_1\rho_1) - \frac{k_1}{\mu_1}\mathrm{H}_{\nu}^{(2)}(k_2\rho_1)\,\mathrm{J}_{\nu}'(k_1\rho_1)\right\}\left\{ \frac{k_2}{\mu_2}\mathrm{J}_{\nu}(k_0\rho_2)\,\mathrm{J}_{\nu}'(k_2\rho_2) - \frac{k_0}{\mu_0}\mathrm{J}_{\nu}(k_2\rho_2)\,\mathrm{J}_{\nu}'(k_0\rho_2)\right\} + \right. \\ & \left.\left\{ \frac{k_2}{\mu_2}\mathrm{J}_{\nu}(k_1\rho_1)\,\mathrm{J}_{\nu}'(k_2\rho_1) - \frac{k_1}{\mu_1}\mathrm{J}_{\nu}'(k_1\rho_1)\,\mathrm{J}_{\nu}(k_2\rho_1)\right\}\left\{ \frac{k_0}{\mu_0}\mathrm{H}_{\nu}^{(2)}(k_2\rho_2)\,\mathrm{J}_{\nu}'(k_0\rho_2) - \frac{k_2}{\mu_2}\mathrm{H}_{\nu}'^{(2)}(k_2\rho_2)\,\mathrm{J}_{\nu}(k_0\rho_2)\right\}\right], \end{aligned} \tag{23}$$

$$\begin{aligned} b_{4n} = {} & -b_{3n} + \frac{\mu_0}{k_0\mathrm{H}_{\nu}^{(2)}(k_0\rho_0)}\frac{\mathrm{J}_{\nu}(k_0\rho_0)}{D(k_0,k_1,k_2,\rho_0,\rho_1,\rho_2)} \\ & \left[\left\{ \frac{k_2}{\mu_2}\mathrm{H}_{\nu}'^{(2)}(k_2\rho_1)\,\mathrm{J}_{\nu}(k_1\rho_1) - \frac{k_1}{\mu_1}\mathrm{H}_{\nu}^{(2)}(k_2\rho_1)\,\mathrm{J}_{\nu}'(k_1\rho_1)\right\}\left\{ \frac{k_0}{\mu_0}\mathrm{J}_{\nu}(k_2\rho_2)\,\mathrm{H}_{\nu}'^{(2)}(k_0\rho_2) - \frac{k_2}{\mu_2}\mathrm{J}_{\nu}'(k_2\rho_2)\,\mathrm{H}_{\nu}^{(2)}(k_0\rho_2)\right\} + \right. \\ & \left.\left\{ \frac{k_2}{\mu_2}\mathrm{J}_{\nu}'(k_2\rho_1)\,\mathrm{J}_{\nu}(k_1\rho_1) - \frac{k_1}{\mu_1}\mathrm{J}_{\nu}'(k_1\rho_1)\,\mathrm{J}_{\nu}(k_2\rho_1)\right\}\left\{ \frac{k_2}{\mu_2}\mathrm{H}_{\nu}^{(2)}(k_0\rho_2)\,\mathrm{H}_{\nu}'^{(2)}(k_2\rho_2) - \frac{k_0}{\mu_0}\mathrm{H}_{\nu}'^{(2)}(k_0\rho_2)\,\mathrm{H}_{\nu}^{(2)}(k_2\rho_2)\right\}\right] \end{aligned} \tag{24}$$

with

$$\begin{aligned} D(k_0,k_1,k_2,\rho_0,\rho_1,\rho_2) = {} & \frac{\rho_0(2\pi-\alpha)}{2I_e\mathrm{j}\omega\mathrm{H}_{\nu}^{(2)}(k_0\rho_0)}\left\{ \mathrm{H}_{\nu}^{(2)}(k_0\rho_0)\,\mathrm{J}_{\nu}'(k_0\rho_0) - \mathrm{H}_{\nu}'^{(2)}(k_0\rho_0)\,\mathrm{J}_{\nu}(k_0\rho_0)\right\} \\ & \left[\left\{ \frac{k_2}{\mu_2}\mathrm{J}_{\nu}'(k_2\rho_1)\,\mathrm{J}_{\nu}(k_1\rho_1) - \frac{k_1}{\mu_1}\mathrm{J}_{\nu}'(k_1\rho_1)\,\mathrm{J}_{\nu}(k_2\rho_1)\right\}\left\{ \frac{k_0}{\mu_0}\mathrm{H}_{\nu}'^{(2)}(k_0\rho_2)\,\mathrm{H}_{\nu}^{(2)}(k_2\rho_2) - \frac{k_2}{\mu_2}\mathrm{H}_{\nu}'^{(2)}(k_2\rho_2)\,\mathrm{H}_{\nu}^{(2)}(k_0\rho_2)\right\} - \right. \\ & \left.\left\{ \frac{k_2}{\mu_2}\mathrm{H}_{\nu}'^{(2)}(k_2\rho_1)\,\mathrm{J}_{\nu}(k_1\rho_1) - \frac{k_1}{\mu_1}\mathrm{J}_{\nu}'(k_1\rho_1)\,\mathrm{H}_{\nu}^{(2)}(k_2\rho_1)\right\}\left\{ \frac{k_0}{\mu_0}\mathrm{H}_{\nu}'^{(2)}(k_0\rho_2)\,\mathrm{J}_{\nu}(k_2\rho_2) - \frac{k_2}{\mu_2}\mathrm{H}_{\nu}^{(2)}(k_0\rho_2)\,\mathrm{J}_{\nu}'(k_2\rho_2)\right\}\right]. \end{aligned} \tag{25}$$

The plane wave excitation ($\rho_0 \to \infty$) can be computed using the asymptotic expansion of $\mathrm{H}_{\nu}^{(2)}(k_0\rho_0)$ for large argument as

$$\mathrm{H}_{\nu}^{(2)}(k_0\rho_0) \simeq \sqrt{\frac{2}{\pi k_0\rho_0}}\mathrm{e}^{-\mathrm{j}\left(k_0\rho_0 - \frac{\nu\pi}{2} - \frac{\pi}{4}\right)}. \tag{26}$$

In this case, the unknown coefficients become

$$a_{1n} = -\frac{k_0k_2\left\{ \mathrm{H}_{\nu}^{(2)}(k_0\rho_2)\,\mathrm{J}_{\nu}'(k_0\rho_2) - \mathrm{H}_{\nu}'^{(2)}(k_0\rho_2)\,\mathrm{J}_{\nu}(k_0\rho_2)\right\}}{\mu_0\mu_2\tilde{D}(k_0,k_1,k_2,\rho_1,\rho_2)}\left\{ \mathrm{H}_{\nu}^{(2)}(k_2\rho_1)\,\mathrm{J}_{\nu}'(k_2\rho_1) - \mathrm{H}_{\nu}'^{(2)}(k_2\rho_1)\,\mathrm{J}_{\nu}(k_2\rho_1)\right\}, \tag{27}$$

$$a_{2n} = \frac{k_0\left\{ \mathrm{H}_{\nu}^{(2)}(k_0\rho_2)\,\mathrm{J}_{\nu}'(k_0\rho_2) - \mathrm{H}_{\nu}'^{(2)}(k_0\rho_2)\,\mathrm{J}_{\nu}(k_0\rho_2)\right\}}{\mu_0\tilde{D}(k_0,k_1,k_2,\rho_1,\rho_2)}\left\{ \frac{k_1}{\mu_1}\mathrm{H}_{\nu}^{(2)}(k_2\rho_1)\,\mathrm{J}_{\nu}'(k_1\rho_1) - \frac{k_2}{\mu_2}\mathrm{H}_{\nu}'^{(2)}(k_2\rho_1)\,\mathrm{J}_{\nu}(k_1\rho_1)\right\}, \tag{28}$$

$$\begin{aligned} a_{3n} = {} & \frac{1}{\tilde{D}(k_0,k_1,k_2,\rho_1,\rho_2)} \\ & \left[\left\{ \frac{k_2}{\mu_2}\mathrm{H}_{\nu}'^{(2)}(k_2\rho_1)\,\mathrm{J}_{\nu}(k_1\rho_1) - \frac{k_1}{\mu_1}\mathrm{H}_{\nu}^{(2)}(k_2\rho_1)\,\mathrm{J}_{\nu}'(k_1\rho_1)\right\}\left\{ \frac{k_0}{\mu_0}\mathrm{H}_{\nu}'^{(2)}(k_0\rho_2)\,\mathrm{J}_{\nu}(k_2\rho_2) - \frac{k_2}{\mu_2}\mathrm{H}_{\nu}^{(2)}(k_0\rho_2)\,\mathrm{J}_{\nu}'(k_2\rho_2)\right\} + \right. \\ & \left.\left\{ \frac{k_2}{\mu_2}\mathrm{H}_{\nu}^{(2)}(k_0\rho_2)\,\mathrm{H}_{\nu}'^{(2)}(k_2\rho_2) - \frac{k_0}{\mu_0}\mathrm{H}_{\nu}'^{(2)}(k_0\rho_2)\,\mathrm{H}_{\nu}^{(2)}(k_2\rho_2)\right\}\left\{ \frac{k_2}{\mu_2}\mathrm{J}_{\nu}'(k_2\rho_1)\,\mathrm{J}_{\nu}(k_1\rho_1) - \frac{k_1}{\mu_1}\mathrm{J}_{\nu}'(k_1\rho_1)\,\mathrm{J}_{\nu}(k_2\rho_1)\right\}\right], \end{aligned} \tag{29}$$

$$b_{2n} = \frac{j\omega \left\{ H_v^{(2)}(k_0\rho_2) J_v'(k_0\rho_2) - H_v'^{(2)}(k_0\rho_2) J_v(k_0\rho_2) \right\}}{\tilde{D}(k_0, k_1, k_2, \rho_1, \rho_2)} \left\{ \frac{k_2}{\mu_2} J_v'(k_2\rho_1) J_v(k_1\rho_1) - \frac{k_1}{\mu_1} J_v'(k_1\rho_1) J_v(k_2\rho_1) \right\}, \quad (30)$$

$$\begin{aligned} b_{3n} = {} & \frac{1}{\tilde{D}(k_0, k_1, k_2, \rho_1, \rho_2)} \\ & \left[\left\{ \frac{k_2}{\mu_2} H_v'^{(2)}(k_2\rho_1) J_v(k_1\rho_1) - \frac{k_1}{\mu_1} H_v^{(2)}(k_2\rho_1) J_v'(k_1\rho_1) \right\} \left\{ \frac{k_2}{\mu_2} J_v(k_0\rho_2) J_v'(k_2\rho_2) - \frac{k_0}{\mu_0} J_v(k_2\rho_2) J_v'(k_0\rho_2) \right\} + \right. \\ & \left. \left\{ \frac{k_2}{\mu_2} J_v(k_1\rho_1) J_v'(k_2\rho_1) - \frac{k_1}{\mu_1} J_v'(k_1\rho_1) J_v(k_2\rho_1) \right\} \left\{ \frac{k_0}{\mu_0} H_v^{(2)}(k_2\rho_2) J_v'(k_0\rho_2) - \frac{k_2}{\mu_2} H_v'^{(2)}(k_2\rho_2) J_v(k_0\rho_2) \right\} \right] \quad (31) \end{aligned}$$

where

$$\begin{aligned} \tilde{D}(k_0, k_1, k_2, \rho_1, \rho_2) = {} & -\frac{2\pi - \alpha}{4j^v \pi} \\ & \left\{ \frac{k_2}{\mu_2} J_v'(k_2\rho_1) J_v(k_1\rho_1) - \frac{k_1}{\mu_1} J_v'(k_1\rho_1) J_v(k_2\rho_1) \right\} \left\{ \frac{k_0}{\mu_0} H_v'^{(2)}(k_0\rho_2) H_v^{(2)}(k_2\rho_2) - \frac{k_2}{\mu_2} H_v'^{(2)}(k_2\rho_2) H_v^{(2)}(k_0\rho_2) \right\} + \\ & \left\{ \frac{k_2}{\mu_2} H_v'^{(2)}(k_2\rho_1) J_v(k_1\rho_1) - \frac{k_1}{\mu_1} J_v'(k_1\rho_1) H_v^{(2)}(k_2\rho_1) \right\} \left\{ \frac{k_2}{\mu_2} H_v^{(2)}(k_0\rho_2) J_v'(k_2\rho_2) - \frac{k_0}{\mu_0} H_v'^{(2)}(k_0\rho_2) J_v(k_2\rho_2) \right\}. \quad (32) \end{aligned}$$

In (27)–(32), all the coefficients are normalized with incident plane wave which is given by

$$E_0 = -I_e \frac{\omega\mu_0}{4} \sqrt{\frac{2j}{\pi k_0 \rho_0}} e^{-jk_0\rho_0}. \quad (33)$$

For $N = 3$ case

$$\begin{aligned} a_{1n} = {} & \left[\frac{k_2 k_3 \left\{ H_v^{(2)}(k_0\rho_3) J_v'(k_0\rho_3) - H_v'^{(2)}(k_0\rho_3) J_v(k_0\rho_3) \right\}}{\mu_2\mu_3 R(k_0, k_1, k_2, k_3, \rho_0, \rho_1, \rho_2, \rho_3)} \right] \\ & \left[\left\{ H_v^{(2)}(k_3\rho_2) J_v'(k_3\rho_2) - H_v'^{(2)}(k_3\rho_2) J_v(k_3\rho_2) \right\} \left\{ H_v'^{(2)}(k_2\rho_1) J_v(k_2\rho_1) - H_v^{(2)}(k_2\rho_1) J_v'(k_2\rho_1) \right\} \right], \quad (34) \end{aligned}$$

$$\begin{aligned} a_{2n} = {} & \left[\frac{k_3 \left\{ \frac{k_1}{\mu_1} H_v^{(2)}(k_2\rho_1) J_v'(k_1\rho_1) - \frac{k_2}{\mu_2} H_v'^{(2)}(k_2\rho_1) J_v(k_1\rho_1) \right\}}{\mu_3 R(k_0, k_1, k_2, k_3, \rho_0, \rho_1, \rho_2, \rho_3)} \right] \\ & \left[\left\{ H_v^{(2)}(k_3\rho_2) J_v'(k_3\rho_2) - H_v'^{(2)}(k_3\rho_2) J_v(k_3\rho_2) \right\} \left\{ H_v'^{(2)}(k_0\rho_3) J_v(k_0\rho_3) - H_v^{(2)}(k_0\rho_3) J_v'(k_0\rho_3) \right\} \right], \quad (35) \end{aligned}$$

$$a_{3n} = A(k_1, k_2, k_3, \rho_1, \rho_2, \rho_3) \frac{\left\{ H_v^{(2)}(k_0\rho_3) J_v'(k_0\rho_3) - H_v'^{(2)}(k_0\rho_3) J_v(k_0\rho_3) \right\}}{R(k_0, k_1, k_2, k_3, \rho_0, \rho_1, \rho_2, \rho_3)}, \quad (36)$$

$$\begin{aligned} a_{4n} = {} & \frac{\mu_0}{k_0} \frac{1}{R(k_0, k_1, k_2, k_3, \rho_0, \rho_1, \rho_2, \rho_3)} \\ & \left[\frac{k_3}{\mu_3} H_v^{(2)}(k_0\rho_3) \left\{ J_v'(k_3\rho_3) A(k_1, k_2, k_3, \rho_1, \rho_2, \rho_3) + H_v'^{(2)}(k_3\rho_3) B(k_1, k_2, k_3, \rho_1, \rho_2, \rho_3) \right\} - \right. \\ & \left. \frac{k_0}{\mu_0} H_v'^{(2)}(k_0\rho_3) \left\{ J_v(k_3\rho_3) A(k_1, k_2, k_3, \rho_1, \rho_2, \rho_3) + H_v^{(2)}(k_3\rho_3) B(k_1, k_2, k_3, \rho_1, \rho_2, \rho_3) \right\} \right], \quad (37) \end{aligned}$$

$$b_{2n} = \frac{k_3 \left\{ \mathrm{H}_v^{(2)}(k_0\rho_3)\,\mathrm{J}_v'(k_0\rho_3) - \mathrm{H}_v'^{(2)}(k_0\rho_3)\,\mathrm{J}_v(k_0\rho_3) \right\}}{\mu_3 R(k_0, k_1, k_2, k_3, \rho_0, \rho_1, \rho_2, \rho_3)}$$
$$\left[\left\{ \mathrm{H}_v^{(2)}(k_3\rho_2)\,\mathrm{J}_v'(k_3\rho_2) - \mathrm{H}_v'^{(2)}(k_3\rho_2)\,\mathrm{J}_v(k_3\rho_2) \right\} \left\{ \frac{k_2}{\mu_2}\mathrm{J}_v'(k_2\rho_1)\,\mathrm{J}_v(k_1\rho_1) - \frac{k_1}{\mu_1}\mathrm{J}_v'(k_1\rho_1)\,\mathrm{J}_v(k_2\rho_1) \right\} \right], \quad (38)$$

$$b_{3n} = B(k_1, k_2, k_3, \rho_1, \rho_2, \rho_3)\,\frac{\left\{ \mathrm{H}_v^{(2)}(k_0\rho_3)\,\mathrm{J}_v'(k_0\rho_3) - \mathrm{H}_v'^{(2)}(k_0\rho_3)\,\mathrm{J}_v(k_0\rho_3) \right\}}{R(k_0, k_1, k_2, k_3, \rho_0, \rho_1, \rho_2, \rho_3)}, \quad (39)$$

$$b_{4n} = \frac{\mu_0}{k_0}\frac{1}{R(k_0, k_1, k_2, k_3, \rho_0, \rho_1, \rho_2, \rho_3)}$$
$$\left[\frac{k_3}{\mu_3}\mathrm{J}_v(k_0\rho_3) \left\{ \mathrm{J}_v'(k_3\rho_3)\,A(k_1, k_2, k_3, \rho_1, \rho_2, \rho_3) + \mathrm{H}_v'^{(2)}(k_3\rho_3)\,B(k_1, k_2, k_3, \rho_1, \rho_2, \rho_3) \right\} - \right.$$
$$\left. \frac{k_0}{\mu_0}\mathrm{J}_v'(k_0\rho_3) \left\{ \mathrm{J}_v(k_3\rho_3)\,A(k_1, k_2, k_3, \rho_1, \rho_2, \rho_3) + \mathrm{H}_v^{(2)}(k_3\rho_3)\,B(k_1, k_2, k_3, \rho_1, \rho_2, \rho_3) \right\} \right], \quad (40)$$

$$b_{5n} = -b_{4n} + \frac{\mu_0 \mathrm{J}_v(k_0\rho_0)}{k_0 \mathrm{H}_v^{(2)}(k_0\rho_0)\,R(k_0, k_1, k_2, k_3, \rho_0, \rho_1, \rho_2, \rho_3)}$$
$$\left[\frac{k_3}{\mu_3}\mathrm{H}_v^{(2)}(k_0\rho_3) \left\{ \mathrm{J}_v'(k_3\rho_3)\,A(k_1, k_2, k_3, \rho_1, \rho_2, \rho_3) + \mathrm{H}_v'^{(2)}(k_3\rho_3)\,B(k_1, k_2, k_3, \rho_1, \rho_2, \rho_3) \right\} - \right.$$
$$\left. \frac{k_0}{\mu_0}\mathrm{H}_v'^{(2)}(k_0\rho_3) \left\{ \mathrm{J}_v(k_3\rho_3)\,A(k_1, k_2, k_3, \rho_1, \rho_2, \rho_3) + \mathrm{H}_v^{(2)}(k_3\rho_3)\,B(k_1, k_2, k_3, \rho_1, \rho_2, \rho_3) \right\} \right] \quad (41)$$

where

$$R(k_0, k_1, k_2, k_3, \rho_0, \rho_1, \rho_2, \rho_3) =$$
$$\frac{\rho_0(2\pi - \alpha)}{2I_e \mathrm{j}\omega \mathrm{H}_v^{(2)}(k_0\rho_0)} \left[\mathrm{H}_v'^{(2)}(k_0\rho_0)\,\mathrm{J}_v(k_0\rho_0) - \mathrm{H}_v^{(2)}(k_0\rho_0)\,\mathrm{J}_v'(k_0\rho_0) \right]$$
$$\left[\left[\frac{k_0 k_3}{\mu_0 \mu_3}\mathrm{H}_v'^{(2)}(k_0\rho_3) \left\{ \mathrm{H}_v^{(2)}(k_3\rho_3)\,\mathrm{J}_v'(k_3\rho_2) - \mathrm{H}_v'^{(2)}(k_3\rho_2)\,\mathrm{J}_v(k_3\rho_3) \right\} + \right. \right.$$
$$\left. \frac{k_3^2}{\mu_3^2}\mathrm{H}_v^{(2)}(k_0\rho_3) \left\{ \mathrm{H}_v'^{(2)}(k_3\rho_2)\,\mathrm{J}_v'(k_3\rho_3) - \mathrm{H}_v'^{(2)}(k_3\rho_3)\,\mathrm{J}_v'(k_3\rho_2) \right\} \right]$$
$$\left[\frac{k_2}{\mu_2}\mathrm{J}_v(k_1\rho_1) \left\{ \mathrm{H}_v^{(2)}(k_2\rho_2)\,\mathrm{J}_v'(k_2\rho_1) - \mathrm{H}_v'^{(2)}(k_2\rho_1)\,\mathrm{J}_v(k_3\rho_2) \right\} + \right.$$
$$\left. \frac{k_1}{\mu_1}\mathrm{J}_v'(k_1\rho_1) \left\{ \mathrm{H}_v^{(2)}(k_2\rho_1)\,\mathrm{J}_v(k_2\rho_2) - \mathrm{H}_v^{(2)}(k_2\rho_2)\,\mathrm{J}_v(k_2\rho_1) \right\} \right] +$$
$$\left[\frac{k_3}{\mu_3}\mathrm{H}_v^{(2)}(k_0\rho_3) \left\{ \mathrm{H}_v^{(2)}(k_3\rho_2)\,\mathrm{J}_v'(k_3\rho_3) - \mathrm{H}_v'^{(2)}(k_3\rho_3)\,\mathrm{J}_v(k_3\rho_2) \right\} + \right.$$
$$\left. \frac{k_0}{\mu_0}\mathrm{H}_v'^{(2)}(k_0\rho_3) \left\{ \mathrm{H}_v^{(2)}(k_3\rho_3)\,\mathrm{J}_v(k_3\rho_2) - \mathrm{H}_v^{(2)}(k_3\rho_2)\,\mathrm{J}_v(k_3\rho_3) \right\} \right]$$
$$\left[\frac{k_2^2}{\mu_2^2}\mathrm{J}_v(k_1\rho_1) \left\{ \mathrm{H}_v'^{(2)}(k_2\rho_1)\,\mathrm{J}_v'(k_2\rho_2) - \mathrm{H}_v'^{(2)}(k_2\rho_2)\,\mathrm{J}_v'(k_2\rho_1) \right\} + \right.$$
$$\left. \left. \frac{k_1 k_2}{\mu_1 \mu_2}\mathrm{J}_v'(k_1\rho_1) \left\{ \mathrm{H}_v'^{(2)}(k_2\rho_2)\,\mathrm{J}_v(k_2\rho_1) - \mathrm{H}_v^{(2)}(k_2\rho_1)\,\mathrm{J}_v'(k_2\rho_2) \right\} \right] \right], \quad (42)$$

and

$$A(k_1,k_2,k_3,\rho_1,\rho_2,\rho_3)=$$

$$\left[\frac{k_2^2}{\mu_2^2}\mathrm{H}_\nu^{(2)}(k_3\rho_2)\,\mathrm{J}_\nu(k_1\rho_1)\left\{\mathrm{H}_\nu^{\prime(2)}(k_2\rho_1)\,\mathrm{J}_\nu^{\prime}(k_2\rho_2)-\mathrm{H}_\nu^{\prime(2)}(k_2\rho_2)\,\mathrm{J}_\nu^{\prime}(k_2\rho_1)\right\}+\right.$$
$$\frac{k_1k_2}{\mu_1\mu_2}\mathrm{J}_\nu^{\prime}(k_1\rho_1)\,\mathrm{H}_\nu^{(2)}(k_3\rho_2)\left\{\mathrm{H}_\nu^{\prime(2)}(k_2\rho_2)\,\mathrm{J}_\nu(k_2\rho_1)-\mathrm{H}_\nu^{(2)}(k_2\rho_1)\,\mathrm{J}_\nu^{\prime}(k_2\rho_2)\right\}+$$
$$\frac{k_1k_3}{\mu_1\mu_3}\mathrm{J}_\nu^{\prime}(k_1\rho_1)\,\mathrm{H}_\nu^{\prime(2)}(k_3\rho_2)\left\{\mathrm{H}_\nu^{(2)}(k_2\rho_1)\,\mathrm{J}_\nu(k_2\rho_2)-\mathrm{H}_\nu^{(2)}(k_2\rho_2)\,\mathrm{J}_\nu(k_2\rho_1)\right\}+$$
$$\left.\frac{k_2k_3}{\mu_2\mu_3}\mathrm{H}_\nu^{\prime(2)}(k_3\rho_2)\,\mathrm{J}_\nu(k_1\rho_1)\left\{\mathrm{H}_\nu^{(2)}(k_2\rho_2)\,\mathrm{J}_\nu^{\prime}(k_2\rho_1)-\mathrm{H}_\nu^{\prime(2)}(k_2\rho_1)\,\mathrm{J}_\nu(k_2\rho_2)\right\}\right], \quad (43)$$

and

$$B(k_1,k_2,k_3,\rho_1,\rho_2,\rho_3)=$$

$$\left[\frac{k_2^2}{\mu_2^2}\mathrm{J}_\nu(k_3\rho_2)\,\mathrm{J}_\nu(k_1\rho_1)\left\{\mathrm{H}_\nu^{\prime(2)}(k_2\rho_2)\,\mathrm{J}_\nu^{\prime}(k_2\rho_1)-\mathrm{H}_\nu^{\prime(2)}(k_2\rho_1)\,\mathrm{J}_\nu^{\prime}(k_2\rho_2)\right\}+\right.$$
$$\frac{k_1k_2}{\mu_1\mu_2}\mathrm{J}_\nu^{\prime}(k_1\rho_1)\,\mathrm{J}_\nu(k_3\rho_2)\left\{\mathrm{H}_\nu^{(2)}(k_2\rho_1)\,\mathrm{J}_\nu^{\prime}(k_2\rho_2)-\mathrm{H}_\nu^{\prime(2)}(k_2\rho_2)\,\mathrm{J}_\nu(k_2\rho_1)\right\}+$$
$$\frac{k_1k_3}{\mu_1\mu_3}\mathrm{J}_\nu^{\prime}(k_1\rho_1)\,\mathrm{J}_\nu^{\prime}(k_3\rho_2)\left\{\mathrm{H}_\nu^{(2)}(k_2\rho_2)\,\mathrm{J}_\nu(k_2\rho_1)-\mathrm{H}_\nu^{(2)}(k_2\rho_1)\,\mathrm{J}_\nu(k_2\rho_2)\right\}+$$
$$\left.\frac{k_2k_3}{\mu_2\mu_3}\mathrm{J}_\nu^{\prime}(k_3\rho_2)\,\mathrm{J}_\nu(k_1\rho_1)\left\{\mathrm{H}_\nu^{\prime(2)}(k_2\rho_1)\,\mathrm{J}_\nu(k_2\rho_2)-\mathrm{H}_\nu^{(2)}(k_2\rho_2)\,\mathrm{J}_\nu^{\prime}(k_2\rho_1)\right\}\right]. \quad (44)$$

Thus, analytical closed form representations are readily available for 2- and 3- layer capped wedge. Without resorting to numerical matrix inversion, these coefficients can be used directly for field calculation.

4. Numerical Simulations

Explicit forms for 2- and 3-layer capped wedge are utilized for numerical simulation of several wedge problems. To observe the influence of different parameters such as radius of the layer (ρ), relative permittivity (ε_r), relative permeability (μ_r), and loss tangent ($\tan\delta$) on the scattered electric field with plane wave excitation, numerical results are presented in Figures 2 through 9. The variation of the scattered field pattern is normalized as $|E_z| / |E_0|$. Analytical expressions for two different wedge angles and different layer properties are computed and compared to PEC-only case. The wedge angles of 10° and 30° are chosen to represent airplane wing geometries, but closed form expressions can be used for any wedge angle. Dielectric layer parameters are chosen from well-known materials.

First, permittivity values of 2-layer geometry with different loss tangents are studied. Magnitude of the electric field is presented in Fig. 2. More than 50 % reduction in magnitude is observed compared to PEC-only case. Small loss tangent did not make any considerable change in the electric field magnitude. Increased loss tangent values, of course, would increase the scattered field. Next, the same wedge shape with the same permittivity but different

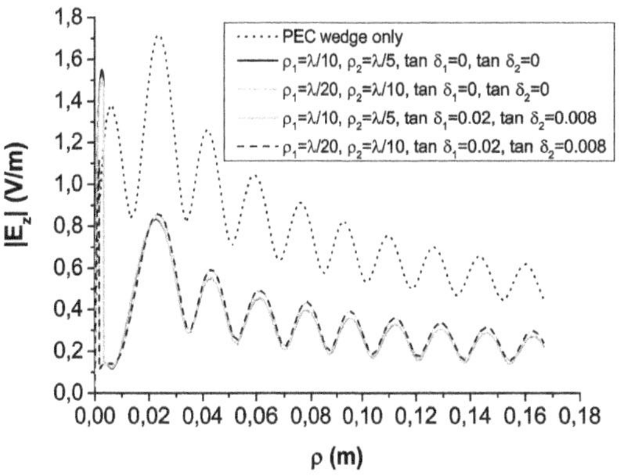

Fig. 2. Comparison of z components of electric field with respect to ρ. The parameters are $N = 2$, $\varepsilon_1 = 4.4$, $\varepsilon_2 = 2.2$, $\alpha = 30°$, $f = 9$ GHz and $\varphi = 180°$.

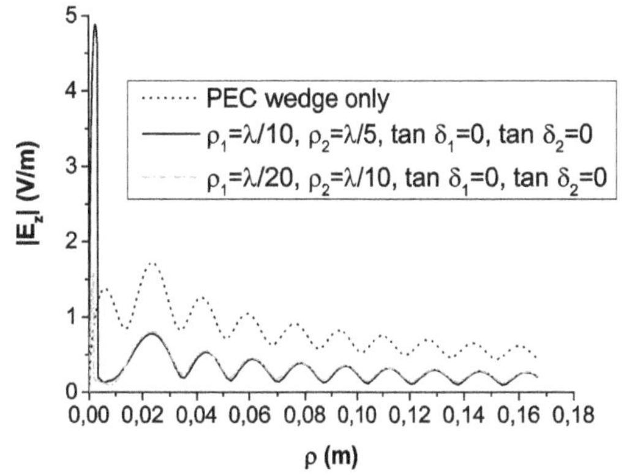

Fig. 3. Comparison of z components of electric field with respect to ρ. The parameters are $N = 2$, $\mu_1 = 7$, $\mu_2 = 4$, $\alpha = 30°$, $f = 9$ GHz and $\varphi = 180°$.

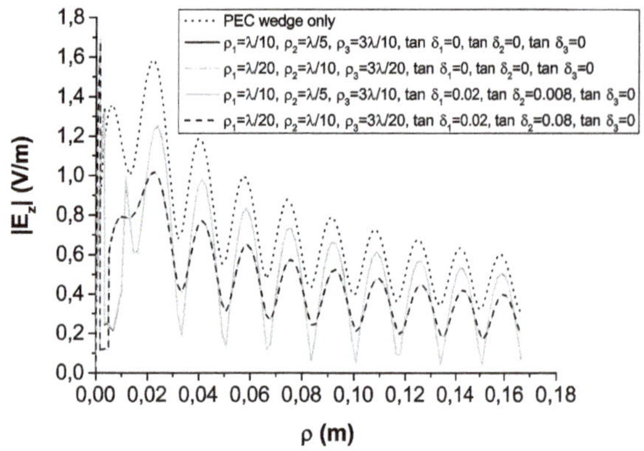

Fig. 4. Comparison of z components of electric field with respect to ρ. The parameters are $N = 3$, $\varepsilon_1 = 6$, $\varepsilon_2 = 4.4$, $\varepsilon_3 = 2.2$, $\alpha = 10°$, $f = 9$ GHz and $\varphi = 180°$.

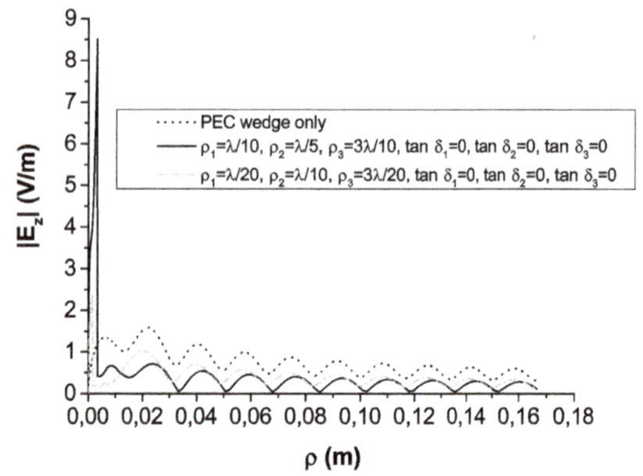

Fig. 5. Comparison of z components of electric field with respect to ρ. The parameters are $N = 3$, $\mu_1 = 11$, $\mu_2 = 7$, $\mu_3 = 4$, $\alpha = 10°$, $f = 9$ GHz and $\varphi = 180°$.

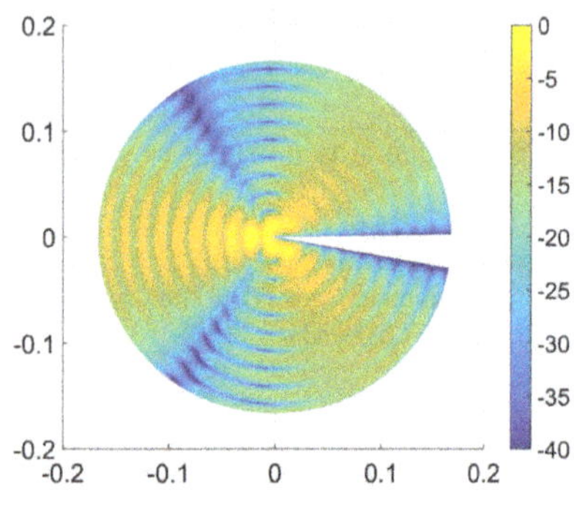

Fig. 6. Variation of $|E_Z|$(dBV/m) for standalone PEC wedge. The parameters are $\alpha = 10°$, $f = 9$ GHz and $\varphi_0 = 180°$.

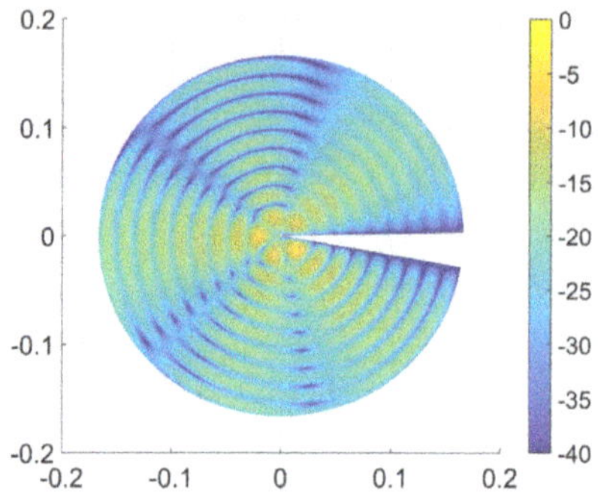

Fig. 7. Variation of $|E_Z|$(dBV/m). The parameters are $N = 2$, $\rho_1 = \lambda : 20$, $\rho_2 = \lambda : 10$, $\tan\delta_1 = 0$, $\tan\delta_2 = 0$, $\alpha = 10°$, $f = 9$ GHz and $\varphi_0 = 180°$.

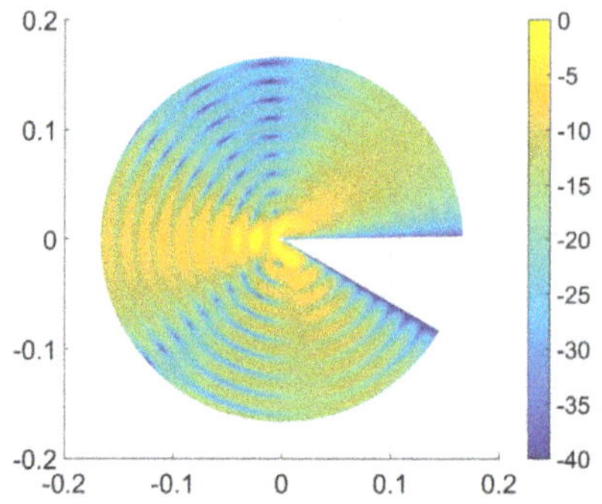

Fig. 8. Variation of $|E_Z|$(dBV/m) for standalone PEC wedge. The parameters are $\alpha = 30°$, $f = 9$ GHz and $\varphi_0 = 180°$.

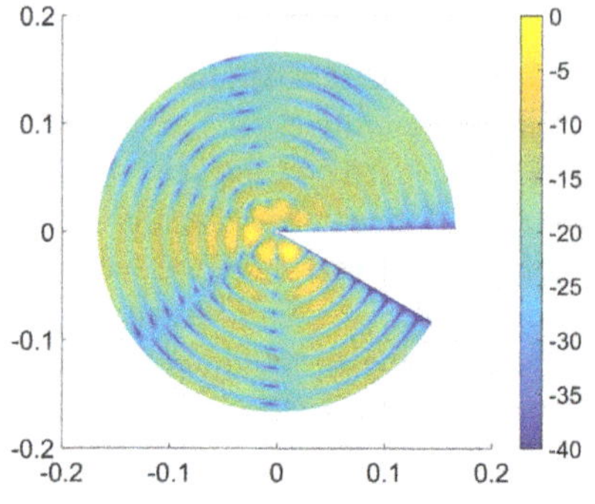

Fig. 9. Variation of $|E_Z|$(dBV/m). The parameters are $N = 3$, $\rho_1 = \lambda : 20$, $\rho_2 = \lambda : 10$, $\rho_3 = 3\lambda : 20$, $\tan\delta_1 = 0$, $\tan\delta_2 = 0$, $\tan\delta_3 = 0$, $\alpha = 30°$, $f = 9$ GHz and $\varphi_0 = 180°$.

permeability values are presented in Fig. 3. Again, the field magnitude was reduced more than half with this configuration. The thickness of the layers did not substantially influence the magnitude. The spike in field magnitude near the tip of the wedge was larger than PEC-only case as field localized in this asymptotic region was enhanced due to permeability of the layer.

Similar analysis was carried out for narrower wedge angle of 10° in Figures 4 and 5. It is observed that the effects of dielectric layers were slightly less than those of 30° wedge angle as expected. As the wedge angle gets smaller, incident plane encounters less cross sectional area of the wedge, which in turn, leads to lowered electric field magnitude in

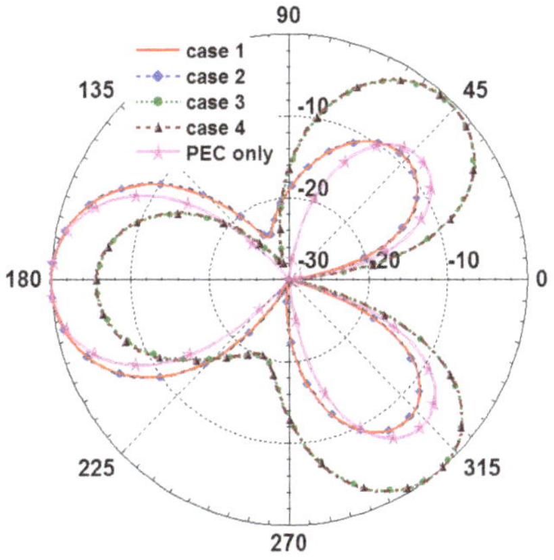

Fig. 10. Modified RCS pattern (dB) for $\alpha = 10°$, $\varphi_0 = 180°$, $\varepsilon_1 = 4.4$ and $\varepsilon_2 = 2.2$, for Cases 1-4 and PEC-only case. Case 1: N = 2, $\rho_1 = \lambda : 10$, $\rho_2 = \lambda : 5$, $\rho_0 = 3\lambda : 5$, $\tan\delta_1 = 0$, $\tan\delta_2 = 0$, Case 2: $N = 2$, $\rho_1 = \lambda : 10$, $\rho_2 = \lambda : 5$, $\rho_0 = 3\lambda : 5$, $\tan\delta_1 = 0.02$, $\tan\delta_2 = 0.008$, Case 3: $N = 2$, $\rho_1 = \lambda : 5$, $\rho_2 = 2\lambda : 5$, $\rho_0 = 6\lambda : 5$, $\tan\delta_1 = 0$, $\tan\delta_2 = 0$, Case 4: $N = 2$, $\rho_1 = \lambda : 5$, $\rho_2 = 2\lambda : 5$, $\rho_0 = 6\lambda : 5$, $\tan\delta_1 = 0.02$, $\tan\delta_2 = 0.008$.

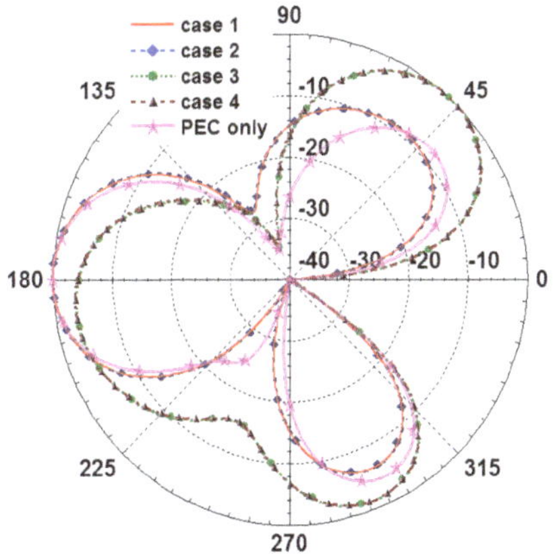

Fig. 11. Modified RCS pattern (dB) for $\alpha = 30°$, $\varphi_0 = 180°$, $\varepsilon_1 = 4.4$ and $\varepsilon_2 = 2.2$, for Cases 1-4 and PEC-only case. Case 1: N = 2, $\rho_1 = \lambda : 10$, $\rho_2 = \lambda : 5$, $\rho_0 = 3\lambda : 5$, $\tan\delta_1 = 0$, $\tan\delta_2 = 0$, Case 2: $N = 2$, $\rho_1 = \lambda : 10$, $\rho_2 = \lambda : 5$, $\rho_0 = 3\lambda : 5$, $\tan\delta_1 = 0.02$, $\tan\delta_2 = 0.008$, Case 3: $N = 2$, $\rho_1 = \lambda : 5$, $\rho_2 = 2\lambda : 5$, $\rho_0 = 6\lambda : 5$, $\tan\delta_1 = 0$, $\tan\delta_2 = 0$, Case 4: $N = 2$, $\rho_1 = \lambda : 5$, $\rho_2 = 2\lambda : 5$, $\rho_0 = 6\lambda : 5$, $\tan\delta_1 = 0.02$, $\tan\delta_2 = 0.008$.

PEC-only case. Lossless dielectric layers were more successful in reducing the scattered electric field.Variation of electric field magnitude at z = 0 plane was also studied for both wedge angles. In Figures 6 and 7, PEC-only case and permittivity varied case were shown, respectively. Reduction in electric field magnitude was greater than 3 dB in almost every direction. PEC-only case for 30° wedge angle is displayed in Fig. 8 and 3-layer case with different permittivity values are shown in Fig. 9.

Far-field analysis of the electric field can be expressed as

$$\sigma' = \lim_{\rho\to\infty} 2\pi\rho \frac{|E_{\text{tot}}|^2}{|E_{\text{inc}}|^2}$$

where σ' denotes modified RCS as total electric field instead of scattered electric field is considered. For 10° and 30° wedge angles, four different cases are compared to PEC only case in Figures 10 and 11, respectively. The incident electric field is assumed from 180° direction to the wedge. It is observed that for monostatic radar applications, Cases 3 and 4 provide 5.8 and 5.7 dB reduction relative to PEC-only case on incident wave direction, whereas Cases 1 and 2 provide almost no reduction but about 1 dB reduction at +/-20° off of incident direction. On opposite side of the incident direction, layered wedge has more RCS compared to PEC-only case as expected. Since Cases 1, 3 and Cases 2, 4 differ only for layer thicknesses with other parameters being equal, layer thicknesses play a critical role in achieving RCS reduction.

5. Conclusions

Layered capped wedge with various thicknesses and dielectric loss values have been studied for electric field magnitude in longitudinal direction. Explicit expressions for 2- and 3-layer geometries are derived rigorously. It is observed that for certain values of cap radii and loss factors, more than 3 dB reduction in scattered electric field is possible with 2- and 3-layer cap geometries in the near field. Similar analysis is also carried out for modified RCS to compare far-field effects of capped wedge and it is observed that almost 6 dB reduction is possible for certain combination of layer thicknesses. The dielectric layer thicknesses were kept small to enable coating on the PEC surface at the target frequency of 9 GHz. This particular frequency is chosen at X-band where most military radars operate. It is shown that with simple dielectic coating on wing-like airplane structures radar cross section can be reduced by more than half compared to PEC-only case.

Acknowledgments

This work was supported by TÜBİTAK (The Scientific and Technological Research Council of TURKEY) under BİDEB postdoctoral scholarship program.

References

[1] MENTZER, J. R. *Scattering and Diffraction of Radio Waves*. 1st ed. New York, NY (USA): Pergamon Press, 1955.

[2] FELSEN, L. B., MARCUVITZ, N. *Radiation and Scattering of Waves*. Piscataway, NJ (USA): IEEE Press, 1994. ISBN: 978-0780310889

[3] HARRINGTON, R. F. *Time-Harmonic Electromagnetics Field*. 1st ed. New York, NY (USA): McGraw & Hill, 1961.

[4] BOWMAN, J. J., SENIOR, T. B. A., USLENGHI, P. L. E. *Electromagnetic and Acoustic Scattering by Simple Shapes*. New York, NY (USA): Hemisphere Publishing, 1987.

[5] REDADAA, S., BOUALLEG, A., MERABTINE, N., et al. Radar cross section study from wave scattering structures. *Semiconductor Physics, Quantum Electronics & Optoelectronics*, 2006, vol. 9, no. 4, p. 71–76.

[6] LEWIN, L., SREENIVASIAH, I. *Diffraction by a Dielectric Wedge, Technology Report*. 191 pages.

[7] MALIUZHINETS, G. D. Excitation, reflection, and emission of surface waves from a wedge with given face impedances. *Soviet Physics Doklady*, 1958, vol. 3, p. 752–755.

[8] FELSEN, L. B. Electromagnetic properties of wedge and cone surfaces with a linearly varying surface impedance. *IRE Transactions on Antennas and Propagation*, Dec. 1959, vol. 7, p. 231–243. DOI: 10.1109/TAP.1959.1144752

[9] ISENLIK, T., YEGIN, K. Paraxial fields of a wedge with anisotropic impedance and perfect electric conductor faces excited by a dipole. *Electromagnetics*, 2010, vol. p. 30, no. 7, 589-608. DOI: 10.1080/02726343.2010.513932

[10] ISENLIK, T., YEGIN, K. Derivations of Green's functions for paraxial fields of a wedge with particular anisotropic impedance faces. *Electromagnetics*, 2013, vol. 33, no. 5, p. 392–412. DOI: 10.1080/02726343.2013.792722

[11] ENGHETA, N., MURPHY, W. D., ROKHLIN, V., et al. The fast multipole method (FMM) for electromagnetic scattering problems. *IEEE Transactions on Antennas and Propagation*, 1992, vol. 40, no. 6, p. 634–641. DOI: 10.1109/8.144597

[12] ERGUL, O., GUREL, L. *The Multilevel Fast Multipole Algorithm (MLFMA) for Solving Large-Scale Computational Electromagnetics Problems*. Piscateway, NJ (USA): Wiley-IEEE Press, 2014. ISBN: 978-1-119-97741-4

About the Authors . . .

Hulya OZTURK was born in 1984 in İstanbul, Turkey. She received her M.Sc. and Ph.D. degree in mathematics from Gebze Technical University, Kocaeli, Turkey, in 2010 and 2015, respectively. Her research interests are in electromagnetic and acoustic wave scattering and propagation.

Korkut YEGIN received the B.Sc. degree in electrical and electronics engineering from Middle East Technical University, Ankara, Turkey, in 1992, the M.Sc. degree in electrical and computer engineering, the M.Sc. degree in plasma physics, and the Ph.D. degree in electrical and computer engineering from Clemson University, Clemson, SC, USA, in 1996, 1998 and 1999, respectively. He is a professor at Ege University Electrical and Electronics Engineering , İzmir, Turkey. His research interests include electromagnetic scattering, VSAT antennas, and UWB radar.

2

Novel Design of Recursive Differentiator Based on Lattice Wave Digital Filter

Richa BARSAINYA, Tarun K. RAWAT

Division of Electronics and Communication, Netaji Subhas Institute of Technology, Sector-3, Dwarka, 110078 New Delhi, India

richa.barsainya@gmail.com, tarundsp@gmail.com

Abstract. *In this paper, a novel design of third and fifth order differentiator based on lattice wave digital filter (LWDF), established on optimizing L_1-error approximation function using cuckoo search algorithm (CSA) is proposed. We present a novel realization of minimum multiplier differentiator using LWD structure leading to requirement of optimizing only N coefficients for Nth order differentiator. The γ coefficients of lattice wave digital differentiator (LWDD) are computed by minimizing the L_1-norm fitness function leading to a flat response. The superiority of the proposed LWDD is evident by comparing it with other differentiators mentioned in the literature. The magnitude response of the designed LWDD is found to be of high accuracy with flat response in a wide frequency range. The simulation and statistical results validates that the designed minimum multiplier LWDD circumvents the existing one in terms of minimum absolute magnitude error, mean relative error (dB) and efficient structural realization, thereby making the proposed LWDD a promising approach to digital differentiator design.*

Keywords

Lattice wave digital filter, digital differentiator, wideband, L_1-CSA, minimum multiplier

1. Introduction

In recent years, digital differentiators are widely applied in various fields of signal processing, image processing, biomedical engineering, radar engineering, control systems, etc. [1–4]. The design and realization of digital differentiator have emerged as an active area of research due to its wide range of applications. Digital differentiators are effectively used to compute the time-derivative of real time and/or stored signal, which necessitates their results to be of high accuracy and its structural realization to be robust. The frequency response of an ideal digital differentiator is given by

$$H_d(\omega) = j\omega \tag{1}$$

where $j = \sqrt{-1}$ and $\omega \in [0, \pi]$ is the normalized frequency. An ideal differentiator has a constant phase response of $\frac{\pi}{2}(\approx 1.57\,\text{rad})$ over the entire Nyquist frequency range. In available literature, several methods have been extensively explored for designing and implementing a digital differentiator. Interpolation and approximation based techniques [5–8] and different optimization techniques are most prevalent approaches used for the designing purpose [9–16]. The optimization based design of digital differentiator can be described as an approximation problem which comprises of four steps. First, a desired ideal frequency response of the digital differentiator is defined. Second, selection of type of system (either FIR or IIR). Third, developing an optimality criterion to approximate the ideal response. Lastly, an application of an optimization method to compute the optimal system coefficients.

In this paper, instead of a direct form of FIR or IIR system, a new and improved class of IIR system is reported due to design constraints of IIR system. IIR systems suffer from stability problem due to its recursive nature, especially when quantization of signal and coefficient is applied. IIR systems also have limitations like sensitivity to wordlength and coefficient round-off errors which make their implementation tricky. Direct form structure of IIR system based Nth order digital differentiator requires $2N + 1$ multiplier coefficients. Due to involvement of large number of multiplications in the filter algorithm, hardware implementation incurs excessive area, delays and power consumption. Analyzing all these limitations and considering the requirement of competent design of the digital differentiator, mathematical modelling based on the lattice wave digital filter system is proposed. The lattice wave digital filter is a specific class of wave digital filter [17]. LWDF structures flaunt many fascinating properties such as low coefficient sensitivity and consequently the low accuracy requirements for register wordlength, higher dynamic range, higher overflow level, lower round-off noise, assurance of stability and good nonlinear properties under finite arithmetic conditions where the effects of rounding, truncation and overflow are present [17–19]. In the past, LWDF structures are used for realizing lowpass-highpass filter, bandpass-bandstop filter and Hilbert transformers [20–22]. Their resulting structures are highly modular, less sensitive and found to have minimum hardware which make them suitable for signal processors and VLSI implementation.

In the literature, many researchers have proposed different wideband recursive differentiators by approximating the ideal differentiator response efficiently in a full Nyquist frequency range. The use of evolutionary algorithms in optimization of digital operators is an exponentially increasing field with usage of different optimization algorithms for obtaining improved models of digital differentiator. These intelligent algorithms are capable of providing optimal results by minimizing any multi-modal error objective functions in less computation time. Simulated annealing (SA) technique is used to optimize the differentiator designed using segment rule [12]. Genetic algorithm (GA) was practiced for designing of second order recursive differentiator and the designs were shown to perform well in terms of both magnitude and phase response [16]. A third order wideband differentiator using linear programming (LP) technique was introduced in [9]. The techniques SA, GA and Fletcher and Powell optimization were incorporated to optimize the interpolated coefficients of first and higher order differentiator [13]. Furthermore, the modified particle swarm optimization (PSO) algorithm based digital differentiator design was presented and compared with the designs of GA, variants of PSO and PSO-GA hybrid techniques [14]. Jain et al. proposed IIR differentiators using minimax and pole, zero, and constant optimization (MPZCO) methods [15]. The objective of these algorithms are to minimize the mean square error approximation by searching for an optimized set of numerator and denominator coefficients of IIR system. Direct form structure of Nth order IIR differentiator requires $2N+1$ multiplier coefficients. It complicates the search as large computational complexity occur due to the increased number of variable to be optimized and the constraints that need to be incorporated to ensure stability of an IIR system. On the contrary, LWDF is completely characterized by a set of coefficients (γ) that have excellent dynamic range and low wordlength requirements and only N number of γ coefficients are to be optimized for the design of Nth order differentiator.

In this paper, CSA technique aims to find the optimal coefficients for LWDD that closely match the ideal counterpart by minimizing the fitness function, that is, L_1-norm error. The L_1-norm based optimality criterion is employed due to its ability to produce the flattest response over wide range of frequency that approaches the ideal one [23]. Thus, in order to design digital differentiator with desired response, the L_1-norm is hybridized with the evolutionary algorithm to find the system coefficients.

In this paper, novelty lies in the fact that the wideband differentiator designed using LWDF system will incorporate all the advantageous properties of LWDF with utilization of minimum hardware and also competent results are gained by using optimization techniques. The success of this combination is validated by the results and comparisons performed in this work. The capability of global search and optimal

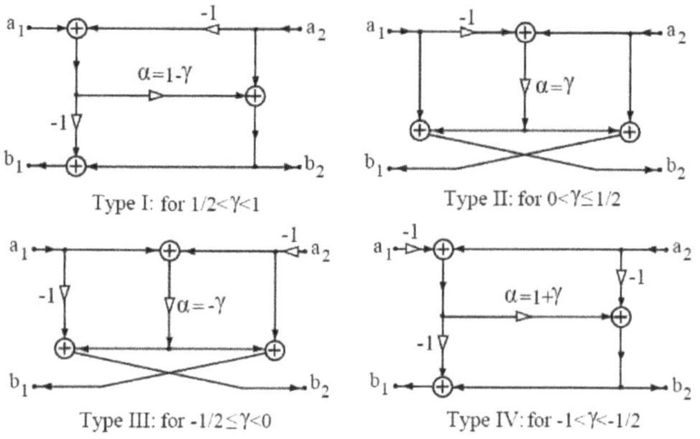

Fig. 1. Signal flow graph of four symmetric two-port adaptors.

robust solution finding feature of cuckoo search algorithm is explored for the design of lattice wave digital differentiator. Thus, the hybrid L_1-CSA is formulated for designing of the LWDD in order to optimize γ coefficients of a LWDF transfer function. The novel structural realization of the proposed 3rd and 5th order lattice wave digital differentiator is also introduced in this paper. The graphical analysis of magnitude response, phase response, magnitude error response and statistical analysis of absolute magnitude error (AME), mean relative error (MRE) in dB and mean phase error (MPE), demonstrates that the proposed minimum multiplier LWDD outperforms the reported differentiators.

The rest of the paper is segmented as follows. Brief overview of the LWDF along with the mathematical articulation is provided in Sec. 2. Section 3 focuses on the formulation of lattice wave digital differentiator design problem using L_1-error function. Section 4 provides the brief overview of the employed optimization algorithm. Simulation results and comparative analysis of the proposed LWDD is provided in Sec. 5 and Sec. 6. Section 7 highlights conclusion of the presented work.

2. Lattice Wave Digital Filter

This section gives a brief overview of lattice wave digital filter structure. Alike WDF, LWDFs are also related to certain analog prototype networks, i.e., lattice networks [17]. LWDF is represented by two parallel branches, which realize all-pass filters. These all-pass filters can be realized by using first- and/or second-order wave digital all-pass structures. These all-pass structures are implemented by using symmetric two-port adaptors and delay elements [18]. An adaptor requires a single multiplication and three additions. The application of adaptor produces an efficient realization in terms of number of multipliers for a given order and have low sensitivity to coefficient quantization. LWDF uses four types of adaptors as its building blocks. The signal flow graphs of four single multiplier symmetric two-port configurations are shown in Fig. 1.

According to [19], the adaptor coefficients γ can be guaranteed to fall in the interval $-1 < \gamma < 1$. Methods to easily calculate these coefficients from the design specifications have been discussed in [19]. The choice of an adaptor structure depends on the value of γ coefficient. The α/γ conversion expressions is shown in Fig. 1. The value α coefficient should always be positive and less than or equal to half i.e., $0 < \alpha \leq \frac{1}{2}$.

3. Problem Formulation

The ideal differentiator in (1) is approximated to the LWDF system. The transfer function of LWDF system is specified as

$$H_{\mathrm{LWDF}}(z) = \frac{1}{a}[H_1(z) - H_2(z)] \tag{2}$$

where $H_1(z)$ and $H_2(z)$ are allpass filter of order P and Q, respectively. In case of lowpass/highpass filters, $P = Q - 1$ or $P = Q + 1$. The overall order of $H_{\mathrm{LWDF}}(z)$ is $P + Q$. $H_1(z)$ and $H_2(z)$ can be expressed in terms of adaptor coefficients as

$$H_1(z) = \frac{-\gamma_0 + z^{-1}}{1 - \gamma_0 z^{-1}} \prod_{l=1}^{m} \frac{-\gamma_{2l-1} + \gamma_{2l}(\gamma_{2l-1} - 1)z^{-1} + z^{-2}}{1 + \gamma_{2l}(\gamma_{2l-1} - 1)z^{-1} - \gamma_{2l-1}z^{-2}}, \tag{3}$$

$$H_2(z) = \prod_{l=m+1}^{m+n} \frac{-\gamma_{2l-1} + \gamma_{2l}(\gamma_{2l-1} - 1)z^{-1} + z^{-2}}{1 + \gamma_{2l}(\gamma_{2l-1} - 1)z^{-1} - \gamma_{2l-1}z^{-2}} \tag{4}$$

where $m = \frac{P-1}{2}$ and $n = \frac{Q}{2}$ and γ is the adaptor coefficient that characterizes LWDF.

In this paper, design of 3rd and 5th order differentiator is considered. The differentiator of even order can also be realized using LWDF, however, in that case complex coefficients are required. In this paper, only real coefficients, odd order differentiator design is considered and the transfer function of 3rd and 5th order LWDF system for designing of differentiator is expressed as

$$H^{\mathrm{3rd}}_{\mathrm{LWDF}}(z) = \frac{1}{a}\left[\left(\frac{-\gamma_0 + z^{-1}}{1 - \gamma_0 z^{-1}}\right) - \left(\frac{-\gamma_1 + \gamma_2(\gamma_1 - 1)z^{-1} + z^{-2}}{1 + \gamma_2(\gamma_1 - 1)z^{-1} - \gamma_1 z^{-2}}\right)\right], \tag{5}$$

$$H^{\mathrm{5th}}_{\mathrm{LWDF}}(z) = \frac{1}{a}\left[\left(\frac{-\gamma_0 + z^{-1}}{1 - \gamma_0 z^{-1}}\right)\left(\frac{-\gamma_1 + \gamma_2(\gamma_1 - 1)z^{-1} + z^{-2}}{1 + \gamma_2(\gamma_1 - 1)z^{-1} - \gamma_1 z^{-2}}\right) - \left(\frac{-\gamma_3 + \gamma_4(\gamma_3 - 1)z^{-1} + z^{-2}}{1 + \gamma_4(\gamma_3 - 1)z^{-1} - \gamma_3 z^{-2}}\right)\right] \tag{6}$$

where a is the scaling factor. In order to obtain the digital differentiator with the desired specification, an objective function in terms of error between the ideal frequency response and LWDF frequency response is developed. The objective function is formulated using L_1-norm to obtain a set of optimized coefficients and can be expressed as

$$||E||_1 = \sum_{\omega} |\mathrm{e}(\omega)| \tag{7}$$

where $||.||$ denotes norm of the function and $\mathrm{e}(\omega) = H_{\mathrm{d}}(\omega) - H_{\mathrm{LWDF}}(\omega)$ is the error objective function. The objective fitness function is minimized iteratively to obtain the optimized γ coefficients of the 3rd and 5th order LWDF (given in (5) and (6), respectively) to design minimum multiplier differentiator with desired specifications. The motivation behind implementing the L_1-norm fitness function is due to the fact that it is capable of delivering flattest response amongst others, such as the L_2 and L_∞-norms [23], [24].

4. Brief Outline of Cuckoo Search Algorithm

In this paper, CSA is used to design minimum multiplier digital differentiator. CSA tries to find the optimal γ coefficients by iteratively minimizing the error objective function, leading to development of the lattice wave digital differentiator. In previous literatures, GA and PSO are most utilized optimization algorithm for designing of optimal recursive wideband differentiators. However, when tested on different benchmark functions, CSA found to be potentially more powerful than the GA and PSO [25].

Cuckoo search algorithm is instigated by the unique breeding behavior of cuckoo bird and the concept of Lévy flights which are observed in some species of birds and animals [25–27]. Cuckoo birds depend on some other bird's nest for hatching their eggs. Cuckoo attempts to determine a nest where the host bird has recently laid eggs. Cuckoo uses this nest to hide its own eggs. If the host bird identifies that the eggs are not its own, it may abandon the nest or either choose to destroy the alien eggs. This heads to the evolution of the cuckoo eggs, which make attempt to mimic the eggs of the host bird.

To apply cuckoo search algorithm, following assumptions are considered:

(i) Cuckoo bird hatches one egg at a time and hide it in the host's nest chosen randomly.

(ii) The nests with the best host environment and with the best eggs will survive and move forward to the next generations.

(iii) The number of host nests is fixed.

(iv) The probability of identifying alien eggs by the host bird is P_{a}.

The Lévy flights are the forward steps taken by living beings, such as birds, insects and animals in search of their food. Lévy flight refers to a series of straight line flights followed by sudden 90° turns. The usage of Lévy flight for choosing a random new nest is the main factor for improving the performance of CSA. To get a new random nest following Lévy flight based formula is used,

$$x_l = x_i + \nu \oplus L\acute{e}vy(\lambda) \tag{8}$$

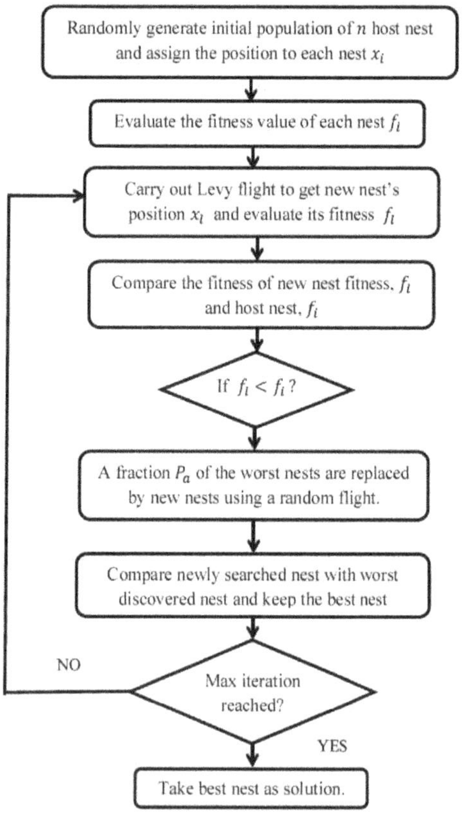

Fig. 2. Flow chart of Cuckoo search algorithm.

Parameters	CSA
Population size	25
Maximum iterations	300
Tolerance	10^{-5}
Lower bound (X_{min})	−1
Upper bound (X_{max})	1
Initial value of all coefficients	0.01
Discovering rate of alien eggs (P_a)	0.25

Tab. 1. Control Parameters of CSA for the optimized LWDD.

Order	γ coefficients	γ range	Adaptor type	α coefficients
3rd	$\gamma_0 = -0.41434$	$-\frac{1}{2} \le \gamma < 0$	III	$\alpha_0 = 0.41434$
	$\gamma_1 = -0.06457$	$-\frac{1}{2} \le \gamma < 0$	III	$\alpha_1 = 0.06457$
	$\gamma_2 = -0.64613$	$-1 < \gamma < -\frac{1}{2}$	IV	$\alpha_2 = 0.35387$
	$a = 0.65775$			
5th	$\gamma_0 = -0.08045$	$-\frac{1}{2} \le \gamma < 0$	III	$\alpha_0 = 0.08045$
	$\gamma_1 = -0.33818$	$-\frac{1}{2} \le \gamma < 0$	III	$\alpha_1 = 0.33818$
	$\gamma_2 = -0.91047$	$-1 < \gamma < -\frac{1}{2}$	IV	$\alpha_2 = 0.08953$
	$\gamma_3 = -0.20044$	$-\frac{1}{2} \le \gamma < 0$	III	$\alpha_3 = 0.20044$
	$\gamma_4 = -0.83884$	$-1 < \gamma < -\frac{1}{2}$	IV	$\alpha_4 = 0.16116$
	$a = 0.64531$			

Tab. 2. Optimized Coefficients for the 3rd and 5th order LWDD using L_1-CSA.

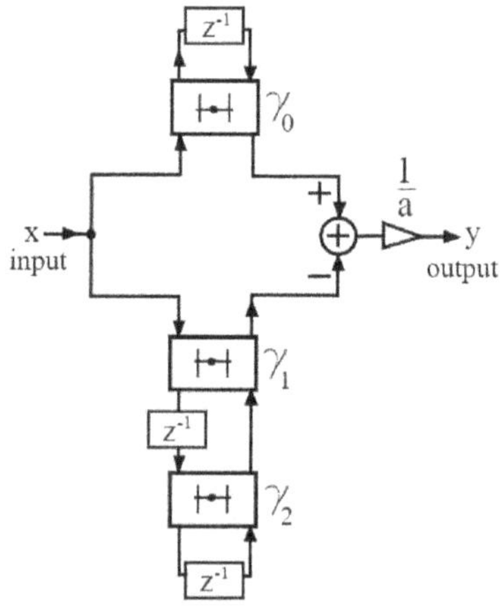

Fig. 3. Structural realization of the 3rd order lattice wave digital differentiator.

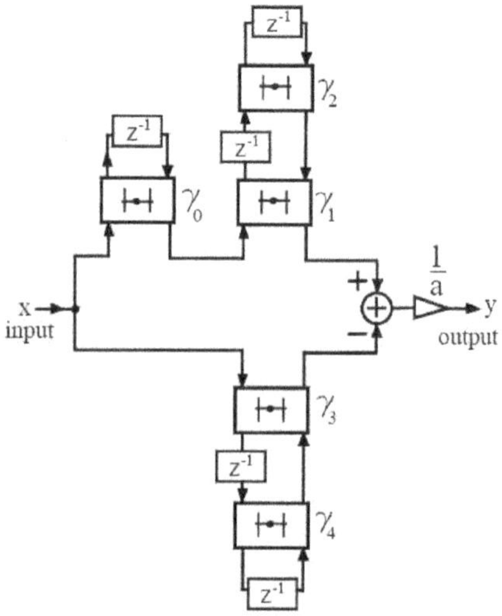

Fig. 4. Structural realization of the 5th order lattice wave digital differentiator.

where v is the step size related to the problem specified, $\oplus$ represents entry wise multiplication and λ is a Lévy flight parameter. This equation represents a random walk which is a Markov chain which means its next step depends on the current location and the transition probability. The implementation steps of CSA are explained with a flow chart presented in Fig. 2.

5. Simulation Analysis

This section shows the design process of lattice wave digital differentiator using CSA and its efficient LWD structural realization with minimum multipliers. To incur the proposed LWDD, simulations are performed in MATLAB on Intel Core i5, 3.20 GHz with 2 GB RAM. The optimal set of CSA parameters for design of minimum multiplier differentiator are reported in Tab. 1. After exhaustive simulation and analysis, the best optimal coefficients for the designed 3rd and 5th order LWDD are calculated by minimizing error fitness function using L_1-CSA are reported in Tab. 2. Best results are reported here after 100 simulation trails with random parameter value selection. The tuning of controlling parameters is a typical task and there exists no definite methodology in the available literature to provide an optimal set of parameter values. So, optimal parameter values can differ for different problems.

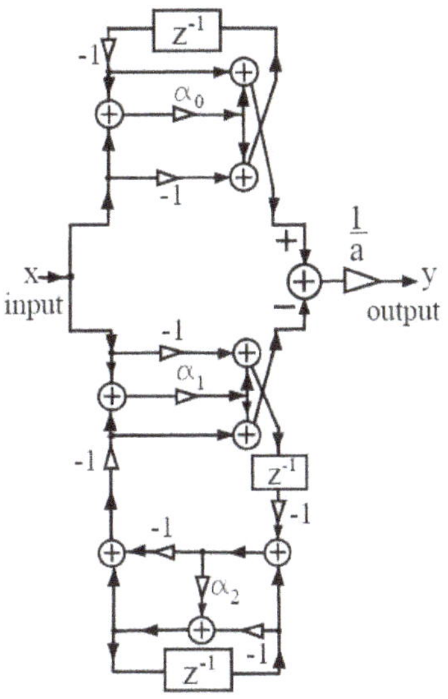

Fig. 5. Signal flow diagram of the 3rd order lattice wave digital differentiator.

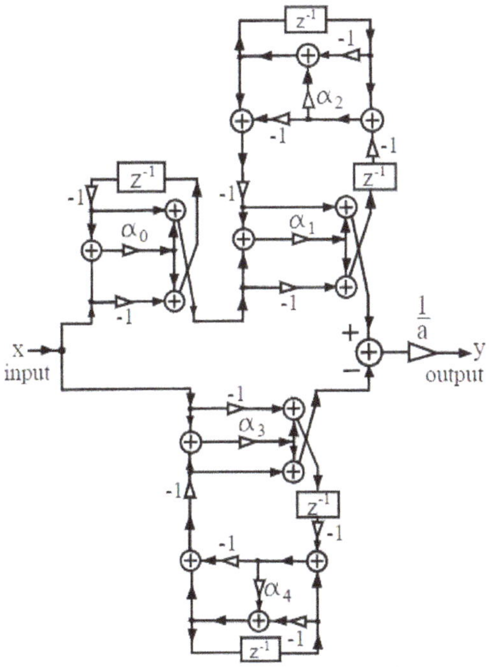

Fig. 6. Signal flow diagram of the 5th order lattice wave digital differentiator.

Table 2 also provides information about the adaptor type chosen depending on the value of γ coefficients and the exact value of adaptor multiplier coefficients, α. The lattice wave digital realization of the designed 3rd order LWDD is shown in Fig. 3. A minimum multiplier design for the 3rd order LWDD is procured where only three adaptors are utilized, where each adaptor comprises of a single multiplier. The lattice wave digital realization of the designed 5th order LWDD is shown in Fig. 4.

A minimum multiplier design for the 5th order LWDD is obtained, thereby reducing the computational requirement or the complexity of hardware in implementation. The signal flow diagram of 3rd and 5th order lattice wave digital differentiator is depicted in Figures 5 and 6, respectively. The absolute magnitude error of 1.4706 and 0.3243 have been achieved for the 3rd and 5th order LWDD optimized using L_1-CSA. The execution time to achieve corresponding values of AME and MPE is 80.4564 second and 106.2131 second, respectively. Furthermore, statistical analysis shows that the mean relative error (dB) of –51.7828 and –64.0481 have been achieved for the 3rd and 5th order LWDD optimized using CSA. The MRE is computed over the complete frequency range using following equation

$$MRE = \frac{1}{N}\sum_{i=1}^{N}\left|\frac{H_{\mathrm{d}}(\omega) - H_{\mathrm{LWDF}}(\omega)}{H_{\mathrm{d}}(\omega)}\right|, \quad 0 \leq \omega \leq \pi. \tag{9}$$

6. Comparison of the Proposed Differentiators with the Existing Ones.

To verify and evaluate the efficiency, the designed LWDD is compared with the existing differentiators. Several existing differentiators procured through different methods are tabulated in Tab. 3 for comparison.

The comparison is performed on the basis of procured magnitude response, magnitude error response and phase curve. The magnitude response plots of 3rd and 5th order LWDDs are obtained by simulating (5)–(6), after substituting the value of obtained optimal γ coefficients. Figures 7 and 8, represent the graphical comparison of magnitude and phase response of the 3rd order LWDD and all the existing 3rd order differentiators mentioned in Tab. 3, along with the ideal response. The magnitude error plot of the designed and reported 3rd order differentiators is shown in Fig. 9.

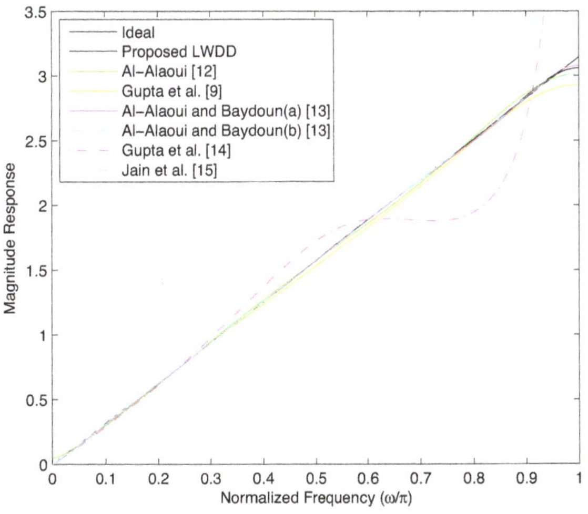

Fig. 7. Magnitude response comparison of the proposed 3rd order LWDD and existing differentiators.

Order	Reference	Method	Numerator coefficient	Denominator coefficient
3rd	Proposed LWDD	CSA	0.53177; 0.8672; –0.8672; –0.53177	1.0000; 1.1022; 0.34957; 0.02675
	Al-Alaoui	3-segment	0.01903; –0.02905; 1.1230; –1.1810	1.0000; 0.1846; –0.001748; 0.03484
	Gupta et al.	LP	1.0000; –1.0000; 0.0000; 0.0000	0.3290; 0.8077; –0.1694; 0.0338
	Al-Alaoui and Baydoun (a)	GA	1.1533; –0.4432; –0.7060; –0.0041	1.0000; 0.7981; 0.0884; 0.0000
	Al-Alaoui and Baydoun (b)	SA	1.1555; –0.3580; –0.7140; –0.0833	1.0000; 0.8662; 0.1612; 0.0028
	Gupta et al.	PSO	0.3237; 1.0000; –0.7133; –0.6124	1.0000; 1.0000; 0.2759; 0.1595
	Jain et al.	MPZCO	1.0000; –0.0243; –0.8315; –0.1436	0.8646; 1.0001; 0.2470; 0.0056
5th	Proposed LWDD	CSA	0.1295; 0.6828; 0.6019; –0.6019; –0.6828; –0.1295	0.6541; 1.3175; 0.9691; 0.3156; 0.0435; 0.0018
	Devate et al.	curve fitting	0.2500; 2.1408; 2.0522; –2.0523; –2.1407; –0.2499	2.2257; 4.1683; 2.6300; 0.6353; 0.0517; 0.0000

Tab. 3. Optimal coefficients of the existing digital differentiators using different design methods.

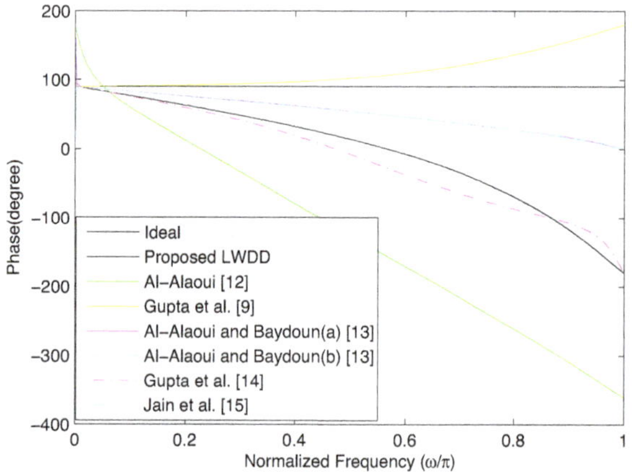

Fig. 8. Phase response comparison of the proposed 3rd order LWDD and existing differentiators.

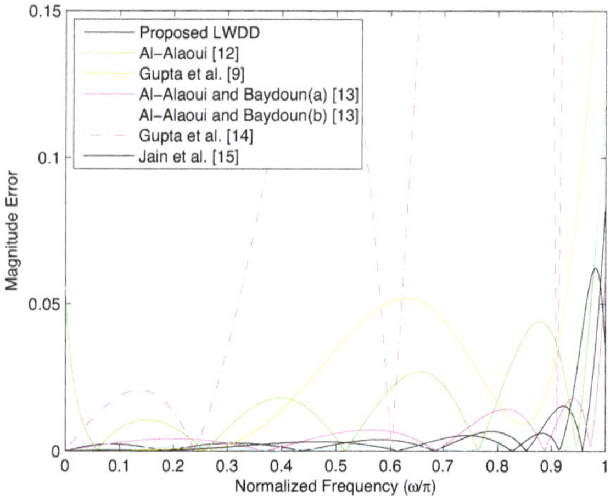

Fig. 9. Magnitude error comparison of the proposed 3rd order LWDD and existing differentiators.

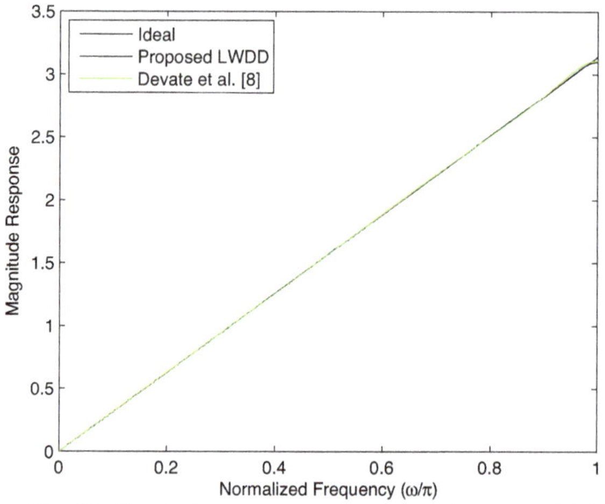

Fig. 10. Magnitude response comparison of the proposed 5th order LWDD and existing differentiator.

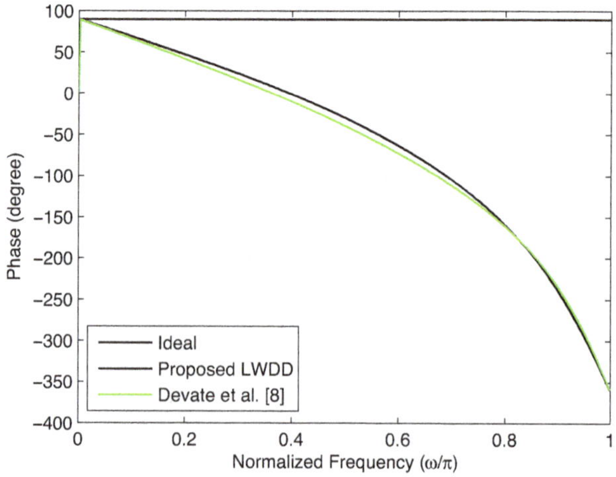

Fig. 11. Phase response comparison of the proposed 5th order LWDD and existing differentiator.

It is discernible that the deviation of magnitude from ideality varies least in case of the proposed LWDD over a wide range of frequency. Figures 10 and 11, represent the graphical comparison of magnitude response and phase response of the 5th order LWDD and the existing 5th order differentiator mentioned in Tab. 3. Further in Fig. 12, the magnitude error plot of the designed and existing 5th order differentiator is shown. All reported figures depict that the proposed minimum multiplier LWDD approximates the magnitude response of the ideal differentiator over wide frequency band in comparison to the reported differentiators.

For the affirmative analysis, absolute magnitude error, mean relative error (dB) and number of multiplier operations are the main parameters taken into consideration to evaluate performance of the designed LWDDs. The value of AME, MRE (dB) and MPE of the proposed minimum multiplier 3rd and 5th order LWDD and different reported differentiators are provided in Tab. 4. The absolute magnitude error is observed to be the lowest (1.4706 and 0.3243, respectively) for the proposed 3rd and 5th order LWDD amongst all reported differentiators. The MRE and mean phase error are evaluated over the complete frequency range. The MPE (rad) is calculated using the following equation

Order	Reference	Absolute magnitude error	Mean relative error (dB)	Mean phase error (rad)
3rd	Proposed LWDD	1.4706	-51.7828	1.4896
	Al-Alaoui [12]	5.5590	-13.3765	3.7851
	Gupta et al. [9]	8.4601	-38.0100	0.4117
	Al-Alaoui and Baydoun (a) [13]	1.8570	-46.9583	0.6726
	Al-Alaoui and Baydoun (b) [13]	1.5120	-52.7418	0.6682
	Gupta et al. [14]	91.4645	-17.8199	1.8253
	Jain et al. [15]	1.5372	-52.9934	0.6632
5th	Proposed LWDD	0.3243	-64.0481	2.5895
	Devate et al. [8]	0.9008	-57.4759	2.6660

Tab. 4. Statistical comparison of the designed LWDD with existing designs.

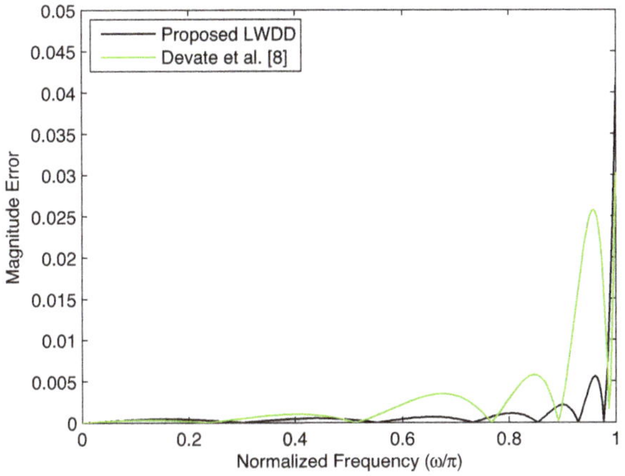

Fig. 12. Magnitude error comparison of the proposed 5th order LWDD and existing differentiator.

$$MPE = \frac{1}{N}\sum_{i=1}^{N}\left|\frac{\angle H_{\rm d}(\omega) - \angle H_{\rm LWDF}(\omega)}{\angle H_{\rm d}(\omega)}\right|, \quad 0 \le \omega \le \pi. \tag{10}$$

Comparative results reveal that the 3rd and 5th order differentiators designed using LWDF possess the lowest MRE (dB) (−51.7828 dB and −64.0481 dB, respectively) amongst all. However, MPE of the proposed 3rd order LWDD is slight higher than that of differentiator designed using GA and MPZCO [13], [15]. Whereas, MPE of the proposed 5th order LWDD is lower than the differentiator proposed by Devate et al. [8]. On the grounds of the above facts, it is inferred that the proposed minimum multiplier wideband LWDD performs better than the reported differentiators by approaching the ideal curve to the maximum and also inherit all the advantageous properties of LWDF.

Transfer function of a recursive differentiator can be realized using different structures. For infinite precision arithmetic all structures produce the same output for a given input, but under finite precision arithmetic different kinds of errors occur. The best structure is a trade-off between many different aspects such as stability, coefficient sensitivity and other quantization effects. Hence, need of the competent structures for differentiation applications is rectified by using lattice wave digital structural realization. Estimation of hardware requirements of the proposed LWDD structure

Order	Differentiator designs	Number of Multipliers
3rd	Proposed LWDD	4
	Al-Alaoui	7
	Gupta et al.	5
	Al-Alaoui and Baydoun(a)	6
	Al-Alaoui and Baydoun(b)	7
	Gupta et al.	5
	Jain et al.	7
5th	Proposed LWDD	6
	Devate et al.	11

Tab. 5. Multiplication operations required by the proposed LWDD and the existing differentiators.

can be done by counting the number of multiplications required for its implementation. The number of multiplication operations required by the proposed 3rd and 5th order differentiators are computed from the signal-flow graph provided in Figures 5 and 6, respectively. Arithmetic complexity of the reported digital differentiators is computed by considering their realization using direct form II structure.

The number of multiplication operations required by the proposed LWDD and the existing differentiators are provided in Tab. 5, which are computed from the signal-flow graph. The number of multiplication operations required by the proposed Nth order LWDD is $N + 1$ multiplication, whereas, the exiting Nth order differentiator structures require $2N + 1$ multiplication. The proposed 3rd and 5th order LWDD require 43 % and 45 % less multipliers as compared to the reported 3rd and 5th order differentiators. This implies that the LWDD yields a smaller arithmetic complexity, in terms of multiplications, compared to the existing ones leading to minimum hardware cost and making it suitable for VLSI implementation. Based on the observations, it is ratified that the proposed minimum multiplier LWDD outperforms all the existing differentiators both in design as well as structural realization.

7. Conclusion

A novel method for designing of minimum multiplier digital differentiator based on lattice wave digital filter and L_1-CSA is proposed. The optimal coefficients of 3rd and 5th order lattice wave digital differentiator are procured by minimizing the L_1-error objective function by means of cuckoo search algorithm. The proposed LWDDs accurately approximate the ideal response with very small AME and MRE over the entire frequency range. Simulation results clearly demonstrate the effective performance of the proposed differentiator over other existing differentiators along with reduced computational complexity. A significant improvement in hardware utilization is achieved by incorporating the lattice wave digital structural realization which makes the proposed LWDD useful in various applications.

References

[1] AL-ALAOUI, M. A. Novel FIR approximations of IIR differentiators with applications to image edge detection. In *Proceedings of the 18th IEEE International Conference on Electronics, Circuits and Systems (ICECS)*. Beirut (Lebanon), 2011, p. 11–14. DOI: 10.1109/ICECS.2011.6122335

[2] LAGUNA, P., THAKOR, N. V., CAMINAL, P., et al. Low-pass differentiators for biological signals with known spectra: Application to ECG signal processing. *IEEE Transactions on bio-medical engineering*, 1990, vol. 37, no. 4, p. 420–425. DOI: 10.1109/10.52350

[3] SKOLNIK, M. I. *Introduction to Radar Systems*. 2nd ed. New York, NY (USA): McGraw & Hill, 1980. ISBN: 0070579091

[4] XU, Y., DAI, T., SYCARA, K., et al. Service level differentiation in multi-robots control. In *Proceedings of the International Conference on Intelligent Robots and Systems (IROS)*. Taipei (China), 2010, p. 2224–2230. DOI: 10.1109/IROS.2010.5649366

[5] AL-ALAOUI, M. A. Novel digital integrator and differentiator. *IET Electronics letters*, 1993, vol. 29, no. 4, p. 376–378. DOI: 10.1049/el:19930253

[6] NGO, N. Q. A new approach for the design of wideband digital integrator and differentiator. *IEEE Transactions on Circuits and Systems II: Express Briefs*, 2006, vol. 53, no. 9, p. 936–940. DOI: 10.1109/TCSII.2006.881806

[7] GUPTA, M., JAIN, M., KUMAR, B. Novel class of stable wideband recursive digital integrators and differentiators. *IET Signal Processing*, 2010, vol. 4, no. 5, p. 560–566. DOI: 10.1049/iet-spr.2009.0030

[8] DEVATE, J., KULKARNI, S. Y., PAI, K. R. Wideband IIR digital integrator and differentiator using curve fitting technique. In *Proceedings of the International Conference on Signal Processing, Communication and Networking (ICSCN)*. 2015, p. 1–4. DOI: 10.1109/ICSCN.2015.7219845

[9] GUPTA, M., JAIN, M., KUMAR, B. Recursive wideband digital integrator and differentiator. *International Journal of Circuit Theory and Applications*, 2011, vol. 39, no. 7, p. 775–782.

[10] UPADHYAY, D. K. Class of recursive wideband digital differentiators and integrators. *Radioengineering*, 2012, vol. 21, no. 3, p. 904–910. ISSN: 1805-9600

[11] UPADHYAY, D. K. Recursive wideband digital differentiators. *IET Electronics letters*, 2010, vol. 46, no. 25, p. 1661–1662. DOI: 10.1049/el.2010.2113

[12] AL-ALAOUI, M. A. Class of digital integrators and differentiators. *IET Signal Processing*, 2011, vol. 5, no. 2, p. 251–260. DOI: 10.1049/iet-spr.2010.0107

[13] AL-ALAOUI, M. A., BAYDOUN, M. Novel wideband digital differentiators and integrators using different optimization techniques. In *Proceedings of the International Symposium on Signals, Circuits and Systems (ISSCS)*. Iasi (Romania), 2013, p. 1–4. DOI: 10.1109/ISSCS.2013.6651225

[14] GUPTA, M., RELAN, B., YADAV, R., et al. Wideband digital integrators and differentiators designed using particle swarm optimisation. *IET Signal Processing*, 2014, vol. 8, no. 6, p. 668–679. DOI: 10.1049/iet-spr.2013.0011

[15] JAIN, M., GUPTA, M., JAIN, N. K. Analysis and design of digital IIR integrators and differentiators using minimax and pole, zero, and constant optimization methods. *ISRN Electronics*, 2013, Article ID: 493973. DOI: 10.1155/2013/493973

[16] GUPTA, M., JAIN, M., JAIN, N. Linear phase second order recursive digital integrators and differentiators. *Radioengineering*, 2012, vol. 21, no. 2, p. 712–717. ISSN: 1805-9600

[17] FETTWEIS, A. Wave digital filters: Theory and practice. *IEEE Proceeding*, 1986, vol. 74, no. 2, p. 270–327. DOI: 10.1109/PROC.1986.13458

[18] YLI-KAAKINEN, J., SARAMÄKI, T. A systematic algorithm for the design of lattice wave digital filters with short-coefficient wordlength. *IEEE Transaction on Circuits and Systems-I*, 2007, vol. 54, no. 8, p. 1838–1851. DOI: 10.1109/TCSI.2007.902513

[19] GAZSI, L. Explicit formulas for lattice wave digital filters. *IEEE Transaction on Circuits and Systems*, 1985, vol. 32, no. 1, p. 68–88. DOI: 10.1109/TCS.1985.1085595

[20] AGGARWAL, M., BARSAINYA, R., RAWAT, T. K. FPGA implementation of Hilbert transformer based on lattice wave digital filters. In *Proceedings of the IEEE Conference on Reliability, Infocom Technologies and Optimization (ICRITO)*. Noida (India), 2015, p. 1–5. DOI: 10.1109/ICRITO.2015.7359331

[21] BARSAINYA, R., AGGARWAL, M., RAWAT, T. K. Multiplierless implementation of quadrature mirror filter. In *Proceedings of the IEEE Conference on Reliability, Infocom Technologies and Optimization (ICRITO)*. Noida (India), 2015, p. 1–6. DOI: 10.1109/ICRITO.2015.7359328

[22] BARSAINYA, R., AGGARWAL, M., RAWAT, T. K. Minimum multiplier implementation of a comb filter using lattice wave digital filter. In *Proceedings of the Annual IEEE India Conference (INDICON)*. New Delhi (India), 2015, p. 1–6. DOI: 10.1109/INDICON.2015.7443491

[23] GROSSMANN, L. D., ELDAR, Y. C. An L_1-method for the design of linear-phase FIR digital filters. *IEEE Transactions on Signal Processing*, 2007, vol. 55, no. 11, p. 5253–5266. DOI: 10.1109/TSP.2007.896088

[24] HASHIM, H. A., EL-FERIK, S., ABIDO, M. A. A fuzzy logic feedback filter design tuned with PSO for L_1 adaptive controller. *Expert Systems with Applications*, 2015, vol. 42, no. 23, p. 9077–9085. DOI: 10.1016/j.eswa.2015.08.026

[25] YANG, X. S., DEB, S. Cuckoo search via Lévy flights. In *Proceedings of the 2009 World Congress on Nature & Biologically Inspired Computing (NaBIC)*. Coimbatore (India), 2009, p. 210–214. DOI: 10.1109/NABIC.2009.5393690

[26] YANG, X. S., DEB, S. Engineering optimisation by cuckoo search. *International Journal of Mathematical Modelling and Numerical Optimisation*, 2010, vol. 1, no. 4, p. 330–343. arXiv: 2010arXiv1005.2908Y

[27] YANG, X. S., DEB, S. Cuckoo search: Recent advances and applications. *Neural Computing and Applications*, 2014, vol. 24, no. 1, p. 169–174. DOI: 10.1007/s00521-013-1367-1

About the Authors . . .

Richa BARSAINYA incurred her B. Tech degree in Electronics and communication Engineering in the year 2010 from Bundelkhand University, Uttar Pradesh. She completed M. Tech in signal processing from Ambedkar Institute of Technology, Guru Gobind Singh Indraprastha University in 2012. Presently, she is pursuing Ph.D. in the field of wave digital filter design from Netaji Subhas Institute of Technology, University of Delhi.

Tarun Kumar RAWAT received his M. Tech and Ph.D. from NSIT, University of Delhi, in 2003 and 2010, respectively. Presently, he is working in NSIT as an Assistant Professor in the Department of Electronics and Communication Engineering department. His teaching and research work include digital signal processing, statistical signal processing, VLSI signal processing and wave digital filter. He has authored/co-authored over 50 research papers and contributed two books: Signal and System and Digital Signal Processing in Oxford University Press.

3

Power Allocation and Low Complexity Detector for Differential Full Diversity Spatial Modulation Using Two Transmit Antennas

Kavishaur DWARIKA [1], *Hongjun XU* [1,2]

[1] School of Engineering, University of KwaZulu-Natal, King George V Avenue, 4041 Durban, South Africa
[2] School of Information and Electronic Engineering, Zhejiang Gongshang University, P.R. China

kavishaur@gmail.com, xuh@ukzn.ac.za

Abstract. *Differential full diversity spatial modulation (DFD-SM) is a differential spatial modulation (DSM) scheme that makes use of a cyclic unitary M-ary phase shift keying (M-PSK) constellation to achieve diversity gains at both the transmitter and receiver. In this paper, we extend the power allocation concept of generalized differential modulation (GDM) to DFD-SM to improve its block error rate (BLER). A novel power allocation scheme is formulated, and its optimum power allocation is derived. An asymptotic upper bound is presented for the new scheme and results are verified through Monte Carlo simulations. It can be seen that for a large enough frame length, the proposed scheme can almost achieve coherent performance. We also propose a low complexity detection scheme for DFD-SM. We evaluate the computational complexity of the maximum-likelihood (ML) detector and compare it to that of the proposed algorithm. It is shown that our scheme is independent of the constellation size. Numerical simulations of the BLER are presented, and it can be seen that the proposed scheme provides near-ML performance throughout the entire signal-to-noise ratio (SNR) range with a complexity reduction of about 55 % and 52 % for one and two receive antennas respectively, in the high SNR region.*

Keywords

Spatial Modulation (SM), differential spatial modulation (DSM), full transmit diversity, maximum-likelihood (ML) decoding, computational complexity

1. Introduction

Spatial Modulation (SM) [1] is an efficient multiple-input multiple-output (MIMO) system which has a low complexity implementation. Coherent SM generally requires full knowledge of the channel state information (CSI), which adds to the complexity of implementing the system at the receiver. Coherent systems are also susceptible to pilot overhead and estimation errors [2]. Non-coherent systems do not require CSI and are thus less complex to be implemented at the receiver, however they do suffer from an error performance penalty when compared to coherent systems. As such, to combat pilot overhead and estimation errors, multiple differentially encoded SM (DSM) systems have been introduced in [3–6].

Bian et al. in [3] introduced the concept of an $N_R \times 2$ DSM system, where N_R is the total number of receive antennas, and 2 is the total number of transmit antennas. In DSM, communication is carried out block-wise. Two antenna matrices are created, which encode the space and time dimensions of the two M-ary phase shift keying (M-PSK) symbols to be transmitted in two time slots. At any given time slot, only one transmit antenna is active. The recursive formula to differentially encode the transmit symbols is introduced. A maximum-likelihood (ML) detector is derived which estimates the transmitted symbols without the need for CSI. The detector searches through a total of $2M^2$ possible combinations in order to find the optimum solution.

Bian et al. in [4] further extended the work of [3] to an $N_R \times N_T$ DSM system, where N_T is the total number of transmit antennas. The design for antenna selection is introduced to accommodate for the increase in the number of antenna configurations. This increases the system's spectral efficiency. The ML detector has to search through a total of $2^{\log_2 \lfloor N_T! \rfloor} M^{N_T}$ possible combinations to find the optimal solution, where $\lfloor \cdot \rfloor$ denotes the floor function. The system's results are compared with that of conventional SM, and it can be seen that DSM only suffers from a 3 dB penalty [4].

Ishikawa in [5] introduced a unified DSM architecture. In order to attain a diversity gain, the number of symbols employed per antenna-index block is a design variable. It can be seen that based on this design, a flexible rate-diversity tradeoff is achieved [5].

Zang et al. in [6] proposed an $N_R \times 2$ DSM scheme which uses a cyclic M-PSK constellation to achieve full diversity, i.e. transmit and receive diversity. In order to achieve

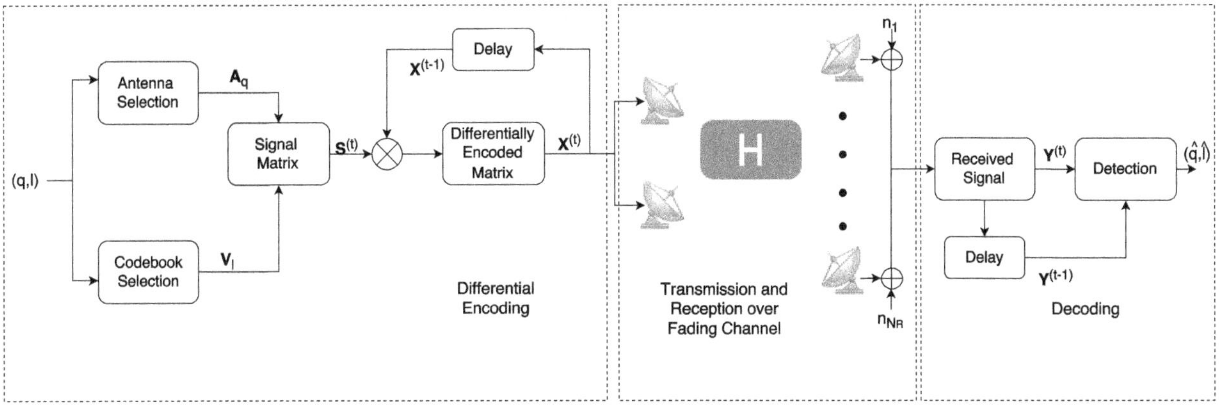

Fig. 1. System model of DFD-SM [6].

a transmit diversity gain; the data rate has to be lowered, which coincides with [5]. The system transmits the same symbol over the two time slots. The ML detector searches a total of $2M$ combinations in order to find the optimal solution.

In order to improve error performance of conventional differential modulation (CDM) a generalized differential modulation (GDM) scheme is introduced in [7], [8]. In GDM, a frame is split up into two parts, namely a reference part and a normal part. Both the reference and normal parts convey information. The reference part differentially encodes the normal part in the current frame and the reference part in the next frame. The system allocates more power to the reference part in order to improve the system's error performance. It can be seen, that for a large enough frame length, the error performance of GDM can almost approach that of coherent detection. The optimal power allocation of GDM for two-way amplify-and-forward relaying [7] differs from that of space-time block codes [8], as it depends on the statistics of the differential modulation scheme. However, the proposed power allocation scheme in [8] can be applied to any differential modulation scheme, as it is only dependent on the structure of the received signal in CDM.

The work of [7], [8] motivates us to extend the power allocation concept of GDM to differential full diversity spatial modulation (DFD-SM) to further improve its error performance. We also propose a low complexity detection algorithm for DFD-SM, as only the ML detector is discussed in literature.

The paper is organized as follows: Section 2 is broken down into 3 subsections. Section 2.1 gives a brief overview of conventional DFD-SM and introduces its system model. Section 2.2 introduces the proposed scheme and Sec. 2.3 discusses the power allocation of the proposed system. In Sec. 3, the optimum power allocation and asymptotic upper bound on the block error rate (BLER) are derived. Section 4 introduces the low complexity detection scheme for conventional DFD-SM and Sec. 5 explores the complexity analysis of the proposed detection scheme against the optimal detector. Section 6 provides the simulation results and discussion and finally, Section 7 concludes the paper.

Notation used in this paper: Bold upper/lower case letters represent matrices/vectors. $(\cdot)^T$, $(\cdot)^H$, and $(\cdot)^*$ represent the transpose, Hermitian and complex conjugate operations respectively. $X(i, j)$ denotes the element located at the ith row and j^{th} column of matrix $\mathbf{X}$ and $\text{Tr}(\mathbf{X})$ denotes the trace operation, which is the sum of all elements on the main diagonal of matrix $\mathbf{X}$. $\Re(z)$ and $\angle z$ denotes the real part and the phase of the complex number z respectively. $\arg\max$ returns the maximum argument passed to it and $(\cdot)!!$ denotes the double factorial operator. The modulo operation is represented as $\bmod(x, y) = x - y\lfloor x/y \rfloor$ and $\text{round}(x)$ rounds x up or down to the nearest integer.

2. System Model

In this section, the system model of conventional DFD-SM is first introduced. Based on GDM, we discuss the system model of the proposed scheme. Finally, we discuss the power allocation of the proposed DFD-SM system.

2.1 Conventional DFD-SM System

The conventional DFD-SM system is represented in Fig. 1 [6]. The system consists of two transmit antennas and N_R receive antennas. Let $\mathbf{H} = \begin{bmatrix}\mathbf{h}_1 & \mathbf{h}_2\end{bmatrix}$ and $\mathbf{N} = \begin{bmatrix}\mathbf{n}_1 & \mathbf{n}_2\end{bmatrix}$ denote the $N_R \times 2$ fading channel matrix and the $N_R \times 2$ additive white Gaussian noise (AWGN) channel matrix, respectively, where $\mathbf{h}_i = [h_{1,i} h_{2,i} \dots h_{N_R,i}]^T$ and $\mathbf{n}_i = [n_{1,i} n_{2,i} \dots n_{N_R,i}]^T$, $i = 1, 2$. The entries of $\mathbf{h}_i$ and $\mathbf{n}_i$ are independent and identically distributed (i.i.d.) complex Gaussian random variables with zero mean and a variance of 0.5 and $\frac{\sigma_n^2}{2}$ per dimension, respectively. The transmitted symbols are drawn from a unit M-PSK constellation. The system's average signal-to-noise ratio (SNR) for conventional DFD-SM is therefore defined as $\bar{\gamma} = \frac{1}{\sigma_n^2}$. In conventional DFD-SM, communication is carried out block-wise. Two antenna index matrices are defined according to [6] as

$$\mathbf{A}_0 = \begin{bmatrix} 1 & 0 \\ 0 & 1 \end{bmatrix}, \mathbf{A}_1 = \begin{bmatrix} 0 & \exp(\mathrm{j}\phi) \\ \exp(\mathrm{j}\phi) & 0 \end{bmatrix} \tag{1}$$

where ϕ is the rotation angle to be optimized to achieve transmit diversity. The codebook $\mathcal{V} = \{\mathbf{V}_0, \mathbf{V}_1, \dots, \mathbf{V}_{M-1}\}$

as seen in Fig. 1, is defined to be the set of M distinct unitary matrices chosen from a cyclic signal constellation whose l^{th} element is of the form $\mathbf{V}_l = \text{diag}\left(\exp\left(\frac{j2\pi u_1 l}{M}\right), \exp\left(\frac{j2\pi u_2 l}{M}\right)\right)$, where the parameters of u_1 and u_2 are optimized to achieve full transmit diversity [6]. $\mathbf{V}_l$ consists of a single information carrying M-PSK symbol over the two symbol durations. It can be seen that only a single antenna is activated for each symbol duration. From the analysis provided in [6], it was found that $\phi = \pi/4$, $u_1 = 1$ and $u_2 = 7$ for $M = 16$, which will be used in this paper.

In the tth block, $\log_2(M) + 1$ information bits are mapped to (q, l) where q, $q \in \{0, 1\}$, indicates the selected antenna activation order matrix $\mathbf{A}_q$ and l, $l \in \{0, 1, \ldots, M-1\}$, indicates the selected unitary matrix $\mathbf{V}_l$ [6]. A $\mathbb{C}^{2\times 2}$ signal matrix $\mathbf{S}^{(t)}$, which encodes the space and time dimensions, is defined as [6]

$$\mathbf{S}^{(t)} = \mathbf{A}_q^{(t)}\mathbf{V}_l^{(t)}. \tag{2}$$

The signal matrix is differentially encoded in a $\mathbb{C}^{2\times 2}$ space-time matrix $\mathbf{X}^{(t)}$ as [6]

$$\mathbf{X}^{(t)} = \mathbf{X}^{(t-1)}\mathbf{S}^{(t)}. \tag{3}$$

In the first block, the differentially encoded matrix is set as $\mathbf{X}^{(0)} = \mathbf{I}_2 = \mathbf{V}_0$, where $\mathbf{I}_2$ represents the 2×2 identity matrix, for simplicity.

The received signal in the tth block is given by

$$\mathbf{Y}^{(t)} = \mathbf{H}^{(t)}\mathbf{X}^{(t)} + \mathbf{N}^{(t)}. \tag{4}$$

Assuming quasi-static fading, $\mathbf{H}^{(t)} = \mathbf{H}^{(t-1)}$, the ML detector can be derived as [6]

$$(\hat{q}, \hat{l}) = \arg \max_{\substack{\hat{q}\in\{0:1\}\\ \hat{l}\in\{0:M-1\}}} \Re\left[\text{Tr}\left(\left(\mathbf{Y}^{(t)}\right)^H \mathbf{Y}^{(t-1)}\mathbf{A}_{\hat{q}}\mathbf{V}_{\hat{l}}\right)\right]. \tag{5}$$

2.2 Proposed DFD-SM System

In this subsection, we extend the power allocation concept of GDM to the conventional DFD-SM system. In [8], the differential detector is considered to have an *estimation-detection* structure. This implies that the previous received block, $\mathbf{Y}^{(t-1)}$, is used as an estimation to the fading channel matrix, $\mathbf{H}^{(t)}$, in order to coherently detect the information conveyed in the current received block, $\mathbf{Y}^{(t)}$ [8]. We exploit this property in our new scheme.

We assume each frame contains $(K + 1)$ blocks. K blocks, defined as normal blocks, will convey information. The first block transmitted in a frame, defined as a reference block, will serve as reference to the next K blocks in the frame. The normal and reference blocks are transmitted with unequal power, with more power allocated to the reference block.

The reference (first) block transmitted in a frame is given by

$$\mathbf{Y}_{\text{ref}} = \mathbf{H}_{\text{ref}}\mathbf{X}_{\text{ref}} + \mathbf{N}_{\text{ref}} \tag{6}$$

where $\mathbf{X}_{\text{ref}} = \mathbf{I}_2 = \mathbf{V}_0$. The entries of $\mathbf{N}_{\text{ref}}$ are i.i.d. complex Gaussian random variables with $\mathcal{CN}(0, \sigma^2_{\text{ref}})$ distribution in the reference block. Thus, the reference block has an average SNR of $\bar{\gamma}_{\text{ref}} = \frac{1}{\sigma^2_{\text{ref}}}$. This block provides the channel estimation for the next K blocks in the frame. We further denote the K received normal blocks as

$$\mathbf{Y}^{(t)}_{\text{norm}} = \mathbf{H}^{(t)}_{\text{norm}}\mathbf{X}^{(t)}_{\text{norm}} + \mathbf{N}^{(t)}_{\text{norm}}, \qquad 1 \le t \le K \tag{7}$$

where $\mathbf{X}^{(t)}_{\text{norm}} = \mathbf{X}_{\text{ref}}\mathbf{S}^{(t)}$. The entries of $\mathbf{N}^{(t)}_{\text{norm}}$ are also i.i.d. complex Gaussian random variables with $\mathcal{CN}(0, \sigma^2_{\text{norm}})$ distribution in the normal block. Thus, the normal block has an average SNR of $\bar{\gamma}_{\text{norm}} = \frac{1}{\sigma^2_{\text{norm}}}$.

For quasi-static fading, $\mathbf{H}_{\text{ref}} = \mathbf{H}^{(t)}_{\text{norm}}$, $1 \le t \le K$, we can re-write the received signal in (7) as

$$\mathbf{Y}^{(t)}_{\text{norm}} = \mathbf{Y}_{\text{ref}}\mathbf{S}^{(t)} - \mathbf{N}_{\text{ref}}\mathbf{S}^{(t)} + \mathbf{N}^{(t)}_{\text{norm}}. \tag{8}$$

The ML detector can now be derived similar to [6] as

$$(\hat{q}, \hat{l}) = \arg \max_{\substack{\hat{q}\in\{0:1\}\\ \hat{l}\in\{0:M-1\}}} \Re\left[\text{Tr}\left(\left(\mathbf{Y}^{(t)}_{\text{norm}}\right)^H \mathbf{Y}_{\text{ref}}\mathbf{A}_{\hat{q}}\mathbf{V}_{\hat{l}}\right)\right]. \tag{9}$$

2.3 Power Allocation

In GDM [7], [8], both schemes allocate a very large portion of power to the reference blocks in the frame. In order to ensure an average transmit power, P at the transmitter, the system needs to abide to the following power constraint [7], [8]

$$P_1 + (L-1)P_2 = LP \tag{10}$$

where P_1 is the power allocated to the reference block, P_2 is the power allocated to the normal block and L is the number of blocks in a frame. The optimal power allocation for each scheme is found by formulating a minimization problem based on (10), as well as the statistics of the modulation scheme. In [7], the power allocation problem is formulated into a function of one variable based on the performance analysis of the system, after which the derivative is taken in order to find the optimal solution, whereas in [8], the Lagrange multiplier method is used to find the optimum power allocation. Similar to GDM, we introduce a type of power allocation to DFD-SM.

Defining the average transmit power constraint (10) in terms of the system's average SNR, $\bar{\gamma}$, for our proposed scheme, we have: $\bar{\gamma}_{\text{ref}} + K\bar{\gamma}_{\text{norm}} = (K+1)\bar{\gamma}$. We propose a novel re-allocation of power scheme. First, we remove a fraction of power, denoted as α, from each of the normal blocks in the frame. This can be represented mathematically, in terms of the system's average SNR, as $\bar{\gamma}_{\text{norm}} = (1-\alpha)\bar{\gamma}$. We then re-allocate this fraction of power from all K normal blocks to the reference block, i.e. $\bar{\gamma}_{\text{ref}} = (1 + K\alpha)\bar{\gamma}$. It can be seen that the power allocated to the reference block is greater than that of the normal blocks, so as to improve channel estimation and hence reduce errors. The proposed scheme does not require any information on the statistics of the system and can therefore be applied to any differential modulation scheme.

3. Optimal Power Allocation and Asymptotic Error Performance Analysis

In this section we find the optimal power allocation for the proposed DFD-SM scheme and then derive an upper bound on the BLER. Following the normal and reference SNRs defined in the previous section, making the variance of noise the subject of the formula, we have

$$\sigma_{\text{ref}}^2 = \frac{1}{(1+K\alpha)\bar{\gamma}} \quad \text{and} \quad \sigma_{\text{norm}}^2 = \frac{1}{(1-\alpha)\bar{\gamma}}. \tag{11}$$

If we analyze the received signal in (8), it is obvious that the signal contains two noise elements comprising of two different variances. We define the effective noise variance, which is the coherent equivalent noise variance, as

$$\sigma_{\text{eff}}^2 = \frac{\sigma_{\text{ref}}^2 + \sigma_{\text{norm}}^2}{2}. \tag{12}$$

We can find the optimal α by minimizing the noise variance, i.e. taking the derivative of (12) with respect to α and equating it to 0. After some algebraic manipulations, we find that

$$\alpha_{\text{opt}} = \frac{\sqrt{K}-1}{\sqrt{K}+K} \tag{13}$$

When selecting the value of K, one should be mindful of the practical implementation of the peak to average power ratio (PAPR) required [8], as well as the block duration for which the channel remains constant.

Based on the optimal power allocation, we can derive the effective SNR and thus further derive the asymptotic BLER. The effective SNR can be derived following (12) as $\bar{\gamma}_{\text{eff}} = \frac{1}{\sigma_{\text{eff}}^2}$. This is found to be

$$\bar{\gamma}_{\text{eff}} = \left[\frac{2(1-\alpha)(1+K\alpha)}{2+(K-1)\alpha}\right]\bar{\gamma}. \tag{14}$$

By providing one reference block with a high SNR, for the K normal blocks, the effective SNR per block is increased.

From [6], the asymptotic upper bound on the BLER for conventional DFD-SM is given by

$$P_{\text{BLER}_{\text{conv}}} \leq \left(\frac{1}{\bar{\gamma}^{2N_R}}\right)\frac{(4N_R-1)!!}{2(4N_R)!!} \left[\sum_{\hat{l}=0}^{M-1}\left(\frac{2}{\Lambda(u_1+u_2,\hat{l},\phi)}\right)^{2N_R} + \sum_{\hat{l}=1}^{M-1}\left(\frac{2}{\Delta(u_1,u_2,\hat{l})}\right)^{2N_R}\right] \tag{15}$$

where $\Lambda(u,l,\phi) = \left|\sin\left(\frac{\pi u l}{M}-\phi\right)\right|$ and $\Delta(u_1,u_2,l) = \left|\sin\left(\frac{\pi u_1 l}{M}\right)\sin\left(\frac{\pi u_2 l}{M}\right)\right|$. Substituting (14) into (15), the upper bound for the proposed power scheme is found to be

$$P_{\text{BLER}_{\text{new}}} \leq \left(\frac{1}{\bar{\gamma}_{\text{eff}}^{2N_R}}\right)\frac{(4N_R-1)!!}{2(4N_R)!!} \left[\sum_{\hat{l}=0}^{M-1}\left(\frac{2}{\Lambda(u_1+u_2,\hat{l},\phi)}\right)^{2N_R} + \sum_{\hat{l}=1}^{M-1}\left(\frac{2}{\Delta(u_1,u_2,\hat{l})}\right)^{2N_R}\right]. \tag{16}$$

4. Low Complexity Detection Scheme

In conventional DFD-SM, the ML detector seen in (5), searches through a total of $2M$ possible combinations made up of all codebook and antenna index elements. The authors in [6] suggest that the high complexity of the detector in the proposed scheme is outweighed by the performance gains of the systems against which it was compared. In DFD-SM, there exists a symmetric relationship between the two symbols contained in each codebook entry. We exploit this relationship and propose a low complexity detection algorithm in this section. In the proposed detection algorithm, we first estimate the received symbols based on the activated antennas. We then estimate which elements of codebook $\mathcal{V}$ was received based on the estimated received symbols. Using these estimates, we reduce the number of elements needed to be tested by the ML detector, thereby reducing the complexity of the conventional scheme. The proposed detection scheme comprises of three steps. In the first two steps, we assume $\mathbf{A}_{\hat{q}} = \mathbf{A}_0$ and $\mathbf{A}_{\hat{q}} = \mathbf{A}_1$ respectively and apply our algorithm for each case. In the final step, we choose the most likely solution.

Step 1: Low Complexity Detection for $\mathbf{A}_0$.
Consider a symbol in an M-PSK constellation, of the form

$$s(l) = \exp\left(\frac{\mathrm{j}2\pi l}{M}\right), \qquad l = 0,1,\ldots,M-1. \tag{17}$$

We will denote the symbols found at element $\mathbf{V}_l$ in codebook $\mathcal{V}$ for DFD-SM as

$$s_{b_0}(l) = \exp\left(\frac{\mathrm{j}2\pi u_1 l}{M}\right) \text{ and } s_{p_0}(l) = \exp\left(\frac{\mathrm{j}2\pi u_2 l}{M}\right). \tag{18}$$

It can be seen that there exists a relationship between the phase of s_{b_0} and s_{p_0}, i.e. if the phase of $s_{b_0}(l) = \frac{2\pi u_1 l}{M}$, the phase of $s_{p_0}(l) = \frac{2\pi u_2 l}{M}$. This relationship can be seen in Fig. 2.

Following (2), the signal matrix is

$$\mathbf{S}^{(t)} = \mathbf{A}_0^{(t)}\mathbf{V}_l^{(t)} = \begin{bmatrix} s_{b_0}(l) & 0 \\ 0 & s_{p_0}(l) \end{bmatrix}. \tag{19}$$

After obtaining the received signal in (4), we solve for the estimation of $s_{b_0}(l)$ and $s_{p_0}(l)$ as

$$\hat{s}_{b_0} = \sum_{i=1}^{N_R} Y^{(t)}(i,1)\left(Y^{(t-1)}(i,1)\right)^*, \tag{20}$$

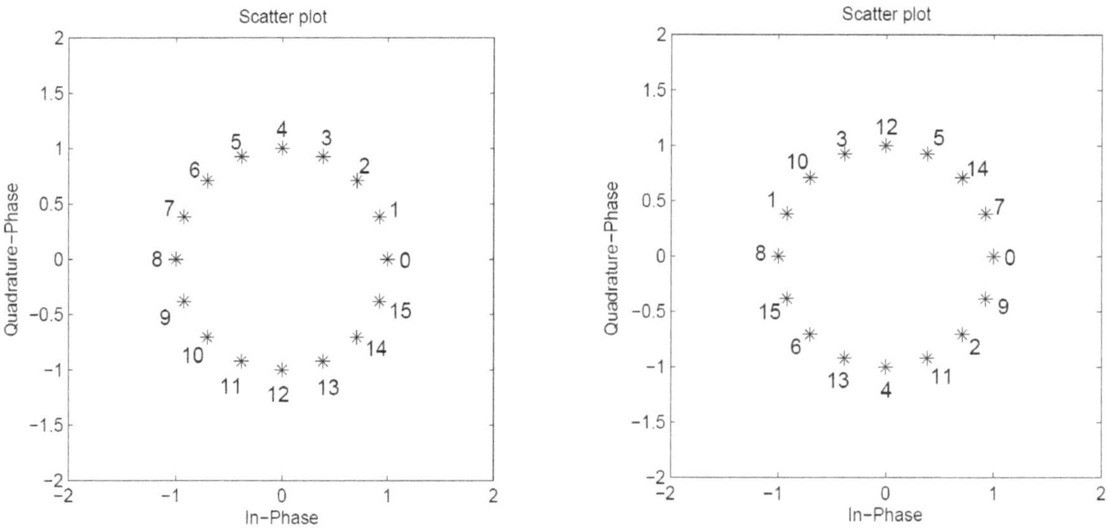

Fig. 2. Constellations of s_{b_0} and s_{p_0} for $u_1 = 1$, $u_2 = 7$ and $M = 16$.

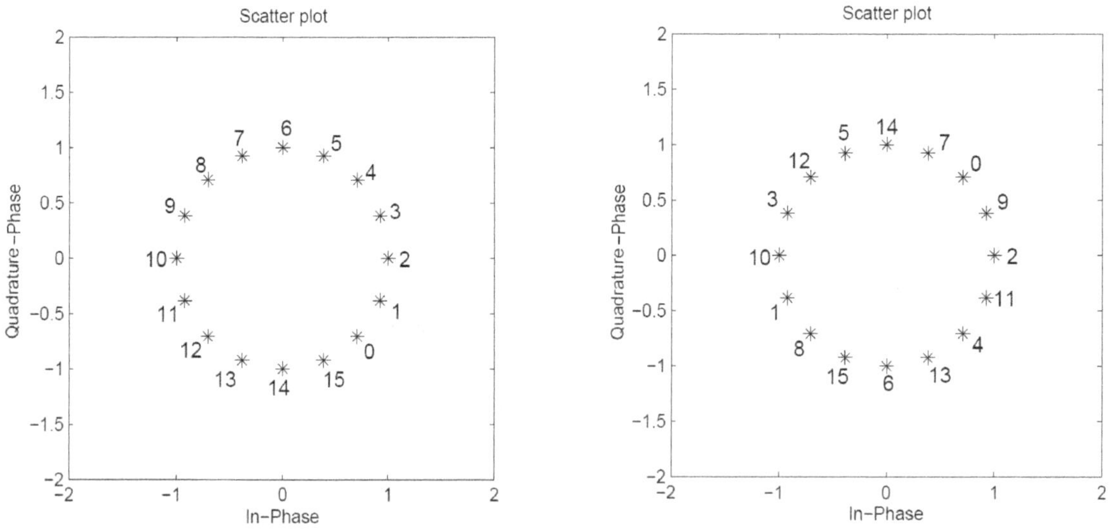

Fig. 3. Constellations of s_{b_1} and s_{p_1} for $u_1 = 1$, $u_2 = 7$, $\phi = \pi/4$ and $M = 16$.

$$\hat{s}_{p_0} = \sum_{i=1}^{N_R} Y^{(t)}(i,2)\left(Y^{(t-1)}(i,2)\right)^* . \tag{21}$$

Let $Q_b = \frac{\angle \hat{s}_{b_0}}{2\pi/M}$ and $Q_p = \frac{\angle \hat{s}_{p_0}}{2\pi/M}$ [9], we can then proceed to find their indices using a modified version of [9]

$$\hat{l}_b = \mod(\text{round}(Q_b), M), \tag{22}$$

$$\hat{l}_p = \mod(\text{round}(Q_p), M). \tag{23}$$

Substituting (22) and (23) into (17), we can solve for the transmitted symbols. Note that the solutions to (22) and (23) are of the form $\hat{l}_b = \mod(u_1 l', M)$, $\hat{l}_p = \mod(u_2 l'', M)$, $\hat{l}_b, \hat{l}_p, l', l'' \in \{0, 1, \ldots, M-1\}$ and l' and l'' are the indices of the codebook. In essence, we are finding the index of codebook $\mathcal{V}$, using the estimated symbols index. The received signal is distorted by the effects of fading and AWGN, and as a result the calculated index may lie in the wrong decision region. To improve the error performance of our detection scheme, we utilize a type of nearest neighbor algorithm. As a result, we add and subtract π/M to both Q_b and Q_p independently in order to more accurately determine its index, i.e.

$$\hat{Q}_{b_1} = Q_b + \frac{\pi}{M} \text{ and } \hat{Q}_{b_2} = Q_b - \frac{\pi}{M}, \tag{24}$$

$$\hat{Q}_{p_1} = Q_p + \frac{\pi}{M} \text{ and } \hat{Q}_{p_2} = Q_p - \frac{\pi}{M}. \tag{25}$$

Thereafter we proceed with using (22) and (23) to obtain our estimates. We will denote the two estimates obtained from $\hat{Q}_{b_i}$ as l_i' and from $\hat{Q}_{p_i}$ as l_i'', $i = 1, 2$, respectively. Note that the solutions of (22) and (23) using (24) and (25) respectively, are of the form $\mod(u_1 l_i', M)$ and $\mod(u_2 l_i'', M)$, respectively. In order to determine the l_i' and l_i'' solutions, we make use of look-up tables, which can be constructed using Fig. 2.

Now if $l' = l''$, it can be observed that the estimates of (22) and (23) lie in the same set. In this case, $\{l'_1, l'_2\} \in \{l''_1, l''_2\}$ and we only need to test two elements of Codebook $\mathcal{V}$

$$(0, \hat{l}_{A_0}) = \arg \max_{\substack{\hat{q}\in\{0\} \\ \hat{l}\in\{l'_1,l'_2\}}} \Re\left[\operatorname{Tr}\left(\left(\mathbf{Y}^{(t)}\right)^H \mathbf{Y}^{(t-1)}\mathbf{A}_0\mathbf{V}_{\hat{l}}\right)\right]. \quad (26)$$

However if $l' \neq l''$, we find that the estimates of (22) and (23) do not lie in the same set. In this case $\{l'_1, l'_2\} \notin \{l''_1, l''_2\}$ and we will have to test four elements of Codebook $\mathcal{V}$

$$(0, \hat{l}_{A_0}) = \arg \max_{\substack{\hat{q}\in\{0\} \\ \hat{l}\in\{l'_1,l'_2,l''_1,l''_2\}}} \Re\left[\operatorname{Tr}\left(\left(\mathbf{Y}^{(t)}\right)^H \mathbf{Y}^{(t-1)}\mathbf{A}_0\mathbf{V}_{\hat{l}}\right)\right]. \quad (27)$$

We store the maximum argument obtained from (26) or (27) as d_1.

Step 2: Low Complexity Detection for $\mathbf{A}_1$.
For $\mathbf{A}_{\hat{q}} = \mathbf{A}_1$, (18) is now modified to include the rotation angle

$$s_{b_1}(l) = \exp\left(\frac{\mathrm{j}2\pi u_1 l}{M} + \mathrm{j}\phi\right) \text{ and } s_{p_1}(l) = \exp\left(\frac{\mathrm{j}2\pi u_2 l}{M} + \mathrm{j}\phi\right). \quad (28)$$

For the case of $\phi = \pi/4$ and $M = 16$, we have $s_{b_1}(l) = \exp\left(\frac{\mathrm{j}2\pi(u_1 l+2)}{16}\right)$ and $s_{p_1}(l) = \exp\left(\frac{\mathrm{j}2\pi(u_2 l+2)}{16}\right)$. This shows that two is just being added to the index of the $\mathbf{A}_0$ case, and is represented in Fig. 3.

Following (2), the signal matrix is

$$\mathbf{S}^{(t)} = \mathbf{A}_1^{(t)}\mathbf{V}_l^{(t)} = \begin{bmatrix} 0 & s_{p_1}(l) \\ s_{b_1}(l) & 0 \end{bmatrix}. \quad (29)$$

Equations (20) and (21) now become

$$\hat{s}_{b_1} = \sum_{i=1}^{N_R} Y^{(t)}(i,1)\left(Y^{(t-1)}(i,2)\right)^*, \quad (30)$$

$$\hat{s}_{p_1} = \sum_{i=1}^{N_R} Y^{(t)}(i,2)\left(Y^{(t-1)}(i,1)\right)^*. \quad (31)$$

We then carry out all steps performed from (22) to (25), using look-up tables constructed from Fig. 3. The ML detector from (26) and (27) now becomes

$$(1, \hat{l}_{A_1}) = \arg \max_{\substack{\hat{q}\in\{1\} \\ \hat{l}\in\{l'_1,l'_2\}}} \Re\left[\operatorname{Tr}\left(\left(\mathbf{Y}^{(t)}\right)^H \mathbf{Y}^{(t-1)}\mathbf{A}_1\mathbf{V}_{\hat{l}}\right)\right], \quad (32)$$

$$(1, \hat{l}_{A_1}) = \arg \max_{\substack{\hat{q}\in\{1\} \\ \hat{l}\in\{l'_1,l'_2,l''_1,l''_2\}}} \Re\left[\operatorname{Tr}\left(\left(\mathbf{Y}^{(t)}\right)^H \mathbf{Y}^{(t-1)}\mathbf{A}_1\mathbf{V}_{\hat{l}}\right)\right], \quad (33)$$

respectively. We store the maximum argument obtained from (32) or (33) as d_2.

Step 3: Detection.
Now based on d_1 and d_2, the greater of the two will provide us with our transmitted bits.

The proposed low complexity detection scheme is for conventional DFD-SM [6]. The algorithm can be modified to be used with the new power allocation scheme by replacing $\mathbf{Y}^{(t-1)}$ with $\mathbf{Y}_{\text{ref}}$ in the above equations, however for the purposes of this paper, we only analyze it for the conventional scheme. The algorithm is summarized below

i. Perform $\mathbf{A}_0$ detection
   a. Solve for $\hat{s}_{b_0}$ and $\hat{s}_{p_0}$ using (20) and (21).
   b. Solve for l' and l'' from (22) and (23).
   c. Obtain $\hat{Q}_{b_i}$ by using (24), then solve for l'_i using (22), $i = 1, 2$.
   d. Obtain $\hat{Q}_{p_i}$ by using (25), then solve for l''_i using (23), $i = 1, 2$.
   e. If $l' = l''$, use the ML detector in (26). Else, use the ML detector in (27).
   f. Save the solution of the ML detector as $(0, \hat{l}_{A_0})$, along with its maximum value as d_1.

ii. Perform $\mathbf{A}_1$ detection
   a. Solve for $\hat{s}_{b_1}$ and $\hat{s}_{p_1}$ using (30) and (31).
   b. Solve for l' and l'' from (22) and (23).
   c. Obtain $\hat{Q}_{b_i}$ by using (24), then solve for l'_i using (22), $i = 1, 2$.
   d. Obtain $\hat{Q}_{p_i}$ by using (25), then solve for l''_i using (23), $i = 1, 2$.
   e. If $l' = l''$, use the ML detector in (32). Else, use the ML detector in (33).
   f. Save the solution of the ML detector as $(1, \hat{l}_{A_1})$, along with its maximum value as d_2.

iii. If $d_1 > d_2$, $(\hat{q}, \hat{l}) = (0, \hat{l}_{A_0})$. Else, $(\hat{q}, \hat{l}) = (1, \hat{l}_{A_1})$

5. Computational Complexity

In this section, we analyse the computational complexity of the proposed scheme and compare it to the optimal detection scheme. We use the concept of computational complexity, as discussed in [10], which is defined as the total number of real-valued multiplications in a given algorithm.

We first derive the computational complexity of the optimal detection scheme found in (5).

i. $\left(\mathbf{Y}^{(t)}\right)^H \mathbf{Y}^{(t-1)}$ is a 2×2 matrix and needs to be computed once. It requires $4N_R$ complex multiplications to be computed, which equates to $16N_R$ real multiplications.

ii. When $\mathbf{A}_{\hat{q}} = \mathbf{A}_0$
   - Each trial of $\mathbf{A}_0\mathbf{V}_{\hat{l}}$ will require 2 multiplications of a complex number by a real number which results in a total of 4 real multiplications.

iii. When $\mathbf{A}_{\hat{q}} = \mathbf{A}_1$

- Each trial of $\mathbf{A}_1\mathbf{V}_{\hat{l}}$ will require 2 multiplications of a complex number by another complex number which results in a total of 8 real multiplications.

iv. Computing $\left(\mathbf{Y}^{(t)}\right)^H \mathbf{Y}^{(t-1)}\mathbf{A}_{\hat{q}}\mathbf{V}_{\hat{l}}$ will require 4 complex multiplications, resulting in 16 real multiplications.

v. The trace and real operations require no multiplications.

The computational complexity for $\mathbf{A}_0$ is $C_{\text{ML1}} = 16N_R + (4+16)M$. The computational complexity for $\mathbf{A}_1$ is $C_{\text{ML2}} = 16N_R + (8+16)M$. Eliminating all common steps between C_{ML1} and C_{ML2}, the total computational complexity of the optimal detector for conventional DFD-SM is $C_{\text{ML}} = C_{\text{ML1}} + C_{\text{ML2}} - 16N_R = 16N_R + 44M$.

For the derivation of the proposed detection scheme's computational complexity, we consider the steps for $\mathbf{A}_0$, unless otherwise stated.

i. Computing both $\hat{s}_{b_0}$ and $\hat{s}_{p_0}$ requires a total of $2N_{\text{R}}$ complex multiplications, translating to $8N_{\text{R}}$ real multiplications.

ii. The calculation of l' and l'' requires 6 real multiplications in total. This is broken down into 2 real multiplications for computing both Q_b and Q_p, and the modulo operation for finding both $\hat{l}_b$ and $\hat{l}_p$ requires a total of 2×2 real multiplications. We treat $2\pi/M$ as a constant. l' and l'' are found from $\hat{l}_b$ and $\hat{l}_p$ respectively, using look-up tables and hence require no multiplications.

iii. The addition and subtraction of π/M (treated as a constant) to Q_b and Q_p (computed in step ii.) requires no multiplications. Solving for l'_1, l'_2, l''_1 and l''_2 using the modulo operation requires a total of 4×2 real multiplications. l'_1, l'_2, l''_1 and l''_2 can be obtained by using look-up tables and hence involve no real multiplications.

iv. We use the ML detector for the last step with a reduced search space. If $l' = l''$, we will require $16N_R + 2(4+16)$ real multiplications, else we will require $16N_R + 4(4+16)$ real multiplications. For $\mathbf{A}_1$ detection it will be $16N_R + 2(8+16)$ and $16N_R + 4(8+16)$ real multiplications respectively.

The computational complexity for $\mathbf{A}_0$ is $C_{\text{proposed1}} = 8N_R + 6 + 8 + 16N_R + \mu(4+16)$, $\mu \in \{2,4\}$. The computational complexity for $\mathbf{A}_1$ is $C_{\text{proposed2}} = 8N_R + 6 + 8 + 16N_R + \mu(8+16)$, $\mu \in \{2,4\}$. Eliminating all the common steps between the two, the total computational complexity is $C_{\text{proposed}} = C_{\text{proposed1}} + C_{\text{proposed2}} - 16N_R = 32N_R + 28 + 44\mu'$, $\mu' \in \{6,8\}$. The proposed scheme will always have either 6 or 8 estimates

Detection scheme	Real-valued multiplications
ML Detector (5)	$16N_R + 44M$
Proposed Detector	$32N_R + 28 + 44\mu'$, $\mu' \in \{6,8\}$

Tab. 1. Complexity order for DFD-SM detection schemes.

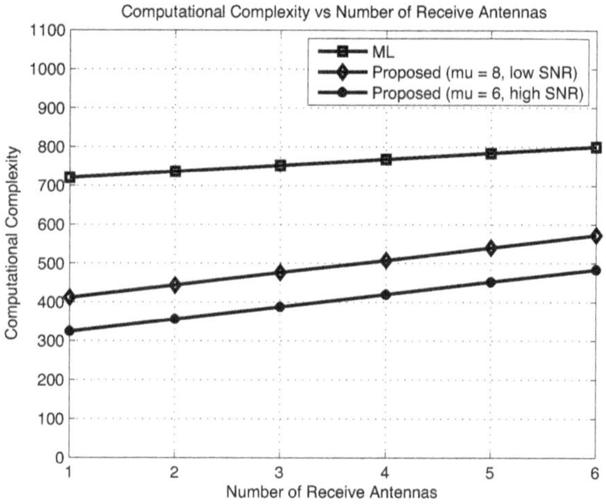

Fig. 4. Computational complexity vs. N_{R} $(M = 16)$.

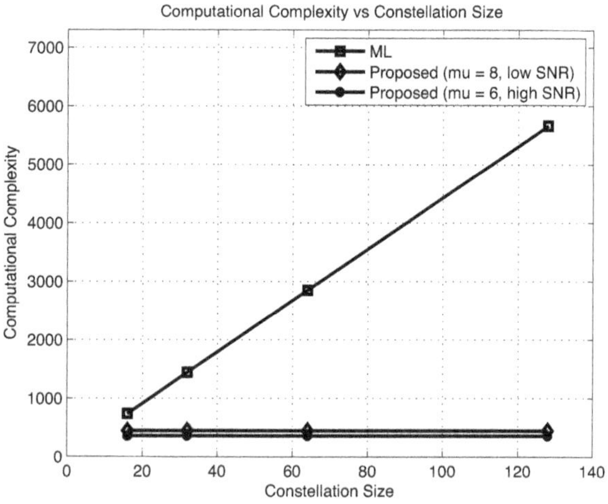

Fig. 5. Computational complexity vs. M $(N_{\text{R}} = 2)$.

and never 4, as a result $\mu' \in \{6,8\}$. We will assume that $\mu' = 6$ in the high SNR region, while $\mu' = 8$ will be in the low SNR region to aid discussion, although this will not always hold true. The computational complexity of both schemes have been summarized in Tab. 1.

Figures 4 and 5 show the computational complexity of the detection schemes as a function of N_{R} and M, respectively. It can be seen from Fig. 4, that our proposed scheme is highly dependent on N_{R} and the computational complexity increases as N_{R} increases for a 16-PSK constellation. However, for a practical number of receive antennas, it can be seen that the proposed scheme still requires fewer real multiplications as compared to the ML detector. Since the proposed algorithm reduces the search space of the optimal detector to either six or eight estimates, it is recommended that proposed detection scheme be used for constellations of order $M \geq 16$. Figure 5 highlights the fact that the proposed detection scheme is independent of the constellation's size M, as the computational complexity of 16,32,64,128-PSK constellations are found for $N_R = 2$. For $M = 16$ and $N_R = 1$, our algorithm demonstrates a 55% reduction in computational complexity in the high SNR region as compared to the optimal detector, and a 43% reduction in the low SNR region. For $M = 16$ and $N_R = 2$, a 52% reduction is realized in the high SNR region, and a 40% reduction in the low SNR region.

6. Simulations

For the simulations, we assumed a quasi-static Rayleigh fading channel. The simulations were performed for one and two receive antennas. Firstly we compared the new power allocation system against conventional DFD-SM in Figures 6 and 7. The upper bound derived in (16), as well as DFD-SM with coherent transmission/detection are also included in Figures 6 and 7. We choose a frame length of $K = 100$ and $K = 500$ for comparison. The BLER is plotted against the average SNR $\bar{\gamma}$ (in dB) for the proposed scheme. At a BLER $= 10^{-4}$, we see that the proposed scheme outperforms the conventional scheme by approximately 2 dB and is shown to be 1 dB behind that of the coherent scheme for $K = 500$. The proposed scheme is seen to obtain a gain of about 0.4 dB when the frame length, K, is increased from 100 to 500. Since α is a function of K, it can be seen that as the frame length increases, the power allocated to the reference block increases. This provides better channel estimation for the normal blocks, and thus better error performance. For a large enough K, the proposed scheme can approach the performance of coherent transmission/detection. The bound of (16) is observed to be tight at high SNR.

We next verify that α_{opt} found in (13) allows for optimal error performance. Using (13), we have $\alpha_{\text{opt}} = 0.0409$ for $K = 500$ and $\alpha_{\text{opt}} = 0.0818$ for $K = 100$. Fig. 8 contains a plot of $P_{\text{BLER}_{\text{new}}}$ (16) as a function of α, at $\bar{\gamma} = 30$ dB for $N_R = 1$ and $\bar{\gamma} = 20$ dB for $N_R = 2$ respectively. From Fig. 8, we observe that the BLER is a minimum when $\alpha = \alpha_{\text{opt}}$.

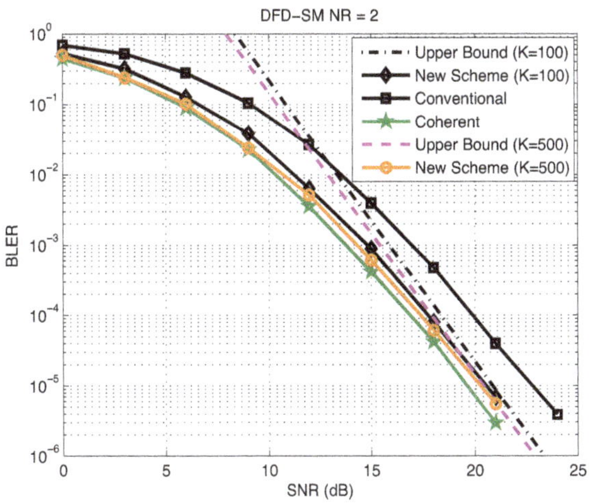

Fig. 7. BLER of conventional, proposed and coherent DFD-SM with $N_R = 2$.

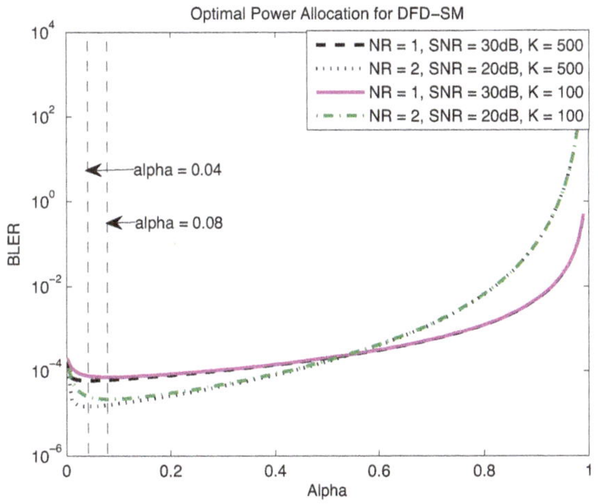

Fig. 8. Theoretically calculated upper bound of BLER vs. α.

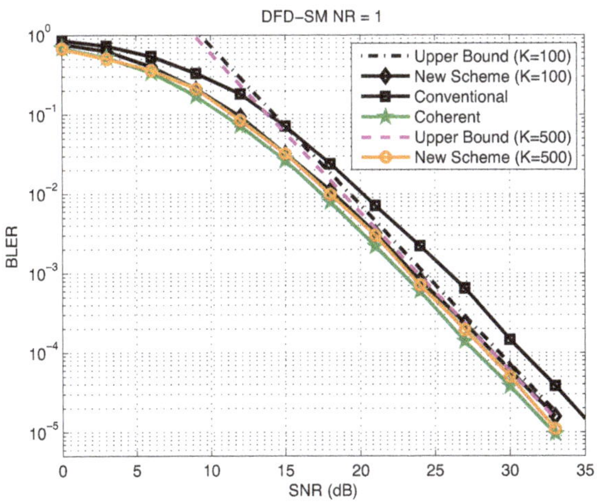

Fig. 6. BLER of conventional, proposed and coherent DFD-SM with $N_R = 1$.

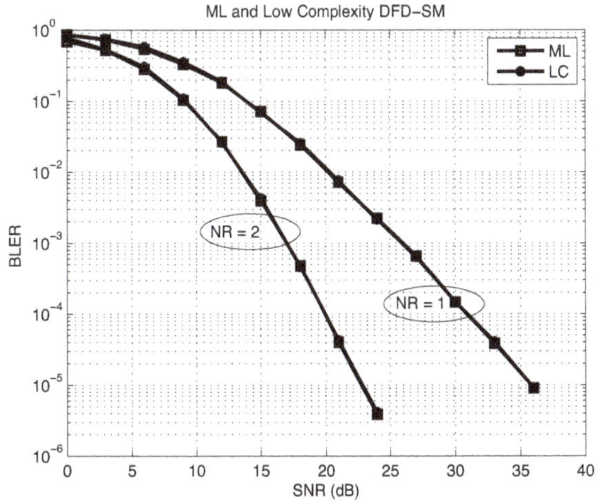

Fig. 9. DFD-SM low complexity and ML detection for $N_R = 1$ and $N_R = 2$.

Finally we compare the performance of our proposed low complexity (LC) detection scheme for conventional DFD-SM against the optimal detector found in (5) in Fig. 9. It is observed that the proposed algorithm provides near-ML performance throughout the entire SNR range.

7. Conclusion

In this paper we have provided a new power allocation scheme for DFD-SM, based on GDM. The optimal power allocation and theoretical upper bound on the BLER were derived. It was shown that the proposed scheme outperforms the conventional scheme and closes the gap between conventional differential detection and coherent detection. A low complexity detection scheme for conventional DFD-SM was also introduced. The computational complexity of the optimal detector and proposed detector were presented, with the proposed scheme providing approximately a 55 % and 52 % complexity reduction for one and two receive antennas, respectively. Numerical simulations show that the proposed scheme provides near-ML performance throughout the entire SNR range.

Acknowledgments

The financial assistance of the National Research Foundation (NRF) towards this research is hereby acknowledged. Opinions expressed and conclusions arrived at, are those of the author and are not necessarily to be attributed to the NRF.

References

[1] MESLEH, R., HAAS, H., SINANOVIC, S., et al. Spatial modulation. *IEEE Transactions on Vehicular Technology*, Jul. 2008, vol. 57, no. 4, p. 2228–2242. DOI: 10.1109/TVT.2007.912136

[2] SUGIURA, S., CHEN, S., HANZO, L. Coherent and differential space time shift keying: A dispersion matrix approach. *IEEE Transactions on Communications*, Nov. 2010, vol. 58, no. 11, p. 3219–3230. DOI: 10.1109/TCOMM.2010.093010.090730

[3] BIAN, Y., WEN, M., CHENG, X., et al. A differential scheme for spatial modulation. In *Proceedings of the 2013 IEEE Global Communications Conference (GLOBECOM)*. Atlanta (USA), 2013, p. 3925–3930. DOI: 10.1109/GLOCOM.2013.6831686

[4] BIAN, Y., CHENG, X., WEN, M., et al. Differential spatial modulation. *IEEE Transactions on Vehicular Technology*, Jul. 2015, vol. 64, no. 7, p. 3262–3268. DOI: 10.1109/TVT.2014.2348791

[5] ISHIKAWA, N., SUGIURA, S. Unified differential spatial modulation. *IEEE Wireless Communications Letters*, Aug. 2014, vol. 3, no. 4, p. 337–340. DOI: 10.1109/LWC.2014.2315635

[6] ZANG, W., YIN, Q., DENG, H. Differential full diversity spatial modulation and its performance analysis with two transmit antennas. *IEEE Wireless Communications Letters*, Apr. 2015, vol. 19, no. 4, p. 677–680. DOI: 10.1109/LCOMM.2015.2403859

[7] FANG, Z., LIANG, F., LI, L., et al. Performance analysis and power allocation for two-way amplify-and-forward relaying with generalized differential modulation. *IEEE Transactions on Vehicular Technology*, Feb. 2014, vol. 63, no. 2, p. 937–942. DOI: 10.1109/TVT.2013.2279856

[8] LI, L., FANG, Z., ZHU. Y., et al. Generalized differential transmission for STBC systems. In *Proceedings of the 2008 IEEE Global Communications Conference (GLOBECOM)*. New Orleans (USA), Dec. 2008, p. 1–5. DOI: 10.1109/GLOCOM.2008.ECP.836

[9] MEN, H., JIN, M. A low complexity ML detection algorithm for spatial modulation systems with MPSK constellation. *IEEE Communications Letters*, Jun. 2014, vol. 18, no. 8, p. 1375–1378. DOI: 10.1109/LCOMM.2014.2331283

[10] XIAO, L., YANG P., LEI X., et al. A low-complexity detection scheme for differential spatial modulation. *IEEE Communications Letters*, Jun. 2015, vol. 19, no. 9, p. 1516–1519. DOI: 10.1109/LCOMM.2015.2448616

About the Authors . . .

Kavishaur DWARIKA received his B.Sc. degree in Electronic Engineering from the University of KwaZulu-Natal, South Africa, in 2014. Since 2015, he is an M.Sc. candidate with the University of KwaZulu-Natal. His research interests include wireless communications and digital signal processing.

Hongjun XU (MIEEE, 07) received the B.Sc. degree from the University of Guilin Technology, Guilin, China, in 1984; the M.Sc. degree from the Institute of Telecontrol and Telemeasure, Shi Jian Zhuang, China, in 1989; and the Ph.D. degree from the Beijing University of Aeronautics and Astronautics, Beijing, China, in 1995. From 1997 to 2000, he was a Postdoctoral Fellow with the University of Natal, Durban, South Africa, and Inha University, Incheon, Korea. He is currently a Professor with the School of Engineering at the University of KwaZulu-Natal, Durban. He is the author of more than 20 journal papers. His research interests include wireless communications and digital systems.

4

DOA Estimation of Uncorrelated, Partly Correlated and Coherent Signals Using Alternating Oblique Projection

Huijun HOU [1], *Xingpeng MAO* [1,2]

[1] School of Electronics and Information Engineering, Harbin Institute of Technology, Harbin 150001, P. R. China
[2] Collaborative Innovation Center of Information Sensing and Understanding, Harbin Institute of Technology, Harbin 150001, P. R. China

mxp@hit.edu.cn

Abstract. *The problem of direction-of-arrival (DOA) estimation is important in array signal processing. To estimate the DOAs of uncorrelated, partly correlated and coherent signals, a new iterative DOA estimation algorithm, named AOP-DOA, is proposed by using alternating oblique projection (AOP). In each iteration, the oblique projection approach is employed to separate the received signals, then the DOA of each separated signal is estimated one after another. After theoretical analysis on the relationship between the proposed AOP-DOA and the conventional alternating projection based maximum likelihood estimator (AP-MLE), an AOP&AP-DOA algorithm, which is a combination of AOP-DOA and AP-MLE, is developed to reduce the computational complexity of AOP-DOA. Extensive experiments validate the effectiveness and complexity of the proposed two algorithms. Particularly, AOP&AP-DOA keeps the merits of AOP-DOA, but exhibits superiority over AOP-DOA in terms of computational complexity when proper adaptive grid refinement strategy is applied.*

Keywords

Oblique projection, direction-of-arrival (DOA) estimation, maximum likelihood (ML), coherent signal

1. Introduction

The problem of direction-of-arrival (DOA) estimation received considerable attention in the fields of radar, communication, sonar, etc. [1], [2]. The maximum likelihood (ML) [3–5] and subspace-based high resolution algorithms, including multiple signal classification (MUSIC) [6] and estimation of signal parameters via rotational invariance techniques (ESPRIT) [7], have been proved to be effective and computationally efficient for uncorrelated and partially correlated signals. The ML algorithms [3–5] are also capable of yielding asymptotically optimal solutions for fully correlated (coherent) signals, but they are computationally intensive since the multivariate nonlinear optimization procedure is involved [8]. The subspace-based MUSIC and ESPRIT are relatively computational simplicity and yield suboptimal solutions, but they are inefficient for coherent signals [9], [10].

To solve the aforementioned problems, several preprocessing algorithms, such as redundancy averaging [11], forward/backward spatial smoothing (FBSS) [12], [13], etc., are developed to decorrelate signals in the array covariance matrix. However, the algorithm of redundancy averaging leads to biased DOA estimates [14], and the FBSS algorithm suffers from performance degradation due to the array aperture loss [15], [16]. On the other hand, iterative algorithms, including the alternating projection (AP) technique [17], [18] and the modified Gauss-Newton technique [19] are investigated to circumvent the multivariate nonlinear optimization procedure. However, the global convergence of AP based ML estimator (AP-MLE) can not be guaranteed, and the Newton-type algorithms also have to be carefully initialized so that the global convergence is achieved [20].

Recently, much attention has been drawn to sparse signal reconstruction (SSR) perspective for DOA estimation [21]. Many SSR algorithms associated with l_p-norm-based ($1 \geq p \geq 0$) convex relaxation [22–25] and sparse Bayesian learning (SBL) [26], [27] are developed. They are capable of handling uncorrelated, partially correlated, and coherent signals, even if in the small sample-size case. Apart from the ML and SSR algorithms, the RELAX algorithm in [28] as well as the iterative adaptive approach for amplitude and phase estimation algorithm in [29] also works well in the same context.

In this paper, an alternating oblique projection (AOP) algorithm is proposed for DOA estimation. The technique of oblique projection (OP) [30] has a distinct advantage in terms of separating signal-of-interest (SOI) in the signal subspace while zero-forcing structured interferences in the interference subspace [31], [32]. In the context of DOA estimation, the OP technique was utilized to develop a scaled version of the MUSIC algorithm [33], and was utilized to integrate a prior known location of several sources into the MUSIC algorithm and circumvent their influences on the estimation of the unknown ones [34]. However, both are not applicable for coherent signals, since they aim at improving the performance of MUSIC for noncoherent signals. In [35–40], for DOA estimation of mixed coherent and noncoherent signals, the OP technique was utilized to eliminate the contributions of noncoherent signals from the data covariance matrix so that only those of coherent signals remain. Then the DOAs of coherent signals are estimated by the algorithm of spatial smoothing, redundancy averaging, etc.. But confined to the decorrelation algorithms, [35–40] were only researched with certain regular arrays (e.g., uniform linear array).

The proposed algorithm is named AOP-DOA, and is able to handle coherent signals. But different to [35–40], the proposed AOP-DOA has no limitation on antenna array geometry since the OP technique is differently used. Herein, AOP-DOA employs OP technique to alternately separate each of the received signals from array measurements. By this means, the received multiple signals are separated into a series of small signal groups, where each group contains one signal. Then the DOA of each signal is obtained by performing single target source localization. Based on AOP-DOA, we also propose an AOP&AP-DOA algorithm, which is a combination of AOP-DOA and AP-MLE, and reduces the computational complexity of AOP-DOA.

The main contributions of this paper are presented as follows.

1) By employing OP technique, a new AOP-DOA algorithm, which is applicable to arbitrary array geometries, is proposed for DOA estimation.

2) On the basis of AOP-DOA, an AOP&AP-DOA algorithm, which could reduce the computational complexity of AOP-DOA, is proposed.

3) The proposed two algorithms can work well with uncorrelated and correlated (including coherent) signals.

The remainder of the paper is organized as follows. Section 2 describes the system model and introduces the basic assumptions. The derivation and discussion of AOP-DOA and AOP&AP-DOA are presented in Sec. 3. In Sec. 4, simulation experiments are presented to evaluate the effectiveness of proposed algorithms, and performance comparisons between our algorithms and the existing algorithms are made. Finally, conclusions are drawn in Sec. 5.

Notations: matrices and vectors are denoted by boldface uppercase and lowercase letters, respectively. $(\cdot)^{\dagger}$, $(\cdot)^{\mathrm{H}}$, $\|\cdot\|$, $(\cdot)^{\perp}$, $E\{\cdot\}$ and $\mathrm{tr}\{\cdot\}$ denote the Moore-Penrose pseudoinverse, Hermitian transposition, l_2-norm, orthogonal complement, statistical expectation and trace operator, respectively. $(\cdot)^{-1}$ and $\mathrm{eig}(\cdot)$ stand for the inverse and eigendecomposition of the bracketed matrix, respectively. Additionally, $\oplus$ represents the direct sum operator, $\mathcal{R}\{\cdot\}$ signifies the range space of the bracketed quantity, and $\hat{x}$ refers to the estimate of x.

2. System Model and Assumptions

Consider an antenna array composed of M sensors with arbitrary array geometry, and assume that K narrowband far-field source signals impinge on the array from distinct locations $\Theta_1, \Theta_2, \cdots, \Theta_K$. Further, the steering vector and the baseband signal waveform of the kth incident signal are denoted as $\mathbf{a}(\Theta_k)$ and $s_k(t)$, respectively. $k = 1, 2, \cdots, K$, and the array output vector at the moment of t can be expressed as [17], [41]

$$\mathbf{x}(t) = \sum_{k=1}^{K} \mathbf{a}(\Theta_k) s_k(t) + \mathbf{n}(t) = \mathbf{A}(\boldsymbol{\Theta})\mathbf{s}(t) + \mathbf{n}(t) \quad (1)$$

where $\mathbf{A}(\boldsymbol{\Theta}) = [\mathbf{a}(\Theta_1), \mathbf{a}(\Theta_2), \cdots, \mathbf{a}(\Theta_K)]$, and $\mathbf{s}(t) = [s_1(t), s_2(t), \cdots, s_K(t)]^{\mathrm{T}}$. $\mathbf{n}(\mathbf{t})$ refers to the additive noise, which is modeled as a zero-mean, temporally and spatially white complex-valued Gaussian random process. Further, the covariance matrix of $\mathbf{n}(t)$ takes the form $\sigma^2\mathbf{I}$, where σ^2 signifies the noise power, and $\mathbf{I}$ denotes the identity matrix.

Using the system model (1), the data covariance matrix is given by

$$\mathbf{R}_x = E\left\{\mathbf{x}(t)\mathbf{x}^{\mathrm{H}}(t)\right\} = \mathbf{A}(\boldsymbol{\Theta})\mathbf{R}_s\mathbf{A}^{\mathrm{H}}(\boldsymbol{\Theta}) + \sigma^2\mathbf{I} \quad (2)$$

where $\mathbf{R}_s = E\left\{\mathbf{s}(t)\mathbf{s}^{\mathrm{H}}(t)\right\}$ denotes the source covariance matrix. Additionally, the following basic assumptions are made:

A1) The set of steering vectors $\{\mathbf{a}(\Theta_1), \mathbf{a}(\Theta_2), \cdots, \mathbf{a}(\Theta_K)\}$ is linearly independent for any set of distinct source locations $\Theta_1, \Theta_2, \cdots, \Theta_K$, where $K < M$.

A2) The narrowband signals $s_1(t), s_2(t), \cdots, s_K(t)$ are zero-mean, and they may be uncorrelated, partially correlated or coherent. In addition, the signals are independent of the noise.

A3) The source number K is known, or it has been estimated by the existing number detection methods (cf. [41], [42] and references therein).

The DOA estimation problem addressed in this paper is to estimate the source locations $\{\Theta_k\}_{k=1}^{K}$ from multiple snapshots $\{\mathbf{x}(t_l)\}_{l=1}^{L}$, where L denotes the number of snapshots.

3. Alternating Oblique Projection for DOA Estimation

AOP-DOA is a data-dependent, alternating least squares (LS) adaptive algorithm based on the relaxed optimization principle ("one parameter at the time") in [43], [44]. The received signals are separated into multiple single signals by using OP technique, and the DOAs of the separated signals are estimated separately. To achieve source separation, the OP matrices in use are closely associated with the source locations. Because of this, AOP-DOA is implemented iteratively, and its cost function is given by [45]

$$\frac{1}{L}\sum_{l=1}^{L}\left\|\mathbf{x}(t_l)-\sum_{k=1}^{K}\mathbf{a}(\Theta_k)s_k(t_l)\right\|^2. \tag{3}$$

3.1 Principle of AOP-DOA

Assume that the signal at Θ_k is to be separated, where $k = 1, 2, \cdots, K$. Correspondingly, the signal $s_k(t)$ is referred to as the SOI, and the remaining $K-1$ received signals at $\{\Theta_1, \cdots, \Theta_{k-1}, \Theta_{k+1}, \cdots, \Theta_K\}$ are relatively referred to as the "interferences". Then, (1) can be rewritten as

$$\mathbf{x}(t) = \mathbf{a}(\Theta_k)s_k(t) + \mathbf{B}_k\mathbf{s}_{\mathbf{B}_k}(t) + \mathbf{n}(t) \tag{4}$$

where $\mathbf{s}_{\mathbf{B}_k}(t) = [s_1(t), \cdots, s_{k-1}(t), s_{k+1}(t), \cdots, s_K(t)]^{\mathrm{T}}$, and

$$\mathbf{B}_k = [\mathbf{a}(\Theta_1), \cdots, \mathbf{a}(\Theta_{k-1}), \mathbf{a}(\Theta_{k+1}), \cdots, \mathbf{a}(\Theta_K)]. \tag{5}$$

According to the basic assumptions claimed in Sec. 2, the signal subspace is given by $\mathcal{R}\{\mathbf{A}(\boldsymbol{\Theta})\} = \mathcal{R}\{\mathbf{a}(\Theta_k)\} \oplus \mathcal{R}\{\mathbf{B}_k\}$, where $\mathcal{R}\{\mathbf{a}(\Theta_k)\}$ and $\mathcal{R}\{\mathbf{B}_k\}$ are disjoint. Therefore, the OP matrix whose range space is $\mathcal{R}\{\mathbf{B}_k\}$ and whose null space contains $\mathcal{R}\{\mathbf{a}(\Theta_k)\}$ can be expressed as [32]

$$\mathbf{E}_{\mathbf{B}_k,\mathbf{a}(\Theta_k)} = \mathbf{B}_k\left(\mathbf{P}^{\perp}_{\mathbf{a}(\Theta_k)}\mathbf{B}_k\right)^{\dagger} \tag{6}$$

where $\mathbf{P}^{\perp}_{\mathbf{a}(\Theta_k)} = \mathbf{I} - \mathbf{a}(\Theta_k)\mathbf{a}^{\dagger}(\Theta_k)$, $\mathbf{E}_{\mathbf{B}_k,\mathbf{a}(\Theta_k)}\mathbf{a}(\Theta_k) = \mathbf{0}$, and $\mathbf{E}_{\mathbf{B}_k,\mathbf{a}(\Theta_k)}\mathbf{B}_k = \mathbf{B}_k$. Naturally, the received SOI can be separated from the array measurements in the following way:

$$\mathbf{y}_k(t) = (\mathbf{I} - \mathbf{E}_{\mathbf{B_k},\mathbf{a}(\boldsymbol{\Theta}_k)})\mathbf{z}(\mathbf{t}) \tag{7}$$

where

$$\mathbf{z}(t) = \mathbf{x}(t) - \mathbf{n}(t) \tag{8}$$

signifies the array output in noiseless scenario, and $\mathbf{y}_k(t)$ is referred to as the "cleaned" observation of the SOI.

Based on (4), (7) and the LS criterion in [46], the SOI can be finally estimated by solving the following minimization problem:

$$\min_{\Theta_k,\{s_k(t_l)\}_{l=1}^{L}} \frac{1}{L}\sum_{l=1}^{L}\left\|\mathbf{y}_k(t_l) - \mathbf{a}(\Theta_k)s_k(t_l)\right\|^2. \tag{9}$$

Minimizing (9) with respect to $s_k(t_l)$, then the optimal solution to this problem yields

$$\hat{s}_k(t_l) = \mathbf{a}^{\dagger}(\Theta_k)\mathbf{y}_k(t_l), \quad l = 1, 2, \cdots, L. \tag{10}$$

Substituting (10) into (9), it follows that the DOA of the SOI can be estimated by solving the following minimization problem:

$$\hat{\Theta}_k = \underset{\Theta_k}{\operatorname{argmin}}\frac{1}{L}\sum_{l=1}^{L}\left\|\mathbf{P}^{\perp}_{\mathbf{a}(\Theta_k)}\mathbf{y}_k(t_l)\right\|^2. \tag{11}$$

In the above manner, all the K received signals can be separated one after another, and the DOA estimate of each separated signal can be achieved separately. Remark that the OP matrix in (7) is irrelevant to the correlations among signals, and it is this property that makes the proposed AOP-DOA being applicable for uncorrelated, partially correlated and coherent signals. But it is noted that the computation of $\mathbf{E}_{\mathbf{B}_k,\mathbf{a}(\Theta_k)}$ requires knowledge of the source locations, whereas none of them are considered to be known a prior. To solve this problem, AOP-DOA is iterative, and it updates $\hat{\Theta}_k^{(i)}$ by using previously estimated values $\{\hat{\Theta}_1^{(i)}, \cdots, \hat{\Theta}_{k-1}^{(i)}, \hat{\Theta}_k^{(i-1)}, \hat{\Theta}_{k+1}^{(i-1)}, \cdots, \hat{\Theta}_K^{(i-1)}\}$, where $\hat{\Theta}_k^{(i)}$ denotes the DOA estimates of Θ_k at the ith iteration (similarly hereinafter), $i = 1, 2, \cdots$.

The relative change of the cost function in (3) between two consecutive iterations determines the convergence of AOP-DOA. Substituting the iteration results of (10) and (11) into (3), it follows that (see the Appendix A)

$$\begin{aligned}&\frac{1}{(M-K)L}\sum_{l=1}^{L}\left\|\mathbf{x}(t_l) - \sum_{k=1}^{K}\mathbf{a}(\Theta_k)s_k(t_l)\right\|^2 - \frac{K}{M-K}\hat{\sigma}^2\\ &= \frac{1}{M-K}\mathrm{tr}\{\mathbf{P}^{\perp}_{\mathbf{A}(\boldsymbol{\Theta})}\hat{\mathbf{R}}_x\} = \hat{\sigma}^2\end{aligned} \tag{12}$$

where $\hat{\mathbf{R}}_x = \frac{1}{L}\sum_{l=1}^{L}\mathbf{x}(t_l)\mathbf{x}^{\mathrm{H}}(t_l)$ is referred to as the sample data covariance matrix which signifies a practical estimation of the data covariance matrix from the samples $\{\mathbf{x}(t_l)\}_{l=1}^{L}$. $\hat{\sigma}^2 = \frac{1}{L}\sum_{l=1}^{L}\|\mathbf{n}(t_l)\|^2$ denotes an estimation of noise power from the samples $\{\mathbf{n}(t_l)\}_{l=1}^{L}$.

Instead of using (3), it is natural to terminate the AOP-DOA iterations by checking the relative change of

$$\frac{1}{M-K}\mathrm{tr}\{\mathbf{P}^{\perp}_{\mathbf{A}(\boldsymbol{\Theta})}\hat{\mathbf{R}}_x\}. \tag{13}$$

The fundamental objective of the proposed AOP-DOA is to find the global optimal solution of the nonlinear LS problem in (3), and it is achieved by separately solving the optimization problem of a series of separable variable pairs $\{\Theta_k, s_k(t)\}_{k=1}^{K}$ (see (9)) and following the relaxed iterative approach in [28](the iterative implementation of AOP-DOA is summarized in Tab. 1, Sec. 3.2). These processes are essentially the same as these of the RELAX estimator in [28], hence the convergence properties of AOP-DOA and RELAX are consistent.

The same as RELAX, for the proposed AOP-DOA, the estimation results of "interferences" are removed from the array measurements, and the parameters of the SOI are estimated based on the cleaned data. The difference of these two algorithms is that AOP-DOA obtain the "cleaned" observation of the SOI from noiseless array output via employing the OP technique, whereas that of RELAX is obtained by directly subtracting the "interferences" from the noisy array measurements. Because of this, the specific implementation of AOP-DOA is quite distinct from that of RELAX (cf. [28]), which will be described in the following subsection.

3.2 Implementation of AOP-DOA

According to Sec. 3.1, at the ith iteration, $\mathbf{y}_k(t)$ is calculated as follows:

$$\mathbf{y}_k^{(i)}(t) = \left(\mathbf{I} - \mathbf{E}_{\hat{\mathbf{B}}_k^{(i)},\mathbf{a}(\hat{\Theta}_k^{(i-1)})}\right)\mathbf{z}(t) \tag{14}$$

where

$$\hat{\mathbf{B}}_k^{(i)} = [\mathbf{a}(\hat{\Theta}_1^{(i)}), \cdots, \mathbf{a}(\hat{\Theta}_{k-1}^{(i)}), \mathbf{a}(\hat{\Theta}_{k+1}^{(i-1)}), \cdots, \mathbf{a}(\hat{\Theta}_K^{(i-1)})]. \tag{15}$$

Further, it follows from (8) and (14) that

$$\begin{aligned}\hat{\mathbf{R}}_{y_k}^{(i)} &= \frac{1}{L}\sum_{l=1}^{L}\mathbf{y}_k^{(i)}(t_l)\,(\mathbf{y}_k^{(i)}(t_l))^{\mathrm{H}} \\ &= (\mathbf{I} - \mathbf{E}_{\hat{\mathbf{B}}_k^{(i)},\mathbf{a}(\hat{\Theta}_k^{(i-1)})})\hat{\mathbf{R}}_z(\mathbf{I} - \mathbf{E}_{\hat{\mathbf{B}}_k^{(i)},\mathbf{a}(\hat{\Theta}_k^{(i-1)})})^{\mathrm{H}}.\end{aligned} \tag{16}$$

where

$$\hat{\mathbf{R}}_z = \frac{1}{L}\sum_{l=1}^{L}\mathbf{z}(t_l)\,\mathbf{z}^{\mathrm{H}}(t_l) = \hat{\mathbf{R}}_x - \hat{\sigma}^2\mathbf{I}. \tag{17}$$

Thus, (11) can be rewritten as

$$\hat{\Theta}_k^{(i)} = \underset{\Theta_k}{\operatorname{argmin}}\,\mathrm{tr}\{\mathbf{P}_{\mathbf{a}(\Theta_k)}^{\perp}\hat{\mathbf{R}}_{y_k}^{(i)}\}. \tag{18}$$

Note that $\mathrm{tr}\{\mathbf{P}_{\mathbf{a}(\Theta_k)}^{\perp}\hat{\mathbf{R}}_{y_k}^{(i)}\} = \mathrm{tr}\{\hat{\mathbf{R}}_{y_k}^{(i)}\} - \mathrm{tr}\{\mathbf{P}_{\mathbf{a}(\Theta_k)}\hat{\mathbf{R}}_{y_k}^{(i)}\}$. Hence, the optimal problem of (18) is equivalent to

$$\hat{\Theta}_k^{(i)} = \underset{\Theta_k}{\operatorname{argmax}}\,\mathrm{tr}\{\mathbf{P}_{\mathbf{a}(\Theta_k)}\hat{\mathbf{R}}_{y_k}^{(i)}\} = \underset{\Theta_k}{\operatorname{argmax}}\,\frac{\mathbf{a}^{\mathrm{H}}(\Theta_k)\hat{\mathbf{R}}_{y_k}^{(i)}\mathbf{a}(\Theta_k)}{\mathbf{a}^{\mathrm{H}}(\Theta_k)\,\mathbf{a}(\Theta_k)}. \tag{19}$$

Using Rayleigh quotient properties in [47], [48], the solution to this problem yields

$$\mathcal{R}\{\mathbf{a}(\hat{\Theta}_k^{(i)}\} = \mathcal{R}\{\mathbf{u}_{\max,k}^{(i)}\} \tag{20}$$

where $\mathbf{u}_{\max,k}^{(i)}$ refers to the unit eigenvector corresponding to the maximum eigenvalue of $\hat{\mathbf{R}}_{y_k}^{(i)}$. Therefore,

$$\hat{\Theta}_k^{(i)} = \underset{\Theta_k}{\operatorname{argmax}}\,|\mathbf{a}^{\mathrm{H}}(\Theta_k)\,\mathbf{u}_{\max,k}^{(i)}|^2. \tag{21}$$

This maximization problem can be solved without much computational effort by means of adaptive grid refinement [23] only around the regions where signals were present in the last iteration.

Using (16), (21) and referring to the relaxed iterative method in [28], the implementation procedure of AOP-DOA is summarized in Tab. 1, where $\mathcal{S}^{(0)}$ represents a rough region of potential source locations (a 1° or 2° uniform sampling usually suffices [23]), and $\mathcal{S}_j^{(i)}$ denotes a refined region around the previous source location $\hat{\Theta}_j^{(i-1)}$, $j = 1, 2, \cdots, K$.

As given in Tab. 1, the initialization step of AOP-DOA is implemented with assumption $\mathbf{a}(\hat{\Theta}_k^{(-1)}) = \mathbf{0}$, where $k = 1, 2, \cdots, K$. At the beginning of each outer loop, $\hat{\Theta}_k^{(0)}$ is determined with (16) and (21) (see lines 5–7). After that, we sequentially redetermine $\hat{\Theta}_1, \hat{\Theta}_2, \cdots, \hat{\Theta}_k$ and iteratively update these DOA estimates until convergence is achieved (see lines 8–18). AOP-DOA stops when all of the K outer loops finished.

Input:
1: measurement data $\mathbf{x}(t_l)$, $l = 1, 2, \cdots, L$
Initialization:
2: $\mathbf{a}(\hat{\Theta}_1^{(-1)}) = \mathbf{0}, \mathbf{a}(\hat{\Theta}_2^{(-1)}) = \mathbf{0}, \cdots, \mathbf{a}(\hat{\Theta}_K^{(-1)}) = \mathbf{0}$
Main Loop:
3: **for** $k = 1, 2, \cdots, K$ **do**
4: $i \leftarrow 0$
5: $\hat{\mathbf{R}}_{y_k}^{(0)} = (\mathbf{I} - \mathbf{E}_{\hat{\mathbf{B}}_k^{(0)},\mathbf{a}(\hat{\Theta}_k^{(-1)})})\hat{\mathbf{R}}_z(\mathbf{I} - \mathbf{E}_{\hat{\mathbf{B}}_k^{(0)},\mathbf{a}(\hat{\Theta}_k^{(-1)})})^{\mathrm{H}}$
6: $\mathbf{u}_{\max,k}^{(0)} \leftarrow \mathrm{eig}(\hat{\mathbf{R}}_{y_k}^{(0)})$
7: $\hat{\Theta}_k^{(0)} = \underset{\Theta_k \in \mathcal{S}^{(0)}}{\operatorname{argmax}}\,|\mathbf{a}^{\mathrm{H}}(\Theta_k)\,\mathbf{u}_{\max,k}^{(0)}|^2$
8: **repeat**
9: $i \leftarrow i + 1$
10: $0 \leftarrow j$
11: **while** $j < k$ **do**
12: $j + 1 \leftarrow j$
13: $\hat{\mathbf{R}}_{y_j}^{(i)} = (\mathbf{I} - \mathbf{E}_{\hat{\mathbf{B}}_j^{(i)},\mathbf{a}(\hat{\Theta}_j^{(i-1)})})\hat{\mathbf{R}}_z(\mathbf{I} - \mathbf{E}_{\hat{\mathbf{B}}_j^{(i)},\mathbf{a}(\hat{\Theta}_j^{(i-1)})})^{\mathrm{H}}$
14: $\mathbf{u}_{\max,j}^{(i)} \leftarrow \mathrm{eig}(\hat{\mathbf{R}}_{y_j}^{(i)})$
15: $\hat{\Theta}_j^{(i)} = \underset{\Theta_j \in \mathcal{S}_j^{(i)}}{\operatorname{argmax}}\,|\mathbf{a}^{\mathrm{H}}(\Theta_j)\,\mathbf{u}_{\max,j}^{(i)}|^2$
16: **end while**
17: **until** (convergence)
18: $\mathbf{a}(\hat{\Theta}_1^{(0)}) \leftarrow \mathbf{a}(\hat{\Theta}_1^{(i)}), \cdots, \mathbf{a}(\hat{\Theta}_j^{(0)}) \leftarrow \mathbf{a}(\hat{\Theta}_j^{(i)})$
19: **end for**
Output:
20: $\hat{\Theta}_1 \leftarrow \hat{\Theta}_1^{(i)}, \hat{\Theta}_2 \leftarrow \hat{\Theta}_2^{(i)}, \cdots, \hat{\Theta}_K \leftarrow \hat{\Theta}_K^{(i)}$

Tab. 1. The AOP-DOA algorithm.

3.3 Relationship Between AOP-DOA and AP-MLE

At the ith iteration, AP-MLE updates the kth DOA estimates as follows [17]:

$$\begin{aligned}\hat{\Theta}_{k,\,\text{AP-MLE}}^{(i)} &= \underset{\Theta_k}{\operatorname{argmin}}\,\mathrm{tr}\{\mathbf{P}_{[\tilde{\mathbf{B}}_k^{(i-1)},\,\mathbf{a}(\Theta_k)]}^{\perp}\hat{\mathbf{R}}_x\} \\ &= \underset{\Theta_k}{\operatorname{argmin}}\,\mathrm{tr}\{\mathbf{P}_{[\tilde{\mathbf{B}}_k^{(i-1)},\,\mathbf{a}(\Theta_k)]}^{\perp}\hat{\mathbf{R}}_z + \hat{\sigma}^2\mathbf{P}_{[\tilde{\mathbf{B}}_k^{(i-1)},\,\mathbf{a}(\Theta_k)]}^{\perp}\} \\ &= \underset{\Theta_k}{\operatorname{argmin}}\,\mathrm{tr}\{\mathbf{P}_{[\tilde{\mathbf{B}}_k^{(i-1)},\,\mathbf{a}(\Theta_k)]}^{\perp}\hat{\mathbf{R}}_z + \hat{\sigma}^2(M-K)\}\end{aligned} \tag{22}$$

where $\tilde{\mathbf{B}}_k^{(i)} = [\mathbf{a}(\hat{\Theta}_1^{(i)}), \cdots, \mathbf{a}(\hat{\Theta}_{k-1}^{(i)}), \mathbf{a}(\hat{\Theta}_{k+1}^{(i)}), \cdots, \mathbf{a}(\hat{\Theta}_K^{(i)})]$. According to Appendix B, $\mathbf{E}_{\mathbf{a}(\Theta_k),\mathbf{B}_k} = \mathbf{P}_{\mathbf{a}(\Theta_k)}\,(\mathbf{I} - \mathbf{E}_{\mathbf{B}_k,\mathbf{a}(\Theta_k)})$ and

$$\begin{aligned}\mathbf{P}_{\mathbf{A}(\Theta)}^{\perp} &= \mathbf{I} - \mathbf{E}_{\mathbf{B}_k,\mathbf{a}(\Theta_k)} - \mathbf{P}_{\mathbf{a}(\Theta_k)}\,(\mathbf{I} - \mathbf{E}_{\mathbf{B}_k,\mathbf{a}(\Theta_k)}) \\ &= \mathbf{P}_{\mathbf{a}(\Theta_k)}^{\perp}\,(\mathbf{I} - \mathbf{E}_{\mathbf{B}_k,\mathbf{a}(\Theta_k)})\,.\end{aligned} \tag{23}$$

Therefore, (22) is equivalent to

$$\hat{\Theta}_{k,\,\text{AP-MLE}}^{(i)} = \underset{\Theta_k}{\operatorname{argmin}}\sum_{l=1}^{L}\frac{\left\|\mathbf{P}_{\mathbf{a}(\Theta_k)}^{\perp}\left(\mathbf{I} - \mathbf{E}_{\tilde{\mathbf{B}}_k^{(i-1)},\mathbf{a}(\Theta_k)}\right)\mathbf{z}(t)\right\|^2}{L}. \tag{24}$$

Recall from (16) and (18) that, at the ith iteration, AOP-DOA updates the kth DOA estimate as follows:

Input:
1: $\{\hat{\Theta}_k\}_{k=1}^{K}$: DOA estimates obtained from AOP-DOA
Initialization:
2: $\hat{\Theta}_1^{(0)} \leftarrow \hat{\Theta}_1, \hat{\Theta}_2^{(0)} \leftarrow \hat{\Theta}_2, \cdots, \hat{\Theta}_K^{(0)} \leftarrow \hat{\Theta}_K$
Main Loop:
3: $i \leftarrow 0$
4: **repeat**
5: $\quad i \leftarrow i+1$
6: $\quad$ **for** $k = 1, 2, \cdots, K$ **do**
7: $\quad\quad \hat{\Theta}_k^{(i)} = \underset{\Theta_k \in \mathcal{S}_k^{(i)}}{\operatorname{argmin}} \operatorname{tr}\{\mathbf{P}^{\perp}_{[\tilde{\mathbf{B}}_k^{(i-1)}, \mathbf{a}(\Theta_k)]} \hat{\mathbf{R}}_x\}$
8: $\quad$ **end for**
9: **until** (convergence)
Output:
10: $\hat{\Theta}_1 = \hat{\Theta}_1^{(i)}, \hat{\Theta}_2 = \hat{\Theta}_2^{(i)}, \cdots, \hat{\Theta}_K = \hat{\Theta}_K^{(i)}$

Tab. 2. The AOP&AP-DOA algorithm.

$$\hat{\Theta}^{(i)}_{k,\text{AOP-DOA}} = \underset{\Theta_k}{\operatorname{argmin}} \operatorname{tr}\{\mathbf{P}^{\perp}_{\mathbf{a}(\Theta_k)} \hat{\mathbf{R}}^{(i)}_{y_k}\}$$
$$= \underset{\Theta_k}{\operatorname{argmin}} \frac{1}{L} \sum_{l=1}^{L} \left\| \mathbf{P}^{\perp}_{\mathbf{a}(\Theta_k)} \left(\mathbf{I} - \mathbf{E}_{\hat{\mathbf{B}}_k^{(i)}, \mathbf{a}(\hat{\Theta}_k^{(i-1)})} \right) \mathbf{z}(t) \right\|^2. \tag{25}$$

Evidently, both AOP-DOA and AP-MLE solve one-dimensional optimal problems. Comparing (24) with (25), it is found that AOP-DOA utilizes previously estimated value $\hat{\Theta}_k^{(i-1)}$ to estimate $\hat{\Theta}_k^{(i)}$, whereas AP-MLE does not.

The results of AOP-DOA can be used to provide a good initial estimate for the main loop of AP-MLE. This method is referred to as AOP&AP-DOA and is summarized in Tab. 2. Instead of adopting a universally fine grid search, the source locations can be accurately estimated via AP-MLE by means of a fine search only around the estimated values of AOP-DOA. By this means, AOP-DOA provides relatively coarse DOA estimates in the first stage (e.g., in the adaptive gird refinement strategy, the lower limit of fine gird spacing (LLFGS) may be set to 0.1°, 0.25°, etc.), and, in the second stage, AP-MLE provides fine DOA estimates (e.g., the LLFGS can be set to 0.01°).

3.4 Computational Complexity

The computational complexity of proposed AOP-DOA is briefly analyzed as follows, where a flop is defined as a complex floating-pint multiplication or addition operation.

The computation of matrix $\hat{\mathbf{R}}_z$ takes about $O(LM^2+M^3)$ flops, where the noise power estimate $\hat{\sigma}^2$ is obtained by averaging the $M-K$ smallest eigenvalues of covariance matrix $\hat{\mathbf{R}}_x$ [6]. The calculation of the OP matrix $\mathbf{E}_{\hat{\mathbf{B}}_k^{(i)}, \mathbf{a}(\hat{\Theta}_k^{(i-1)})}$ (or $\mathbf{E}_{\hat{\mathbf{B}}_j^{(i)}, \mathbf{a}(\hat{\Theta}_j^{(i-1)})}$) requires roughly $O((K-1)M^2)$ flops, and that of the matrix $\hat{\mathbf{R}}^{(i)}_{y_k}$ (or $\hat{\mathbf{R}}^{(i)}_{y_j}$) requires roughly $O(M^3)$ flops. The eigendecomposition of $\hat{\mathbf{R}}^{(i)}_{y_k}$ (or $\hat{\mathbf{R}}^{(i)}_{y_j}$) requires approximately $O(M^3)$ flops. Suppose that $\mathcal{S}^{(0)}$ and $\mathcal{S}_j^{(i)}$ consist of N_0 and N_i potential source locations, respectively. Then the number of flops roughly required to solve the maximization problem in line 7, Tab. 1 and line 15, Tab. 1 is $O(MN_0)$ and $O(MN_i)$, respectively. In addition, the computation of the cost function in (13) requires roughly $O(M^3)$ flops.

Therefore, the computational complexity of AOP-DOA is roughly $O(LM^2+M^3+K(2M^3+(K-1)M^2+MN_0)+\sum_{k=1}^{K}\sum_{i=1}^{n_k}((2M^3+(K-1)M^2+MN_i)k+M^3))$ in total, where n_k denotes the iteration number included in the kth outer loop. In particular, the total computational complexity is approximately $O(LM^2+K(M^3+MN_0)+\sum_{k=1}^{K}\sum_{i=1}^{n_k}(M^3+MN_i)k)$ flops, when $N_0 \gg M > K$, which occurs often in practical applications.

The proposed AOP&AP-DOA can be viewed as a combination of AOP-DOA and AP-MLE. Thus the computational complexity of AOP&AP-DOA consists of two parts, where the part of AP-MLE additionally takes about $O(M^3K\tilde{N}_i\, n_{AP})$ flops. n_{AP} denotes the iteration number, and it is supposed that $\mathcal{S}_k^{(i)}$ (line 7, Tab. 2) consists of $\tilde{N}_i$ potential source locations.

The computational complexities of AOP-DOA and AOP&AP-DOA are summarized in Tab. 3. Remark that the major computational complexity is proportional to $\sum_{i=1}^{n_k} N_i$ and N_0, when M, K and L are fixed. Therefore, the employed adaptive grid refinement strategy not only determines the computational accuracy, but also influences the computational cost. Compared with AOP-DOA, in Sec. 4, we show that AOP&AP-DOA could have a lower computational load when proper adaptive grid refinement strategy is applied.

Algorithm	Complexity
AOP-DOA	$O\big(LM^2+K(M^3+MN_0)\big)$ $+O\big(\sum_{k=1}^{K}\sum_{i=1}^{n_k}(M^3+MN_i)k\big)$
AOP&AP-DOA	$O\big(M^3K\tilde{N}_i\, n_{AP}+LM^2+K(M^3+MN_0)\big)$ $+O\big(\sum_{k=1}^{K}\sum_{i=1}^{n_k}(M^3+MN_i)k\big)$

Tab. 3. Computational complexity of the proposed algorithms.

4. Simulation Results

Extensive experiments are presented to evaluate the effectiveness of proposed AOP-DOA and AOP&AP-DOA. The stochastic Cramér-Rao bound (CRB) [3] and the existing high resolution DOA estimation algorithms, including AP-MLE [17], RELAX [28], MUSIC [6], l_1-SVD [23] and perturbed SBL (PSBL) [27] are utilized for performance comparison.

The root mean square error (RMSE) of DOA estimates is defined as

$$\text{RMSE} = \sqrt{\frac{1}{KM_c}\sum_{m=1}^{M_c}\sum_{k=1}^{K}\left(\hat{\Theta}_{m,k}-\Theta_k\right)^2} \tag{26}$$

where M_c is the number of trials, $\hat{\Theta}_{m,k}$ denotes the estimate of the kth DOA Θ_k in the mth trial. All the simulations are carried out on a uniform linear array with $M = 6$ sensors and half-wavelength sensor spacing. Unless otherwise stated, each of the simulations is evaluated by $M_c = 200$ independent trials, and $L = 100$ snapshots are employed.

The initialization condition for PSBL is set in the same way as [27]. l_1-SVD is considered with SeDuMi packages [49], and the regularization parameter is set to 0.625. For l_1-SVD, RELAX, AOP-DOA and AOP&AP-DOA, the adaptive grid refinement strategy [23] is employed, where the spacing of the initial coarse grid is 2° and the LLFGS is set to 0.01°, unless otherwise stated. For MUSIC and AP-MLE, we use 0.01° uniform sampling of the spatial location of the sources. All the simulations are performed using MATLAB 2013b running on a computer with a 2.3 GHz Intel Quad-Core processor and 12GB RAM, under Windows 8.1. The computational complexity of an algorithm is evaluated in terms of the average CPU processing time [27].

4.1 Discussion: Complexity and Accuracy

In Sec. 3.4, the complexities of the proposed algorithms are analyzed, and it is shown that, when the number of snapshots L, array size M and source number K are fixed, the computational complexity is jointly proportional to the iteration number and the grid size which is considered in the adaptive grid refinement strategy.

This subsection intends to evaluate the complexities of proposed AOP-DOA and AOP&AP-DOA by employing different adaptive grid refinement strategies. Two equal-power uncorrelated signals impinging from $[-5.12°, 5.37°]$ are considered in this experiment, and the RMSE is utilized to evaluate the precision of DOA estimates. Both the CPU time and the average number of iterations are analyzed at each evaluated signal-to-noise ratio (SNR). In each simulation, the number of iterations of AOP&AP-DOA is the sum of the iteration numbers that required in the two stages DOA estimation.

Firstly, the proposed AOP-DOA is investigated. Three different LLFGS are considered, and the SNR is varied from -15 dB to 25 dB. Fig. 1 shows the statistical results. It is seen that AOP-DOA exhibits a satisfying performance when different LLFGS is used. Moreover, AOP-DOA can coincide well with the CRB at a moderate to high SNR when proper adaptive grid refinement strategy is employed. But we also should note that a fine grid search requires a heavy computational cost.

In the next, the proposed AOP&AP-DOA is evaluated. Since the DOA estimation of AOP&AP-DOA consists of two stages, we consider different LLFGS in the first stage. But the LLFGS in the second stage is fixed at 0.01°. Other settings remain the same as previous experiment. Figure 2 shows the statistical results of AOP&AP-DOA, and it is indicated that AOP&AP-DOA is able to coincide well with the CRB at a moderate to high SNR. But AOP&AP-DOA consumes different time, when different adaptive grid refinement strategies are applied. The running time is proportional to the LLFGS in the first stage. With proper adaptive grid refinement strategy, the convergence speed of AOP&AP-DOA is fast. In the following simulations, the LLFGS in the first stage of AOP&AP-DOA is fixed at 0.1° and that in the second stage is fixed at 0.01°.

Besides, when comparing Fig. 1 with Fig. 2, it can be seen that AOP-DOA may show a competitive estimation precision with a high SNR, but AOP&AP-DOA can own a lower running time. In particular, a fine grid search leads to a large number of iterations. But for a fixed grid search approach, with the increase of SNR, the numbers of iterations of AOP-DOA and AOP&AP-DOA are stable or approximately stable, especially when the incident signals dominate the background noise. The same is true for the running time under different SNR.

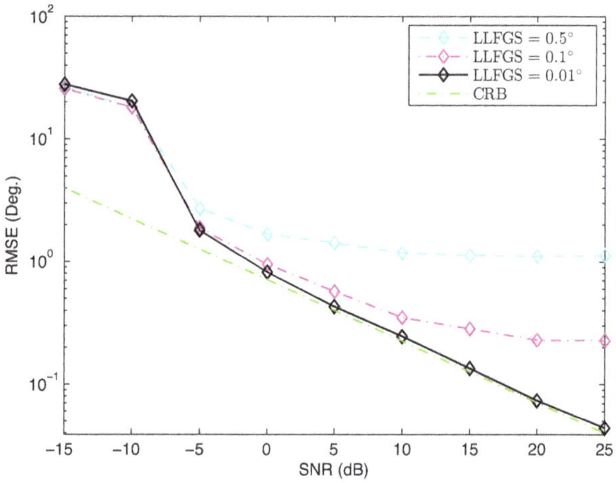

(a) RMSE versus SNR

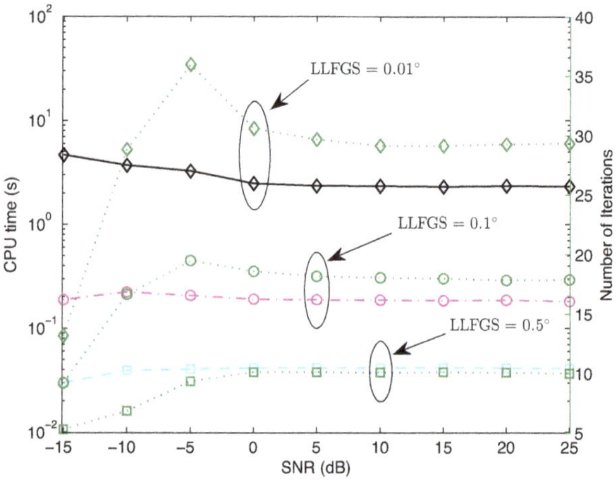

(b) CPU time versus SNR

Fig. 1. RMSE and CPU time comparison of the proposed AOP-DOA for two uncorrelated sources impinging from $[-5.12°, 5.37°]$ when three different LLFGS are considered. The SNR ranges from -15 dB to 25 dB.

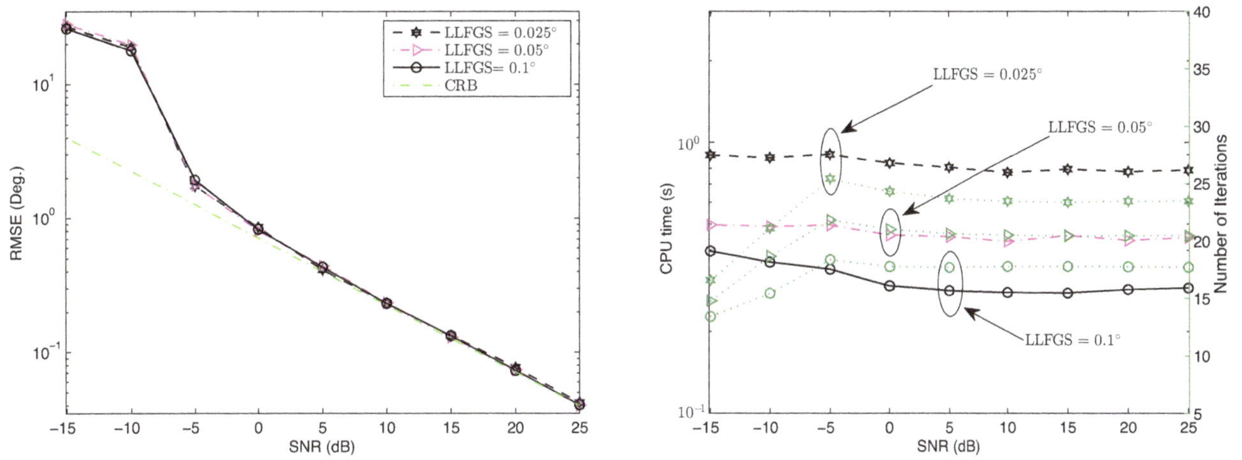

(a) RMSE versus SNR

(b) CPU time versus SNR

Fig. 2. RMSE and CPU time comparison of the proposed AOP&AP-DOA for two uncorrelated sources impinging from $[-5.12°, 5.37°]$ when the LLFGS in the first stage ranges from $0.025°$ to $0.1°$. The LLFGS in the second stage is fixed at $0.01°$, and the SNR ranges from -15 dB to 25 dB.

4.2 RMSE and CPU Time Comparisons

This subsection presents detailed performance analyses to illustrate the superiority of the proposed AOP-DOA and AOP&AP-DOA in cases of uncorrelated and correlated(including coherent) signals. For each algorithm, the computational complexity and the estimation accuracy are employed for performance comparison, and they are evaluated by the running time and the RMSE, respectively.

Figure 3 depicts the estimation accuracy of the DOA estimates and the time usage of each comparing algorithm when two uncorrelated signals are considered. We assume the signals impinge on the array from $[-5.12°, 5.37°]$, and they have equal SNR ranging from -15 dB to 20 dB. The results of Fig. 3(a) show that the proposed AOP-DOA and AOP&AP-DOA, as well as RELAX and AP-MLE, exhibit a satisfying estimation precision, whereas PSBL and l_1-SVD not coincide well with the CRB. It is of interest to note that MUSIC also shows a high precision when the SNR is high enough, but it fails to hold a superresolution at a lower SNR. Fig. 3(b) indicates that AP-MLE consumes the maximum running time, while the proposed AOP-DOA and AOP&AP-DOA algorithms take the relatively lower one. This is because the conventional AP-MLE always adopts a universally fine grid search, whereas our algorithms utilize fine grid search only around the regions where signals are present. Besides, the computational load of RELAX and the proposed AOP-DOA is approximately at the same level, and is higher than that of the proposed AOP&AP-DOA. The running time of PSBL fluctuates remarkably when the SNR changes, but that of the other algorithms are not. To illustrate this further, Fig. 3(c) presents the iteration numbers of PSBL, RELAX and the proposed AOP-DOA and AOP&AP-DOA algorithms at each SNR. It is seen that the number of iterations of AOP&AP-DOA is the smallest, while the number of iterations of RELAX and AOP-DOA are approximately the same though the latter is slightly smaller. Particularly, the iteration number of PSBL decreases fast when the SNR increases, and that's the reason the CPU time of PSBL fluctuates (the complexity per iteration of PSBL is identical [27]). These phenomena are common in the following simulations, hence the curve of the number of iterations is omitted, and the similar phenomena are not explained repeatedly.

Figure 4 considers two uncorrelated signals with different angular separations. We assume $\Theta_1 = 5.37°$ and $\Theta_2 = 5.37° - \Delta$, where Δ denotes the source separation and it ranges from $2°$ to $24°$. The SNR is fixed at 10 dB. It is seen that the proposed AOP-DOA and AOP&AP-DOA have a distinct advantage against spatially adjacent signals, and both are high resolution algorithms. Compared with AP-MLE, the proposed algorithms exhibit the similar estimation precision but consume much less running time. Compared with PSBL and RELAX, the proposed algorithms exhibit a lower RMSE when the angular separation is small, and the time usage of the proposed algorithms decreases rapidly when the angular separation largely increases. A large angular separation may lead to a faster convergence speed of an iterative algorithm, but it is not the main factor that determines the computational complexity of MUSIC and l_1-SVD (cf. [6], [45], [23]). The CPU time of MUSIC and l_1-SVD has tiny changes for different angular separations. Hence, as the angular separation becomes largely, the time usage of the proposed AOP-DOA could be lower than that of the MUSIC and l_1-SVD.

Both Fig. 5 and Fig. 7 consider the case of correlated signals. We assume two signals impinge on the array from directions $[-15.12°, 5.37°]$. In Fig. 5, the correlation coefficient of the signals is set to 0.5, and we vary the SNR from -15 dB to 20 dB. In Fig. 7, the SNR is fixed at 15 dB, but we vary the correlation coefficient of the signals from 0 to 1. The simulation results illustrated in Fig. 5 and Fig. 7

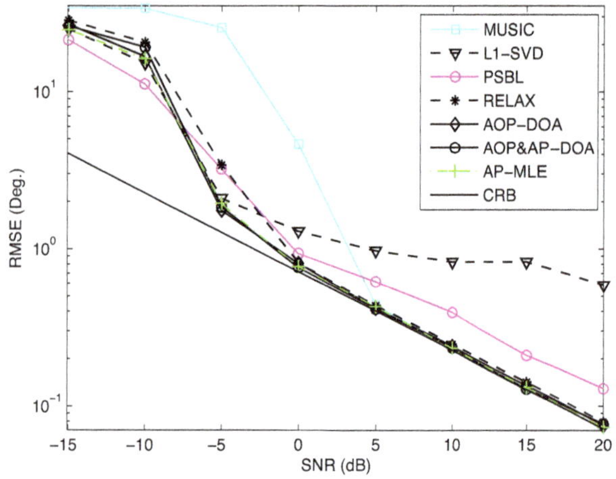

(a) RMSE versus SNR

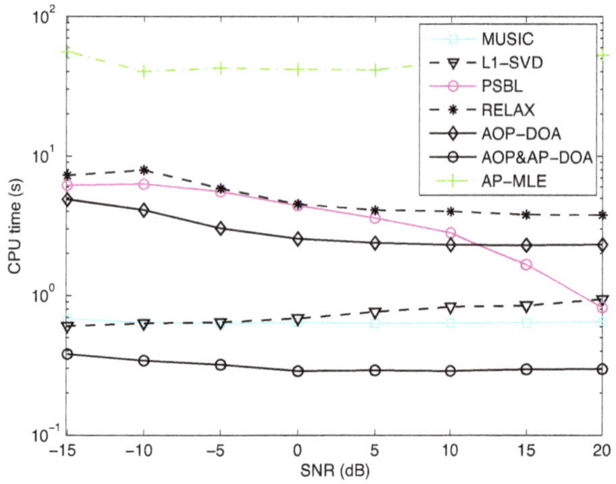

(b) CPU time versus SNR

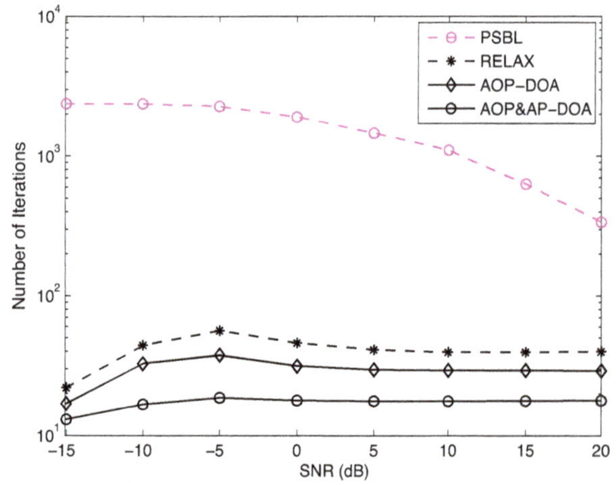

(c) Number of iterations at each SNR

Fig. 3. RMSE and CPU time comparison for two uncorrelated sources impinging from $[-5.12°, 5.37°]$.

show that the proposed AOP-DOA and AOP&AP-DOA can achieve satisfactory accuracies when the signals are partially or fully correlated. The DOA estimation performance of each algorithm deteriorates along with the increased correlation coefficient, and the MUSIC fails when the signals

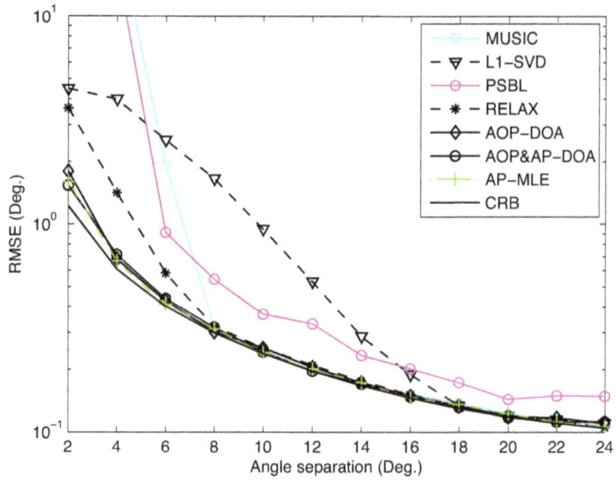

(a) RMSE versus angular separation

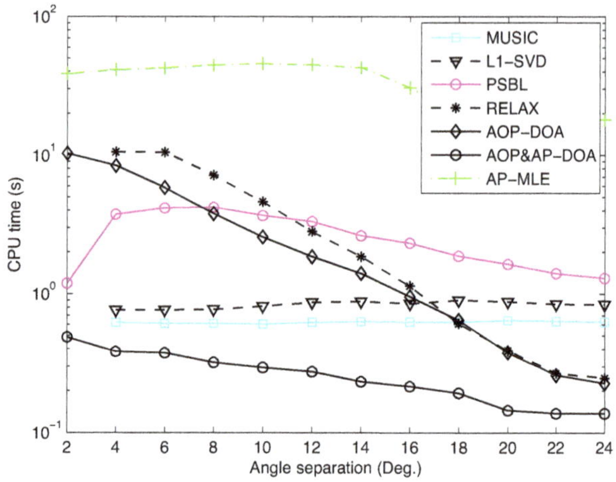

(b) CPU time versus angular separation

Fig. 4. RMSE and CPU time comparison for two uncorrelated sources impinging from $[5.37°, 5.37° - \Delta]$, where the angular separation Δ ranges from 2° to 24° and the SNR is fixed at 10 dB.

are coherent. But the proposed AOP-DOA and AOP&AP-DOA, as well as AP-MLE and RELAX, are able to coincide well with the CRB, and they could have a smaller RMSE than PSBL and l_1-SVD. In addition, the proposed two algorithms have a much lower computational cost when compared with AP-MLE, even if the sources are correlated. The increased correlation coefficient of signals makes AP-MLE, RELAX, AOP-DOA and AOP&AP-DOA need more iterations (running time) to achieve convergence, whereas the complexities of MUSIC, l_1-SVD and PSBL are insensitive to the variations of the correlation property. Hence the MUSIC, l_1-SVD and PSBL show a stable running time, and this causes their CPU time curves to have intersections.

Figure 6 considers the DOA estimation for a mixture of uncorrelated and coherent signals. The incident signals are a group of two coherent signals from $[-15.12°, 5.37°]$ and an uncorrelated signal from 42.95°. The SNR ranges from -5 dB to 20 dB. The simulation results depicted in Fig. 6

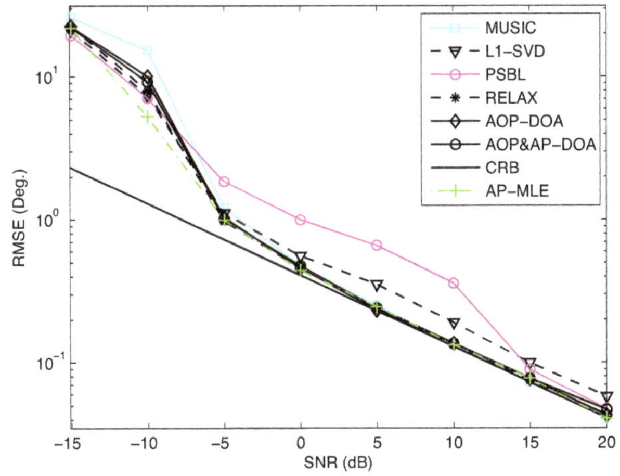

(a) RMSE versus SNR

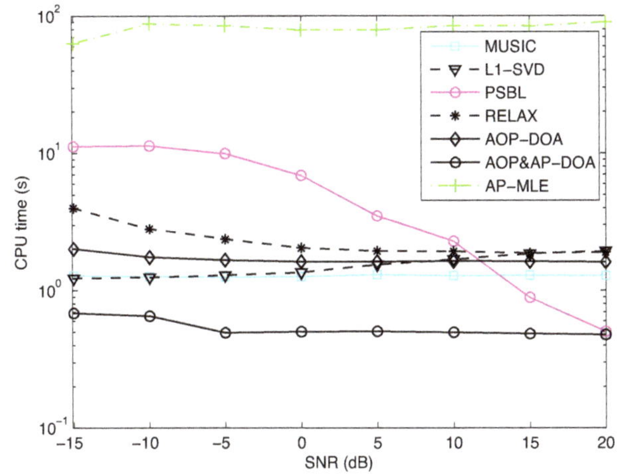

(b) CPU time versus SNR

Fig. 5. RMSE and CPU time comparison for two partially correlated sources impinging from [−15.12°, 5.37°], where the correlation coefficient of the signals is set to 0.5.

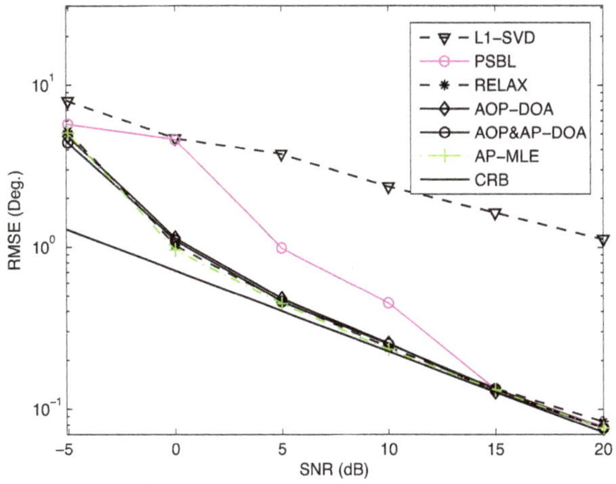

(a) RMSE versus SNR

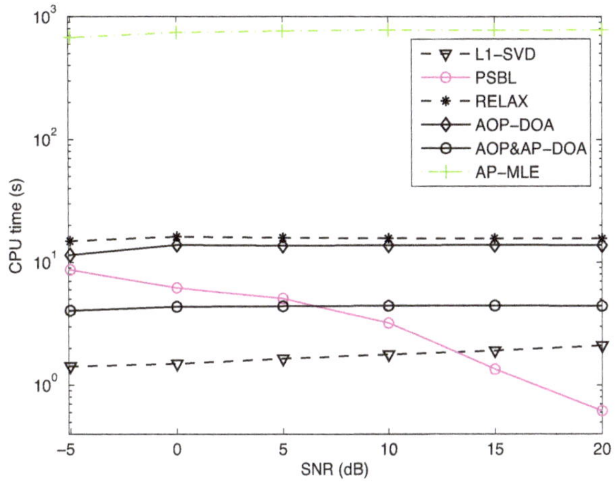

(b) CPU time versus SNR

Fig. 6. RMSE and CPU time comparison for a mixture of uncorrelated and coherent signals, where a group of two coherent signals from [−15.12°, 5.37°] and an uncorrelated signal from 42.95°.

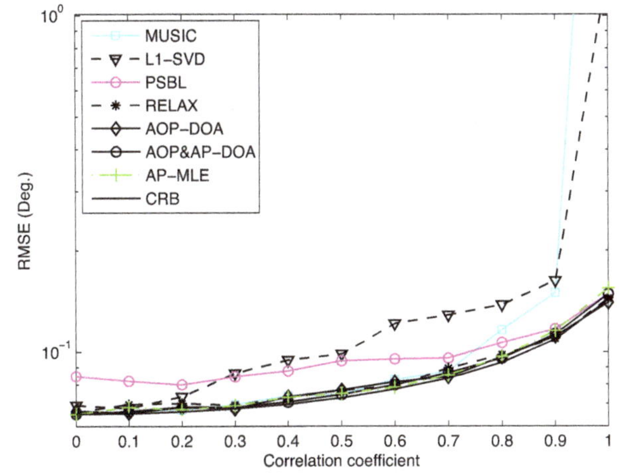

(a) RMSE versus correlation coefficient

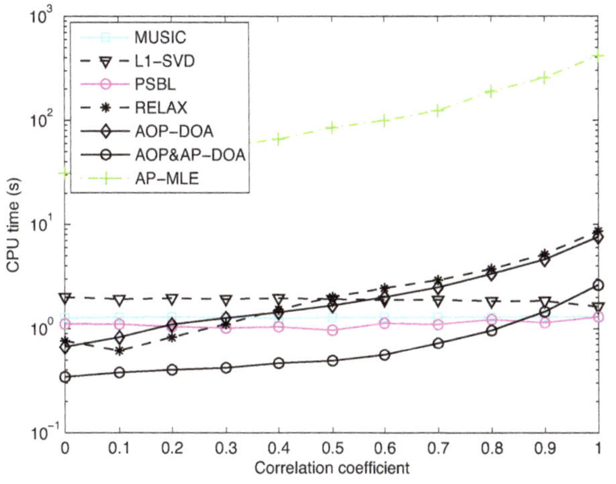

(b) CPU time versus correlation coefficient

Fig. 7. RMSE and CPU time comparison for two sources impinging from [−15.12°, 5.37°], where the SNR is fixed at 15 dB, but the sources' correlation coefficient varies.

show that the proposed AOP-DOA and AOP&AP-DOA algorithms, as well as the RELAX and AP-MLE, exhibit a satisfying estimation performance when uncorrelated and fully correlated signals coexist (remark that the MUSIC algorithm fails in this case, thus it is omitted here). On the whole, when compared with PSBL, the proposed algorithms have a better estimation performance and a more stable computational load under different SNR. Compared with l_1-SVD, the proposed algorithms can have a more superior estimation performance, although their computational loads are slightly higher in this scenario. In addition, as we have already verified in the previous simulations, the proposed AOP-DOA and AOP&AP-DOA have a much lower computational load when compared with AP-MLE.

4.3 Probability of Success Comparison

This subsection intends to evaluate the success probabilities of the proposed AOP-DOA and AOP&AP-DOA algorithms. The DOA estimation of K signals is considered successful if and only if $|\hat{\Theta}_1 - \Theta_1|, |\hat{\Theta}_2 - \Theta_2|, \cdots, |\hat{\Theta}_K - \Theta_K|$ are less than a user-selected parameter ϵ, which is set to be 1° in the following simulations.

Figure 8 considers two uncorrelated signals with different angular separations. The signals impinge on the array from $\Theta_1 = 5.37°$ and $\Theta_2 = 5.37° - \Delta$, and the source separation Δ ranges from 2° to 20°. In addition, the SNR is fixed at 10 dB.

Figure 9 considers three signals, where we assume a group of two correlated signals with correlation coefficient r impinge on the array from $[-15.12°, 5.37°]$ while an uncorrelated signal impinges on the array from 42.95°. The correlation coefficient r is set to 0.9 and 1 in Fig. 9(a) and Fig. 9(b), respectively. Other settings remain the same as the last experiment considered in Sec. 4.2.

Both Fig. 8 and Fig. 9 show that the proposed AOP-DOA and AOP&AP-DOA algorithms have a relatively higher success probability when compared with l_1-SVD, PSBL and MUSIC. In addition, the success probability of the proposed AOP&AP-DOA is able to coincide well with that of the AP-MLE, and is not lower than that of the RELAX. Remark that when signals are coherent, i.e., $r = 1$, the RMSE of l_1-SVD is relatively higher than the threshold ϵ (cf. Fig. 6(a)). So the success probability of l_1-SVD is low. Meanwhile, the MUSIC is inefficient for identifying DOAs of coherent signals, hence it fails.

Figure 10 tests the proposed AOP-DOA and AOP&AP-DOA algorithms by using a uniform circular array (UCA), where we consider azimuth-only DOA estimation and assume a group of two coherent signals impinging from [−15.12°, 5.37°] and an uncorrelated signal impinging from 42.95°. The SNR ranges form −15 dB to 20 dB. $M = 15$ and the UCA is considered with two different radii. It can be seen that the proposed two algorithms are applicable to UCA, and this is because they have no limitation on antenna array geometry.

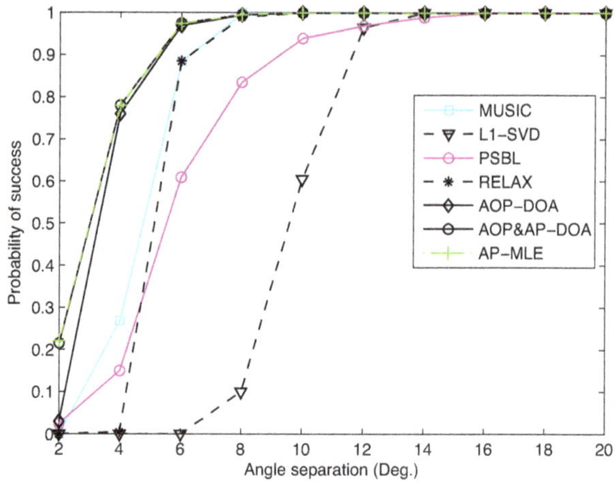

Fig. 8. Probability of success comparison for two uncorrelated signals impinging from $[5.37°, 5.37° - \Delta]$, where Δ ranges from 2° to 20° and the SNR is fixed at 10 dB.

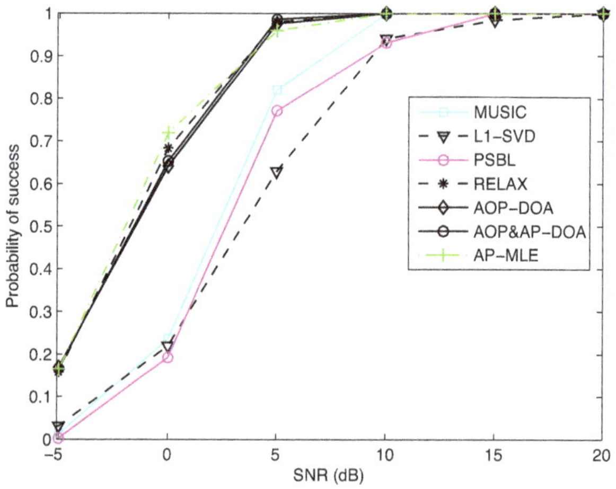

(a) Probability of success versus SNR when $r = 0.9$

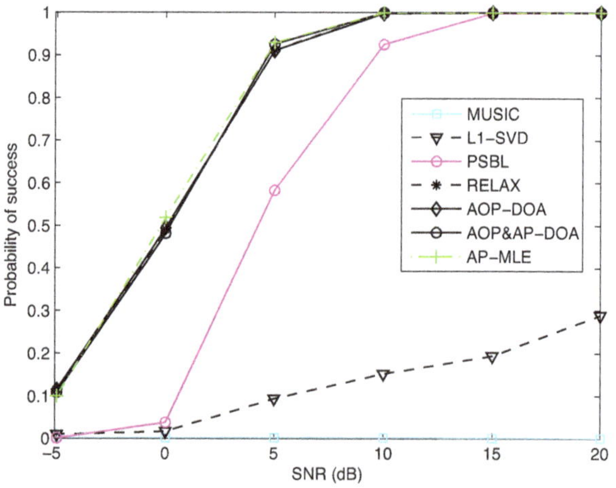

(b) Probability of success versus SNR when $r = 1$

Fig. 9. Probability of success comparison for a mixture of uncorrelated and correlated signals, where a group of two correlated signals with correlation coefficient r impinging from $[-15.12°, 5.37°]$ and an uncorrelated signal impinging from 42.95°.

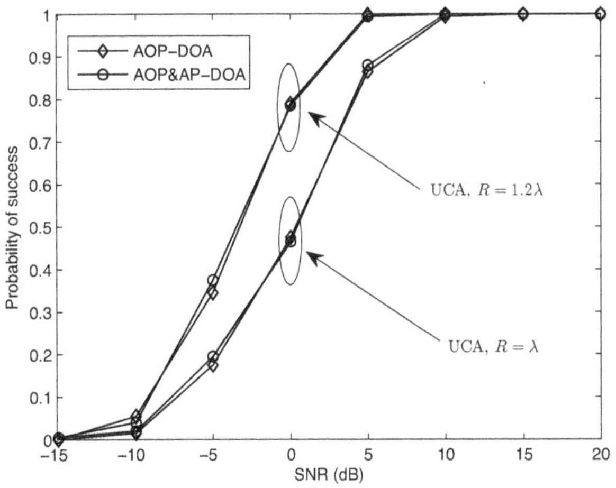

Fig. 10. Probability of success comparison of the proposed AOP-DOA and AOP&AP-DOA for a mixture of uncorrelated and coherent signals by using a UCA, where R and λ denote array radius and wavelength, respectively.

5. Conclusion

In this paper, a new iterative AOP-DOA algorithm is proposed for the DOA estimation of uncorrelated and correlated signals. To reduce the computational complexity of AOP-DOA, an algorithm named AOP&AP-DOA, which is a combination of AOP-DOA and AP-MLE, is also proposed.

Both AOP-DOA and AOP&AP-DOA are able to estimate DOAs of uncorrelated, partially correlated and coherent signals. The simulation results indicate that they exhibit excellent performance in terms of accuracy, success probability and complexity, when compared with MUSIC, l_1-SVD, PSBL, RELAX and AP-MLE. Firstly, for coherent signals, AOP-DOA has a distinct advantage over MUSIC in terms of DOA estimation since the latter algorithm is not applicable to this case. Secondly, compared with l_1-SVD and PSBL, AOP-DOA presents better estimation accuracy within moderate to high SNR region and presents a relatively higher success probability in low SNR region. Thirdly, the complexities of AOP-DOA and RELAX are approximately presented at the same level, but when the angular separation is small, for uncorrelated signals, the accuracy and success probability of AOP-DOA are higher than those of RELAX. Lastly, when compared with AP-MLE, the proposed AOP-DOA not only exhibits a much lower computational complexity but also shows competitive success probability and accuracy.

Compared with AOP-DOA, the AOP&AP-DOA not only keeps the merits of AOP-AOP, but also can be implemented with a much lower computational cost when proper adaptive grid refinement strategy is applied. In addition, both AOP-DOA and AOP&AP-DOA are not restricted by the array geometry. Extensive experiments have been undertaken to verify our analysis and demonstrate the performance of our algorithms in cases of uncorrelated and correlated signals as compared to the existing ones. Note that only the white noise scenario is analyzed in this paper. Future work includes extension of the proposed algorithms to general noise environments [50], [20].

Acknowledgments

This work was supported in part by the National Natural Science Foundation of China (Grants No. 61171180 and No. 61371100) and the Fundamental Research Funds for the Central Universities (Grants No. HIT. MKSTISP. 2016 13 and No. HIT. MKSTISP. 2016 26).

References

[1] KRIM, H., VIBERG, M. Two decades of array signal processing research: the parametric approach. *IEEE Signal Processing Magazine*, 1996, vol. 13, no. 4, p. 67–94. DOI: 10.1109/79.526899

[2] GAN, L., LUO, X. Direction-of-arrival estimation for uncorrelated and coherent signals in the presence of multipath propagation. *IET Microwaves, Antennas and Propagation*, 2013, vol. 7, no. 9, p. 746–753. DOI: 10.1049/iet-map.2012.0659

[3] STOICA, P., NEHORAI, A. Performance study of conditional and unconditional direction-of-arrival estimation. *IEEE Transactions on Acoustics, Speech, and Signal Processing*, 1990, vol. 38, no. 10, p. 1783–1795. DOI: 10.1109/29.60109

[4] STOICA, P., SHARMAN, K. C. Maximum likelihood methods for direction-of-arrival estimation. *IEEE Transactions on Acoustics, Speech, and Signal Processing*, 1990, vol. 38, no. 7, p. 1132–1143. DOI: 10.1109/29.57542

[5] SHEINVALD, J., WAX, M., WEISS, A. J. On maximum-likelihood localization of coherent signals. *IEEE Transactions on Signal Processing*, 1996, vol. 44, no. 10, p. 2475–2482. DOI: 10.1109/78.539032

[6] SCHMIDT, R. O. Multiple emitter location and signal parameter estimation. *IEEE Transactions on Antennas and Propagation*, 1986, vol. 34, no. 3, p. 276–280. DOI: 10.1109/TAP.1986.1143830

[7] ROY, R., KAILATH, T. ESPRIT-estimation of signal parameters via rotational invariance techniques. *IEEE Transactions on Acoustics, Speech, and Signal Processing*, 1989, vol. 37, no. 7, p. 984–995. DOI: 10.1109/29.32276

[8] XIN, J. M., SANO, A. Computationally efficient subspace-based method for direction-of-arrival estimation without eigendecomposition. *IEEE Transactions on Signal Processing*, 2004, vol. 52, no. 4, p. 876–893. DOI: 10.1109/TSP.2004.823469

[9] ZOLTOWSKI, M., HABER, F. A vector space approach to direction finding in a coherent multipath environment. *IEEE Transactions on Antennas and Propagation*, 1986, vol. 34, no. 9, p. 1069–1079. DOI: 10.1109/TAP.1986.1143956

[10] TAO, H., XIN, J., WANG, J., et al. Two-dimensional direction estimation for a mixture of noncoherent and coherent signals. *IEEE Transactions on Signal Processing*, 2015, vol. 63, no. 2, p. 318–333. DOI: 10.1109/TSP.2014.2369004

[11] LINEBARGER, D. A. Redundancy averaging with large arrays. *IEEE Transactions on Signal Processing*, 1993, vol. 41, no. 4, p. 1707–1710. DOI: 10.1109/78.212750

[12] PILLAI, S. U., KWON, B. H. Forward/backward spatial smoothing techniques for coherent signal identification. *IEEE Transactions on Acoustics, Speech, and Signal Processing*, 1989, vol. 37, no. 1, p. 8–15. DOI: 10.1109/29.17496

[13] WAX, M., SHEINVALD, J. Direction finding of coherent signals via spatial smoothing for uniform circular arrays. *IEEE Transactions on Antennas and Propagation*, 1994, vol. 42, no. 5, p. 613–620. DOI: 10.1109/8.299559

[14] INDUKUMAR, K. C., REDDY, V. U. A note on redundancy averaging. *IEEE Transactions on Signal Processing*, 1992, vol. 40, no. 2, p. 466–469. DOI: 10.1109/78.124962

[15] QIAN, C., HUANG, L., ZENG, W. J., et al. Direction-of-arrival estimation for coherent signals without knowledge of source number. *IEEE Sensors Journal*, 2014, vol. 14, no. 9, p. 3267–3273. DOI: 10.1109/JSEN.2014.2327633

[16] RAHAMIM, D., TABRIKIAN, J., SHAVIT, R. Source localization using vector sensor array in a multipath environment. *IEEE Transactions on Signal Processing*, 2004, vol. 52, no. 11, p. 3096–3103. DOI: 10.1109/TSP.2004.836456

[17] ZISKIND, I., WAX, M. Maximum likelihood localization of multiple sources by alternating projection. *IEEE Transactions on Acoustics, Speech, and Signal Processing*, 1988, vol. 36, no. 10, p. 1553–1560. DOI: 10.1109/29.7543

[18] OH, S. K., UN, C. K. Efficient realisation of alternating projection algorithm for maximum likelihood direction finding. *Electronics Letters*, 1989, vol. 25, no. 20, p. 1325–1326. DOI: 10.1049/el:19890885

[19] VIBERG, M., OTTERSTEN, B., KAILATH, T. Detection and estimation in sensor arrays using weighted subspace fitting. *IEEE Transactions on Signal Processing*, 1991, vol. 39, no. 17, p. 2436–2449. DOI: 10.1109/78.97999

[20] LI, M. H., LU Y. L. Maximum likelihood DOA estimation in unknown colored noise fields. *IEEE Transactions on Aerospace and Electronic Systems*, 2008, vol. 44, no. 3, p. 1079–1090. DOI: 10.1109/TAES.2008.4655365

[21] MASSA, A., ROCCA, P., OLIVERI, G. Compressive sensing in electromagnetics – a review. *IEEE Antennas and Propagation Magazine*, 2015, vol. 57, no. 1, p. 224–238. DOI: 10.1109/MAP.2015.2397092

[22] SAHOO, S. K., MAKUR, A. Signal recovery from random measurements via extended orthogonal matching pursuit. *IEEE Transactions on Signal Processing*, 2015, vol. 63, no. 10, p. 2572–2581. DOI: 10.1109/TSP.2015.2413384

[23] MALIOUTOV, D., CETIN, M., WILLSKY, A. S. A sparse signal reconstruction perspective for source localization with sensor arrays. *IEEE Transactions on Signal Processing*, 2005, vol. 53, no. 8, p. 3010–3022. DOI: 10.1109/TSP.2005.850882

[24] YIN, J., CHEN, T. Direction-of-arrival estimation using a sparse representation of array covariance vectors. *IEEE Transactions on Signal Processing*, 2011, vol. 59, no. 9, p. 4489–4493. DOI: 10.1109/TSP.2011.2158425

[25] HYDER, M. M., MAHATA, K. Direction-of-arrival estimation using a mixed $l_{2,0}$ norm approximation. *IEEE Transactions on Signal Processing*, 2010, vol. 58, no. 9, p. 4646–4655. DOI: 10.1109/TSP.2010.2050477

[26] CARLIN, M., ROCCA, P., OLIVERI, G., et al. Directions-of-arrival estimation through bayesian compressive sensing strategies. *IEEE Transactions on Antennas and Propagation*, 2013, vol. 61, no. 7, p. 3828–3838. DOI: 10.1109/TAP.2013.2256093

[27] WU, X., ZHU, W. P., YAN, J. Direction of arrival estimation for off-grid signals based on sparse bayesian learning. *IEEE Sensors Journal*, 2016, vol. 16, no. 7, p. 2004–2016. DOI: 10.1109/JSEN.2015.2508059

[28] LI, J., ZHENG, D. M., STOICA, P. Angle and waveform estimation via RELAX. *IEEE Transactions on Aerospace and Electronic Systems*, 1997, vol. 33, no. 3, p. 1077–1087. DOI: 10.1109/7.599338

[29] YARDIBI, T., LI, J., STOICA, P., et al. Source localization and sensing: a nonparametric iterative adaptive approach based on weighted least squares. *IEEE Transactions on Aerospace and Electronic Systems*, 2010, vol. 46, no. 1, p. 425–443. DOI: 10.1109/TAES.2010.5417172

[30] BEHRENS, R. T., SCHARF, L. L. Signal processing applications of oblique projection operators. *IEEE Transactions on Signal Processing*, 1994, vol. 42, no. 6, p. 1413–1424. DOI: 10.1109/78.286957

[31] BOYER, R. Oblique projection for source estimation in a competitive environment: algorithm and statistical analysis. *Signal Processing*, 2009, vol. 89, no. 12, p. 2547–2554. DOI: 10.1016/j.sigpro.2009.04.023

[32] SCHARF, L. L., MCCLOUD, M. L. Blind adaptation of zero forcing projections and oblique pseudo-inverses for subspace detection and estimation when interference dominates noise. *IEEE Transactions on Signal Processing*, 2002, vol. 50, no. 12, p. 2938–2946. DOI: 10.1109/TSP.2002.805245

[33] MCCLOUD, M. L., SCHARF, L. L. A new subspace identification algorithm for high-resolution DOA estimation. *IEEE Transactions on Antennas and Propagation*, 2002, vol. 50, no. 10, p. 1382–1390. DOI: 10.1109/TAP.2002.805244

[34] BOYER, R., BOULEUX, G. Oblique projections for direction-of-arrival estimation with prior knowledge. *IEEE Transactions on Signal Processing*, 2008, vol. 56, no. 4, p. 1374–1387. DOI: 10.1109/TSP.2007.909348

[35] TAO, H., XIN, J. M., WANG, J. S., et al. Oblique projection based enumeration of mixed noncoherent and coherent narrowband signals. *IEEE Transactions on Signal Processing*, 2016, vol. 64, no. 16, p. 4282–4295. DOI: 10.1109/TSP.2016.2548994

[36] TAO, H., XIN, J. M., WANG, J. S., et al. Two-dimensional direction estimation for a mixture of noncoherent and coherent signals. *IEEE Transactions on Signal Processing*, 2015, vol. 63, no. 2, p. 318–333. DOI: 10.1109/TSP.2014.2369004

[37] XU, X., YE, Z., Zhang Y., et al. A deflation approach to direction of arrival estimation for symmetric uniform linear array. *IEEE Antennas and Wireless Propagation Letters*, 2006, vol. 5, no. 1, p. 486–489. DOI: 10.1109/LAWP.2006.886304

[38] SI, W. J., ZHAO P. J., QU, Z. Y., et al. Real-valued DOA estimation for a mixture of uncorrelated and coherent sources via unitary transformation. *Digital Signal Processing*, 2016, vol. 58, p. 102–114. DOI: 10.1016/j.dsp.2016.07.024

[39] CUI, Y., LIU, K., WANG, J. Direction-of-arrival estimation for coherent GPS signals based on oblique projection. *Signal Processing*, 2012, vol. 92, no. 1, p. 294–299. DOI: 10.1016/j.sigpro.2011.07.014

[40] XU, X., YE, Z., PENG, J. Method of direction-of-arrival estimation for uncorrelated, partially correlated and coherent sources. *IET Microwaves, Antennas and Propagation*, 2007, vol. 1, no. 4, p. 949–954. DOI: 10.1049/iet-map:20070006

[41] LU, Z., ZOUBIR, A. M. Source enumeration in array processing using a two-step test. *IEEE Transactions on Signal Processing*, 2015, vol. 63, no. 10, p. 2718–2727. DOI: 10.1109/TSP.2015.2414894

[42] HUANG, L., XIAO, Y., LIU, K., et al. Bayesian information criterion for source enumeration in large-scale adaptive antenna array. *IEEE Transactions on Vehicular Technology*, 2016, vol. 65, no. 5, p. 3018–3032. DOI: 10.1109/TVT.2015.2436060

[43] ROUBICEK, T., KRUZIK, M. Adaptive approximation algorithm for relaxed optimization problems. In *Fast Solution of Discretized Optimization Problems*. Birkhäuser (Basel), 2001, p. 242–254. DOI: 10.1007/978-3-0348-8233-0_18

[44] THEODORIDIS, S., SLAVAKIS, K., YAMADA, I. Adaptive learning in a world of projections. *IEEE Signal Processing Magazine*, 2011, vol. 28. no. 1, p. 97–123. DOI: 10.1109/MSP.2010.938752

[45] HURT, N. E. Maximum likelihood estimation and MUSIC in array localization signal processing: a review. *Multidimensional Systems and Signal Processing*, 1990, vol. 1, no. 3, p. 279–325. DOI: 10.1007/BF01812401

[46] GOLUB, G. H., PEREYRA, V. The differentiation of pseudo-inverses and nonlinear least squares problems whose variables separate. *SIAM Journal on Numerical Analysis*, 1973, vol. 10, no. 2, p. 413–432. DOI: 10.1137/0710036

[47] CIRRINCIONE, G., CIRRINCIONE, M., HERAULT, J., et al. The MCA EXIN neuron for the minor component analysis. *IEEE Transactions on Neural Networks*, 2002, vol. 13, no. 1, p. 160–187. DOI: 10.1109/72.977295

[48] MANGASARIAN, O. L., WILD, E. W., Multisurface proximal support vector machine classification via generalized eigenvalues. *IEEE Transactions on Pattern Analysis and Machine Intelligence*, 2006, vol. 28, no. 1, p. 69–74. DOI: 10.1109/TPAMI.2006.17

[49] STURM, J. F. Using SeDuMi 1.02, a MATLAB toolbox for optimization over symmetric cones. *Optimization Methods and Software*, 1999, vol. 11, no. 1–4, p. 625–653. DOI: 10.1080/10556789908805766

[50] LIAO, B., CHAN, S. C., HUANG, L., et al. Iterative methods for subspace and DOA estimation in nonuniform noise. *IEEE Transactions on Signal Processing*, 2016, vol. 64, no. 12, p. 3008–3020. DOI: 10.1109/TSP.2016.2537265

About the Authors . . .

Huijun HOU was born in Anhui, China in 1987. He received the M.S. degree in Information and Communication Engineering from Harbin Institute of Technology, Harbin, China in 2012. He is currently working towards the Ph.D. degree at Harbin Institute of Technology. His research interests include array signal processing and anti-jamming techniques.

Xingpeng MAO (corresponding author) was born in Liaoning, China in 1972. He received his Ph.D. degree from Harbin Institute of Technology, Harbin, China in 2004. He is currently a doctoral advisor and full professor in School of Electronic and Information Engineering, Harbin Institute of Technology, China. His research interests include radar systems and signal processing.

Appendix A: Proof of (12)

After AOP-DOA achieves convergence, it follows from (10) and (11) that

$$s_k(t) = \mathbf{a}^{\dagger}(\Theta_k)\mathbf{y}_k(t). \tag{27}$$

Substituting (7) into (27), then we have

$$s_k(t) = \mathbf{a}^{\dagger}(\Theta_k)(\mathbf{I} - \mathbf{E}_{\mathbf{B}_k,\mathbf{a}(\Theta_k)})\,\mathbf{z}(t) \tag{28}$$

Substituting (28) into (3), it follows that

$$\begin{aligned}&\frac{1}{L}\sum_{l=1}^{L}\left\|\mathbf{x}(t_l) - \sum_{k=1}^{K}\mathbf{a}(\Theta_k)s_k(t_l)\right\|^2\\ &= \frac{1}{L}\sum_{l=1}^{L}\left\|\mathbf{x}(t_l) - \sum_{k=1}^{K}\mathbf{P}_{\mathbf{a}(\Theta_k)}(\mathbf{I} - \mathbf{E}_{\mathbf{B}_k,\mathbf{a}(\Theta_k)})\,\mathbf{z}(t_l)\right\|^2\end{aligned} \tag{29}$$

According to Appendix B, we have

$$\mathbf{P}_{\mathbf{A}(\mathbf{\Theta})} = \sum_{k=1}^{K}\mathbf{P}_{\mathbf{a}(\Theta_k)}(\mathbf{I} - \mathbf{E}_{\mathbf{B}_k,\mathbf{a}(\Theta_k)}). \tag{30}$$

Consequently, (29) can be rewritten as

$$\frac{1}{L}\sum_{l=1}^{L}\left\|\mathbf{x}(t_l) - \sum_{k=1}^{K}\mathbf{a}(\Theta_k)s_k(t_l)\right\|^2 = \sum_{l=1}^{L}\frac{\left\|\mathbf{x}(t_l) - \mathbf{P}_{\mathbf{A}(\mathbf{\Theta})}\mathbf{z}(t_l)\right\|^2}{L} \tag{31}$$

Substituting (1) into (31), then we have

$$\begin{aligned}&\frac{1}{L}\sum_{l=1}^{L}\left\|\mathbf{x}(t_l) - \sum_{k=1}^{K}\mathbf{a}(\Theta_k)s_k(t_l)\right\|^2\\ &= \frac{1}{L}\sum_{l=1}^{L}\left\|\mathbf{P}^{\perp}_{\mathbf{A}(\mathbf{\Theta})}\mathbf{z}(t_l) + \mathbf{n}(t_l)\right\|^2 = \mathrm{tr}\left\{\hat{\mathbf{R}}_n\right\} = M\hat{\sigma}^2\end{aligned} \tag{32}$$

where $\hat{\mathbf{R}}_n = \frac{1}{L}\sum_{l=1}^{L}\mathbf{n}(t_l)\mathbf{n}^{\mathrm{H}}(t_l) = \hat{\sigma}^2\mathbf{I}$.

Substituting (8) into (31), then we have

$$\begin{aligned}&\frac{1}{L}\sum_{l=1}^{L}\left\|\mathbf{x}(t_l) - \sum_{k=1}^{K}\mathbf{a}(\Theta_k)s_k(t_l)\right\|^2\\ &= \frac{1}{L}\sum_{l=1}^{L}\left\|\mathbf{P}^{\perp}_{\mathbf{A}(\mathbf{\Theta})}\mathbf{x}(t_l) + \mathbf{P}_{\mathbf{A}(\mathbf{\Theta})}\mathbf{n}(t_l)\right\|^2 = \mathrm{tr}\left\{\mathbf{P}^{\perp}_{\mathbf{A}(\mathbf{\Theta})}\hat{\mathbf{R}}_x\right\}+\\ &\quad\mathrm{tr}\left\{\mathbf{P}_{\mathbf{A}(\mathbf{\Theta})}\hat{\mathbf{R}}_n\right\} = \mathrm{tr}\left\{\mathbf{P}^{\perp}_{\mathbf{A}(\mathbf{\Theta})}\hat{\mathbf{R}}_x\right\} + K\hat{\sigma}^2.\end{aligned} \tag{33}$$

It follows from (32) and (33) that

$$\begin{aligned}&\frac{1}{(M-K)L}\sum_{l=1}^{L}\left\|\mathbf{x}(t_l) - \sum_{k=1}^{K}\mathbf{a}(\Theta_k)s_k(t_l)\right\|^2 - \frac{K\hat{\sigma}^2}{M-K}\\ &= \frac{1}{M-K}\mathrm{tr}\left\{\mathbf{P}^{\perp}_{\mathbf{A}(\mathbf{\Theta})}\hat{\mathbf{R}}_x\right\} = \hat{\sigma}^2.\end{aligned} \tag{34}$$

Appendix B: Proof of (30)

It follows from (6) and [30] that

$$\begin{aligned}&\sum_{k=1}^{K}\mathbf{E}_{\mathbf{a}(\Theta_k),\mathbf{B}_k}\mathbf{A}(\mathbf{\Theta})\\ &= \sum_{k=1}^{K}\left[\mathbf{E}_{\mathbf{a}(\Theta_k),\mathbf{B}_k}\mathbf{a}(\Theta_1),\cdots,\mathbf{E}_{\mathbf{a}(\Theta_k),\mathbf{B}_k}\mathbf{a}(\Theta_K)\right]\\ &= \mathbf{A}(\mathbf{\Theta}).\end{aligned} \tag{35}$$

Thus, it holds that $(\sum_{k=1}^{K}\mathbf{E}_{\mathbf{a}(\Theta_k),\mathbf{B}_k} - \mathbf{I})\mathbf{A}(\mathbf{\Theta}) = \mathbf{0}$, and $(\sum_{k=1}^{K}\mathbf{E}_{\mathbf{a}(\Theta_k),\mathbf{B}_k} - \mathbf{I})\mathbf{A}(\mathbf{\Theta})\mathbf{A}^{\dagger}(\mathbf{\Theta}) = \mathbf{0}$. Then,

$$\mathbf{P}_{\mathbf{A}(\mathbf{\Theta})} = \sum_{k=1}^{K}\mathbf{E}_{\mathbf{a}(\Theta_k),\mathbf{B}_k}\mathbf{P}_{\mathbf{A}(\mathbf{\Theta})}. \tag{36}$$

Note that $\mathbf{P}_{\mathbf{a}(\Theta_k)}\mathbf{P}^{\perp}_{\mathbf{A}(\mathbf{\Theta})} = \mathbf{0}$, $\mathbf{E}_{\mathbf{a}(\Theta_k),\mathbf{B}_k}\mathbf{E}_{\mathbf{B}_k,\mathbf{a}(\Theta_k)} = \mathbf{0}$ and $\mathbf{P}_{\mathbf{A}(\mathbf{\Theta})} = \mathbf{E}_{\mathbf{a}(\Theta_k),\mathbf{B}_k} + \mathbf{E}_{\mathbf{B}_k,\mathbf{a}(\Theta_k)}$. It follows that $\mathbf{E}_{\mathbf{a}(\Theta_k),\mathbf{B}_k} = \mathbf{P}_{\mathbf{a}(\Theta_k)}\mathbf{E}_{\mathbf{a}(\Theta_k),\mathbf{B}_k} = \mathbf{P}_{\mathbf{a}(\Theta_k)}(\mathbf{I} - \mathbf{E}_{\mathbf{B}_k,\mathbf{a}(\Theta_k)})$, and

$$\mathbf{E}_{\mathbf{a}(\Theta_k),\mathbf{B}_k}\mathbf{P}_{\mathbf{A}(\mathbf{\Theta})} = \mathbf{E}_{\mathbf{a}(\Theta_k),\mathbf{B}_k}. \tag{37}$$

Substituting (37) into (36), then

$$\mathbf{P}_{\mathbf{A}(\mathbf{\Theta})} = \sum_{k=1}^{K}\mathbf{E}_{\mathbf{a}(\Theta_k),\mathbf{B}_k} = \sum_{k=1}^{K}\mathbf{P}_{\mathbf{a}(\Theta_k)}(\mathbf{I} - \mathbf{E}_{\mathbf{B}_k,\mathbf{a}(\Theta_k)}). \tag{38}$$

5

Distributed Matching Algorithms: Maximizing Secrecy in the Presence of Untrusted Relay

Bakhtiar ALI[1], *Nida ZAMIR*[1], *Soon Xin NG*[2], *Muhammad Fasih Uddin BUTT*[1]

[1]Department of Electrical Engineering, COMSATS Institute of Information Technology Islamabad, Pakistan
[2]Department of Electronics and Computer Science, University of Southampton, Southampton SO17 1BJ, United Kingdom

bakhtiar_ali@comsats.edu.pk, nida.zamir@comsats.edu.pk, sxn@ecs.soton.ac.uk, fasih@comsats.edu.pk

Abstract. *In this paper, we propose a secrecy sum-rate maximization based matching algorithm between primary transmitters and secondary cooperative jammers in the presence of an eavesdropper. More explicitly, we consider an untrusted relay scenario, where the relay is a potential eavesdropper. We first show the achievable secrecy regions employing a friendly jammer in a cooperative scenario with employing an untrusted relay. Then, we provide results for the secrecy regions for two cases, one where we consider that there is no direct link between the source and the destination, for the second case we consider that in addition to the relay link we also have a direct link between the source and destination. Furthermore, a friendly jammer helps to send a noise signal during the first phase of the cooperative transmission, for securing the information transmitted from the source. In our matching algorithm, the selected cooperative jammer or the secondary user, is rewarded with the spectrum allocation for a fraction of time slot from the source which is the primary user. The Conventional Distributed Algorithm (CDA) and the Pragmatic Distributed Algorithm (PDA), which were originally designed for maximising the user's sum rate, are modified and adapted for maximizing the secrecy sum-rate for the primary user. Instead of assuming perfect modulation and/or perfect channel coding, we have also investigated our proposed schemes when practical channel coding and modulation schemes are invoked.*

Keywords

Physical layer security, spectrum matching, game theory, spectrum sharing, cognitive radio networks

1. Introduction

The boom in communication technology has brought about revolutionary changes in our lives. This growth along with its benefits has some demerits or challenges since we will be dealing with a huge amount of data coming through billions of connected devices. Current mobile systems would not be able to keep up sufficient provision of providing privacy and security due to the ever-growing number of customers. Enabling technologies like 5G are required for supporting future wireless systems having a large number of devices communicating at ultra high data rates with extreme low latency. 5G technologies such as heterogeneous networks, where lots of devices with different operating systems and protocols will be collaborating or cooperating, will make the problem of privacy and security even more challenging [1–4]. Similarly, Internet of Things (IoT) will be dealing with devices with limited hardware, low complexity and strict energy constraints which presents unique security challenges [5]. In these wireless environments, devices have limited capabilities and are not controlled by a central control station. Hence, the implementation of computationally intensive cryptographic techniques may be challenging. Motivated by these deliberations, substantial research work have been investigating the use of physical layer as a means to develop low-complexity and effective wireless security mechanisms. Such techniques are grouped under the umbrella of Physical Layer Security (PLS) [6]. More explicitly, friendly jamming is a promising PLS technique, which employs cooperating nodes to transmit artificial noise [7–10].

Most of the work in security considers eavesdropper as an outside entity, while authors in [11] presented the motivation for using an untrusted relay for the transmission of information from the source to the destination. They demonstrated that if an untrusted relay is asked to relay information towards the destination, the secrecy rates achieved are higher as compared to the case when the relay is only considered as an eavesdropper. In [12] a link adaptation with untrusted relay assignment framework for cooperative communications is proposed by utilizing arbitrary number of relays for reliable information transfer while ensuring secrecy at the relays. The authors in [13] consider a two-user interference relay channel with the aim to secure the messages from either destinations, as well as the untrusted relay, without the presence of direct link between either of the users. In [14] the authors present the secrecy rates for a dual-hop amplify and forward (AF) multiple-input multiple-output (MIMO) relay network. More specifically, a joint destination based cooperative jamming and joint source, relay and destination precoding based secrecy rate maximization problem is formulated in [14], where simple closed form expressions for asymptotic secrecy rate in high signal to noise ratio (SNR) regime is also presented. In [15] a destination-assisted jamming for secure communication between a source and a destination via a wireless energy harvesting untrusted relay node is proposed. In [16] the authors defined an achievable secrecy rate region, using random binning at the sources and utilizing the compress and forward relaying strategy with the help of cooperative jamming from both destinations. They also derived a genie-aided outer bound on the secrecy rate region. The drawback of friendly jamming is that there have to be dedicated jamming nodes which are willing to share their resources with the nodes that are not related to the jammer except for the case where the jammer is the destination. In contrast, a two way resource sharing would be more practical where the jammer will assist in jamming the source signal from the eavesdropper and in return will gain access to the channel for transmission of its own information.

A game theoretic based friendly jamming mechanism is proposed in [17], where source-destination communication is secured by a Cooperative Jammer (CJ), which is then compensated by the source's spectrum hence enabling the jammer to transmit its own information towards its destination. In [18] a Stackelberg game for maximizing the source and jammer's utility subject to maximum jamming power at the jammer is presented. They provided a uniform pricing algorithm for maximizing the secrecy rate of the system. In [19] the authors proposed a model where each user can act as a data source as well as a friendly jammer. They formalized a coalition game based cooperation for their proposed model. Furthermore, a "merge and split rules", based distributed algorithm was proposed, where the dual-identity nodes can mutually affect and cooperate into disjoint independent coalitions for maximizing the total secrecy capacity participating users. The authors in [20] proposed a cooperative framework to enhance security in a multiple eavesdropper scenario. A game theoretic incentive mechanism was proposed to stimulate the partners to participate into cooperation.

Most of the work in game theoretic based jamming also considers the eavesdropper as an outside entity [17–20]. Here, we present user cooperation based PLS by employing a CJ to provide security by transmitting an artificial noise towards an untrusted relay. Firstly we provide the achievable secrecy regions for such a scenario for two different cases i.e. with and without the Source (S) - Destination (D) link. Then a user cooperation based game theoretic matching algorithm is presented based on the Conventional Distributed Algorithm (CDA) of [21] for maximizing the secrecy provided by the CJ. The CJ is then compensated for its service by the provision of the source's spectrum for a limited amount of time. The downside of the CDA was that the Primary Users (PU) which in this case are S compete among themselves for matching with the best possible Secondary User (SU) which is the CJ. Another matching algorithm called Pragmatic Distributed Algorithm (PDA) was presented in [22] where this competition was eliminated by introducing another game where the PUs participate in a round robin rotation manner for acquiring the best possible SU/CJ. In that way each of the PUs will gain access to its best possible SU for at least one round. Liang et. al. in [22] also provide results for Adaptive Turbo Trellis Coded Modulation (ATTCM) for her algorithm showing the practicality of the algorithm. For our system we also use PDA based cooperative jamming and compare it with CDA based cooperative jamming using idealistic situation where the system can operate at the capacity of the Continuous-Input Continuous-Output Memoryless Channel (CCMC) and that of the Discrete-Input Continuous-Output Memoryless Channel (DCMC) [23]. However these assume perfect modulation and/or perfect coding. For a more realistic scenario, we involve a realistic Self Concatenated Convolutional Coding (SECCC) [24] based scheme. More explicitly, SECCC is a low complexity, flexible and bandwidth-efficient coding scheme which involves only a single encoder and a single decoder. For higher code rates, puncturing can be used but it has a comparable performance to the Turbo Codes.

In this contribution, we present achievable secrecy regions for the scenario where there is a weak link between the source and the destination, with the aid of an untrusted relay. We show results for the two cases i.e., with and without the direct link between the source and destination[1]. Secondly we provide a secrecy maximization framework where we considered the following cooperative distributed matching algorithms which are based on the adaptation of the PDA and CDA of [22]:

1. Secure Pragmatic Distributed Algorithm (S-PDA) which maximizes the secrecy sum-rate for the participating primary nodes.

2. Secure Conventional Distributed Algorithms (S-CDA) for secrecy maximization for the participating primary nodes.

The centralized matching algorithm is also investigated as a comparison to our proposed schemes.

The organization of the paper is as follows. In Sec. 2, we present the friendly jamming based PLS and provide results for the achievable secrecy regions. Based on the results from Sec. 2, we introduce game theoretic secrecy maximization mechanism to further enhance secrecy of the participating nodes in Sec. 3. Finally, we present the conclusion for our findings in Sec. 4.

2. Friendly Jamming Based PLS

Our network includes a source (S) and destination (D) pair, with an untrusted relay (R) and a friendly cooperative jammer (CJ). We consider an AF based network where the relay amplifies and forwards the composite signal resulting from the mixture of S signal mixed with the noise signal

[1]Part of this work was presented at the 24th European Signal Processing Conference (EUSIPCO 2016), Budapest, Hungary, 29th Aug- 2nd Sept 2016.

from the CJ. The noise signal being transmitted from the CJ is assumed to be known at the destination. We assume that there is a weak direct link between the S and D therefore the transmission by S is assisted by an untrusted relay. We consider two scenarios which are further elaborated below.

2.1 Cooperative Jamming Without S-D Direct Link

For our first scenario we consider that the only link available is the link with the untrusted relay and no S to D direct link is included for communication, as shown in Fig. 1. We consider a single antenna system with half duplex operation. The channel between terminal i and terminal j is considered to be a Rayleigh fading channel denoted by h_{ij} and w represents the additive white Gaussian noise (AWGN) at each receiver input with zero mean and variance of σ_w^2 and unilateral power spectral density $N_0 = 2\sigma_w^2$ watts per hertz. The total transmit power is limited by P.

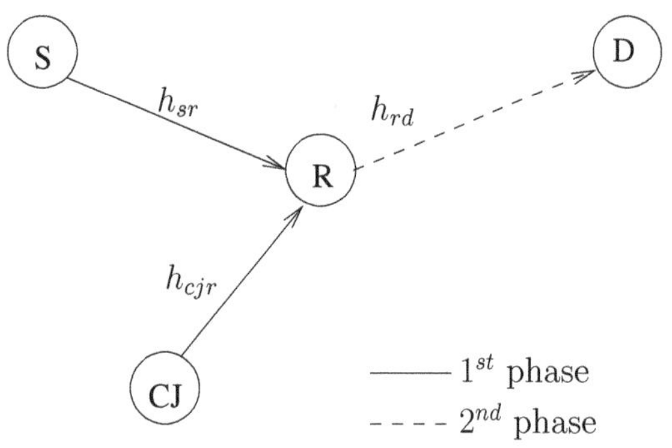

Fig. 1. Cooperative jamming system model (first scenario).

As seen in Fig. 1, S transmits the source signal x_s with power αP and a single selected CJ sends artificial noise η_{CJ}, with power $(1-\alpha)P$ which is known to D, where $\{0 \le \alpha \le 1\}$ is power distribution variant. Therefore we can write the signal received at R as:

$$y_r = h_{sr}\sqrt{\alpha P}x_s + h_{CJr}\sqrt{(1-\alpha)P}\eta_{CJ} + w_r \quad (1)$$

where w_r is the additive noise with unilateral power spectral density $N_0 = 2\sigma_w^2$ watts per hertz at R. After this, R amplifies and forwards the received signal y_r towards D, therefore the signal received at D is given as

$$y_d = h_{rd}\eta_r y_r + w_d \quad (2)$$

where w_d is the Additive White Gaussian Noise (AWGN) at D and the amplification factor is given by

$$\eta_r = \sqrt{\frac{P}{\alpha P|h_{sr}|^2 + (1-\alpha)P|h_{CJr}|^2 + N_0}}. \quad (3)$$

We can calculate the SNR γ_D at D as follows:

$$\gamma_D = \frac{\alpha\gamma_{rd}\gamma_{sr}}{\gamma_{rd} + \alpha\gamma_{rd} + (1-\alpha)\gamma_{CJr} + 1} \quad (4)$$

where γ_{sr} is the SNR for the S to R link, γ_{rd} is the SNR for the R to D link and γ_{CJr} is the SNR for the CJ to R links. Similarly, from (1) we can derive the SNR γ_r at R as follows:

$$\gamma_r = \frac{\alpha\gamma_{sr}}{(1-\alpha)\gamma_{CJr} + 1}. \quad (5)$$

Consequently, the achievable rates $\bar{R}_D$ at D and $\bar{R}_r$ at R will be calculated as

$$\begin{aligned}\bar{R}_D &= \frac{1}{2}\log(1+\gamma_D) \\ &= \frac{1}{2}\log\left(1 + \frac{\alpha\gamma_{rd}\gamma_{sr}}{\gamma_{rd} + \alpha\gamma_{rd} + (1-\alpha)\gamma_{CJr} + 1}\right), \end{aligned} \quad (6)$$

$$\bar{R}_r = \frac{1}{2}\log(1+\gamma_r) = \frac{1}{2}\log\left(1 + \frac{\alpha\gamma_{sr}}{(1-\alpha)\gamma_{CJr} + 1}\right). \quad (7)$$

Finally, the secrecy rate $\bar{R}_s$ of the system is given by

$$\bar{R}_s = \bar{R}_D - \bar{R}_r. \quad (8)$$

2.2 Cooperative Jamming with S-D Direct Link

For our second scenario we include the S to D direct link for our communication. Again the assumption is that the S to D direct link is weak and no communication is possible through this link without the help from the relay. The message signal is transmitted in two phases, where in the first phase a message is broadcasted by the S in parallel to the noise signal being broadcasted by the CJ and in the second phase which is the relaying phase, the relay amplifies the signal it received during the first phase and forwards it to the D, Fig. 2 shows the system model for our second scenario.

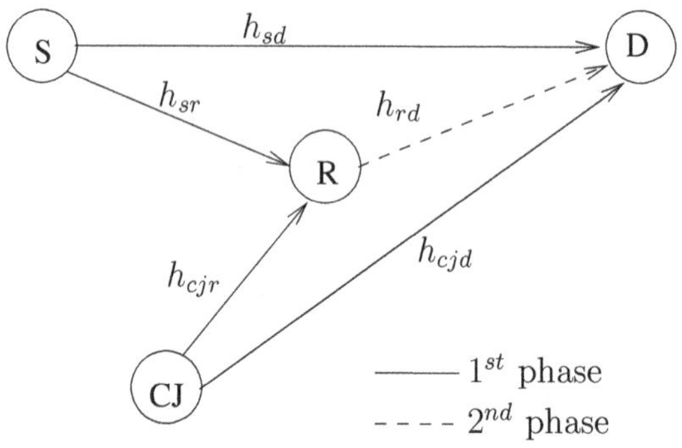

Fig. 2. Cooperative jamming system model (second scenario).

During the 1st phase, S transmits the source signal x_s with power αP and a single selected jammer CJ sends artificial noise η_{CJ}, with power $(1-\alpha)P$ which is known to the D, where $\{0 \le \alpha \le 1\}$ is the power distribution factor. Therefore the signals received at R and D are, given by

$$y_r = h_{sr}\sqrt{\alpha P}x_s + h_{CJr}\sqrt{(1-\alpha)P}\eta_{CJ} + w_r \quad (9)$$

and

$$y_d^{(1)} = h_{sd}\sqrt{\alpha P}x_s + h_{CJd}\sqrt{(1-\alpha)P}\eta_{CJ} + w_d^{(1)} \quad (10)$$

where w_r and $w_d^{(1)}$ represent the AWGN at R and D during the 1st phase, respectively. Then, R amplifies and forwards the received signal y_r during the 2nd phase, where the received signal at D can be expressed as

$$y_{\mathrm{d}}^{(2)} = h_{\mathrm{rd}}\eta_{\mathrm{r}}y_{\mathrm{r}} + w_{\mathrm{d}}^{(2)} \tag{11}$$

where $w_{\mathrm{d}}^{(2)}$ represents the AWGN at D during this phase and the amplification factor may be written as

$$\eta_{\mathrm{r}} = \sqrt{\frac{P}{\alpha P|h_{\mathrm{sr}}|^2 + (1-\alpha)P|h_{\mathrm{CJr}}|^2 + N_0}}. \tag{12}$$

By substituting equations (9) and (12) into (11), we get

$$\begin{aligned} y_{\mathrm{d}}^{(2)} &= h_{\mathrm{rd}}\eta_{\mathrm{r}}(h_{\mathrm{sr}}\sqrt{\alpha P}x_{\mathrm{s}} + h_{\mathrm{CJr}}\sqrt{(1-\alpha)P}\eta_{\mathrm{CJ}} + w_{\mathrm{r}}) + w_{\mathrm{d}}^{(2)} \\ &= \eta_{\mathrm{r}}h_{\mathrm{rd}}h_{\mathrm{sr}}\sqrt{\alpha P}x_{\mathrm{s}} + \eta_{\mathrm{r}}h_{\mathrm{rd}}h_{\mathrm{CJr}}\sqrt{(1-\alpha)P}\eta_{\mathrm{CJ}} \\ &+ \eta_{\mathrm{r}}h_{\mathrm{rd}}w_{\mathrm{r}} + w_{\mathrm{d}}^{(2)}. \end{aligned} \tag{13}$$

The composite signal at the destination after removing the known signals can be written as:

$$\begin{aligned} y_{\mathrm{d}} &= h_{\mathrm{sd}}\sqrt{\alpha P}x_{\mathrm{s}} \\ &+ \sqrt{\frac{\alpha}{\alpha P|h_{\mathrm{sr}}|^2 + (1-\alpha)P|h_{\mathrm{CJr}}|^2 + N_0}}Ph_{\mathrm{rd}}h_{\mathrm{sr}}x_{\mathrm{s}} \\ &+ \sqrt{\frac{P}{\alpha P|h_{\mathrm{sr}}|^2 + (1-\alpha)P|h_{\mathrm{CJr}}|^2 + N_0}}h_{\mathrm{rd}}w_{\mathrm{r}} + w_{\mathrm{d}} \end{aligned} \tag{14}$$

where $w_{\mathrm{d}} = w_{\mathrm{d}}^{(1)} + w_{\mathrm{d}}^{(2)}$, we can calculate the SNR at D as:

$$\gamma_{\mathrm{D}} = \frac{\alpha^2\gamma_{\mathrm{sd}}\gamma_{\mathrm{sr}} + \alpha(1-\alpha)\gamma_{\mathrm{CJr}}\gamma_{\mathrm{sd}} + \alpha\gamma_{\mathrm{sd}} + \alpha\gamma_{\mathrm{rd}}\gamma_{\mathrm{sr}}}{\gamma_{\mathrm{rd}} + \alpha\gamma_{\mathrm{rd}} + (1-\alpha)\gamma_{\mathrm{CJr}} + 1} \tag{15}$$

where γ_{sd} is the SNR for the S to D link, γ_{sr} is the SNR for the S to R link, γ_{rd} is the SNR for the R to D link and γ_{CJr} is the SNR for the CJ to R links. Similarly, from (9) we can derive the SNR at R as:

$$\begin{aligned} \gamma_{\mathrm{r}} &= \frac{|\sqrt{\alpha P}h_{\mathrm{sr}}|^2}{|\sqrt{(1-\alpha P)}h_{\mathrm{CJr}}|^2 + 1} \\ &= \frac{\alpha\gamma_{\mathrm{sr}}}{(1-\alpha)\gamma_{\mathrm{CJr}} + 1}. \end{aligned} \tag{16}$$

Consequently, the achievable rates at D and R are given as:

$$\bar{R}_{\mathrm{D}} = \frac{1}{2}\log(1+\gamma_{\mathrm{D}}) \tag{17}$$

and

$$\bar{R}_{\mathrm{r}} = \frac{1}{2}\log(1+\gamma_{\mathrm{r}}) = \frac{1}{2}\log\left(1 + \frac{\alpha\gamma_{\mathrm{sr}}}{(1-\alpha)\gamma_{\mathrm{CJr}} + 1}\right). \tag{18}$$

The secrecy rate $\bar{R}_{\mathrm{s}}$ of the system can be calculated as

$$\bar{R}_{\mathrm{s}} = \bar{R}_{\mathrm{D}} - \bar{R}_{\mathrm{r}}. \tag{19}$$

2.3 Secrecy Regions

Figures 3, 4 and 5 presents the secrecy rates for both the cases when operating at the CCMC capacity and DCMC capacity as well as when practical adaptive SECCC is employed, respectively. The overall code rate of the SECCC encoder can be calculated as $R_{\mathrm{eq}} = \frac{R_1}{2\times R_2}$. Table 1 shows the different code rates used with their corresponding throughput as well as the mode switching thresholds of SECCC and DCMC.

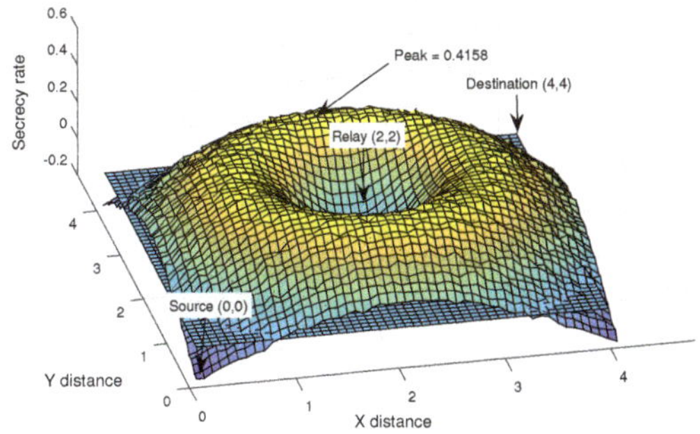

a Without S-D link

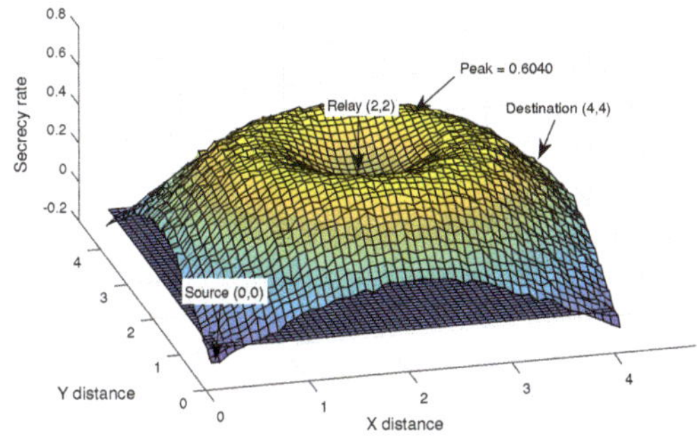

b With S-D link

Fig. 3. Secrecy regions based on the CCMC capacity.

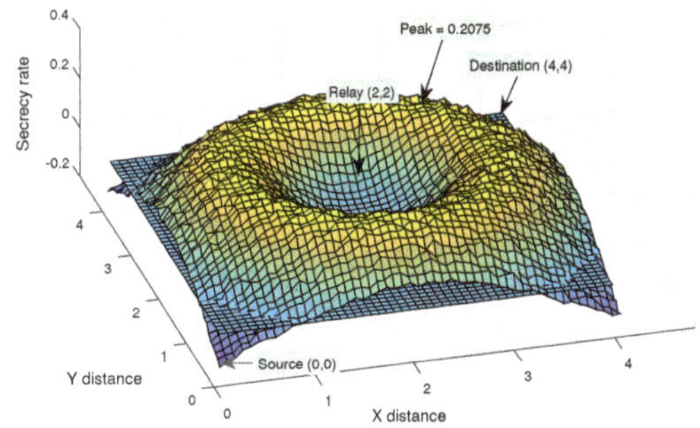

a Without S-D link

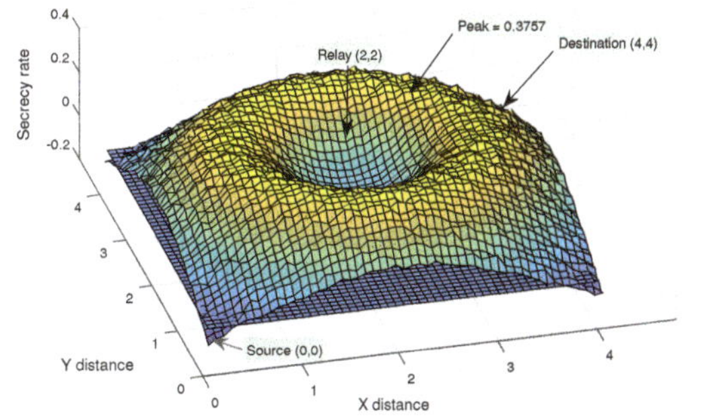

b With S-D link

Fig. 4. Secrecy regions based on the DCMC capacity as detailed in Tab. 1.

Mode	R_1	R_2	R_{eq}	Throughput	DCMC SNR	SECCC SNR @ BER = 10^{-5}
4QAM	1/2	1/2	1/2	1	2	5
8QAM	1/2	1/2	1/2	1.5	5	8
16QAM	1/2	1/2	1/2	2	8	11
32QAM	1/2	5/12	3/5	3	12.5	17
64QAM	1/2	3/8	4/6	4	16	21
256QAM	1/2	8/20	5/8	5	20	24

Tab. 1. Adaptive SECCC mode table.

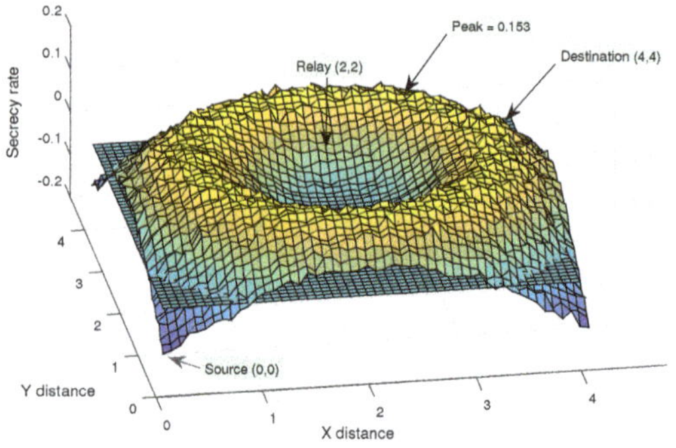

a Without S-D link

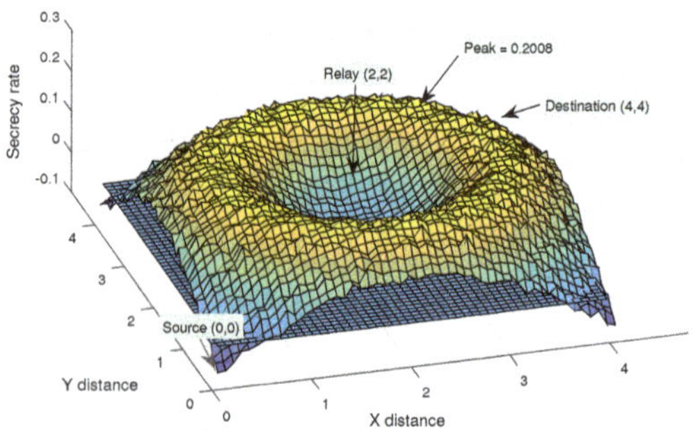

b With S-D link

Fig. 5. Secrecy regions based on the adaptive SECCC scheme, as detailed in Tab. 1.

Here we present the secrecy sum-rate in a 3D space where the $\{X, Y\}$ axis represents the coordinates of the jammer, while the secrecy sum-rate is given on the Z axis for a specific jammer on that specific location of the jammer. The X and Y axis range from $\{X, Y\} = \{0, 0\}$ $\{X, Y\} = \{4, 4\}$ where the distance is represented in km in distance. In our simulations we placed S at $\{X, Y\} = \{0, 0\}$, R at $\{X, Y\} = \{2, 2\}$ and D at $\{X, Y\} = \{4, 4\}$. The path gain [25] is given by: $G_{ij} = \left(\frac{d_{sd}}{d_{ij}}\right)^n$, where d_{ij} is the distance between terminal i and terminal j. Furthermore we have $\bar{h}_{ij} = \sqrt{G_{ij}} h_{ij}$, where h_{ij} is the Rayleigh fading channel coefficients between terminal i and terminal j.

In our simulations we have used $\alpha = 0.9$, while the pathloss exponent is set to $n = 4$. We see an improved secrecy sum-rate when a direct link is considered between source and destination. Secrecy rate improves as the CJ is placed closer to the relay. We can observe a ring shaped region in Fig. 3a around the relay where the secrecy is maximum. If the CJ is placed closer than that ring then we see a degraded performance due to the amplification factor at the relay due to higher jamming interference. The secrecy sum rate is higher if we include the S-D direct link as compared to the case where the S-D link is not considered which can be observed from figures 3, 4 and 5. Figure 3 shows the peak secrecy rates are 0.4158 and 0.6040 for the CJ without S-D and the CJ with S-D scenario when operating at the CCMC capacity which assumes a perfect modulation and a perfect channel code were used. In Fig. 4 the peak secrecy rates are 0.2075 and 0.3757 for the CJ without S-D and the CJ with S-D scenario when operating at the DCMC capacity which assumes a perfect channel code was used. Finally, in Fig. 5 the peak secrecy rates are 0.153 and 0.2008 for the CJ without S-D and the CJ with S-D scenario when a perfect adaptive SECCC scheme is employed.

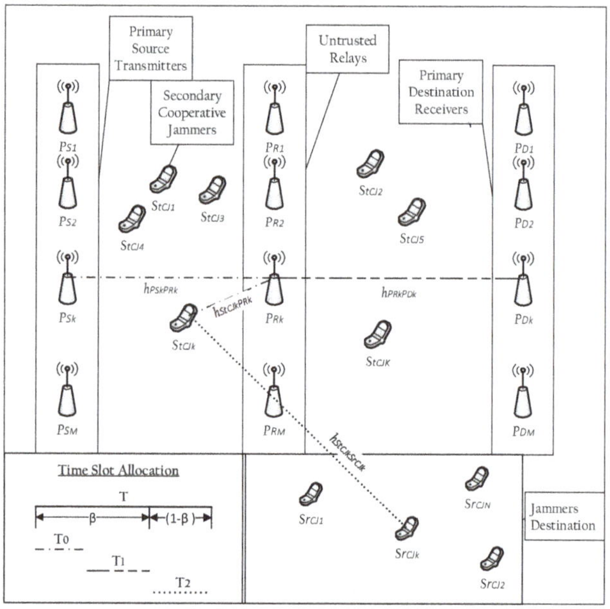

Fig. 6. Cooperative jamming system model.

3. Game Theoretic Secrecy Maximization

We consider an Amplify-and-Forward (AF) based network similar to the model presented in the previous section, but we now consider our network to consist of M pairs of Primary Source transmitters (P_S), Untrusted relays (P_R) and Primary Destination receivers (P_D) ($\{P_{S_m}, P_{R_m}, P_{D_m}\}_{m=1}^{M}$) with the mth pair having a secrecy rate requirement of greater than zero and K pairs of SU Cooperative Jammer transmitters (St_{CJ}) and SU Cooperative Jammer receivers (Sr_{CJ}) ($\{St_{CJ_k}, Sr_{CJ_k}\}$, k=1,...,K) with k^{th} pair having a sum rate requirement of $\bar{R}_{CJ_{k,req}}$ in an overlay cognitive radio network environment. Without loss of generality we assume that each $\{P_{S_m}, P_{D_m}\}$ pair has its unique untrusted relay P_{R_m} which is placed at the center of the $\{P_{S_m}, P_{D_m}\}$ pair. We assume that the S-D link is included for our communication as presented in Sec. 2.2. Each P_S offers a limited duration spectral access as a reward, which is mutually agreed upon, to the St_{CJ} in exchange for its service to securing its unique $\{P_{S_m}, P_{D_m}\}$ pair by transmitting an artificial noise towards its specific P_{R_m}. Our system model can be viewed in Fig. 6 with specific time allocation factors T_0, T_1 and T_2. Furthermore $\beta_{m,k}$ is the time allocation fraction which will be mutually agreed upon by the primary and secondary users based upon the spectrum matching algorithms, discussed in later sections, where $0 < \beta_{m,k} < 1$. During the time interval $T_0 = \beta_{m,k}/2$, P_{S_m} will be broadcasting its signal for P_{R_m} and P_{D_m}, while during the same interval St_{CJ_k} will be sending a noise/jamming signal to block the information leakage at the relay P_{R_m}. The time interval $T_1 = \beta_{m,k}/2$ is only dedicated to P_{R_m} which sends the mixed signal that it has received from S and CJ during interval T_0 to its destination P_{D_m}. Finally during time slot $T_2 = (1 - \beta_{m,k})$, St_{CJ_k} will send its own information towards its destination Sr_{CJ_k}.

The secrecy sum rate for the primary $\{P_{S_m}, P_{R_m}, P_{D_m}\}_{m=1}^{M}$ pair using secondary St_{CJ_k} can be calculated from (17) and (18) as:

$$\bar{R}_{s_{m,k}} = \bar{R}_{D_{m,k}} - \bar{R}_{r_{m,k}}. \tag{20}$$

The achievable rate at m^{th} D using k^{th} CJ is given by:

$$\bar{R}_{D_{m,k}} = \frac{\beta_{m,k}}{2}\log(1 + \gamma_{D_{m,k}}) \tag{21}$$

where $\gamma_{D_{m,k}}$ is given as

$$\frac{\alpha^2\gamma_{sd_m}\gamma_{sr_m} + \alpha(1-\alpha)\gamma_{CJr_{m,k}}\gamma_{sd_m} + \alpha\gamma_{sd_m} + \alpha\gamma_{rd_m}\gamma_{sr_m}}{\gamma_{rd_m} + \alpha\gamma_{sr_m} + (1-\alpha)\gamma_{CJr_{m,k}} + 1}$$

where $\gamma_{sd_m} = \gamma_{PU}|h_{sd_m}|^2 d_{sd_m}^{-n}$, $\gamma_{sr_m} = \gamma_{PU}|h_{sr_m}|^2 d_{sr_m}^{-n}$ and $\gamma_{rd_m} = \gamma_{PU}|h_{rd_m}|^2 d_{rd_m}^{-n}$ are the SNRs, h_{sd_m}, h_{sr_m} and h_{rd_m} are the Rayleigh fading channel coefficients while d_{sd_m}, d_{sr_m} and d_{rd_m} are the distances between the mth S-D, S-R and R-D links, respectively. $\gamma_{CJr_{m,k}} = \gamma_{SU}|h_{CJr_{m,k}}|^2 d_{CJr_{m,k}}^{-n}$ is the SNR with $h_{CJr_{m,k}}$ being the Rayleigh fading channel coefficients while $d_{CJr_{m,k}}$ is the distance between the kth CJ and m^{th} R. The achievable rate at m^{th} R using k^{th} CJ is given as:

$$\bar{R}_{r_{m,k}} = \frac{\beta_{m,k}}{2}\log\left(1 + \frac{\alpha\gamma_{sr_m}}{(1-\alpha)\gamma_{CJr_{m,k}} + 1}\right). \tag{22}$$

The achievable sum-rate for the secondary $\{St_{CJ_k}, Sr_{CJ_k}\}$ pair can be computed as

$$\bar{R}_{CJ_{m,k}} = (1 - \beta_{m,k})\log(1 + \gamma_{CJ_k}) \tag{23}$$

where γ_{CJ_k} is the SNR for $\{St_{CJ_k}, Sr_{CJ_k}\}$ pair. Each P_{S_m} has a list of all the St_{CK_k} which can provide a secrecy sum-rate of greater than zero in a descending order, denoted as $PULIST_m$. Similarly each St_{CJ_k} has a list of all the P_{S_m} that can provide a sum-rate greater than or equal to its minimum rate requirement in a descending order, denoted as $SULIST_k$. Based on this system model we present two secrecy maximization algorithms i.e, the Secure CDA and the Secure PDA which will be elaborated further.

3.1 Secure CDA

CDA was proposed in [21] for maximizing the sum-rate for PU. In our case we modified the algorithm to maximize the secrecy sum-rate while SU benefit in terms of limited duration spectrum access. The algorithm as seen in Algorithm 1 starts by the construction of the $PULIST_m$ by each of the PU, which is the set of all the SUs that provide secrecy of greater than the minimum required by the PU. The list is made in the descending order so that the first entry in the list is an SU which provides the highest secrecy as obtained in Sec. 2.2 of all and so on. Similarly each SU will also create its own $SULIST_k$ which is the set of all the PUs that provide a sum-rate of greater or equal to the minimum required in descending order. In the Secure-CDA each PU offers a limited time allocation factor $\beta_{m,k}$ to the first SU in its $PULIST_m$ in exchange for its service to provide secrecy to the PU. Matching will be made if the offerer PU is present in the $SULIST_k$ of the kth SU. If that PU is not present in the $SULIST_k$, match will not be made and the PU will enhance its time allocation factor for that specific SU by decreasing the time allocation factor by ϵ and the PULIST will be updated accordingly. In this fashion each PU will be making an offer to its desired SU and will try to match with its desired SU. Matching will be broken if any SU which is already matched receives a better offer in terms of its sum-rate from another PU. In that case previous match will be broken and the SU will be matched to the new PU. In this way the algorithm will continue until all the PUs are matched to their desired SUs or until no further matchings are possible.

3.2 Secure PDA

PDA was proposed in [22] also for maximizing the sum-rate for the PU. PDA was better than CDA in terms of the PU sum-rate as it catered for the losses endured by the CDA due to competing primary nodes for acquiring the best SU in terms of secrecy maximization obtained in Sec. 2.2. In Secure-CDA the PUs compete with each other for the acquisition of their desired SU by trying to out-bid their rivals. Due to this competition among the PUs, they end up compromising their secrecy rate. The SUs upon receiving a better

Algorithm 1 Secure CDA

Require: $\bar{R}_{\mathrm{CJ}_{m,k}} \geq 1 \wedge 0 < \beta_{m,k} < 1$
Ensure: $\bar{R}_{\mathrm{S}_{m,k}} > 0$
1: **Initialization**
2: Set matchlist for the set of P_{S_m} to be matched (i.e.)$\{1, \ldots, M\}$.
3: Set the initial TS allocation to $\beta_{init} = 0.99$, and set the step size of TS increment to $\tau = 0.05$.
4: Construct $PULIST_m = \{St_{\mathrm{CK}_k} | \bar{R}_{\mathrm{S}_{m,k}} > 0\}$ and $SULIST_k = \{P_{\mathrm{S}_m} | \bar{R}_{\mathrm{CJ}_{m,k}} \geq 1\}$ in descending order, where $m = \{1, \ldots, M\}$ and $k = \{1, \ldots, K\}$.
5: Set $j = 1$ for the first transmission.
6: **Do the matching for the jth transmission.**
7: P_{S_m} offers $\beta_{m,k}$ to the first SU in its preference list St_{CJ_k}.
8: If P_{S_m} is not in the preference list of St_{CJ_k} then decrease the TS allocation to $\beta_{m,k} = \beta_{m,k} - \tau$ and update both $PULIST_m$ and $SULIST_k$.
9: If P_{S_m} is in the preference list of St_{CJ_k}, then St_{CJ_k} and P_{S_m} are matched.
10: If St_{CJ_k} is already matched to $P_{S_{curr}}$
11: If the P_{S_m} is higher up in the $SULIST_m$ than $P_{S_{curr}}$, then rematch St_{CJ_k} to P_{S_m}.
12: Else decrease the TS allocation to $\beta_{m,k} = \beta_{m,k} - \tau$ and update $PULIST_m$ and $SULIST_k$.
13: If no more matchings are possible then goto step 6.
14: If $j = k$, then the algorithm ends

offer from another PU, breaks the previous match and creates a new match until it receives another better offer. Secure-PDA on the other hand discourages competition among the PUs in their matching, which results in better performance for the PUs in terms of their sececy. It does so by introducing a game where all the PUs prioritize their acquisition of SU in a round robin rotation basis. In Secure-PDA the PUs in addition to their $PULIST_m$, prepare a priority list known as the $ALIST_1 = \{P_{S_1}, P_{S_2}, \ldots, P_{S_M}\}$ which is the PUs priority list for acquiring the best SU. After the first round the $ALIST_i$ is updated in a round robin rotation manner and the new list will be $ALIST_2 = \{P_{S_M}, P_{S_1}, P_{S_2}, \ldots, P_{S_{(M-1)}}\}$ and so on. Therefore, there will be m rounds of our game where each PU will be able to match with its best SU for atleast 1 round. During each round the PU which is at the top of the $ALIST_i$ will have the priority to chose from the available SUs. The PU will make an offer of a limited time allocation factor $\beta_{m,k}$ to the first SU in its $PULIST_m$. If that PU is also present in the $SULIST_k$ of the SU, the match is made. Otherwise the PU will decreasing the time allocation factor by ϵ and update the $PULIST_m$ and make another offer to the first SU in its updated $PULIST_m$. After the first PU is matched, second PU from the $ALIST_i$ will try to make a match with the best SU from the remaining unmatched SUs. In this way the algorithm will continue until all the PUs are matched or until no further matches are possible. During the next round the $ALIST_{i+1}$ will be updated as listed in $ALIST_2$ stated above and new matchings will be made. In this way matchings will be made which will last for at-least m rounds. The detailed PDA algorithm is presented in Algorithm 2.

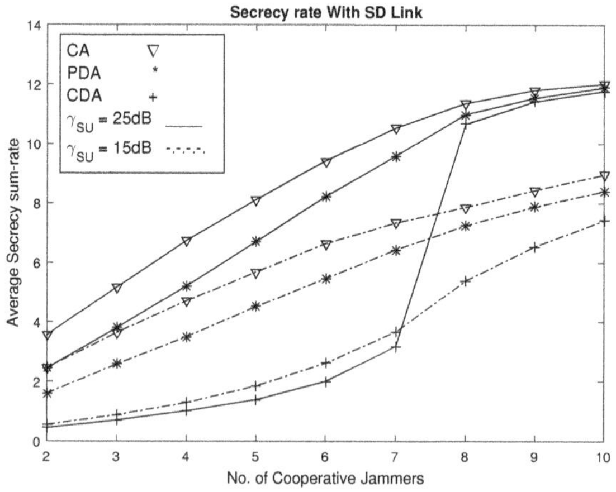

a Secrecy rate, $\alpha = 0.9$ and $\gamma_{\mathrm{PU}} = 20$ dB

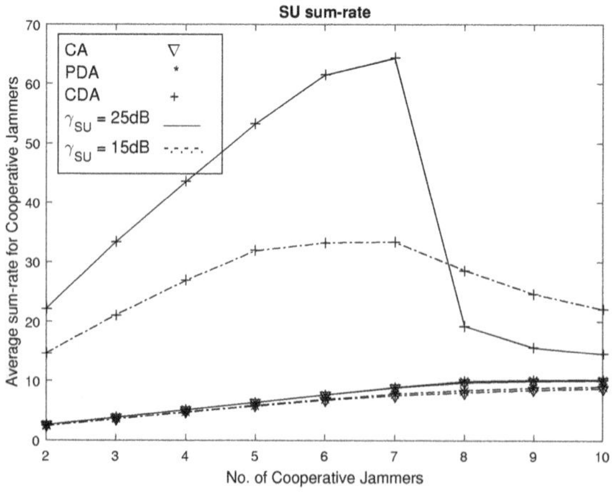

b SU rate, $\alpha = 0.9$ and $\gamma_{\mathrm{PU}} = 20$ dB

Fig. 7. AF based relaying when operating at the CCMC capacity, the number of primary transmitter, relay and receiver pairs is 8.

3.3 Results

We investigated the secrecy sum rate and SU sum rate for our system model for Amplify and Forward (AF) based

Algorithm 2 Secure PDA

Require: $\bar{R}_{\mathrm{CJ}_{m,k}} \geq 1 \wedge 0 < \beta_{m,k} < 1$
Ensure: $\bar{R}_{\mathrm{S}_{m,k}} > 0$
1: **Initialization**
2: Set up the first priority list $ALIST_1 = \{P_{\mathrm{S}_1}, P_{\mathrm{S}_2}, \ldots, P_{\mathrm{S}_M}\}$.
3: Set $i = 1$ for the first round.
4: **Matching for the ith round**
5: Set the initial TS allocation to $\beta_{init} = 0.99$, and set the step size of TS increment to $\tau = 0.05$.
6: Construct $PULIST_m = \{St_{\mathrm{CK}_k} | \bar{R}_{\mathrm{S}_{m,k}} > 0\}$ and $SULIST_k = \{P_{\mathrm{S}_m} | \bar{R}_{\mathrm{CJ}_{m,k}} \geq 1\}$ in descending order, where $m = \{1, \ldots, M\}$ and $k = \{1, \ldots, K\}$.
7: Set $j = 1$ for the first transmission.
8: Do the matching for jth transmission.
9: Find the corresponding P_{S_m} for transmission, based on the $ALIST_i$ (i.e) jth element of $ALIST_i$
10: P_{S_m} selects the best available St_{CJ_k} from its $PULIST$ and offer a time slot $\beta_{m,k}$
11: If P_{S_m} is in the preference list of St_{CJ_k} then St_{CJ_k} and P_{S_m} are matched.
12: If P_{S_m} is not in the preference list of St_{CJ_k} then decrease the TS allocation to $\beta_{m,k} = \beta_{m,k} - \tau$ and update both $PULIST_m$ and $SULIST_k$.
13: If $PULIST_m$ is empty then P_{S_m} is left unmatched.
14: Set $j = j + 1$ and goto step 8 until $j = K$.
15: **Set $i = i + 1$ and goto step 4 for the next round, until $i = K$**

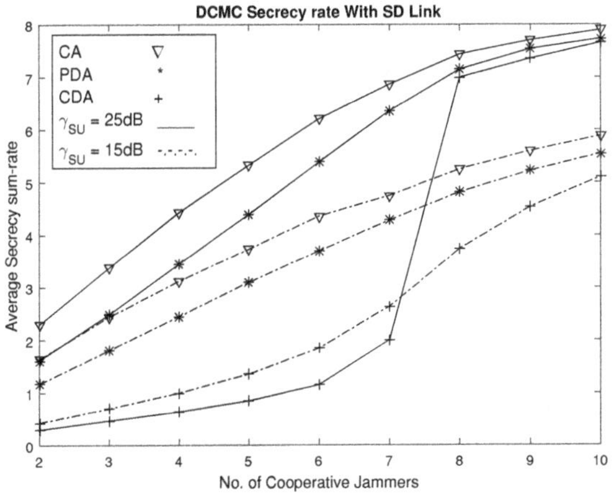

a Secrecy rate, $\alpha = 0.9$ and $\gamma_{\mathrm{PU}} = 20$ dB

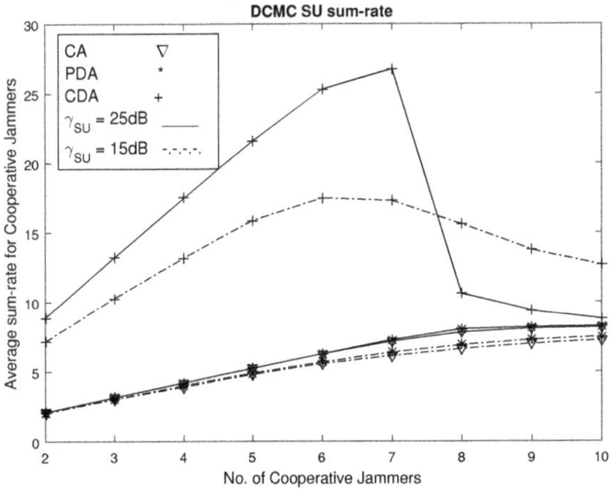

b SU rate, $\alpha = 0.9$ and $\gamma_{\mathrm{PU}} = 20$ dB

Fig. 8. AF based relaying when operating at the DCMC capacity, the number of primary transmitter, relay and receiver pairs is 8.

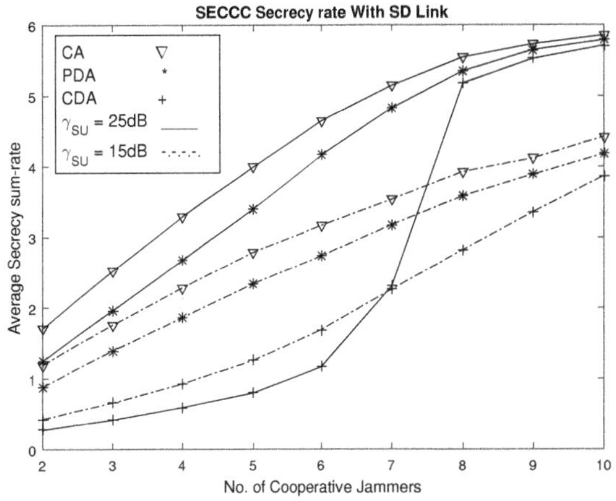

a Secrecy rate, $\alpha = 0.9$ and $\gamma_{\mathrm{PU}} = 20$ dB

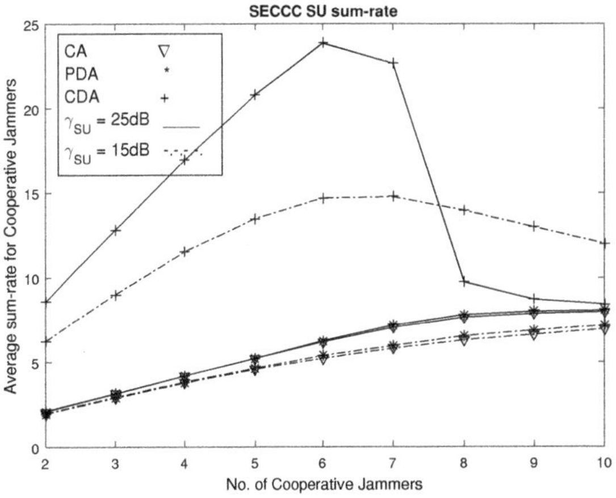

b SU rate, $\alpha = 0.9$ and $\gamma_{\mathrm{PU}} = 20$ dB

Fig. 9. AF based relaying when a practical adaptive SECCC scheme is employed, the number of primary transmitter, relay and receiver pairs is 8.

relaying, when operating at the CCMC capacity, DCMC capacity and when practical adaptive SECCC scheme is used. The performance of the DCMC and SECCC based scheme relies on the SNR thresholds and throughputs in Tab. 1. The results for the Secure Conventional Distributed Algorithm (S-CDA) and Secure Pragmatic Distributed Algorithm (S-PDA) for our system model are shown in Fig. 7 in comparison to the Centralized Algorithm (CA) for secrecy maximization. We consider $M = 8$ with $\gamma_{\text{PU}} = 20$ dB and $K = \{2, 3, \ldots, 10\}$ with $\gamma_{\text{SU}} = \{15, 25\}$ dB. The power distribution factor α is kept at 0.9 while the pathloss exponent n is kept at 4. All the results indicate a superior secrecy sum rate for the S-PDA system which is comparable to that of the CA scheme, while for $K \geq 7$ we see that the S-CDA and S-PDA almost perform similar for the case where the γ_{SU} is kept at 25 dB. The SU sum rate is better for the S-CDA system because of the competition amongst the PUs for acquiring the best SU. As the number of SUs/CJs increases we see a rise in the secrecy sum rate and a decrease in the SU sum rate which again is due to high competition when the number of SUs/CJs is less as compared to the case when the number of SUs/CJs is higher and a decrease in competition is witnessed. Similarly we see an opposite trend in the secrecy sum-rate for the S-CDA where the secrecy sum-rate is lower when we have less number of CJs as all the PUs compete with each other for matching with the CJ, while the secrecy sum-rate increases close to the S-PDA when the number of CJs is equal or greater than the number of PUs due to lesser competition among the PUs for matching with CJs. S-PDA system on the other hand does not have competition amongst the PUs by including them in a round robin rotation based game which encourages the PUs not to acquire the SU which have been assigned to another PU, hence we see a stable increase in the secrecy sum rate as more SU CJs are available, while the SU sum rate will always be closer to their minimum rate requirements.

4. Conclusion

In this paper, we first investigated the secrecy rate regions for friendly jamming in a cooperative network where communication was assisted by an untrusted relay. A friendly jammer was used for providing secrecy by transmitting a noise signal in parallel to the source signal. We investigated the secrecy rate when assuming idealistic performance operating at the CCMC and DCMC capacities, as well as when a practical adaptive SECCC coding scheme was invoked. It was observed that the secrecy rate can be maximized if the jammer is at a certain distance from the relay. We then further proposed the novel S-PDA and S-CDA schemes for maximizing the secrecy based on the cognitive radio approach. More explicitly, selected jammers were rewarded with a limited access to the spectrum for their service in providing the secrecy. The proposed S-PDA and S-CDA schemes were further compared with the CA and it was shown that the S-PDA provides maximum secrecy when the number of jammers is less than the number of primary sources. By contrast, the S-CDA provides a better sum-rate for the jammers and a reduced secrecy for the sources as compared to those of the S-PDA.

References

[1] YANG, N., WANG, L., GERACI, G., et al. Safeguarding 5g wireless communication networks using physical layer security. *IEEE Communications Magazine*, 2015, vol. 53, no. 4, p. 20–27. DOI: 10.1109/MCOM.2015.7081071

[2] ZHONG, Z., PENG, J., LUO, W., et al. A tractable approach to analyzing the physical-layer security in k-tier heterogeneous cellular networks. *China Communications*, 2015, vol. 12, p. 166–173. DOI: 10.1109/CC.2015.7386165

[3] LI, S., XU, L. D., ZHAO, S. The internet of things: A survey. *Information Systems Frontiers*, 2015, vol. 17, no. 2, p. 243–259. DOI: 10.1007/s10796-014-9492-7

[4] WANG, H. M., ZHENG, T. X., YUAN, J., et al. Physical layer security in heterogeneous cellular networks. *IEEE Transactions on Communications*, 2016, vol. 64, no. 3, p. 1204–1219. DOI: 10.1109/TCOMM.2016.2519402

[5] MUKHERJEE, A. Physical-layer security in the internet of things: Sensing and communication confidentiality under resource constraints. *Proceedings of the IEEE*, 2015, vol. 103, no. 10, p. 1747–1761. DOI: 10.1109/JPROC.2015.2466548

[6] SAAD, W., ZHOU, X., DEBBAH, M., et al. Wireless physical layer security. *IEEE Communications Magazine*, 2015, vol. 53, no. 12, p. 18. DOI: 10.1109/MCOM.2015.7355560

[7] ELETREBY, R., RAHBARI, H., KRUNZ, M. Supporting phy-layer security in multi-link wireless networks using friendly jamming. In *Proceedings of the IEEE Global Communications Conference (GLOBECOM)*. 2015, p. 1–6. DOI: 10.1109/GLOCOM.2015.7417141

[8] BERGER, D. S., GRINGOLI, F., FACCHI, N., et al. Friendly jamming on access points: Analysis and real-world measurements. *IEEE Transactions on Wireless Communications*, 2016, vol. 15, no. 9, p. 6189-6202. DOI: 10.1109/TWC.2016.2581165

[9] SHEN, W., LIU, Y., HE, X., et al. No time to demodulate - fast physical layer verification of friendly jamming. In *Proceedings of the IEEE Military Communications Conference (MILCOM)*. 2015, p. 653–658. DOI: 10.1109/MILCOM.2015.7357518

[10] XING, H., WONG, K. K., CHU, Z., et al. To harvest and jam: A paradigm of self-sustaining friendly jammers for secure af relaying. *IEEE Transactions on Signal Processing*, 2015, vol. 63, no. 24, p. 6616–6631. DOI: 10.1109/TSP.2015.2477800

[11] HE, X., YENER, A. Cooperation with an untrusted relay: A secrecy perspective. *IEEE Transactions on Information Theory*, 2010, vol. 56, no. 8, p. 3807–3827. DOI: 10.1109/TIT.2010.2050958

[12] KHODAKARAMI, H., LAHOUTI, F. Link adaptation with untrusted relay assignment: Design and performance analysis. *IEEE Transactions on Communications*, 2013, vol. 61, no. 12, p. 4874–4883. DOI: 10.1109/TCOMM.2013.111513.120888

[13] ZEWAIL, A. A., YENER, A. The two-hop interference untrusted-relay channel with confidential messages. In *Proceedings of the IEEE Information Theory Workshop - Fall (ITW)*. 2015, p. 322–326. DOI: 10.1109/ITWF.2015.7360788

[14] XIONG, J., CHENG, L., MA, D., et al. Destination aided cooperative jamming for dual-hop amplify-and-forward MIMO untrusted relay systems. *IEEE Transactions on Vehicular Technology*, 2015, vol. 65, no. 9, p. 7274-7284. DOI: 10.1109/TVT.2015.2490099

[15] KALAMKAR, S. S., BANERJEE, A. Secure communication via a wireless energy harvesting untrusted relay. *IEEE Transactions on Vehicular Technology*, 2016, vol. 66, no. 3, p. 2199–2213. DOI: 10.1109/TVT.2016.2572960

[16] ZEWAIL, A. A., NAFEA, M., YENER, A. Multi-terminal networks with an untrusted relay. In *Proceedings of the 52nd Annual Conference on Communication, Control, and Computing (Allerton)*. 2014, p. 895–902. DOI: 10.1109/ALLERTON.2014.7028549

[17] STANOJEV, I., YENER, A. Improving secrecy rate via spectrum leasing for friendly jamming. *IEEE Transactions on Wireless Communications*, 2013, vol. 12, no. 1, p. 134–145. DOI: 10.1109/TWC.2012.120412.112001

[18] QU, J., CAI, Y., LU, J., et al. Power allocation based on stackelberg game in a jammer-assisted secure network. In *Proceedings of the International Conference on Cyberspace Technology (CCT 2013)*. 2013, p. 347–352. DOI: 10.1049/cp.2013.2150

[19] ZHILING, H., BAOYUN, W. Coalition formation game of dual-identity nodes for improving phy security of wireless networks. In *Proceedings of the 27th Chinese Control and Decision Conference (2015 CCDC)*. 2015, p. 3658–3662. DOI: 10.1109/CCDC.2015.7162560

[20] ZHANG, N., CHENG, N., LU, N., et al. Partner selection and incentive mechanism for physical layer security. *IEEE Transactions on Wireless Communications*, 2015, vol. 14, no. 8, p. 4265–4276. DOI: 10.1109/TWC.2015.2418316

[21] BAYAT, S., LOUIE, R. H. Y., VUCETIC, B., et al. Dynamic decentralised algorithms for cognitive radio relay networks with multiple primary and secondary users utilising matching theory. *Transactions on Emerging Telecommunications Technologies*, 2013, vol. 24, p. 486–502. DOI: 10.1002/ett.2663

[22] LIANG, W., NG, S. X., FENG, J., et al. Pragmatic distributed algorithm for spectral access in cooperative cognitive radio networks. *IEEE Transactions on Communications*, 2014, vol. 62, no. 4, p. 1188–1200. DOI: 10.1109/TCOMM.2014.030214.130326

[23] NG, S. X., HANZO, L. On the MIMO channel capacity of multidimensional signal sets. *IEEE Transactions on Vehicular Technology*, 2006, vol. 55, no. 2, p. 528–536. DOI: 10.1109/TVT.2005.863357

[24] BUTT, M. F. U., NG, S. X., HANZO, L. Self-concatenated code design and its application in power-efficient cooperative communications. *IEEE Communications Surveys Tutorials*, 2012, vol. 14, no. 3, p. 858–883. DOI: 10.1109/SURV.2011.081511.00104

[25] OCHIAI, H., MITRAN, P., TAROKH, V. Design and analysis of collaborative diversity protocols for wireless sensor networks. In *Proceedings of the 60th IEEE Vehicular Technology Conference (VTC2004-Fall)*. 2004, p. 4645–4649. DOI: 10.1109/VETECF.2004.1404971

About the Authors ...

Bakhtiar ALI received his M.Sc. degree in Personal and Mobile Radio Communications from Lancaster University, UK in September 2008. He is currently pursuing a doctoral degree at COMSATS Institute of Information Technology, Islamabad, Pakistan. His current research interests include the radio resource management in cooperative cognitive radio networks, space time block coding, cooperative communications, physical layer security, game theory and the study of future radio communications systems, i.e., 5G.

Nida ZAMIR received her Bachelor's degree in Electrical Engineering with specialization in Telecommunications from COMSATS Institute of Information Technology (CIIT), Islamabad, Pakistan in 2014. She is currently pursuing her MS in Electrical Engineering from the same institution. Her current research interests include channel coding, physical layer security and game theory.

Soon Xin NG received the B.Eng. degree (First class) in electronics engineering and the Ph.D. degree in wireless communications from the University of Southampton, Southampton, U.K., in 1999 and 2002, respectively. From 2003 to 2006, he was a postdoctoral research fellow working on collaborative European research projects known as SCOUT, NEWCOM and PHOENIX. Since August 2006, he has been a member of academic staff in the School of Electronics and Computer Science, University of Southampton. He is involved in the OPTIMIX and CONCERTO European projects as well as the IUATC and UC4G projects. He is currently an Associate Professor of Telecommunications with the University of Southampton. He has authored over 200 papers and co-authored two John Wiley/IEEE Press books in his research field. His research interests include adaptive coded modulation, coded modulation, channel coding, space-time coding, joint source and channel coding, iterative detection, OFDM, MIMO, cooperative communications, distributed coding, quantum error correction codes and joint wireless-and-optical-fiber communications. He is a Chartered Engineer and a Fellow of the Higher Education Academy in the UK.

Muhammad Fasih Uddin BUTT received his B.E. degree from National University of Sciences & Technology (NUST), Pakistan in 1999. He received his M.E. degree from Center for Advanced Studies in Engineering, UET Taxila, Pakistan with specialization in Digital Communication/Computer Networks in 2003 and his Ph.D. degree from Communications Research Group, School of Electronics and Computer Science, University of Southampton, U.K in June 2010. Currently he is working as Assistant Professor in the Department of Electrical Engineering, COMSATS Institute of Information Technology (CIIT), Islamabad, Pakistan where he has been serving as an academic since 2002. His research interests include channel coding, iterative detection, cooperative cognitive radio networks, mm Wave radio-over-fiber technologies, energy harvesting and physical layer security. He has published over 25 research papers in various reputed journals and conference proceedings.

Approximate Circuits in Low-Power Image and Video Processing: The Approximate Median Filter

Lukas SEKANINA, Zdenek VASICEK, Vojtech MRAZEK

Faculty of Information Technology, IT4Innovations Centre of Excellence, Brno University of Technology, Czech Republic
{sekanina, vasicek, imrazek}@fit.vutbr.cz

Abstract. *Low power image and video processing circuits are crucial in many applications of computer vision. Traditional techniques used to reduce power consumption in these applications have recently been accompanied by circuit approximation methods which exploit the fact that these applications are highly error resilient and, hence, the quality of image processing can be traded for power consumption. On the basis of a literature survey, we identified the components whose implementations are the most frequently approximated and the methods used for obtaining these approximations. One of the components is the median image filter. We propose, evaluate and compare two approximation strategies based on Cartesian genetic programming applied to approximate various common implementations of the median filter. For filters developed using these approximation strategies, trade-offs between the quality of filtering and power consumption are investigated. Under conditions of our experiments we conclude that better trade-offs are achieved when the image filter is evolved from scratch rather than a conventional filter is approximated.*

Keywords

Approximate computing, circuit design, evolutionary computation, image filter

1. Introduction

An efficient implementation of computer vision algorithms is crucial for many smart embedded systems such as traffic control systems, driver assistant systems, production line inspection systems, and robotics. However, providing high-quality outputs in these applications is usually associated with high computation cost and non-trivial requirements on energy. In order to meet real-time constraints and cope with limited power budget, image and video processing algorithms are often accelerated in application-specific integrated circuits (ASIC) or field programmable gate arrays (FPGAs). If additional energy consumption reduction is requested because of, for example, very limited energy available in remote sensors, mobile or wearable devices, the *circuit approximation* is one of the most promising approaches to deliver a suitable solution.

Approximate computing [1] exploits the fact that many applications (image and video processing in particular) are highly error resilient. If occasional errors are acceptable by the users – which is possible because the users as consumers of the outputs of these applications are often unable to recognize small imperfections in images or video sequences – implementations of these applications can be simplified. The goal is to create such an implementation which shows the best trade-off between the error, performance and power consumption. Approximate computing has been progressively developed in recent 5 years and influenced the way how energy efficient computer systems (ranging from tiny battery powered devices via common desktop computers to supercomputers) are now constructed and operated.

In this paper, we focus on approximate circuits that are used in image and video processing applications. On the basis of a literature survey, we identified the components whose implementations are the most frequently approximated and the methods used for obtaining these approximations. One of the components is the median-outputting circuit (median for short) which is typically employed to filter out undesired artefacts (such as shots) in digital images.

As the circuit approximation problem can be formulated as a multi-objective optimization problem (with error, performance and power consumption as objectives), various ad hoc and heuristic methods have been introduced to solve it. In our previous work, we have developed circuit approximation methods [2], [3] based on Cartesian genetic programming (CGP) which is a branch of evolutionary algorithms capable of designing and optimizing digital circuits.

Unfortunately, the quality of approximation methods has been compared in the literature only rarely (see a survey of associated methodological problems in [2]); in most cases only parameters of approximate adders and multipliers were compared [4]. In this paper, we compare two approximation strategies based on CGP applied to approximate various common implementations of the median filter. The first strategy starts with an exact median filter implementation and tries to remove some circuit components (comparators) and reconnect the remaining ones in such a way that the error of filtering is minimized. The second strategy employs CGP to evolve the image filter from scratch; only on the basis of the training data supplied during the evolution. The goal is to demonstrate how different approximation strategies can influence the trade-offs that are obtained between the quality of filtering and power consumption for target circuits.

The rest of the paper is organized as follows. The research area of approximate computing is introduced in Sec. 2. Section 3 deals with a survey of circuits that were approximated for purposes of power consumption reduction in image and video processing applications. Various aspects of the approximation strategies used to obtain desired approximations have been analyzed. Section 4 is devoted to our case study – approximate circuits for image filtering. We present conventional implementations of image filters, CGP as the method used to perform desired approximations and two different approximation strategies based on CGP. Results are summarized in Sec. 5. Conclusions are given in Sec. 6.

2. Approximate Computing

The concept of approximation has been well established in computer science and engineering for decades. For example, a paper with the title "Approximate signal processing" was published in 1997 [5]. However, new problems emerged in the last decade that stimulated new research in applying approximation techniques, but in a slightly different context than before.

In particular, problems with high power density of integrated circuits led to the end of Dennard scaling, i.e. simultaneous doubling the number of transistors on a chip, increasing operation frequency and reducing Vdd have no longer worked together. The coming era of "dark silicon", when many transistors are available on a chip, but cannot be used at the same time on high operating frequency because of thermal issues, has forced us to rethink the basic design principles of computer-based systems [6]. As conventional power reduction techniques such as dynamic voltage-frequency scaling and power gating do not scale sufficiently and alternative post CMOS technologies are not widely adopted, the only solution seems to be to relax the requirement on precise computing across the computer stack.

In approximate computing, the requirement of exact equivalence between the specification and all implementations levels is relaxed in order to reduce power consumption or improve other system parameters such as performance [1], [7].

The approximation can be conducted at the level of software as well as hardware. Mittal [1] discusses a wide spectrum of approximation techniques which include precision scaling, loop perforation, load value approximation, memorization, task dropping/skipping, memory access skipping, using different SW/HW versions, refresh rate reducing in memory, inexact read/write and relaxed synchronization. In the case of digital circuit approximation, voltage over-scaling and functional approximation are the most popular techniques. In the case of *voltage over-scaling*, the circuit is supplied with lower Vdd than nominal, which reduces power consumption, but introduces errors for those inputs whose processing requires attending the critical path of the design. In the case of *functional approximation*, a slightly different function is implemented with respect to the original one, provided that the error is acceptable and key system parameters are improved. The errors induced by approximation are measured using various error metrics such as the average error, error probability, and worst case error.

The approximate solution is usually obtained by a heuristic procedure that modifies the original implementation. In the case of software approximation, programmers can typically declare which parts of the program can be computed approximately and specialized compiler and optimizer then preform requested approximations (see, e.g., EnerJ [8]). In the case of hardware approximation, either general-purpose or circuit-specific approximation methods have been applied. While the aim of general-purpose approximation methods (e.g. SALSA [9], AXILOG [10], ASLAN [11], ABACUS [12], CGP [2], [3]) is to automatically approximate any circuit regardless of its structure, the circuit-specific methods are focused on a rather specific class of circuits (such as adders or multipliers [4]).

3. Approximate Circuits for Image and Video Processing

Based on the analysis of 12 image and data processing applications, Chippa et al. showed that about 83% runtime is spent in error resilient computation kernels that are suitable for approximation [7]. The most dominant kernels were the dot product computation and distance computation. The fact that image and video processing circuits are good targets for circuit approximation can be documented by dozens of papers dealing with this topic in the literature.

It has to be noticed that elementary arithmetic circuits (such and adders and multipliers) are often approximated independently of a potential application. The objective is to create a general-purpose library of approximate implementations showing different trade-offs between power consumption and error. Jiang et al. [4] provided a detailed survey of approaches developed in this direction. In this paper, we will deal with approximate adders or multipliers only if they have been applied in an approximate implementation of image or video processing system.

Approximate circuits are also crucial in energy efficient implementations of image and video processing systems (image classifiers, object detectors) based on (deep) neural networks (DNN). As this is rather a specific area [13], [14], we will not consider it in our survey table, but provide a brief introduction in this paragraph. In DNNs, approximations were introduced at levels of the data type quanti-

Application Type	Ref.	Module	Approx. Component	Approx. Method	Approx. Level	Platform
Filter	[26]	Median	Comparators/Network	Ad hoc	transistor	ASIC
	[24]	Median	Median	CGP	RTL	ASIC
	[27]	Median	Median	CGP	RTL	FPGA
	[31]	Gaussian	Multiplier	Ad hoc	gate	ASIC
	[27]	Gaussian	Adder, Multiplier	CGP	gate	FPGA
	[27]	Sobel	Adder	CGP	gate	FPGA
	[32]	Sobel	OpenCL code	truncation	RTL	FPGA
	[11]	Sobel	Sobel	ASLAN	gate	ASIC
	[10]	Sobel	Verilog code	AXILOG	RTL	ASIC
Metrics	[16]	SAD	SAD	Logic Isolation	gate	ASIC
	[9]	SAD	SAD	SALSA	gate	ASIC
	[16]	EUD	EUD	LogicIsolation	gate	ASIC
Transforms	[16]	DCT	DCT	Logic Isolation	gate	ASIC
	[16]	FFT	FFT	Logic Isolation	gate	ASIC
	[9]	DCT	DCT	SALSA	gate	ASIC
	[32]	DCT	OpenCL code	truncation	RTL	FPGA
JPEG	[31]	JPEG	Multiplier	Ad hoc	gate	ASIC
	[29]	DCT	Adder	Ad hoc	transistor	ASIC
	[34]	DCT	Adder/Multiplier	truncation/MINPS	full adders	ASIC
MPEG	[12]	Block Matching	Verilog code	ABACUS	RTL	ASIC
	[11]	DCT	DCT	ASLAN	gate	ASIC
	[33]	DCT	Adder/Subtractor	Ad hoc	gate	ASIC
	[33]	ME	Adder/Subtractor	Ad hoc	gate	ASIC
HEVC	[28]	SAD in ME	Adder	Ad hoc	gate	ASIC, FPGA
	[30]	DCT	DCT	Ad hoc	gate	ASIC
	[19]	DCT	Adder/Subtractor	CGP	gate	ASIC

SAD (Sum of Absolute Differences) ME (Motion Estimation) MINPS (Mixed Integer Nonlinear Problem Solver)
DCT (Discrete Cosine Transform) FFT (Fast Fourier Transform) EUD (Euclidean distance)

Tab. 1. Circuits approximated in the area of image and video processing.

zation, microarchitecture (e.g. neurons insignificantly contributing to the quality of outputs can be removed), training algorithm (an iterative process which can be stopped when good enough results are obtained), the multiply-accumulate-transform circuits (where the design of approximate multipliers and adders for DNN applications represents an independent topic [15], [16]), and memory cells and architecture (where, e.g., less significant bits can be stored in energy efficient, but less reliable memory cells [17]). An ultra-low power deep learning ASIC for IoT was implemented on a single chip, capable of performing 374 GOPS/W and consuming less than 300 µW. However, performance of this solution is limited as it operates at 3.9 MHz only [18]. While specific low-power electronic circuits can be developed in ASICs (see, e.g., specialized on-chip memory cells and architecture in [18]) to minimize power consumption of DNN, the optimization of an FPGA solution has to be focused on microarchitecture and memory subsystem organization that are composed of (fixed and pre-defined) FPGA primitives.

In our survey, we will primarily focus on functional approximation which is less technology dependent and provides more predictable errors than voltage over scaling. The survey is based on representative papers published in 2011 – 2017 on key relevant conferences and in journals.

The result of our survey is presented in the form of table: Table 1 shows that the papers included into the survey are organized according to the Application Type, where the following major application types were identified: Filters, Metrics, Transforms, image compression (JPEG), and video (de)coders (according to MPEG and HEVC standards). In these Application Types, we investigated:

- what is approximated, i.e. whether the approximation is performed at the level of components (such as adders, multipliers, comparators) or modules (such as filters, DCT and FFT created using these components),
- how the approximation is conducted, i.e. whether an ad hoc or general purpose method is taken,
- what is the level of abstraction, where an approximation is conducted, i.e. whether circuits are approximated at the transistor, gate, register-transfer (RT) or behavioral source code level, and
- target platform, i.e. an ASIC or FPGA.

It can be seen that less complex applications such as image filters can be holistically approximated as a single system. In the case of more complex applications, the design is firstly decomposed and selected components then undergo the approximation process. Some of them can even be removed to further reduce power consumption. The approximation is predominately conducted at the gate level, but there are tools (such as AXILOG, ABACUS and GRATER) in which requirements on the approximation are specified directly at the source code (RT or behavioral) level. The actual approximations are then performed internally by the tool during the synthesis and netlist optimization.

It remains unclear what is the best performing approximation approach in the area of image and video processing. Unfortunately, approximate solutions have been only compared with exact solutions, but almost never with other competitive approximate solutions.

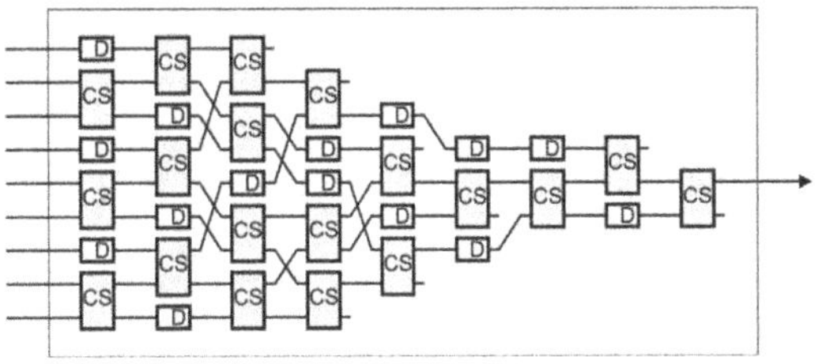

Fig. 1. Pipelined implementation of 9-input median filter consisting of compare-and-swap (CS) blocks and registers (D). All CS blocks contain the output register.

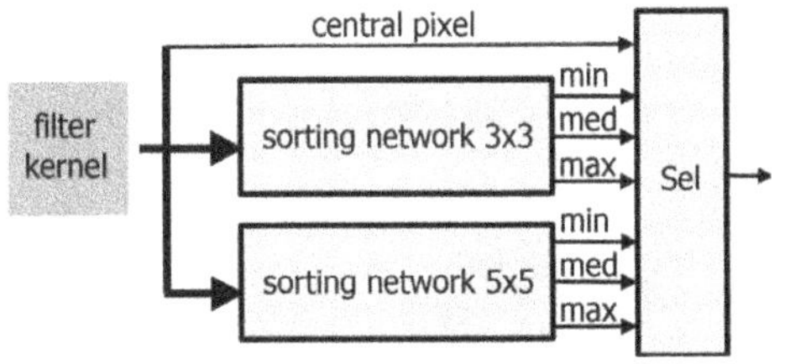

Fig. 2. Adaptive median filter internally computing minimum, maximum and median over kernels with 3×3 and 5×5 pixels and determining the output value using Selector.

4. Case Study

The purpose of this case study is to compare the impact of two fundamentally different approximation strategies on the quality and power consumption of a selected module of an image processing system. We decided to approximate the circuits implementing the shot noise image filter. The approximations will be conducted by CGP which proved to be highly competitive with respect to other circuit approximation methods [3], [19].

Section 4.1 provides a brief overview of conventional implementations of median filters and their extensions. CGP is then introduced in Sec. 4.2. Two approximation strategies (AS) are proposed in Sec. 4.3: (AS1) CGP is employed to approximate circuit implementations of the considered filters. (AS2) CGP is used to holistically evolve desired image filters from scratch.

4.1 Median Filters

Conventional implementations of shot noise elimination filters are usually based on calculating the median over the pixels belonging to the filtering window.

The *median filter* (MF) is a special case of order statistic filters which may be implemented in several different ways [20]. In this paper, we will consider a pipelined implementation based on a median network which is suitable for high-performance applications. The median network consists of a sequence of *compare-and-swap* operations (Fig. 1). Each compare-and-swap (CS) operation acts as a small 2-input sorting network which produces a sorted sequence by outputting the minimum and maximum of the input values.

The *weighted median filter* is an extension of the common median filter, which gives more weight to some values within the filtering window. The *center weighted median filter* (CWMF) represents a special case in which only the central value of the window is counted with additional weight [21]. Compared to the median filter, this modification can preserve more details along the horizontal and vertical directions while suppressing additive white and/or impulsive-type noise.

The median filters uniformly replace the value of every pixel of the filtered image by the median of its neighbors. Consequently, in addition to the removal of noisy pixels, these filters also remove desirable details and thus smudge the resulting image. In order to address this problem, more advanced concepts were introduced. The *adaptive median filter* (AMF) represents a multi-level approach which tries to detect and subsequently replace corrupted pixels only [22]. At each level, filtering windows of different sizes are utilized. Usually, two levels working with the 3×3 and 5×5 filtering window, respectively, are sufficient to obtain a very good image quality (Fig. 2). Hardware implementation consists of two median filters, circuitry that determines minimal and maximal values for each filter window, delay buffers to compensate different latency of median filters and simple logic.

4.2 Evolutionary Approximation

CGP [23] is a form of genetic programming in which each candidate solution is modeled using a two-dimensional array of $n_c \times n_r$ programmable n_a-input/n_b-output nodes whose functions are taken from a set G. The circuit utilizes n_i primary inputs and n_o primary outputs. A unique address is assigned to all primary inputs and to the outputs of all nodes to define an addressing system enabling circuit topologies to be specified (Fig. 3). As no feedback connections are allowed in the basic version of CGP, only combinational circuits can be created. Each candidate circuit is represented using the so-called chromosome which contains $n_c \times n_r \times (n_a + n_b) + n_o$ integers. The $(n_a + n_b)$ integers specify one programmable node: n_a integers specify destination addresses for its inputs and n_b integers determine the function codes from G. All possible legal chromosomes constitute the search space.

The search is usually performed using a simple $(1 + \lambda)$ evolutionary algorithm. In this algorithm, every new population consists of the best individual of the previous population and its λ offspring created using a mutation operator. This operator randomly modifies up to h randomly selected genes (integers) of the chromosome. The search is typically terminated after generating a given number of populations.

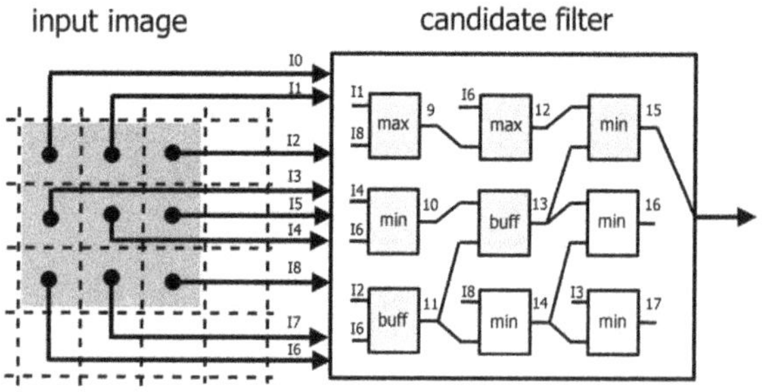

Fig. 3. An example of a simple filtering circuit (with filtering window 3×3 pixels) represented in CGP with parameters: $n_i = 9$, $n_o = 1$, $n_c = 3$, $n_r = 3$, $n_a = 2$, $n_b = 1$, G = {buffer (0), min (1), max(2)}. Nodes 14, 16 and 17 are inactive. Chromosome: 1,8,2; 4,6,1; 2,6,0; 6,9,2; 10,11,0; 8,11,1; 12,13,1; 13,14,1; 13,14,1; 15.

In order to evaluate the population, each candidate solution is evaluated using the so-called fitness function. As the problem is in principle multi-objective (error versus power consumption or area), a suitable multi-objective optimization algorithm has to be taken [2], [3]. While the circuit area on a chip can be easily estimated by summing the areas of components involved in the circuit, the error computation is more time demanding (see next sections).

4.3 Approximation Strategies based on CGP

Two strategies are compared in this case study:

AS1: Since the median filter is implemented as a network of compare-and-swap operations, an obvious approximation strategy is to remove some of them and reconnect the remaining ones in such a way that the error of filtering is minimized. We propose to seed CGP with the best known implementations of median filters and evolve approximate median filters containing fewer comparators than needed in the fully functional implementation. The fitness is constructed according to [24]. AS1, therefore, works at the level of comparators.

AS2: The whole image filtering function is evolved with CGP from scratch. The function set G contains all suitable two-input components, not only the minimum and maximum functions. CGP thus holistically develops a new image filter with the aim to minimize the error of filtering on the training data. Following the approach developed for the evolutionary design of image filters [25], the error is measured by means of the *mean absolute error* (MAE) between the outputs O_f produced by a candidate filter and reference (golden) outputs O_g for a given *training data set*, formally:

$$MAE = \frac{1}{K}\sum_{i=1}^{K} |O_f(i) - O_g(i)| \qquad (1)$$

where K is the number of filtered pixels.

As this approach is not biased by a conventional solution (median filter), there is a chance to discover an implementation showing better filtering properties and lower power consumption.

It has to be noticed that these approximation strategies differ from the approximate median filters proposed in the literature because: paper [26] utilizes approximate transistor-level circuits to implement the comparators (our comparators are always exact) and papers [3], [27] do not initialize CGP with existing median implementations, but rather evolve approximate circuits form scratch using insufficient resources.

5. Results

This section presents the setup used to perform desired approximations, parameters of evolved circuits and a comparison of approximate and original filters in terms of power consumption, area and filtering quality. In order to obtain parameters of evolved filters, we described the filters in VHDL and synthesized them using Synopsys Design compiler with 45 nm PDK. The filters were implemented as pipelined circuits with 8 bit operands. The goal of the synthesis was to produce implementations operating at least at 1 GHz. Section 5.1 deals with the implementation cost of conventional and approximate filters. The filtering quality is compared in Sec. 5.2.

5.1 Implementation Cost

Conventional (Exact) Filters: Table 2 summarizes the synthesis results for various median filters discussed in Sec. 4.1 – median filter operating on 3×3 (5×5) filter window denoted as MF9 (MF25), center weighted median filter operating on 3 × 3 pixels with the weight equal to 3 (CWMF9), and adaptive median filter (AMF25). While MF9 consists of 19 compare-and-swap operations (ops), AMF25 requires nearly ten times more operations. Each compare-and-swap operation is implemented using an 8-bit magnitude comparator and two 8-bit multiplexers. For each filter, the number of compare-and-swap operations, total power consumption and occupied area are presented. Contributions to power and area are given separately for registers and logic. The delay is intentionally omitted in all tables because timing constrains were met in all cases.

The key observation is that logic consumes less than 20% of the total power consumption. This is due to the pipeline nature of the circuits. The area on a chip increases with the increasing complexity (i.e. with the number of compare-and-swap operations) of the filters. As expected,

		power [mW]			**area [μm² × 10³]**		
filter	**ops**	**total**	**regs**	**logic**	**total**	**regs**	**logic**
MF9	19	6.8	80%	19%	7.8	65%	34%
CWMF9	28	12.1	81%	18%	13.6	64%	35%
MF25	99	45.0	86%	13%	37.9	50%	49%
AMF25	182	58.1	85%	14%	52.8	53%	46%

Tab. 2. Results of synthesis for conventional filters.

the common median filter operating with 3×3 pixels is the cheapest solution. If we extend the filter window to 5×5, the power consumption increases more than 6 times and the area on a chip increases nearly 5 times. The adaptive median filter represents the most complex and power-demanding filter in our study. Its power consumption is more than 8 times higher compared to MF9. The implementation costs of CWMF is between MF9 and MF25 since CWMF9 is, in fact, a median network with 11 inputs whose three inputs are connected to the central pixel of the filter window. The power as well as the area on a chip are doubled compared to MF9.

Filters Approximated with AS1: In order to obtain approximate median filters, CGP was seeded with the known optimal implementations of 9-input, 11-input and 25-input median networks exhibiting the minimal number of compare-and-swap operations. CGP operated with $n_a = n_b = 2$, $\lambda = 20$, $h = 5$, and 10^7 (6.10^5 respectively) generations were produced for 9-input (25-input, respectively) circuits. The function set contained 8-bit compare-and-swap functions and identity function. The error was determined as the position distance with respect to the exact median according to [24]. The goal of CGP was to minimize the position error under constrained area (experimented with max. 20% – 95% area of the exact implementation). As the statistical evaluation of this type of evolutionary design has been performed in the literature [24], we will report just the best evolved solutions.

Several hundreds of approximate implementations were produced by CGP in total. We identified ten Pareto-dominant solutions for each type of filter and synthesized them using Synopsys Design compiler to obtain their electrical parameters. Table 3 summarizes the total number of operations, power consumption and area for selected approximate filters. The obtained reduction with respect to the (exact) original solution is included in the 'red.' columns.

Table 3 shows that pruning of the number of compare-and-swap operations and their rearranging enables to significantly reduce not only the area on a chip but also the power consumption. The filtering quality will be reported in Sec. 5.2. For example, 9-input approximate median filter MF9 #9 exhibits a 75% reduction in power consumption and a 69% reduction in the area compared to the accurate optimal implementation. Overall, more than 75% of power budget is due to switching activity of registers. The majority of the area on a chip is utilized by registers.

Table 3 also includes parameters of approximate adaptive median filters. These approximate filters were obtained by replacing the exact 9-input median and 25-input median with their selected approximate implementations. The rest of the circuitry remained unchanged. Three variants of approximate adaptive median filter are presented – AMF25 #19, AMF25 #79 and AMF25 #99. The first variant consists of the exact 9-input approximate median network MF9, the second of approximate MF9 #7 and third employs MF9 #9. In all cases, approximate MF25#9 is employed. The approximate

	ops		power [mW]			area [µm^2 × 10^3]		
filter	**total**	**red.**	**total**	**red.**	**regs**	**total**	**red.**	**regs**
MF9								
#5	15	21%	4.6	31%	78%	5.6	27%	68%
#7	12	36%	3.0	55%	76%	4.1	48%	71%
#9	8	57%	1.7	75%	75%	2.4	69%	73%
CWMF9								
#5	25	10%	8.7	27%	80%	10.3	24%	66%
#7	19	32%	6.9	43%	82%	7.7	42%	63%
#9	13	53%	3.6	70%	78%	4.5	66%	68%
MF25								
#6	64	35%	32.5	27%	89%	26.4	30%	45%
#8	50	49%	20.1	55%	87%	17.7	53%	51%
#9	42	57%	14.5	67%	81%	16.4	56%	64%
AMF25								
#19	125	31%	27.6	52%	81%	31.4	40%	64%
#79	118	35%	23.9	58%	81%	27.7	47%	64%
#99	114	37%	22.5	61%	81%	26.0	50%	64%

Tab. 3. Results of synthesis for filters approximated in AS1.

AMFs occupy nearly half of the area and achieve up to 61% power saving with respect to AMF.

Filters Approximated with AS2: CGP started with a randomly generated initial population and used two-input 8-bit functions (such as minimum, maximum, addition, absolute difference, conditional assignment) and other settings ($n_c = 7$, $n_r = 9$, $n_a = 2$, $n_b = 1$, $\lambda = 7$, $h = 15$) according to [25]. All filters were evolved using fitness function (eq. 1) and appropriate training and golden images consisting of 384×256 pixels, i.e. $K = 98,304$.

Parameters of the best performing filters evolved under AS2 are summarized in Tab. 4. Two noise-specific filters are included in our comparison – a salt-and-pepper noise filter (denoted EVO #1) and a random-valued impulse noise filter (denoted EVO #2). Please, refer to Sec. 5.2 for details dealing with noise description. Both filters operate on the filter window consisting of 5×5 pixels. EVO #1 consists of 27 8-bit components (including 17 min/max functions) and occupies approximately the same area as MF9 but consumes about 50% more power. This is an interesting result because it operates on nearly three times higher number of inputs. EVO #2 is a more complex circuit having 33 8-bit components (including 20 min/max functions). Considering the fact that both filters have 25 inputs, they exhibit significantly lower implementation cost and power compared to MF25. Their filtering quality will be discussed in Sec. 5.2.

In order to improve the quality of output images, an ensemble of filters is often employed. In this evaluation, a bank of filters was constructed using 3 best filters evolved for each type of noise [25] (i.e. BNK #1 for salt-and-peper and BNK #2 for random shot noise). As Fig. 4 shows these filters

filter	power [mW]			area [$\mu m^2 \times 10^3$]		
	total	regs	logic	total	regs	logic
EVO #1	10.2	86%	13%	7.5	55%	44%
BNK #1	39.7	90%	9%	25.5	45%	54%
EVO #2	16.1	85%	14%	11.8	55%	44%
BNK #2	52.5	90%	9%	33.3	44%	55%

Tab. 4. Results of synthesis for filters approximated in AS2.

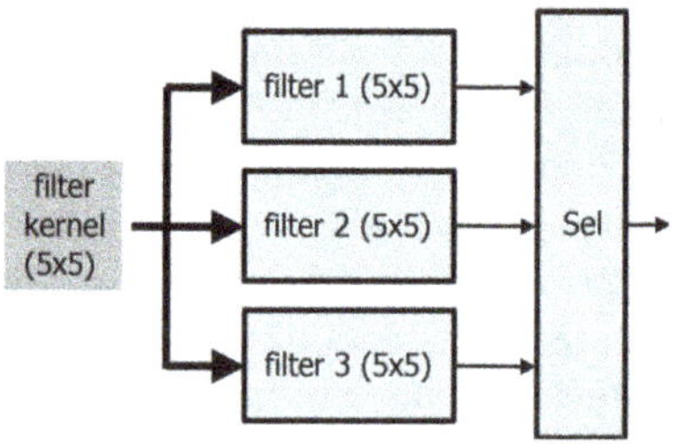

Fig. 4. Bank of filters composed of 3 different evolved filters.

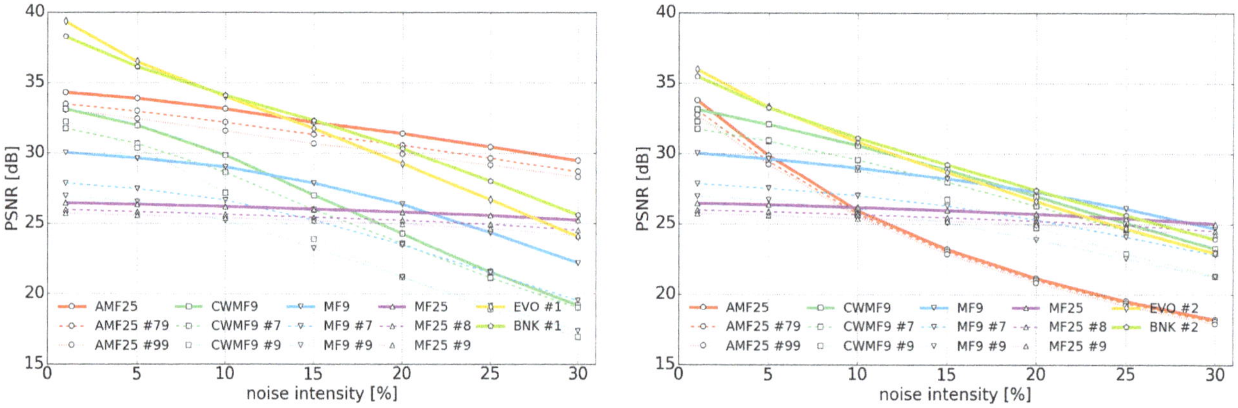

Fig. 5. Mean PSNR on 30 test images and different noise intensities obtained for conventional and approximate filters: salt-and-pepper noise (left) and impulse noise (right).

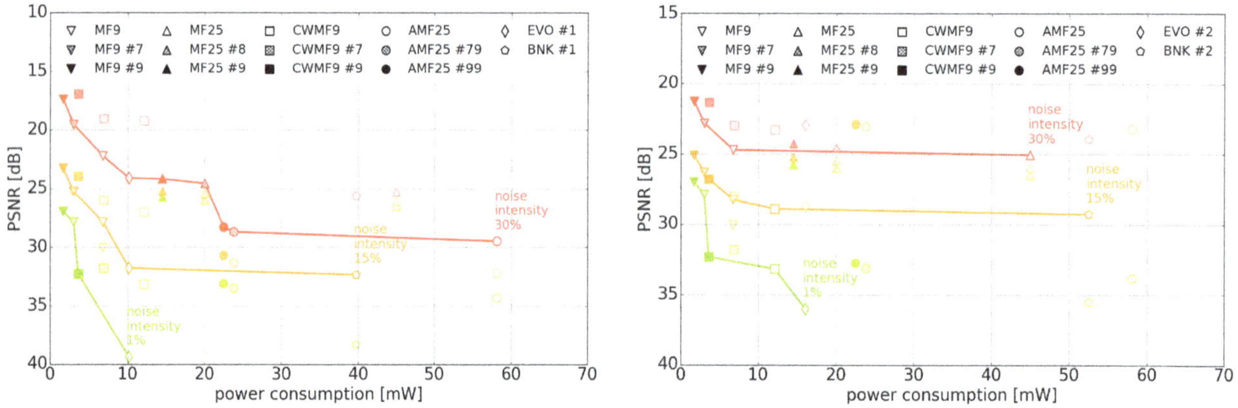

Fig. 6. Mean PSNR and power consumption of selected image filters: salt-and-pepper noise (left) and impulse noise (right).

operate in parallel over the filter window. If the majority of the filters of the bank indicates that the processed pixel is a shot, then the median value is calculated from the outputs of these filters and sent to the primary output of the bank. Otherwise, the original value of the processed pixel is sent to the primary output of the bank. While BNK #1 occupies significantly lower area on a chip compared to MF25 or AMF25, BNK#2 is comparable to AMF25.

5.2 Filtering Quality

The quality of the proposed filters was evaluated using a set of 30 test images corrupted by means of two common types of noise – salt-and-pepper noise and random shot noise. While the salt-and-pepper noise removal represents a typical benchmark problem which can be satisfactory addressed using adaptive median filter, the random shot noise removal is known to be a significantly harder problem. The reason is that the values of noisy pixels for salt-and-pepper noise are equal to either 0 or 255. In the case of the random shot noise, a noisy pixel can gain an arbitrary value from the whole range (i.e. 0 – 255). Therefore, it is more difficult to detect this noise because the deviation of a noise pixel can be very close to its original value.

Figure 5 shows the filtering quality of common and approximate filters in terms of the mean peak signal-to-noise ratio (PSNR). As 7 noise intensities (ranging from 1% to

30%) were considered, every filter was, in fact, applied to 2 (noise type) × 30 (images) × 7 (noise intensity) = 420 images. Resulting trade-offs between power consumption and filtering quality for noise intensity 1%, 15% and 30% are illustrated in Fig. 6.

The most interesting observations are as follows. Regarding the filtering quality, the mean PSNR indicates that filters evolved in AS2 significantly outperform other filters especially if the noise intensity is lower (15-20% depending on the noise type). For highly corrupted images, the bank of evolved filters can be employed to even improve the quality of filtering.

AMF performs well on salt-and-pepper noise, but it is rather poor for random shot noise; however, it is a very expensive solution. When approximate filters are introduced to AMF, the mean PSNR remains practically the same as for AMF25. The output quality depends mainly on the quality of MF9 (see the resulting PSNR for AMF25 #79 and AMF25 #99), but the difference is below 1 dB even when MF9 #9 consisting of eight compare-and-swap operations was employed. Anyway, approximate versions of AMF significantly reduced power consumption of the original AMF. It has to be emphasized that filters evolved in AS2 still consume only around 50% of the power budget of AMF25#19.

CWMF9 and its approximations provide very good results for random shot noise. Hence, CWMF9 (or CWMF#7 having a slightly worse PSNR) seems to be a solution of the first choice for energy constrained applications because it provides 25% benefit in power compared to EVO#2.

If low power consumption is the key design objective then approximate versions of MF9 show the best trade-off.

6. Conclusions

On the basis of the literature survey, we reported approximate circuits and approximation methods that have been applied in the area of image and video processing. We observed that the approximations are conducted at different levels of abstraction (from transistors via gates to RT) and focused either on the whole modules (such as filters or DCT) or elementary components (such as adders and multipliers) of these modules. In addition to ad hoc approximation methods, many general-purpose approximation methods have been used. Only in rare cases the approximation methods were mutually compared in terms of quality of produced approximate circuits.

In order to investigate the impact of approximation methods on the quality of resulting approximate circuits, the median circuit approximation problem was chosen. We compared two CGP-based approximation strategies based on removing and rearranging some components (AS1) and complete redesigning of the circuit (AS2). Three conventional median-based circuits (MF, CWMF, and AMF) were included to our study. The approximations were performed for two types of noise and evaluated for 7 noise intensities.

As all circuits were implemented as pipelined structures operating at least at 1 GHz, the approximation and optimization was focused on obtaining the best trade-offs between power consumption and filtering error. In the case of AS1, approximate circuits consistently show slightly worse filtering quality, but significantly reduced power consumption with respect to their exact counterparts. The best trade-offs were obtained with AS2, i.e. when CGP was not biased by conventional designs and could deliver new well-optimized filtering structures.

We can conclude that complete resynthesizing of the circuit rather than approximating a conventional solution provides better trade-offs, especially if good filtering quality is desired. While this approach (SA2) was applicable to image filtering circuits, it is not currently applicable for complex circuits (such as the whole HEVC coder) because the design process based on CGP is not fully scalable. Improving its scalability will be one of our future research objectives.

Acknowledgments

This work was supported by Czech science foundation project GA16-17538S.

References

[1] MITTAL, S. A survey of techniques for approximate computing. *Journal ACM Computing Surveys (CSUR)*, 2016, vol. 48, no. 4, p. 62:1–62:33. DOI: 10.1145/2893356

[2] MRAZEK, V., HRBACEK, R., VASICEK, Z., et al. Evoapprox8b: Library of approximate adders and multipliers for circuit design and benchmarking of approximation methods. In *Proceedings of the Design, Automation & Test in Europe Conference & Exhibition (DATE)*. 2017, p. 258–261. DOI: 10.23919/DATE.2017.7926993

[3] VASICEK, Z., SEKANINA, L. Evolutionary approach to approximate digital circuits design. *IEEE Transactions on Evolutionary Computation*, 2015, vol. 19, no. 3, p. 432–444. DOI: 10.1109/TEVC.2014.2336175

[4] JIANG, H., LIU, C., LIU, L., et al. A review, classification and comparative evaluation of approximate arithmetic circuits. *ACM Journal on Emerging Technologies in Computing Systems*, 2017, p. 1–37.

[5] NAWAB, S. H., OPPENHEIM, A. V., CHANDRAKASAN, A. P., et al. Approximate signal processing. *Journal of VLSI signal processing systems for signal, image and video technology*, 1997, vol. 15, no. 1, p. 177–200. DOI: 10.1023/A:1007986707921

[6] MARKOV, I. L. Limits on fundamental limits to computation. *Nature*, 2014, vol. 512, p. 147–154. DOI: 10.1038/nature13570

[7] CHIPPA, V. K., CHAKRADHAR, S. T., ROY, K., et al. Analysis and characterization of inherent application resilience for approximate computing. In *Proceedings of the 50th Annual Design Automation Conference (DAC'13)*. 2013, p. 1–9. DOI: 10.1145/2463209.2488873

[8] SAMPSON, A., DIETL, W., FORTUNA, E., et al. EnerJ: Approximate data types for safe and general low-power computation. In *Proceedings of the 32nd ACM SIGPLAN Conference on Programming Language Design and Implementation*. 2011, p. 164–174. DOI: 10.1145/1993498.1993518

[9] VENKATARAMANI, S., SABNE, A., KOZHIKKOTTU, V. J., et al. SALSA: Systematic logic synthesis of approximate circuits. In *Proceedings of the 49th Annual Design Automation Conference (DAC'12)*. 2012, p. 796–801. DOI: 10.1145/2228360.2228504

[10] YAZDANBAKHSH, A., MAHAJAN, D., THWAITES, B., et al. Axilog: Language support for approximate hardware design. In *Proceedings of the Design, Automation and Test in Europe (DATE)*. 2015, p. 1–6. DOI: 10.7873/DATE.2015.0513

[11] RANJAN, A., RAHA, A., VENKATARAMANI, S., et al. ASLAN: Synthesis of approximate sequential circuits. In *Proceedings of the Conference on Design, Automation and Test in Europe (DATE'14)*. 2014, p. 1–6. DOI: 10.7873/DATE.2014.377

[12] NEPAL, K., HASHEMI, S., TANN, H., et al. Automated high-level generation of low-power approximate computing circuits. *IEEE Transactions on Emerging Topics in Computing*, 2016, no. 1, p. 1–13. DOI: 10.1109/TETC.2016.2598283

[13] ESMAEILZADEH, H., SAMPSON, A., CEZE, L., et al. Neural acceleration for general-purpose approximate programs. *Communications of the ACM*, 2015, vol. 58, no. 1, p. 105–115. DOI: 10.1145/2589750

[14] PANDA, P., SENGUPTA, A., SARWAR, S. S., et al. Invited – cross-layer approximations for neuromorphic computing: From devices to circuits and systems. In *Proceedings of the 2016 53nd ACM/EDAC/IEEE Design Automation Conference (DAC)*. 2016, p. 1–6. DOI: 10.1145/2897937.2905009

[15] MRAZEK, V., SARWAR, S. S., SEKANINA, L., et al. Design of power-efficient approximate multipliers for approximate artificial neural networks. In *Proceedings of the IEEE/ACM International Conference on Computer-Aided Design*. 2016, p. 811–817. DOI: 10.1145/2966986.2967021

[16] SHUBHAM JAINA, S. V., RAGHUNATHANC, A. Approximation through logic isolation for the design of quality configurable circuits. In *Proceedings of the Design, Automation & Test in Europe Conference and Exhibition (DATE)*. 2016, p. 1–6. DOI: 10.3850/9783981537079_0416

[17] SRINIVASAN, G., WIJESINGHE, P., SARWAR, S. S., et al. Significance driven hybrid 8t-6t sram for energy-efficient synaptic storage in artificial neural networks. In *Proceedings of the Design, Automation Test in Europe Conference Exhibition (DATE)*. 2016, p. 151–156. DOI: 10.3850/9783981537079 0909

[18] BANG, S., WANG, J., LI, Z., et al. 14.7 A 288µw programmable deep-learning processor with 270KB on-chip weight storage using non-uniform memory hierarchy for mobile intelligence. In *Proceedings of the IEEE International Solid-State Circuits Conference*. 2017, p. 250–251. DOI: 10.1109/ISSCC.2017.7870355

[19] VASICEK, Z., MRAZEK, V., SEKANINA, L. Towards low power approximate DCT architecture for HEVC standard. In *Proceedings of the Design, Automation & Test in Europe Conference & Exhibition (DATE)*. 2017, p. 1576–1581. DOI: 10.23919/DATE.2017.7927241

[20] HUANG, T., YANG, G., TANG, G. A fast two-dimensional median filtering algorithm. *IEEE Transactions on Acoustics, Speech, and Signal Processing*, 1979, vol. 27, no. 1, p. 13–18. DOI: 10.1109/TASSP.1979.1163188

[21] KO, S., LEE, Y. Center weighted median filters and their applications to image enhancement. *IEEE Transactions on Circuits and Systems*, 1991, vol. 15, p. 984–993. DOI: 10.1109/31.83870

[22] HWANG, H., HADDAD, R. Adaptive median filters: New algorithms and results. *IEEE Transactions on Image Processing*, 1995, vol. 4, no. 4, p. 499–502. DOI: 10.1109/83.370679

[23] MILLER, J. F. *Cartesian Genetic Programming*. Springer-Verlag, 2011. DOI: 10.1007/978-3-642-17310-3

[24] VASICEK, Z., MRAZEK, V. Trading between quality and non-functional properties of median filter in embedded systems. *Genetic Programming and Evolvable Machines*, 2017, vol. 18, no. 1, p. 45–82. DOI: 10.1007/s10710-016-9275-7

[25] VASICEK, Z., BIDLO, M., SEKANINA, L. Evolution of efficient real-time non-linear image filters for FPGAs. *Soft Computing*, 2013, vol. 17, no. 11, p. 2163–2180. DOI: 10.1007/s00500-013-1040-8

[26] MONAJATI, M., FAKHRAIE, S. M., KABIR, E. Approximate arithmetic for low-power image median filtering. *Circuits, Systems, and Signal Processing*, 2015, vol. 34, no. 10, p. 3191–3219. DOI: 10.1007/s00034-015-9997-4

[27] VASICEK, Z., MRAZEK, V., SEKANINA, L. Evolutionary functional approximation of circuits implemented into FPGAs. In *Proceedings of the IEEE Symposium Series on Computational Intelligence, Evolvable Systems (SSCI ICES)*. 2016, p. 1–8. DOI: 10.1109/SSCI.2016.7850173

[28] EL-HAROUNI, W., REHMAN, S., PRABAKARAN, B. S., et al. Embracing approximate computing for energy-efficient motion estimation in high efficiency video coding. In *Proceedings of the Design, Automation & Test in Europe Conference & Exhibition (DATE)*. 2017, p. 1384–1389. DOI: 10.23919/DATE.2017.7927209

[29] GUPTA, V., MOHAPATRA, D., RAGHUNATHAN, A., et al. Low-power digital signal processing using approximate adders. *IEEE Transactions on Computer-Aided Design of Integrated Circuits and Systems*, 2017, vol. 32, no. 1, p. 124–137. DOI: 10.1109/TCAD.2012.2217962

[30] JRIDI, M., MEHER, P. A scalable approximate dct architectures for efficient HEVC compliant video coding. *IEEE Transactions on Circuits and Systems for Video Technology*, 2016, p. 1–10. DOI: 10.1109/TCSVT.2016.2556578

[31] KULKARNI, P., GUPTA, P., ERCEGOVAC, M. D. Trading accuracy for power in a multiplier architecture. *Journal of Low Power Electronics*, 2011, vol. 7, no. 4, p. 490–501. DOI: 10.1166/jolpe.2011.1157

[32] LOTFI, A., RAHIMI, A., YAZDANBAKHSH, A., et al. Grater: An approximation workflow for exploiting data-level parallelism in FPGA acceleration. In *Proceedings of the Design, Automation & Test in Europe Conference & Exhibition (DATE)*. 2016, p. 1279–1284. DOI: 10.3850/9783981537079_0805

[33] RAHA, A., JAYAKUMAR, H., RAGHUNATHAN, V. A power efficient video encoder using reconfigurable approximate arithmetic units. In *Proceedings of the 27th International Conference on VLSI Design and 13th International Conference on Embedded Systems*. 2014, p. 324–329. DOI: 10.1109/VLSID.2014.62

[34] SNIGDHA, F. S., SENGUPTA, D., HU, J., et al. Optimal design of JPEG hardware under the approximate computing paradigm. In *Proceedings of the 53rd Annual Design Automation Conference (DAC)*. 2016, p. 106:1–106:6. DOI: 10.1145/2897937.2898057

About the Authors ...

Lukáš SEKANINA received all his degrees (Ing. in 1999, Ph.D in 2002) from Brno University of Technology, Czech Republic. He was awarded with the Fulbright scholarship to work with NASA Jet Propulsion Laboratory at Caltech in 2004. Prof. Sekanina was a visiting professor with CEI UPM Madrid (2012), Pennsylvania State University, Erie (2001) and visiting researcher with Department of Informatics, University of Oslo (2001). He has served as an associate editor of IEEE Transactions on Evolutionary Computation (2011-2014), Genetic Programming and Evolvable Machines

Journal and International Journal of Innovative Computing and Applications. Prof. Sekanina (co)authored over 150 papers mainly on evolutionary design and evolvable hardware and 1 patent. He is currently a full professor and Head of the Department of Computers Systems at Faculty of Information Technology, Brno University of Technology.

Zdeněk VAŠÍČEK received the Ing. and Ph.D. degrees in electrical engineering and computer science from the Faculty of Information Technology, Brno University of Technology, Brno, Czech Republic, in 2006 and 2012. He is an Associate Professor with the Faculty of Information Technology, Brno University of Technology. His research interests include evolutionary design and optimization of complex digital circuits and systems. He has (co)authored over 40 conference/journal papers focused on evolvable hardware and hardware design. His work was awarded with Silver (2011) and Gold (2015) medal at HUMIES.

Vojtěch MRÁZEK received the Ing. degree in computer science and engineering from the Faculty of Information Technology, Brno University of Technology, Czech Republic in 2014. Currently, he is a PhD student at the Faculty of Information Technology with Evolvable Hardware Group. He has (co)authored over 10 conference/journal papers focused on evolvable hardware and hardware design. His research interests are evolvable hardware, power-aware design, approximate computing and genetic programming.

7

A 28-nm 32 Kb SRAM For Low-V_{MIN} Applications Using Write and Read Assist Techniques

Satyendra KUMAR [1], *Kaushik SAHA* [2], *Hariom GUPTA* [1]

[1] Dept. of Electronics and Communication Engineering, Jaypee Institute of Information Technology - Noida, India
[2] Samsung R & D Institute India - Delhi, India

satyena.kumar@gmail.com, ksaha_2000@yahoo.com, hariom.gupta@jiit.ac.in

Abstract. *In this paper new write and read assist techniques, reduced coupling signal negative bitline (RCS-NBL) and low power disturbance noise reduction (LP-DNR) of 6T static random-access memory (SRAM) to improve its minimal supply voltage (V_{MIN}), have been presented. To observe the improvements in V_{MIN} and power consumption of SRAM with the help of proposed assist techniques, a 32 Kb capacity SRAM, with 128 words of 256 bits width, is designed and simulated in 28-nm bulk CMOS technology. New RCS-NBL scheme, shows an improvement in SRAM write V_{MIN} by 295 mV and also reduces overstress on pass transistor (PG) of the selected bitcell by 40 mV. Proposed LP-DNR scheme demonstrates an improvement in SRAM read V_{MIN} by 35 mV and also shows a saving of the power loss in the existing DNR scheme during the read access which occurs due to continuous flow of current from the cross coupled latch to the discharge block path after the bitlines have settled. The static power consumption of this SRAM macro is improved by 48.9 % and 11.7 % while dynamic power by 91.7 % and 8.1 % with the help of proposed write and read assist techniques respectively. Area overheads of these proposed RCS-NBL and LP-DNR assist techniques for this macro are less than 0.79 % and 3.70 % respectively.*

Keywords

Low voltage, low power, SRAM, process variation, write assist, read assist, disturbance noise reduction (DNR)

1. Introduction

In advanced nanometer CMOS technologies, reduction of minimal supply voltage (V_{MIN}) and chip-area are the primary concerns of SRAM design. The reduction in V_{MIN} of the SRAM cell for scaled devices is limited because of the local threshold voltage variations resulting from random dopant fluctuations and lithographic-dependent patterns have been increasing [1]. Also, with the increased threshold voltage variations in scaled transistors the access-disturbance margin (ADM) [2] and write margin (WM) [3] of the SRAM bitcell have been degrading. Process variations make SRAM design less predictable and controllable, moreover the SRAM design space in terms of prediction and control degrades further as supply voltage (V_{DD}) scales down [5]. Meanwhile, to improve the data stability of the bitcell, dual supply voltage schemes have been suggested [6], [7]. These schemes use higher supply voltage for bitcell array and lower supply voltage for peripheral blocks to improve the ADM of an SRAM. Write margin (WM) of an SRAM bitcell, has been improved by pushing selected bitline to negative voltage or by decreasing the cell voltage [2]. Now a days SRAM read and write assist techniques are widely used approaches to lower the V_{MIN} of an SRAM [8]. Firstly, with the help of read and write assist techniques SRAM stability and write ability have been increased from their minimum respective levels required for proper read and write in SRAM without any assist techniques, which allows us to reduce the corresponding SRAM V_{MIN} until the SRAM stability and write ability touch their respective original levels. In this paper, we present 28-nm bulk CMOS technology based 32 Kb 6T SRAM, featuring low V_{MIN} with new write and read assist techniques. The focus has been to reduce the V_{MIN} of SRAM since it is one of the most effective approaches to reduce dynamic as well as static power of SRAM.

Remaining part of this paper is organized as follows. Section 2 describes the conventional SRAM assist schemes. Section 3 elaborates the proposed reduced coupling signal negative bitline (RCS-NBL) scheme. Section 4 discusses the proposed low power disturbance noise reduction (LP-DNR) scheme. Section 5 deals with the impact of process variation on various SRAM parameters. Section 6 demonstrates the implementation and simulation results. Finally, we conclude in Sec. 7.

2. SRAM Assist Schemes

As shown in Fig. 1 conventional SRAM assist techniques are categorized into write and read assist schemes.

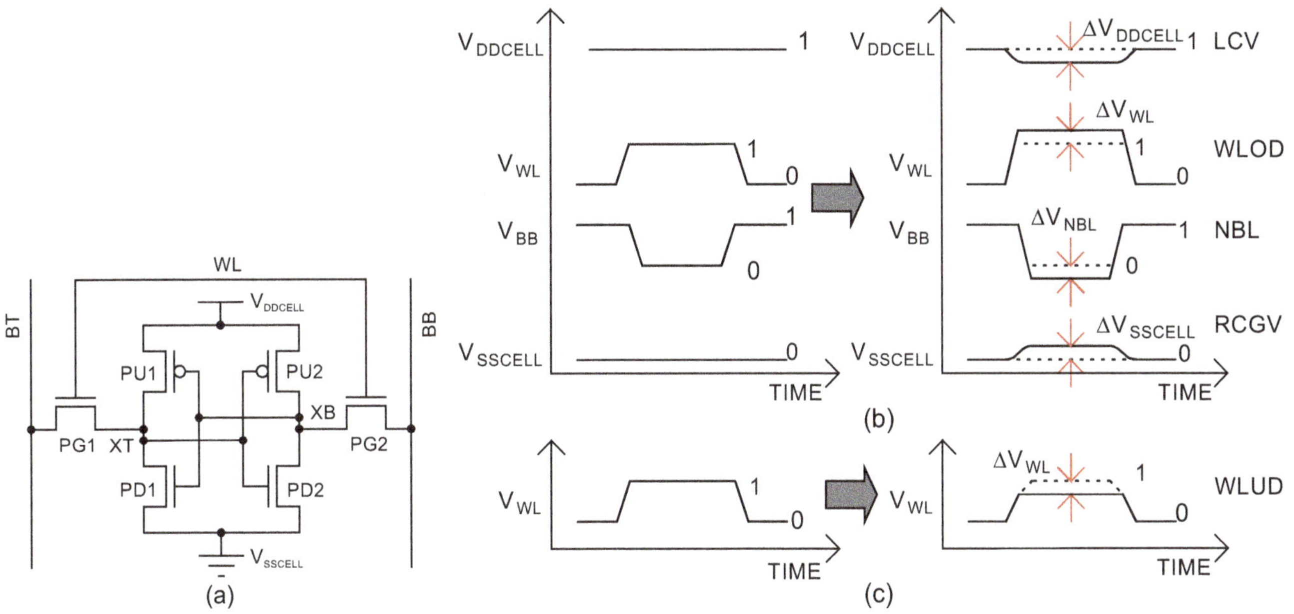

Fig. 1. (a) SRAM bitcell. (b) Timing diagram of write-assist techniques LCV, WLOD, NBL [2] and RCGV technique (c) Timing diagram of read-assist technique WLUD.

2.1 Write Assist Schemes

The techniques which aid the bitcell in changing the state during write operation are called write assist techniques and now these techniques are widely used in most low power SRAMs. The basic idea behind the write assist scheme is to decrease the ratio of strength of pull-up transitor to pass transistor of an SRAM cell when the wordline (WL) of the cell is enabled for write operation.

Conventional SRAM write assist schemes are categorized into following four techniques depending on the approach used to lower the ratio of pull-up to pass transistor strength of an SRAM cell during the write operation. Negative bitline (NBL) scheme, Wordline overdrive (WLOD) scheme, Lowering cell V_{DD} voltage (V_{DDCELL}) (LCV) scheme [2] and Raising cell ground voltage (V_{SSCELL}) (RCGV) scheme. Figure 1(a) shows the schematic of an 6T SRAM bitcell and Fig. 1(b) shows these conventional write assist techniques to enhance the write ability of the SRAM bitcell. In NBL scheme the selected bitline is pushed to negative voltage during write operation, which results in an increase of V_{GS} of the corresponding pass transistor hence the strength of this pass transistor has been enhanced, this improves the WM of the cell. While in WLOD scheme the strength of pass transistor is increased by boosting the WL voltage i.e. the gate voltage of the pass transistor. In LCV scheme, the strength of pull-up device is reduced by lowering the source voltage (V_{DDCELL}) of pull-up devices while keeping wordline voltage (V_{WL}) at V_{DD}. In RCGV scheme also, the same idea is used to weaken the pull-up device, but in this case it is achieved by weakening the pull-up gate voltage instead of the source voltage, which is realized by raising the V_{SSCELL}, during the write operation.

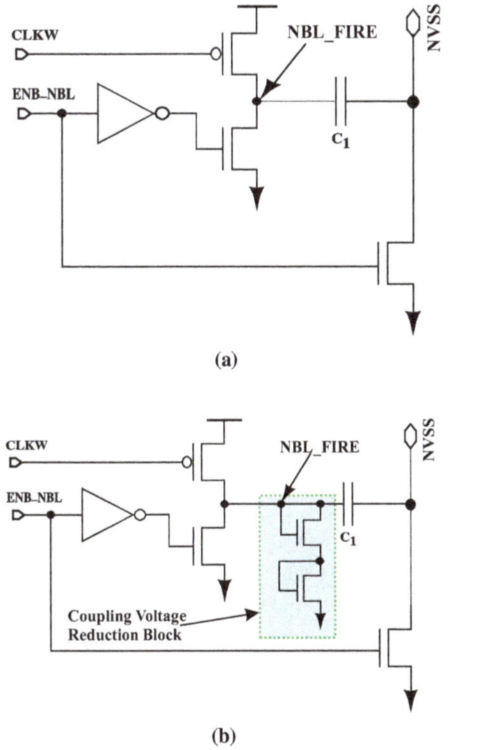

Fig. 2. Schematic of (a) Capacitive coupling signal (CCS) circuit (b) Proposed Reduced coupling signal (RCS) circuit.

2.2 Read Assist Schemes

The read disturb problem can be mitigated by adding a dedicated read port to isolate the bitcell internal nodes from the bitlines, but these resulting 7T, 8T, 9T and 10T SRAM bitcells [9–13] occupy larger area. The access-disturbance margin (ADM) can be improved by reducing the amount of charge injection from the pre-charged bitline to the ‘0’ node of the active bitcell. This can be achieved by reducing the strength of pass transistor and/or bitline capacitance with slow WL rise [14], [15]. Figure 1(c) shows conventional SRAM read assist scheme, WL underdrive (WLUD), used to improve the ADM of an SRAM bitcell by reducing the strength of the pass transistor.

3. Proposed Reduced Coupling Signal Negative Bitline Scheme

In this work, NBL technique is being used as write assist scheme, since this is the most effective technique to reduce the SRAM V_{MIN} [16]. Also, this technique shows highest WM without reducing the ADM of the half selected bitcells [2]. To realize NBL write assist scheme, capacitive coupling signal (CCS) approach shown in Fig. 2(a) is being used, which generates negative voltage at the bitline supposed to get down for write operation. In this scheme ENB_NBL signal propagates to NBL_FIRE signal as the falling edge of ENB_NBL signal triggers the fall of NBL_FIRE signal from the voltage level it was sitting before. The NBL_FIRE signal is coupled to negative bias (NVSS) signal through capacitor C_1. The pre-charged voltage level of NBL_FIRE signal before it starts to fall is one of the key parameters to determine the negative bias (NVSS) voltage level, which means that for higher V_{DD} operation the generated voltage level of NVSS signal will be more negative. As shown in Fig. 3(a) with the help of write driver and column multiplexer, NVSS signal is applied to the selected bitline. Thus, coupling technique produces higher negative bitline bias level for higher V_{DD} operation but there are two main disadvantages of higher negative voltage level at bitline, one stability concern of the half-selected bitcells in the same column and the other one is overstress on pass transistor of the selected bitcell, which is connected to this negative biased bitline, as for this pass transistor V_{GS} is too large, and this overstress condition is also getting worse for high V_{DD} operation [1].

To address these two issues, reduced coupling signal (RCS) circuit as shown in Fig. 2(b) is proposed here as write assist scheme. In this scheme, the voltage level of NBL_FIRE signal is reduced with the help of coupling voltage reduction block shown in this figure.

With the reduction in voltage level of NBL_FIRE signal, the negative bias (NVSS) signal level is reduced and hence negative bitline voltage level is reduced. In this work, 32 Kb SRAM with column mux (CM) = 8 i.e. an SRAM with physical rows (PR)=128 and physical columns (PC)=256 has been simulated with CCS-NBL, proposed RCS-NBL and state of the art SCS-NBL [1] write assist schemes. In CM=8 configuration 8-columns of bitcells are muxed with single write driver and hence single write assist block as shown in Fig.3(a). Simulated waveforms are plotted in Fig. 3(b) and 3(c) for comparison of CCS-NBL with RCS-NBL scheme, and RCS-NBL with SCS-NBL schemes

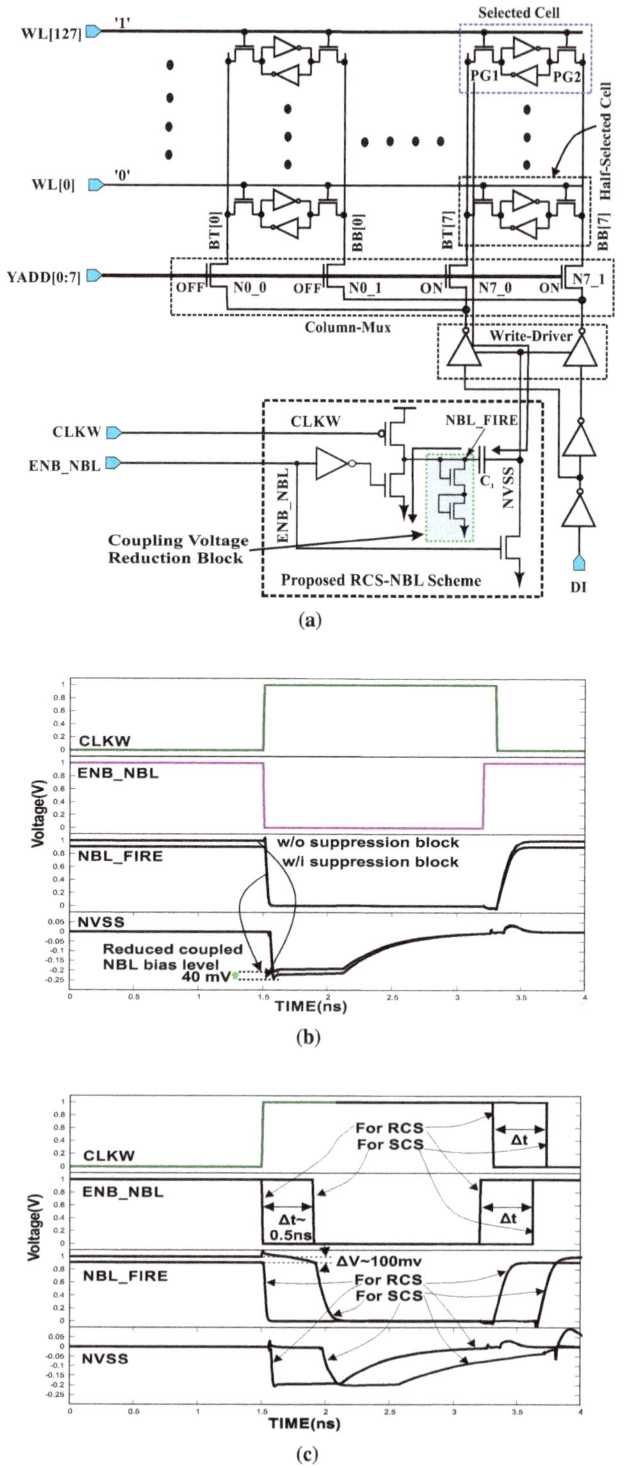

Fig. 3. (a) Proposed RCS-NBL write assist scheme (b) Simulation waveforms of CCS-NBL and RCS-NBL write assist schemes (c) Simulation waveforms of RCS-NBL and SCS-NBL write assist schemes.

respectively. Figure 3(b) and 3(c) demonstrate that the negative voltage level of NVSS is reduced by 40 mV with the help of RCS-NBL scheme as well as SCS-NBL scheme, and which has resulted in reduction of negative bitline voltage by same amount, this reduction will be more significant for higher V_{DD} operation. Thus RCS-NBL and SCS-NBL schemes address both the issues mentioned above. However, as shown in Fig. 3(c) proposed RCS-NBL scheme also improves the performance of write operation of the bitcell as compared to SCS-NBL scheme, because in this scheme the generation of NVSS signal can be triggered at the same time when CLKW signal rises from '0' to '1' while in the case of SCS-NBL scheme generation of NVSS signal will be initiated after Δt time to get same assist level as in the case of RCS-NBL scheme. With 6σ process variation, simulation results shown in Fig. 4 demonstrate the improvement in SRAM write V_{MIN} with NBL techniques. In this figure, blue plot shows the required bitline voltage to write the bitcell, red, green and black plots represent coupled NBL voltage levels with CCS-NBL, proposed RCS-NBL and SCS-NBL techniques, respectively. Points A and B represent SRAM write V_{MIN} without and with write assist techniques respectively. It can be observed from this figure that the proposed RCS-NBL as well as the SCS-NBL write assist technique demonstrate improvement in SRAM write V_{MIN} by 295 mV.

4. Proposed Low Power Disturbance Noise Reduction (LP-DNR) Scheme

Read assist scheme WLUD improves the stability of half-selected bitcells but degrades both the read and write performances of selected bitcell [4]. Further WLUD scheme also shows rise in access time of bitcell with reducing V_{MIN}, then to improve the V_{MIN}, disturbance noise reduction (DNR) scheme was proposed as read assist scheme [2]. In this scheme, both the bitlines are lowered simultaneously, before the WL is activated to reduce the level of noise injection to the bitcell nodes, by discharging through clamping and discharge blocks. With lowering the bitlines voltage level at WL enabled time ADM increases to a certain level and thus the bitlines voltage level at which ADM is getting its maximum value, is defined as bitline safe-voltage level (V_{SAFE}).

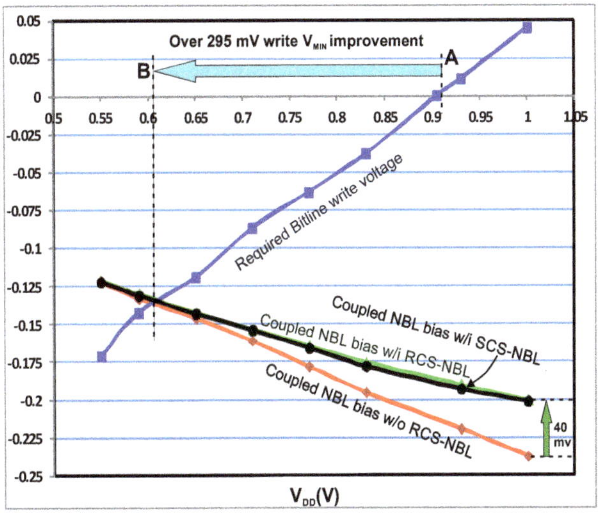

Fig. 4. Simulated write V_{MIN} for the NBL schemes.

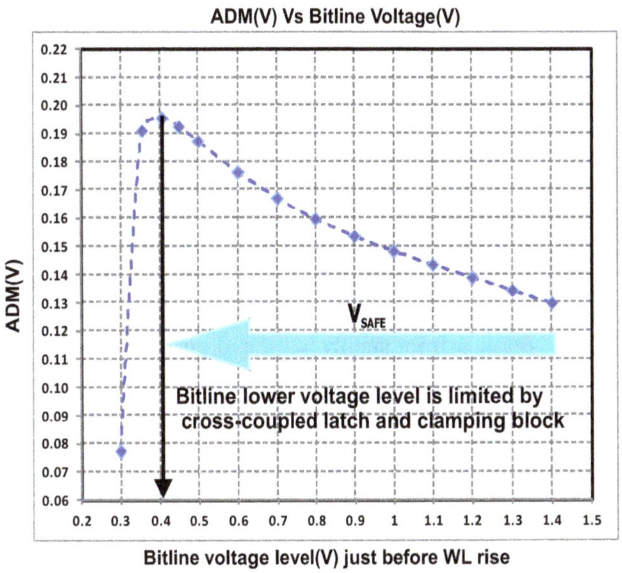

Fig. 5. Simulation result of ADM(V) versus bitline voltage level (V) for 6σ process variation.

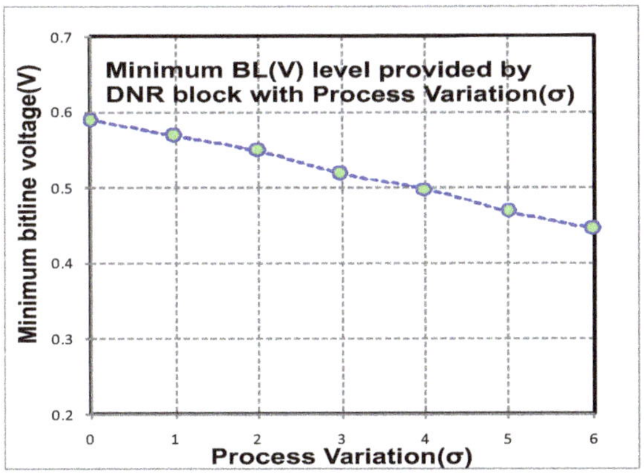

Fig. 6. Simulated minimum bitline voltage(V) at WL rise time with process variation(σ).

However, ADM degrades drastically as bitline voltage level goes below V_{SAFE} [2]. Figure 5 shows the plot of simulated results of ADM versus bitline voltage level at WL enabled time with 6σ process variation for the bitcell being used in this work and these results also endorse the same trend of steep degradation of ADM with bitline voltage level below V_{SAFE}. Thus to ensure enough data stability of the selected bitcells, proposed circuit must provide bitline voltage level at WL enable timing, more than or equal to V_{SAFE} but not below it. Hence here for 6σ process variation cross-coupled latch, clamping and discharge circuits have been designed to keep bitline lowest voltage level to V_{SAFE}. Figure 6 shows the simulated results of biltline voltage level provided by DNR block with process variation for $V_{DD} = 1.0$ V. These results demonstrate that DNR block being used in this work is providing bitline voltage level which is more than or equal to V_{SAFE} as indicated in Fig. 5. In DNR scheme, after the lowering of bitlines, as WL is activated bitcell read current (I_{Read}) from bitline to cell node storing '0' , in addition to bitline discharge current pulls down the corresponding bitline low enough to put ON one of the two PMOSs of cross-coupled latch for which this bitline is acting as its gate. Thus as per the action of cross-coupled latch, this bitline goes down to '0' and other one goes to ($V_{SAFE}+\Delta V$). PMOS which is ON keeps one of the two bitlines at ($V_{SAFE}+\Delta V$) for the period WL enabled time to the time at which the bitline is pre-charged to V_{DD} again. For this period, pass transistor of the bitcell connected to the same bitline supplies current to it since this pass transistor is working very close to the subthreshold region. Thus, during this time period, a sum of PMOS ON current and pass transistor subtheshold current, I_{LOSS}, is drawn from supply V_{DD} and pushed to discharge block through the pair of clamping PMOS devices connected to the bitline settled at ($V_{SAFE}+\Delta V$).

To save this loss of power contributed by I_{LOSS}, DNR circuit [2] has been modified and resulting circuit shown in Fig. 7(a) is proposed here as a new low power disturbance noise reduction (LP-DNR) read assist scheme. Fig.7(b) shows the timing diagram of proposed LP-DNR circuit for an access operation of the bitcell. As shown in this figure after WL is enabled, I_{Read} in addition to discharge current pulls the bitline BT enough low to turned-on PMOS P10 of I_{LOSS} path shut-off circuit, subsequently shut-off signal is activated (shut-off='1') with LP_DNR='1'. Thus during SHUT_PCH window, active shut-off signal turns off P2 & P3 devices, which results in stop of the flow of I_{Loss} but in case of DNR scheme this flow continues throughout SHUT_PCH window. In proposed LP-DNR scheme, as the flow of I_{LOSS} is blocked by turning off P2 device, the bitline BB starts to charge towards V_{DD} while BT remains at 0V, finally bitlines, BB & BT are synchronized with cell data, hence neither '1' noise nor '0' noise can inject to the cell which results in highest stability of the bitcell. Also, with this synchronization of bitlines with cell data the active current drawn from power supply V_{DD} is diminished.

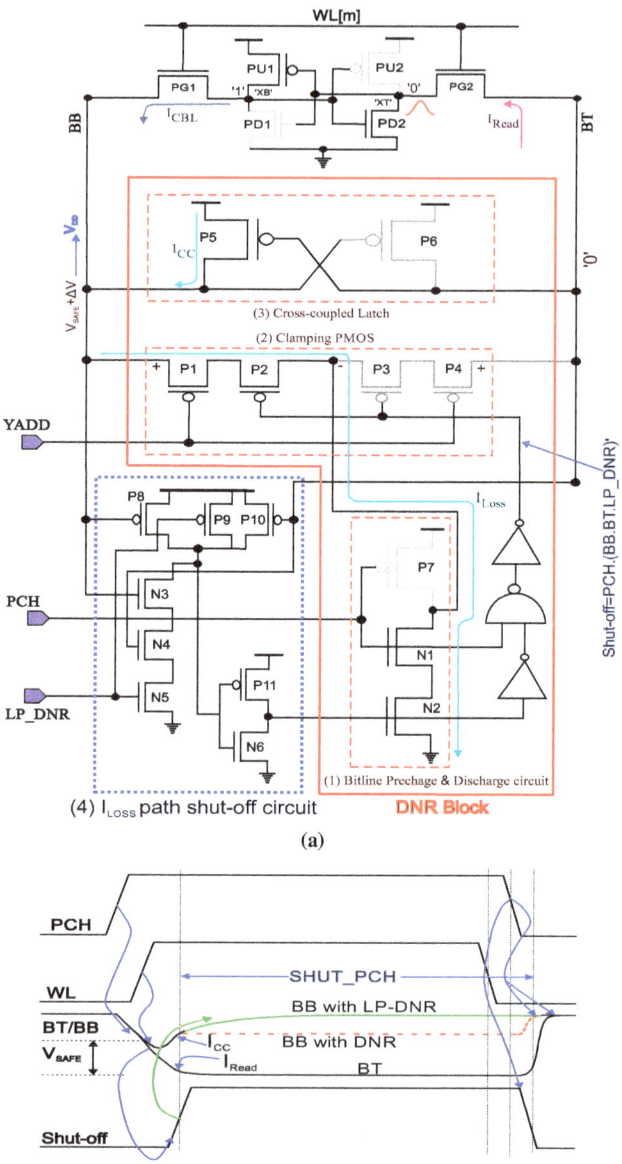

Fig. 7. (a) Proposed LP-DNR circuit (b) Timing diagram of the proposed LP-DNR read assist scheme.

5. Impact of Process Variability

In modern technologies intrinsic device variability of scaled devices dominates the traditional (worst-case) overall process spread that is generally used to determine the design window for the digital design community [17]. Transistor threshold voltage standard deviation σV_T can be used to represent the intrinsic device variability, which is expressed as follows [18]

$$\sigma V_T = 3.19 \times 10^{-8} \frac{t_{ox} . N_A^{0.401}}{\sqrt{L_{eff} W_{eff}}} \quad [V]. \qquad (1)$$

To simulate various SRAM parameters, with the impact of process variation, σV_T for all the concerned devices have been obtained using (1).

The simulation results for 6.5σ weak bitcell, shown in Fig. 8 demostrate the need of LP-DNR circuit to combat the access disturbance of the cell. For SRAM read operation bitlines are pre-charged to V_{DD} before the access of the cell, and access of the cell is obtained by turning ON pass transistors (PG1,PG2) of the cell by enabling WL. Figure 8(a) shows the simulated waveforms for read operation, as the bitcell is accessed, noise from BT will be injected to the cell storage node XT storing '0' data before this access of the cell, thus XT node voltage rises to the voltage level which is sufficient to flip the cell, which results in functional failure. Further, the required voltage level at XT to flip the cell is also degrading as the cell gets weaker due to process variation. Figure 8(b) shows that the bitlines are lowered by LP-DNR circuit before the access of the cell hence the level of noise injection from BT to bitcell node XT storing '0' data before the access of the cell, is reduced which results in successful read operation of the cell.

Simulation results shown in Fig. 9, demonstrate the improvement in read V_{MIN} with the help of LP-DNR and DNR schemes respectively. In this figure, green curve shows the

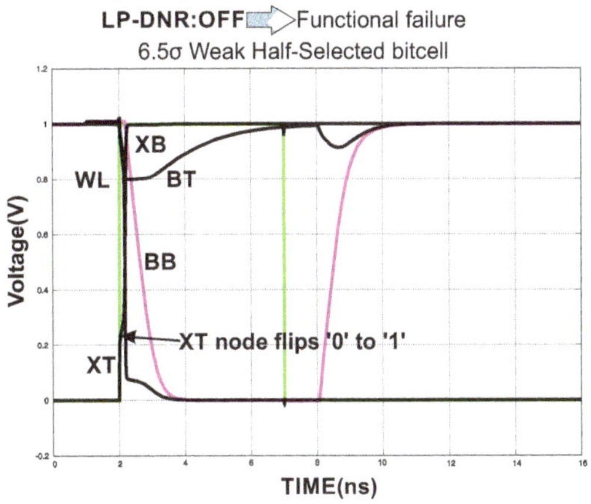

(a)

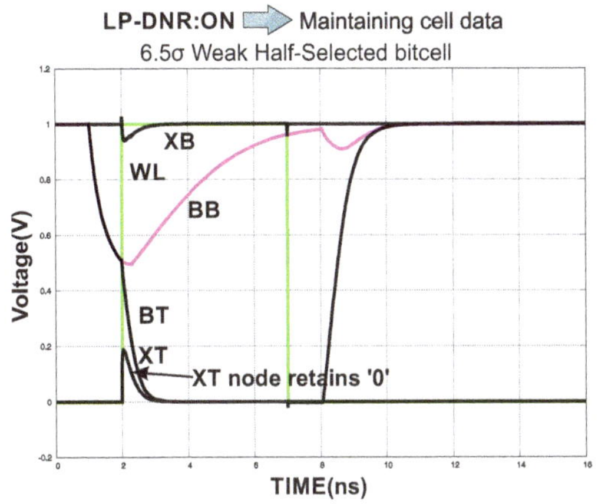

(b)

Fig. 8. LP-DNR effect on disturbance failure for weak bitcell (a) Simulated waveform without LP-DNR (b) Simulated waveform with LP-DNR.

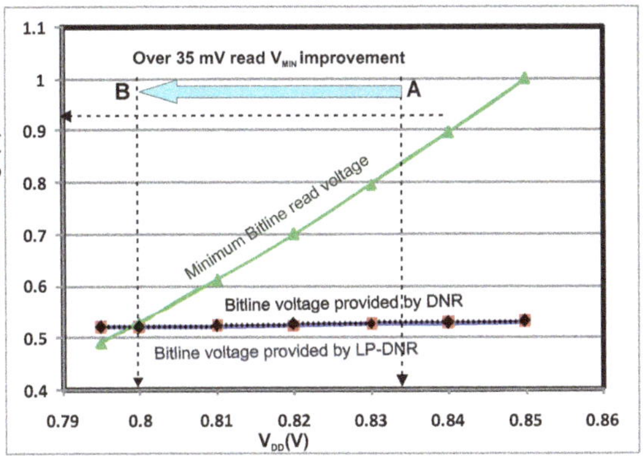

Fig. 9. Simulated read V_{MIN} for LP-DNR and DNR schemes.

minimum bitline voltage for non-destructive read operation of the cell, blue and red curves represent the bitline voltage level provided by LP-DNR and DNR schemes respectively. Here points A and B are representing SRAM read V_{MIN} without and with read assist techniques, respectively. It can be observed from these results that with the help of both the proposed as well as the DNR scheme, SRAM read V_{MIN} has improved by 35 mV.

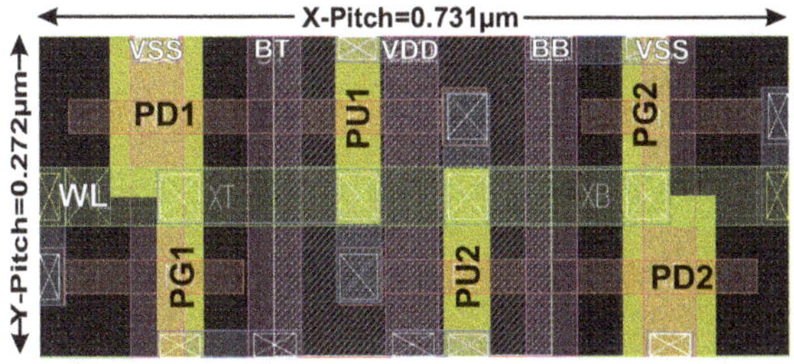

(a)

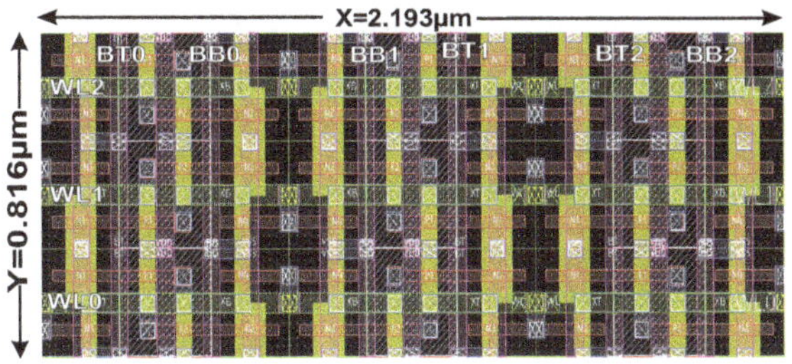

(b)

Fig. 10. Layout of 6T SRAM bitcell (a). Layout of 3×3 miniarray for 6T SRAM cell (b).

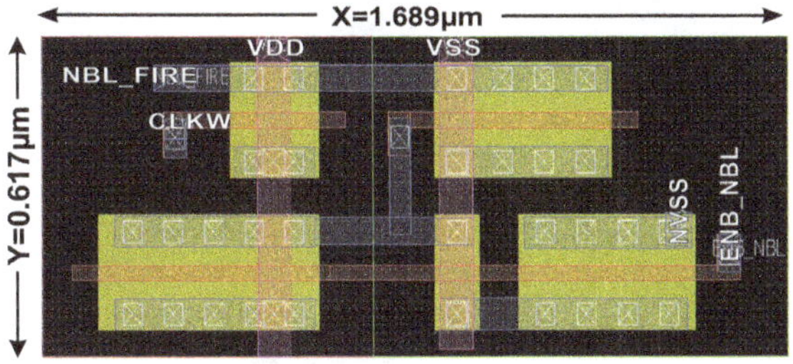

(a)

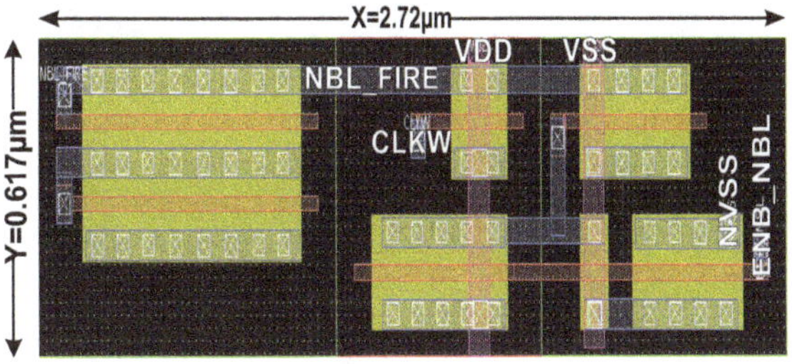

(b)

Fig. 11. Layout of Capacitive coupling signal (CCS) circuit (a) and proposed Reduced coupling signal (RCS) circuit (b).

6. Implementation and Simulation Results

In this paper, all the layouts have been carried out using Synopsys Custom Designer Layout Editor (CDLE) and verified for DRC and LVS checks using IC Validator for 28-nm CMOS technology node while the RC parasitic extraction has been done using Synopsys Tool STAR-RC for the same technology node. Layouts of 6T SRAM bitcell as well as 3×3 miniarray are shown in Fig. 10(a) and (b).

Parasitics of centred 6T-SRAM bitcell of miniarray have been deduced from the extracted netlist of miniarray. Figures 11(a) and (b) show the layouts of capacitive coupling signal (CCS) circuit and proposed reduced coupling signal (RCS) circuit respectively. Layouts of DNR and LP-DNR circuits are as shown in Fig. 12(a) and (b). In this work, the simulation setup for 32 Kb SRAM with CM=8 , is made using center decode architecture as shown in Fig. 13. To get the loading netlist for this SRAM instance each bitcell in array is replaced by its extracted netlist deduced from miniarray as discussed above. To observe the impact of proposed assist schemes with respect to no assist circuits and with assist circuits, extracted netlists for existing assist circuits, CCS and DNR have been used. For the proposed assist circuits RCS-NBL and LP-DNR, extracted netlists of the respective blocks have been used.

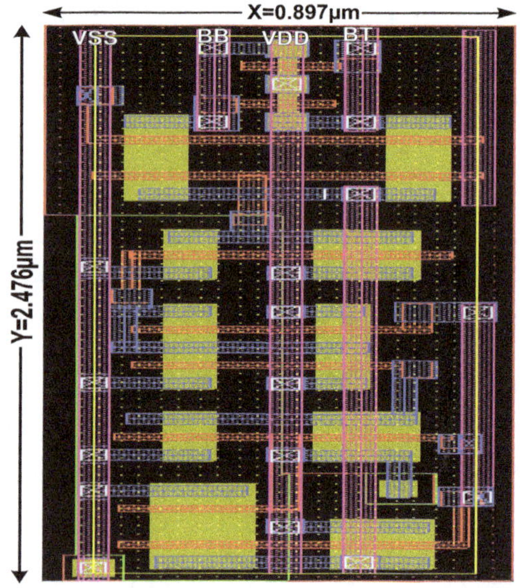

(a)

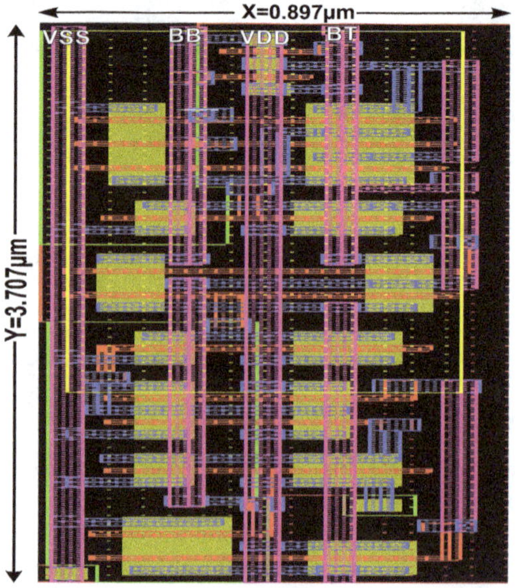

(b)

Fig. 12. Layout of DNR circuit (a). Proposed LP-DNR circuit (b).

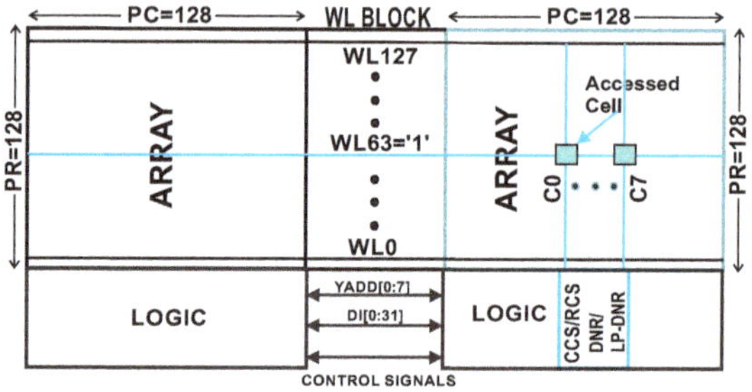

Fig. 13. Architecture used for simulation 32 Kb SRAM.

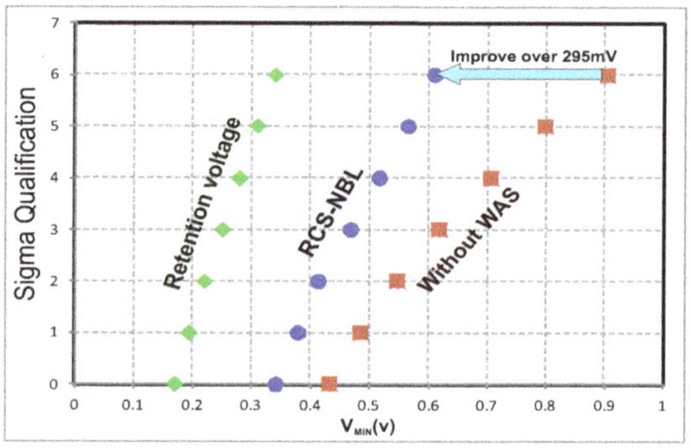

Fig. 14. Simulated SRAM V_{MIN} with process variation(σ).

Figure 14 depicts the sigma-qualification of SRAM. In this figure, red square, blue dot and green rhombus represent SRAM V_{MIN} to qualify for a particular level of process variation(σ) for no write assist scheme, with proposed write assist scheme (RCS-NBL) and for retention mode, respectively.

Power numbers along with V_{MIN} of above mentioned SRAM macro, collected with the help of simulations, for respective write assist and read assist schemes as well as without assist schemes are shown in Tab. 1 (a) and (b) respectively.

SCHEME	V_{MIN}(V)	POWER(µW)	
		DYNAMIC	STATIC
W/O WRITE ASSIST	0.905	975	22
CCS-NBL	0.61	75.2	9.82
RCS-NBL	0.61	81.3	11.3
State of the art (SCS-NBL)	0.61	76.2	9.82

(a)

SCHEME	V_{MIN}(V)	POWER(µW)	
		DYNAMIC	STATIC
W/O READ ASSIST	0.834	311	26.31
LP-DNR	0.799	286	23.20
State of the art (DNR)	0.799	334	23.21

(b)

Tab. 1. Dynamic and static power-numbers along with V_{MIN} for (a) write assist schemes (b) read assist schemes.

SCHEME	V_{MIN}[V]		DYNAMIC POWER RATIO	STATIC POWER RATIO
	W/O ASSIST	W/I ASSIST		
RCS-NBL	0.905	0.61	91.7% ↓	48.9% ↓
LP-DNR	0.834	0.799	8.1% ↓	11.7% ↓

Tab. 2. Static power-improvement with assist circuits.

7. Conclusions

We have proposed new write and read assist circuits, RCS and LP-DNR respectively, to improve SRAM V_{MIN} for 28-nm bulk CMOS technology. Simulation results with 6σ process variation for 32 Kb SRAM CM=8 macro, demonstrate that SRAM write $V_{MIN}$is lowered to 0.61 V, which is an improvement of 295 mV with respect to without any write assist scheme.

SRAM read V_{MIN} has been lowered to 0.799 V, which is an improvement of 35 mV as compared to without any read assist scheme. Table 2 summarizes the improvements in the respective SRAM V_{MIN}, dynamic power and static power consumption for this SRAM macro with the help of corresponding assist techniques.

LP-DNR scheme also improves dynamic power consumption by 16.8 % as compared to DNR scheme, while in the case of RCS-NBL scheme dynamic power consumption degrades by 6.3 % as compared with SCS-NBL write assist scheme. Over stress on PG transistor of selected bitcell as well as the stability issue of half-selected bitcells in the existing CCS-NBL scheme especially at higher operating voltages has been addressed in the proposed RCS-NBL scheme at the cost of power degradation during the short duration of clock inactive window. Area overheads of these proposed RCS-NBL and LP-DNR assist techniques for this macro are less than 0.79 % and 3.70 % respectively.

References

[1] CHEN, Y.-H., CHAN, W.-M., WU, W.-C., et al. A 16 nm 128 Mb SRAM in high-k metal-gate FinFET technology with write-assist circuitry for low-vmin applications. *IEEE Journal of Solid-State Circuits*, 2015, vol. 50, no. 1, p. 170–177. DOI: 10.1109/JSSC.2014.2349977

[2] SONG, T., RIM, W., JUNG, J., et al. A 14 nm FinFET 128 Mb SRAM with v enhancement techniques for low-power applications. *IEEE Journal of Solid-State Circuits*, 2015, vol. 50, no. 1, p. 158–169. DOI: 10.1109/JSSC.2014.2362842

[3] KUMAR, S., TIKKIWAL, V., GUPTA, H. Read SNM free SRAM cell design in deep submicron technology. In *Proceedings of the International Conference on Signal Processing and Communication*. 2013, p. 375–380. DOI: 10.1109/ICSPCom.2013.6719816

[4] NII, K., YABUUCHI, M., TSUKAMOTO, Y., et al. A 45-nm bulk CMOS embedded SRAM with improved immunity against process and temperature variations. *IEEE Journal of Solid-State Circuits*, 2008, vol. 43, no. 1, p. 180–191. DOI: 10.1109/JSSC.2007.907998

[5] GROSSAR, E., STUCCHI, M., MAEX, K., et al. Read stability and write-ability analysis of SRAM cells for nanometer technologies. *IEEE Journal of Solid-State Circuits*, 2006, vol. 41, no. 11, p. 2577–2588. DOI: 10.1109/JSSC.2006.883344

[6] YAMAOKA, M., OSADA, K., ISHIBASHI, K. 0.4-V logic-library-friendly SRAM array using rectangular-diffusion cell and delta-boosted-array voltage scheme. *IEEE Journal of Solid-State Circuits*, 2004, vol. 39, no. 6, p. 934–940. DOI: 10.1109/JSSC.2004.827796

[7] ZHANG, K., BHATTACHARYA, U., CHEN, Z., et al. A 3-GHz 70-mb SRAM in 65-nm CMOS technology with integrated column-based dynamic power supply. *IEEE Journal of Solid-State Circuits*, 2006, vol. 41, no. 1, p. 146–151. DOI: 10.1109/JSSC.2005.859025

[8] CHANDRA, V., PIETRZYK, C., AITKEN, R. On the efficacy of write-assist techniques in low voltage nanoscale SRAMs. In *Proceedings of the Design, Automation & Test in Europe Conference & Exhibition (DATE)*. 2010, p. 345–350. DOI: 10.1109/DATE.2010.5457179

[9] TAKEDA, K., HAGIHARA, Y., AIMOTO, Y., et al. A read-static-noise-margin-free SRAM cell for low-VDD and high-speed applications. *IEEE Journal of Solid-State Circuits*, 2006, vol. 41, no. 1, p. 113–121. DOI: 10.1109/JSSC.2005.859030

[10] CHANG, L., FRIED, D. M., HERGENROTHER, J., et al. Stable SRAM cell design for the 32 nm node and beyond. In *Proceedings of the Digest of Technical Papers, Symposium on VLSI Technology*. 2005 , p. 128–129. DOI: 10.1109/.2005.1469239

[11] LIU, Z., KURSUN, V. Characterization of a novel nine-transistor SRAM cell. *IEEE Transactions on Very Large Scale Integration (VLSI) Systems*, 2008, vol. 16, no. 4, p. 488–492. DOI: 10.1109/TVLSI.2007.915499

[12] SHIBATA, N., KIYA, H., KURITA, S., et al. A 0.5-V 25-MHz 1-mW 256-kb MTCMOS/SOI SRAM for solar-power-operated portable personal digital equipment - sure write operation by using step-down negatively overdriven bitline scheme. *IEEE Journal of Solid-State Circuits*, 2006, vol. 41, no. 3, p. 728–742. DOI: 10.1109/JSSC.2005.864124

[13] KUMAR, S., SAHA, K., GUPTA, H. Run time write detection in SRAM. In *Proceedings of the International Conference on Signal Processing and Communication*. 2015, p. 328–333. DOI: 10.1109/ICSPCom.2015.7150671

[14] NII, K., YABUUCHI, M., TSUKAMOTO, Y., et al. A 45-nm singleport and dual-port SRAM family with robust read/write stabilizing circuitry under DVFS environment. In *Proceedings of the IEEE Symposium on VLSI Circuits*. 2008, p. 212–213. DOI: 10.1109/VLSIC.2008.4586011

[15] TAKEDA, K., SAITO, T., ASAYAMA, S., et al. Multi-step word-line control technology in hierarchical cell architecture for scaled-down high-density SRAMs. In *Proceedings of the IEEE Symposium on VLSI Circuits*. 2010, p. 101–102. DOI: 10.1109/VLSIC.2010.5560336

[16] ZIMMER, B., TOH, S. O., VO, H., et al. SRAM assist techniques for operation in a wide voltage range in 28-nm CMOS. *IEEE Transactions on Circuits and Systems II: Express Briefs*, 2012, vol. 59, no. 12, p. 853–857. DOI: 10.1109/TCSII.2012.2231015

[17] TUINHOUT, H. Impact of parametric fluctuations on performance and yield of deepsubmicron technologies. In *Proceedings of the 32nd European Solid-State Device Research Conference*. 2002, p. 95–102. DOI: 10.1109/ESSDERC.2002.194879

[18] ASENOV, A., SAINI, S. Suppression of random dopant-induced threshold voltage fluctuations in sub-0.1- μm MOSFET's with epitaxial and δ-doped channels. *IEEE Transactions on Electron Devices*, 1999, vol. 46, no. 8, p. 1718–1724. DOI: 10.1109/16.777162

About the Authors . . .

Satyendra KUMAR was born in Muzaffarnagar, India. He obtained his B.E. and M.Tech. degrees from Indian Institute of Technology, Roorkee, India, in 1998 and 2002 respectively. He is currently working as Assistant Professor in the Department of Electronics and Communication Engineering at Jaypee Institute of Information Technology, Noida, India. Previously, he has worked in the semiconductor industry on a number of projects for development and characterization of 90nm, 65nm and 55 nm memory compilers for different foundries like TSMC, IBM, UMC and others. His research interests include SRAM Design For Low Power Applications.

Kaushik SAHA was born in Kolkata, India. He received his B.Tech., M.Tech. and Ph.D. degrees from Indian Institute of Technology, Delhi, India, in 1987, 1989 and 1996 respectively. In 1996 he joined STMicroelectronics Ltd., as a designer of semiconductor memories. At present he is Chief Technology Officer, Advanced Software Team, Samsung R&D Institute India, Delhi. His research interests are in the areas of Advanced Processor Architectures and Low voltage SoC design. He is also associated with the Indian Institute of Technology, Delhi, India, in the capacity of Adjunct Faculty, teaching advanced topics in VLSI design. He has published a number of papers in reputed journals.

Hariom GUPTA was born in Agra, India. He obtained his B.E. in Electrical Engineering from the Government Engineering College, Jabalpur, India. He received his M.E. and Ph.D. from Indian Institute of Technology, Roorkee in 1975 and 1980, respectively. Presently he is working as Director, Jaypee Institute of Information Technology, Noida, India. He has published over 300 research papers and 35 technical reports. His total citations are over 2300. He has supervised 31 Ph.Ds.

8

Guidelines on the Switch Transistors Sizing Using the Symbolic Description for the Cross-Coupled Charge Pump

Jan MAREK, Jiri HOSPODKA, Ondrej SUBRT

Dept. of Circuit Theory, Czech Technical University in Prague, Technická 2, 166 27 Praha, Czech Republic

{marekj20, hospodka, subrto}@fel.cvut.cz

Abstract. *This paper presents a symbolic description of the design process of the switch transistors for the cross-coupled charge pump applications. Discrete-time analog circuits are usually designed by the numerical algorithms in the professional simulator software which can be an extremely time-consuming process in contrast to described analytical procedure. The significant part of the pumping losses is caused by the reverse current through the switch transistors due to the continuous-time voltage change on the main capacitors. The design process is based on the analytical expression of the time response characteristics of the pump stage as an analog system with using BSIM model equations. The main benefit of the article is the analytical transistors sizing formula so that the maximum voltage gain is achieved. The diode transistor is dimensioned for the pump requirements, as the maximal pump output ripple voltage, current, etc. The characteristics of the proposed circuit have been verified by simulation in ELDO Spice. Results are valid for N-stage charge pump and also applicable for other model equations as PSP, EKV.*

Keywords

Time response characteristics, reverse current, cross-coupled charge pump, BSIM model, high-voltage

1. Introduction

Charge pumps are switched-capacitor circuits that transport charge between main capacitors to create a higher output voltage. They are used to supply low-power circuits that require relatively high input voltage, for example, EEPROM memories.

The advanced architectures of the modern integrated two–phase charge pump are based on the elimination of the threshold voltage of the active components (Dickson charge pump [1]), that decreases all node voltages. The static charge pumps [2–4] realize the charge transport through the switch transistors, which are controlled by the output voltage from the next stage as is shown in Fig. 1. When the logic levels of the two-phase clock signal are $\phi = 0$ (low) and $\bar{\phi} = V_{DD}$ (high) and assuming the correct function of the charge pump,

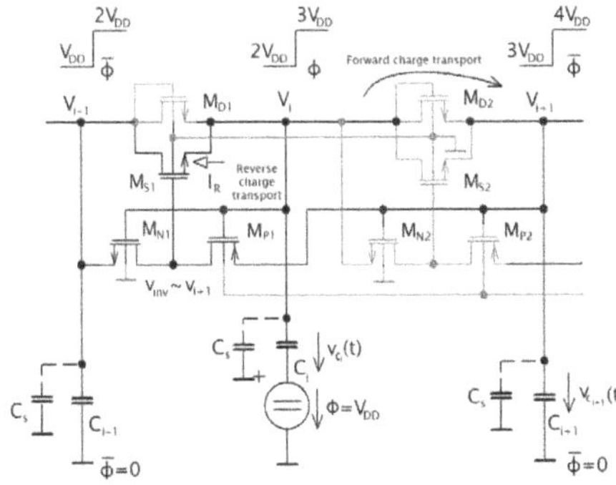

Fig. 1. Reverse current through the switch transistor in the cross-coupled charge pump.

then switch transistor M_{S1} is fully ON, node voltages V_{i-1} and V_{i+1} are pumped up to $2V_{DD}$ and $3V_{DD}$ respectively, and main capacitor connected to node i is charged to $2V_{DD}$. The voltage drop between two nodes is theoretically determined by the saturation voltage of the switch (MOSFET) at the end of the charge transport. Conversely, when $\phi = V_{DD}$ and $\bar{\phi} = 0$, M_{S1} must be OFF [4], so that capacitor connected to node i can be pumped up to $2V_{DD}$ and forward charge transport between nodes i and $i+1$ can be realized. The problem of the reverse current occurs at the same time ($\phi = V_{DD}$ and $\bar{\phi} = 0$) because the voltage difference $v_{inv} - v_{i-1}$ may be higher than threshold voltage $V_{TH_{MS1}}$. Now, the MOSFET electrodes Drain and Source are mutually exchanged and M_{S1} is not OFF in spite of the expectation.

The discharge–reverse current i_R, which flows through the switch is undesirable because it decreases the pump voltage gain. The reverse current of the diode transistor is practically zero. The cross–coupled charge pump contains the inverter that controls the switch at the time intervals defined by the clock signal. This topology allows achieving higher efficiency compared with the static charge pump. However, the problem of reverse current still exists [4], [5].

Simulation results show a strong dependence of the pump voltage gain on the strength of the switch transistor(s). Design of discrete-time analog circuits including charge pump circuits represent the fundamental problem, which relates to the solution of the part steps of the design algorithm. The following three key steps are necessary for a successful design: circuit model, simulation and evaluation of the simulation results. Only transient analyses are allowable. It is a fundamental difference of approach compared to analog circuits [5]. The experimental part including simulation of real properties of the cross-coupled charge pump [5] and their comparison with other architectures (Dickson charge pump, a static charge pump) has been done [1–4], [6], [7]. However, a design process of the circuit has not been known yet. General description methods of discrete-time analog circuits have been published in many books and research papers [5], [8], [9]. Well-known description methods are insufficient because they do not consider the relevant properties (nonidealized structure), that are typical for the behavior of the charge pumps. Optimization is usually circuitous process due to many iterations to achieve of the required parameters (static, dynamic). The different access to solve this task will be offered in this article. The symbolic description of the design process of the cross-coupled charge pump stage as an analog block for high-voltage application will be discussed to find an analytical expression for width and length of the diode and switch transistors so that the voltage gain of the N-stage pump will be maximal. The pump elements (switch, diode, capacitors,..) usually have same parameters at all pump stages. Long channel MOSFET is provided due to high bias voltages (drain-source) in the circuit. Sizing of the switch transistor will be designed so that the reverse current will be suppressed. Sizing of the diode transistor is related to the optimization of pump parameters such as maximation of the load current, minimalization of the output ripple voltage.

The DC characteristics of the pump stage will be firstly found. Because the transistors are operating in strong inversion region, the simplified BSIM model [10] can be used for this purpose. The main part contains an analytical description of the time response characteristics, which are applied in the real circuit. The switch transistors ratio W_s/L_s is set, so that their equivalent resistance value is a compromise between the charge/forward and discharge/reverse current. The ratio is calculated for the worst case of bias voltages because this resistance is nonlinear. The equivalent diode resistance is determined by the change of the pump output voltage. The derived formulas are verified by simulation in ELDO Spice. The effort is to find mentioned solution without using the numerical optimization procedure. The created model including the dominant real properties points to an alternative way to N-stages charge pump draft (static, dynamic parameters). The strong inversion operating region of the MOSFET is expected, in which the behavior of the MOSFET models is correct [11] compared with the real measured curves (BSIM, EKV, PSP, etc.) in the specified technology process.

2. The Static Model of the Pump Stage

One stage of the cross-coupled charge pump is shown in Fig. 2. The drain current of the each MOSFETs is controlled by the input voltage V_{in}. Adjustable DC source voltage is used for analysis instead of the main capacitor in real circuits. All other DC voltages in the diagram are referenced to the ground.

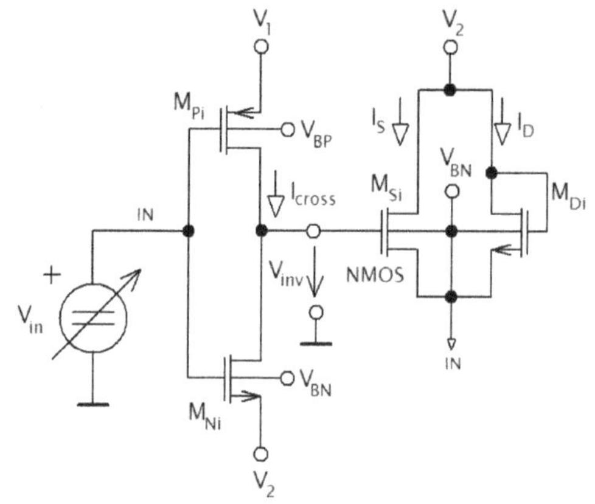

Fig. 2. Diagram of the cross-coupled charge pump stage.

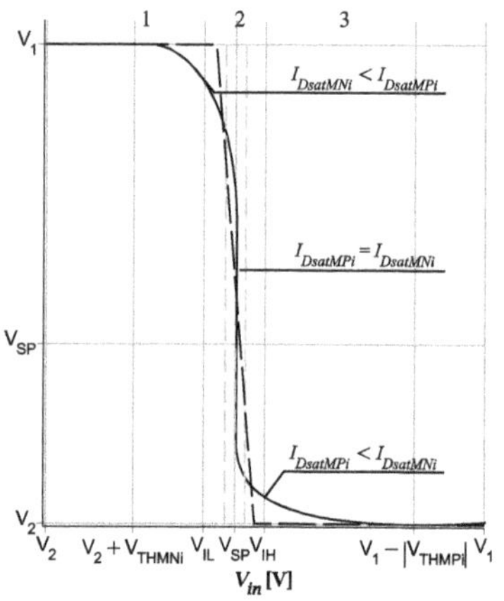

Fig. 3. The voltage transfer characteristics of the CMOS inverter for long channel MOSFET and its linearization.

It is supposed that all transistors are operating in strong inversion region. Hence, the power supply range of the inverter must be adequately high to turn on both of the transistors M_{Ni} and M_{Pi} in the interval $V_{in} \in \langle V_{IL}, V_{IH} \rangle$, see Fig. 3. Pump voltage gain of two stages labeled $G_v = V_1 - V_2$, must be greater than the sum of the threshold voltages of these transistors, labeled $V_{\mathrm{TH}_{\mathrm{MNi}}}$, $V_{\mathrm{TH}_{\mathrm{MPi}}}$. The switch transistor must be ON, when the output inverter is at a high logic level, i.e. $v_{inv} = V_1$,

$$V_{\mathrm{TH}_{\mathrm{MSi}}} < G_v > V_{\mathrm{TH}_{\mathrm{MNi}}} + |V_{\mathrm{TH}_{\mathrm{MPi}}}| \,. \tag{1}$$

The output voltage of the inverter $V_{inv} = f(V_{in})$ is setting the drain current of the switch transistor M_{Si}, labeled I_S. The CMOS inverter voltage transfer characteristic [12–14] is derived based on the fact that the drain current of both MOSFETs must be equal for each of the operating regions.

The complex expression of the voltage transfer characteristics is not necessary for the practical results. Considering the electrical field in structure is much less than critical electrical field [10], then "long channel" can be defined as

$$L_{\mathrm{eff}} \gg V_{\max} \frac{\mu_{\mathrm{eff}}}{2 v_{\mathrm{sat}}}, \tag{2}$$

where L_{eff} is the effective channel length [10], [12], $V_{\max}$ is the maximal bias voltage (drain-source, gate-source), μ_{eff} is effective mobility and v_{sat} is the saturation velocity [10].

Then, the transfer part of the characteristics is well linearized, as it is shown in Fig. 3. Input voltages between the limit values V_{IL} and V_{IH} do not define the valid output logic level [12]. Inverter cross current I_{cross} is the maximal in the switching point when $V_{IN} = V_{inv}$ [12]. Analytical equation of V_{SP} is derived in [5]. Now, an analytical estimation of the voltage transfer characteristics has the following form:

$$V_{out}(V_{in}) \approx \begin{cases} V_1, & V_{in} \le V_{IL}, \\ \frac{V_1 - V_2}{V_{IL} - V_{IH}}(V_{IN} - V_{IL}) + V_1, & V_{IH} > V_{in} > V_{IL}, \\ V_2, & V_{in} \ge V_{IH}. \end{cases} \tag{3}$$

The slope of the transition part is determined by the Early voltage of MOSFETs, V_{ADIBL}, which is proportional to the voltage V_{GS} and relationship only contains the model parameters of PDIBLC2 and PDIBLCB [10]. So that, difference $V_{IH} - V_{IL}$ is very small, $V_{IL} \rightarrow V_{SP}$, $V_{IH} \rightarrow V_{SP}$ and the wide of the transition part is negligibly small (zero in the ideal case).

The voltage on the capacitor is changing continuously from 0 to the supply voltage V_2 in the passive time interval and the voltage at the terminal "IN" may be theoretically doubled in the active interval of the clock signal, i.e. $V_{IN} = 2V_2$. Thus, the drain current I_S through the switch transistor M_{Si} (and its orientation) will be analyzed in the interval of the input voltage $V_{in} \in \langle 0, V_1 \rangle$ (the D,S pins of the switch transistor are not distinguished in the scheme). The direction of the forward–"charging" current (that is required) matches the orientation in the scheme. The control voltages configuration for the setting both the forward (I_{S_F}) and reverse current of the drain (I_{S_R}) of the M_{Si} transistor are shown in Tab 1.

Parameter	Value	
I_S	> 0	≤ 0
V_{in}	$V_{in} \in \langle 0, V_2 \rangle$	$V_{in} \in \langle V_2, V_1 \rangle$
$V_{DS_{MSi}}$	$V_2 - V_{in}$	$V_{in} - V_2$
$V_{GS_{MSi}}$	$V_{inv} - V_{in} = V_1 - V_{in}$	$V_{inv}(V_{in}) - V_2$
$V_{SB_{MSi}}$	$V_{in} - V_{BN}$	$V_2 - V_{BN}$

Tab. 1. M_{Si} transistor control voltages configuration for the setting of I_S.

Respecting the condition (1), transistor M_{Si} is always ON in the interval $V_{in} \in \langle 0, V_2 \rangle$. The bulk both of the M_{Ni} and M_{Si} transistors is connected to the same bias voltage V_{BN} (usually to the ground) and $V_{SB_{MSi}} \le V_{SB_{MNi}}$, then $V_{TH_{MSi}}(V_{SB}) \le V_{TH_{MNi}}(V_{SB})$ in the same technology process. Moreover, this transistor is operating in the triode region, as it is shown bellow. In this case, the condition $V_{DS} < V_{DS_{sat}}$ is valid, where the saturation voltage is calculated from [10], [15]

$$V_{DS_{sat}} = \frac{V_{GS} - V_{TH}}{A_{bulk}(V_{GS}, V_{SB})}. \tag{4}$$

Substituting the specific values from Tab. 1 into the $V_{DS_{sat}}$ expression, following inequality is obtained:

$$V_2 < \frac{V_1 - V_{in} - V_{TH_{MSi}}(V_{in}, V_{BN})}{A_{bulk_{MSi}}(V_1, V_{in}, V_{BN})} + V_{in}. \tag{5}$$

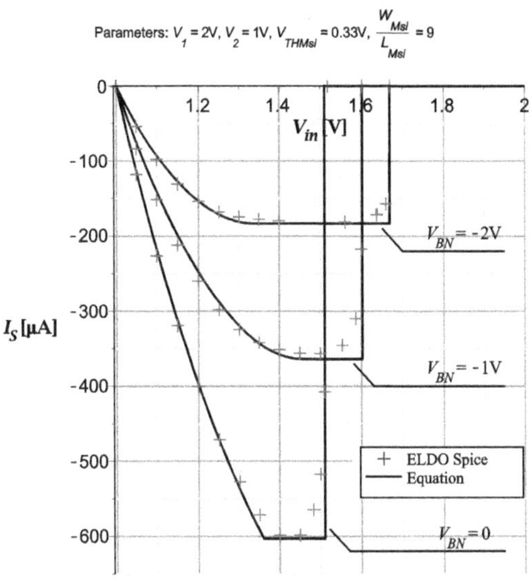

Fig. 4. Reverse current of the switch transistor vs. input voltage.

Expression on the right hand sight of (5) must also satisfy the condition (1). Considering the worst case of the threshold voltage, $V_{TH_{MSi}} = V_{TH_{MNi}}$, then

$$|V_{TH_{MPi}}| + V_{in} > \underbrace{\left(1 - \frac{1}{A_{bulk_{MSi}}}\right)}_{\le 0}(V_1 - V_{TH_{MSi}} - V_{in}). \tag{6}$$

Saturation voltage $V_{DS_{sat}}$ can be approximated by the function $V_{GS} - V_{TH}$ near the point $V_{GS} = V_{TH}$ (long channel MOSFET is provided). However, real saturation voltage is greater than function expressed in (4) for higher voltage V_{GS}, $V_{DS_{sat}} > V_{GS} - V_{TH}$, i.e. $A_{bulk} < 1$, for $V_{GS} \gg V_{TH}$. Subsequently, the inequality (6) is always true.

The drain current direction is changed, and it is controlled by the constant gate-source voltage $V_1 - V_2$, while $V_2 < V_{in} \le V_{IL}$. The gate-source voltage decreasing quickly in the interval $\langle V_{IL}, V_{IH} \rangle$, while a change of the drain-source voltage is negligible. Hence, the drain current achieves the maximal value at point V_{IL} and transistor is abruptly switched off after exceeding the switching point. Neglecting the transition part of the inverter transfer characteristic, drain current can be considered the constant in the interval $V_{in} \in \langle V_{IL}, V_{SP} \rangle$. Total current I_S is given by the following formula:

$$I_S(V_{in}) \approx \begin{cases} I_{DS0_F}, & V_{in} \in \langle 0, V_2 \rangle, \\ I_{S_R}, & V_{in} \in \langle V_2, V_{SP} \rangle, \\ 0, & V_{in} \in \langle V_{SP}, V_1 \rangle. \end{cases} \tag{7}$$

Current I_{S_R} is calculated on the basis of the two following cases:

- if $V_{IL} > V_2 + V_{DSsat_{MSi}}$, then

$$I_{S_R}(V_{in}) \approx \begin{cases} I_{DS0_R}, & V_{in} \in \langle V_2, V_2 + V_{DS_{sat}} \rangle \\ I_{D_{sat0R}}, & V_{in} \in \langle V_2 + V_{DS_{sat}}, V_{SP} \rangle, \end{cases} \tag{8}$$

- if $V_{IL} \le V_2 + V_{DSsat_{MSi}}$, then

$$I_{S_R}(V_{in}) \approx \begin{cases} I_{DS0_R}, & V_{in} \in \langle V_2, V_{IL} \rangle \\ I_{DS0_R}|_{V_{GS}=V_1-V_2, V_{DS}=V_{IL}-V_2}, & V_{in} \in \langle V_{IL}, V_{SP} \rangle, \end{cases} \tag{9}$$

where I_{DS0} is the drain current in triode region and $I_{D_{sat0}}$ is the drain current in saturation region at $V_{DS} = V_{DS_{sat}}$. The reverse current waveform for the both cases is shown in Fig. 4. The source-bulk voltage is the parameter.

The drain current of the M_{Di} transistor is zero in the reverse configuration due to shorted gate and source electrodes,

$$I_D(V_{in}) = \begin{cases} I_{Dsat0}, & V_{in} \in \langle 0, V_2 - V_{TH_{MDi}} \rangle, \\ 0, & \text{otherwise.} \end{cases} \quad (10)$$

3. Time Response Characteristics

Step response is a typical characteristic situation in the switched-capacitor circuits. Step response characteristics of the circuits are shown in Fig. 6 and 8. The time-varying voltage on the main capacitor to the clock signal will be found for both the forward and reverse configuration to determining pumping losses. The extreme values of the bias voltage have been chosen for the following optimization process. The time domain method must be used for the calculation due to the nonlinearity behavior of this system. It is also necessary to define the next conditions for the analysis process:

- parasitic capacitances are negligibly small compared with the main pumping capacitors, $C_s \ll C_i$.
- rise time and fall time delay of the clock signal and propagation delays of the inverter are very short compared to the charge/discharge time of the main capacitors.
- leakage currents of all the components are neglected.
- settling time of the switches is zero.

The main capacitor is charged, when the gate of the switch transistor is connected to high output voltage level of the inverter V_{inv} = "H" = V_1, the drain is connected to the input stage voltage V_2 and the main capacitor is connected to ground. This situation is shown in Fig. 5.

When the switches S_1, S_2 and S_3 are ON at $t = 0$, the current flowing through the capacitors i_{c_F} is supplied both of the transistor until the capacitor voltage does not exceed the value V_{0_F} at time t_{0_F}, see Fig. 5. Total current i_{c_F} is given by

$$i_{c_F}(t) = \begin{cases} i_s(t) + i_d(t), & \text{for } 0 < t \le t_{0_F}, \\ i_s(t), & \text{for } t > t_{0_F}, \\ 0, & \text{otherwise.} \end{cases} \quad (11)$$

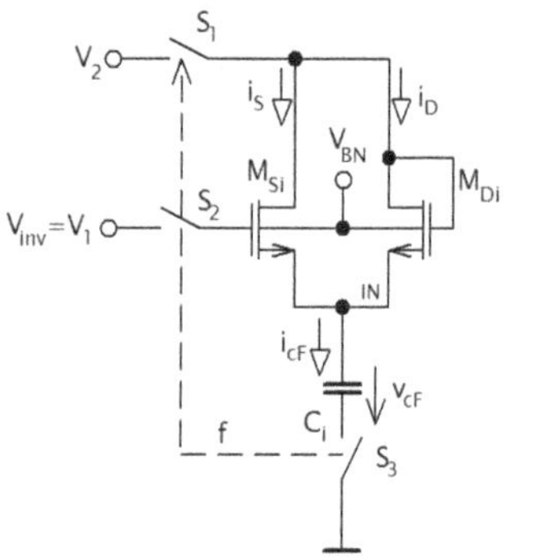

Fig. 5. Configuration for the charge of the main capacitor.

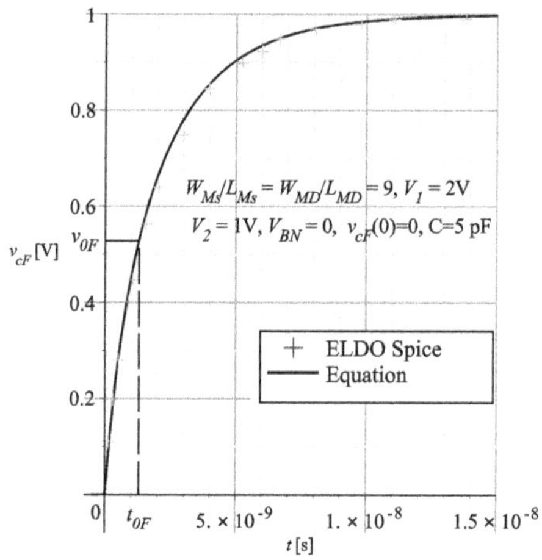

Fig. 6. Time response characteristics of the circuit from Fig. 5.

The voltage on the capacitor is equal to V_2 in steady state and the particular value of the voltage V_{0_F} can be derived from

$$V_{0_F} = V_2 - V_{TH_{MDi}}(v_{SB}) \quad (12)$$

where source-bias voltage is equal to v_{c_F} ($V_{BN} = 0$). Dependence of the threshold voltage on the V_{SB} voltage (body effect [10], [12], [13], [15]) is given by

$$V_{TH} = V_{TH0} + K_{1ox}\sqrt{\phi_s - V_{BS}} - K_1\sqrt{\phi_s} - K_{2ox}V_{BS} \quad (13)$$

where V_{TH0} is threshold voltage at zero bias voltages, ϕ_s is the surface potential and K_1, K_2 are body effect coefficients (model parameters). Combining (12) and (13), the instantaneous value of the voltage in which the transistor M_{Di} will be OFF, is calculated from

$$V_{0_F} = \frac{V_2 + K_1 - V_{TH0_{MDi}}}{K_{2ox} + 1} + \frac{1}{2}\frac{K_{1ox}\left(K_{1ox} - \sqrt{\gamma}\right)}{(K_{2ox} + 1)^2} \quad (14)$$

where

$$\gamma = 4\phi_s (K_{2ox} + 1)^2 + 4 (K_{2ox} + 1)\psi,$$
$$\psi = \left(K_1\sqrt{\phi_s} + V_2 - V_{TH0_{MDi}} + K_{1ox}^2\right).$$

Substituting the voltage v_{cF} in the static model for V_{in} and using equations for the drain current [10], [12], [13], [15], time response characteristic is found by the solving of following differential equation

$$\int \frac{C}{i_{cF}} \mathrm{d}v_c = t + \mathrm{IC} \quad (15)$$

with the initial condition $v_c(t_0) = v_{c_0}$ for each of the intervals, as it is shown in (11). The drain current equation is the composite function (NF) in the form

$$i_c = \mathrm{f}\left[v_{ds}(t), v_{gs}(t), v_{TH}(v_c(t)), A_{bulk}(v_c(t)), \mu_{eff}(v_c(t))\right],$$

consequently, the analytical solution would be unreasonably complicated for practical design. Thus, the estimation is done providing the constant nested functions v_{TH}, A_{bulk} and μ_{eff}

according this criteria: When $t \leq t_{0_F}$, the voltage v_{C_F} change in time is approximately same as at the beginning of the transient process. Contrariwise, when $t > t_{0_F}$ and $i_D = 0$ the characteristic curve is approximated by the nested function values which would acquire in the steady state.

The same principle is also used for the reverse configuration, as it is shown in Fig. 7. Bias voltages are listed in Tab. 2.

Condition		Index of NF.		$\mathbf{V_{in}[V]}$	
		M_{D_i}	M_{S_i}	M_{D_i}	M_{S_i}
$i_s > 0$	$t < t_{0_F}$	D0	S0	$v_{C_F}(0_+)$	$v_{C_F}(0_+)$
	$t \geq t_{0_F}$	X	S	X	V_2
$i_s < 0$	$t < t_{0_R}$	X	SR	X	$v_{C_R}(0_+)$
	$t \geq t_{0_R}$	X	S	X	V_2

Tab. 2. Bias voltages of the nested functions (NF) V_{TH}, A_{bulk} and u_{eff}.

Therefore, solving of (15) can be only found by integrating the voltage square $[v_{gs}(t) - V_{TH}]^2$, eventually $v_{ds}(t)$ and $v_{ds}^2(t)$ for triode region. The time-varying voltage v_{C_F} for the forward configuration is given by

$$v_{C_F}(t) = \begin{cases} V_{C_F}(0), & \text{for } t \leq 0 \\ \dfrac{\sqrt{C_1 C_2} \cdot \tan\left[\frac{(t+\mathrm{IC_{F1}})\sqrt{C_1 C_2}c_{oxe}}{2L \cdot C A_{bulk_{D0}}}\right] - C_4}{\mu_{eff_{D0}} W_{M_S} + C_1\left(A_{bulk_{S0}} - 2\right)}, & \text{for } 0 < t < t_{0_F} \\ \dfrac{C_6 \mathrm{e}^{\frac{t+\mathrm{IC_{F2}}}{C_7}} - V_2}{\mathrm{e}^{\frac{t+\mathrm{IC_{F2}}}{C_7}}(A_{bulk_S} - 2) - 1}, & \text{for } t \geq t_{0_F}, \end{cases} \tag{16}$$

where L is channel length and c_{oxe} is electrical oxide capacitance. Integration constants, labeled $\mathrm{IC_1}$ and $\mathrm{IC_2}$, are generally calculated from the initial conditions that are substituted into (15) – Cauchy's equation:

$$\mathrm{IC_1} = \frac{2LCA_{bulk_{D0}}}{\sqrt{C_1 C_2} c_{oxe}} \arctan(\zeta), \tag{17}$$

$$\mathrm{IC_2} = C \frac{V_{DSsat_{Ms}}}{I_{Dsat0_{Ms}}}|_{V_{in}=V_2} \ln(|\lambda|) - t_0, \tag{18}$$

where $\zeta = \frac{v_{c_0}\left[\mu_{eff_{D0}} W_{M_S} + C_1\left(A_{bulk_{S0}} - 2\right) + C_4\right]}{\sqrt{C_1 C_2}}$, $\lambda = \frac{A_{bulk_S}\left[V_2 - v_{c_0}\right] - 2\left[V_1 - V_{TH_S} - v_{c_0}\right]}{V_2 - v_{c_0}}$, $\mathrm{IC_{F1}} = \mathrm{IC_1}|_{v_{c_0} = V_{C_F}(0_+)}$, for $0 < t < t_0$ and $\mathrm{IC_{F2}} = \mathrm{IC_2}|_{t_0 = t_{0F}, v_{c_0} = V_{C_F}(0_+)}$, for $t \geq t_0$.

Using the voltage V_{0_F} in (18), the initial time t_{0_F} is given by

$$t_{0_F} = \mathrm{IC_{F1}}|_{v_{c_0} = V_{0_F}} - \mathrm{IC_{F1}}|_{v_{c_0} = V_{C_F}(0_+)}, \quad V_{C_F}(0_+) < V_{0_F}. \tag{19}$$

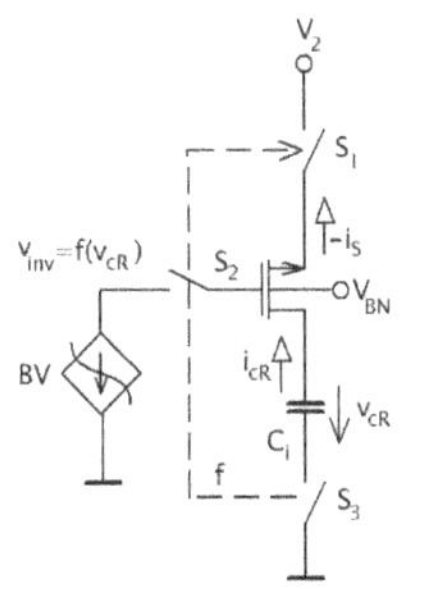

Fig. 7. Configuration for the discharge of the main capacitor

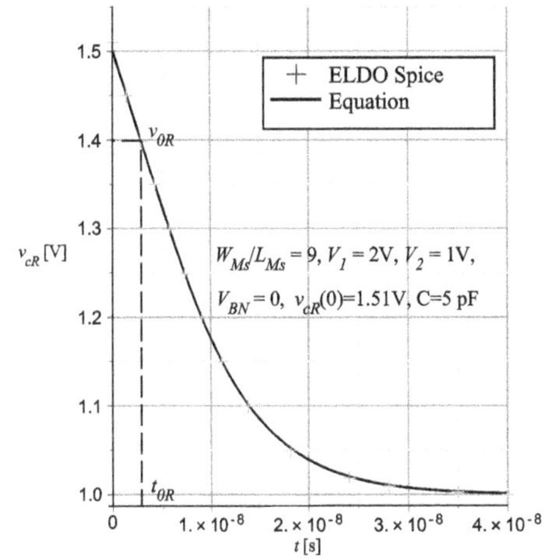

Fig. 8. Time response characteristics of the circuit from Fig. 7.

Coefficients C_1, C_2, C_3, C_4 and C_5 are calculated from:

$$\begin{aligned} C_1 &= -A_{bulk_{D0}} \mu_{eff_{S0}} W_{M_S}, \\ C_2 &= -C_1 (V_1 - V_2 - V_{TH_{S0}})^2 - C_3, \\ C_3 &= \mu_{eff_{S0}} W_{M_D} (2V_1 - 2V_2 - A_{bulk_{S0}} V_{TH_{S0}}), \\ C_4 &= -C_1 (A_{bulk_{S0}} V_2 - V_1 - V_2 + V_{TH_{S0}}) - C_5, \\ C_5 &= \mu_{eff_{D0}} W_{M_D} (V_2 - V_{TH_{D0}}), \\ C_6 &= A_{bulk_S} V_2 - 2(V_1 - V_{TH_S}), \\ C_7 &= -C \frac{V_{DSsat_{MS}}}{I_{Dsat0_{MS}}}|_{V_{in}=V_2}. \end{aligned}$$

Discharge of the main capacitor is shown in Fig. 7. The CMOS inverter is modeled by the voltage source BV controlled by the time-varying voltage v_{CR}. The switch transistor is ON after the switches S_1, S_2, S_3 are closed at $t = 0$ and the capacitor C_i was charged on the value in the interval of the voltages $v_{C_R}(0) \in (V_2, V_{SP})$.

The initial condition $v_{C_R}(0) \in (V_{IL}, V_{SP})$ will be considered to a complete description of time response characteristics. Then, the main capacitor is firstly discharged by the constant current I_S until the voltage of BV achieves V_1 at time $t = t_{0_R}$,

$$i_{C_R}(t) = \begin{cases} I_{S_R}, & \text{for } 0 < t \leq t_{0_R} \\ i_s(t), & \text{for } t \geq t_{0_R}. \end{cases} \tag{20}$$

Value of the constant current I_{S_R} for $V_{in} = v_{C_R}(0_+)$ follows from the static model, parameters for $i_s(t)$ are mentioned in Tab. 2 (index SR).

The default differential equation for each of the time interval is the same as in the previous case,

$$v_{C_R}(t) = \begin{cases} v_{C_R}(0), & \text{for } t \leq 0, \\ -\frac{I_{S_R}}{C} t + v_{C_R}(0), & \text{for } 0 < t \leq t_{0_R}, \\ \dfrac{V_2\left(A_{bulk_S} - \mathrm{e}^{\frac{t+\mathrm{IC_R}}{C_5}}\right) + 2\left(V_1 - V_2 - V_{TH_S}\right)}{A_{bulk_S} - \mathrm{e}^{\frac{t+\mathrm{IC_R}}{C_5}}}, & \text{for } t > t_{0_R}. \end{cases} \tag{21}$$

Because the voltages are equal to V_2 in steady state for both the configurations, the coefficients in the exponential functions are also the same. The integration constant IC_R can be easily expressed as

$$IC_R = IC_2|_{v_{C0}=V_{C_R}(0),t_0=t_{0_R}} \tag{22}$$

and the point t_{0_R} is given by

$$t_{0_R} = \begin{cases} \frac{C}{I_{S_R}}\left[v_{C_R}(0) - V_{IL}\right], & \text{for } V_{SP} \geq v_{C_R}(0) > V_{IL}. \\ 0, & \text{otherwise.} \end{cases} \tag{23}$$

4. Minimization of the Pumping Losses

The sizing of the switch transistor will be discussed in this part. The main criterion of the optimal pump design is based on the maximum voltage gain at the end of each phase of the clock signal, as it is shown in Fig. 9. Sizing of the switch transistor M_{Si} can be set, so that the voltage gain is greater than $2V_2$, better V_{max} at point $T/2$. The transistor length is determined from condition (2) and the width is determined based on the following condition:

$$\max\{V_{C_F}(W_{MS}) + V_{C_R}(W_{MS})\}|_{t=T/2,t>t_0}. \tag{24}$$

The optimal width $W_{MS_{opt}}$ will be searched while using the limit initial conditions to satisfy the worst case that can be taken into account in the real circuit. Using the condition (24) and (16), (21) then the following equality is true:

$$\frac{dV_{C_F}}{dW_{MS}}|_{t=T/2,t>t_{0_F}} = -\frac{dV_{C_R}}{dW_{MS}}|_{t=T/2,t>t_{0_R}},$$
$$v_{C_F}(0_+) = 0,\ v_{C_R}(0_+) = V_{SP}, \tag{25}$$

and it is giving desired value of the width at the known clock frequency. However, the optimal point can be estimated even in a simpler way. Both the time response characteristics $v_{C_F}(t)$ and $v_{C_R}(t)$ in the intervals $t > t_{0_F}$ and $t > t_{0_R}$ are compared to each other via its linearization in the initial time, as it is shown in Fig. 10.

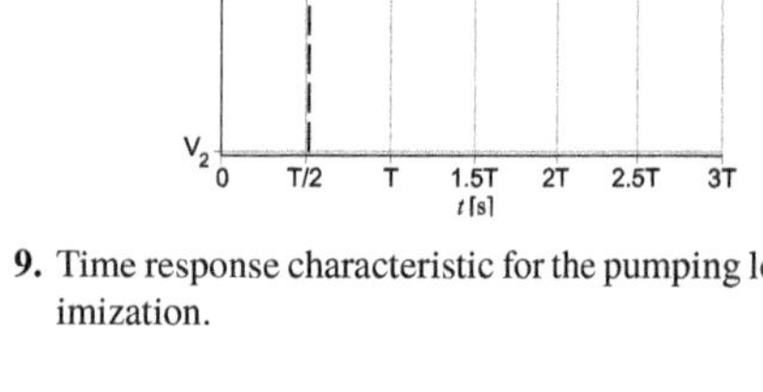

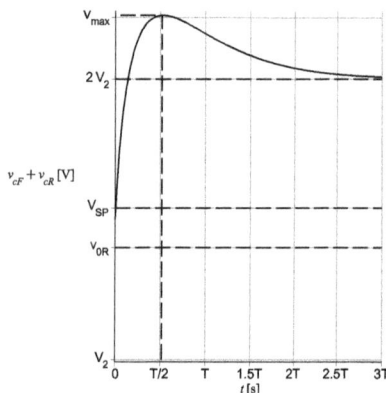

Fig. 9. Time response characteristic for the pumping losses minimization.

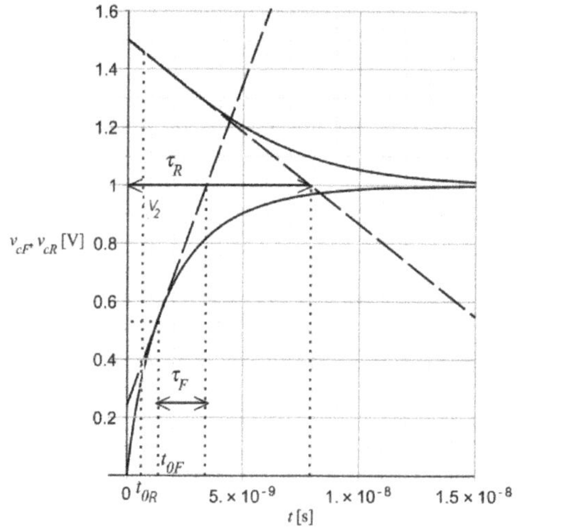

Fig. 10. Linearization of the time response characteristics.

Providing the linear change voltage as in the initial time, the transient process would be terminated at time τ. This parameter is equivalent to the time constant, but it is a function of the bias voltages (is not constant), unlike the first order linear systems. It is derived from the first order Taylor approximation,

$$\tau(v) = \frac{v_{C_\infty} - v_C|_{t=t_0}}{\dot{v_C}|_{t=t_0}}. \tag{26}$$

If the capacitor is charged from the initial value $v_c(t_0)$ to value in the steady state v_{c_∞}, then the voltage on the capacitor reaches the value $\Delta v_c = \alpha\,(v_{C_\infty} - v_{c_0})$ during time $t = k\tau$. Parameters $1 > \alpha > 0$ and $k \geq 0$ are multiple constants. The specific values of α and k parameters calculated for exponential function of $v_{C_R}(t)$, are shown in Tab. 3.

k	0.75	0.9	1	1.5	2	3	5
α	0.5	0.6	0.63	0.76	0.84	0.93	0.98

Tab. 3. Relationship between parameters k and α calculated from (16).

The increase of the pump stage voltage must not fall bellow the value $\Delta V_{max} \geq V_2$ during half of the period, as it is shown in Fig. 7. Discharge time through the parameter τ_R primarily determines the amount of the pumping losses and $\tau_R > \tau_F$, thus

$$\{\Delta v_{C_F} + \Delta v_{C_R} \approx V_2\}|_{t=k\cdot\tau_R+t_{0_F}=T/2}. \tag{27}$$

Thence the parameter α is given by

$$\alpha = \frac{V_2}{V_{SP}}. \tag{28}$$

Parameter τ_R can be calculated from (26), however it is approximately given by the reverse current I_{S_R},

$$\tau_R(W_{M_S}) \approx \frac{(V_{SP} - V_2)C}{I_{S_R}|_{V_{inv}=V_1,V_{in}=V_{IL}}}. \tag{29}$$

Using the condition (27) and respecting the time value t_{0_F} from (19), in which the M_{D_i} transistor is OFF, then the found width $W_{M_{S_{opt}}}$ is given by

$$W_{M_{S_{opt}}} = \frac{kC(V_{SP} - V_2)}{T\hat{I}_{S_R}} + \frac{\hat{I}_{S_R}(Tc_{t_0} + 2a_{t_0}) + \sqrt{D_{t_0}}}{2T\hat{I}_{S_R}b_{t_0}} \tag{30}$$

where

$$a_{t0} = \frac{2C \cdot A_{\mathrm{bulkD0}} W_{\mathrm{M_D}} L \mu_{\mathrm{effD0}} V_{0\mathrm{F}}}{c_{\mathrm{oxe}}},$$
$$b_{t0} = -A_{\mathrm{bulkD0}} \mu_{\mathrm{effD0}} \mu_{\mathrm{effSR}} W_{\mathrm{M_D}} \left(V_{t_a} + V_{t_b} V_{t_c}\right),$$
$$c_{t0} = \left(\mu_{\mathrm{effD0}} W_{\mathrm{M_D}}\right)^2 \left(-V_2 + V_{\mathrm{THSR}}\right) V_{0\mathrm{F}} + \\ + \left(\mu_{\mathrm{effD0}} W_{\mathrm{M_D}}\right)^2 \left(-V_2 + V_{\mathrm{THSR}}\right)^2,$$
$$D_{t0} = \left[2kC \cdot b_{t0}(V_2 - V_{\mathrm{SP}}) - \hat{I}_{\mathrm{SR}} \left(Tc_{t_0} + 2a_{t_0}\right)\right]^2 + \\ +16kCa_{t_0} b_{t_0} \hat{I}_{\mathrm{SR}} (V_2 - V_{\mathrm{SP}})$$

and

$$V_{t_a} = V_{\mathrm{THSR}} \left(V_{\mathrm{THSR}} A_{\mathrm{bulkSR}} - 2V_1 + 2V_2\right),$$
$$V_{t_b} = V_2 \left(A_{\mathrm{bulkSR}} - 1\right) - V_1 + V_{\mathrm{THSR}},$$
$$V_{t_c} = V_{0\mathrm{F}} - 2\left(V_2 - V_{\mathrm{THSR}}\right).$$

Parameter k is selected from Tab. 3 based on the parameter α from (28), $\hat{I}_{\mathrm{SR}}$ is the drain current calculated for the unity width $\left(\hat{I}_{\mathrm{SR}} = I_{\mathrm{SR}}/W\right)$.

In case the multiple of the time constant satisfies the session $k\tau_{\mathrm{R}} \gg t_{0\mathrm{F}}$, (30) now becomes to

$$W_{\mathrm{M_{S_{opt}}}} \approx \frac{2kC(V_{\mathrm{SP}} - V_2)}{T\hat{I}_{\mathrm{SR}}}. \tag{31}$$

5. Sizing of the "Diode" Transistor

Analysis results show that dynamic properties are not practically dependent on the sizing of the transistor $M_{\mathrm{D_i}}$ in the wide range of the ratio W/L. It only needs to be adequately dimensioned for the pump output load current i_{L} in steady state. After the clock signal $\bar{\phi}$ goes to H logic level (corresponds to V_{DD}), the output voltage starts from the initial value $V_{\mathrm{out,av}} - V_{\mathrm{r}}/2$ and can theoretically achieve the maximum value $V_{\mathrm{out,max}}$ during $T/2$. $V_{\mathrm{out,av}}$ is the required average value of the output voltage and V_{r} is the peak value of the ripple voltage. The situation is shown in Fig. 11. The transistor $M_{\mathrm{D}_{N+1}}$ is on in the active interval of the clock signal.

Time response characteristics will be firstly determined. Providing the capacitive character of the load impedance, the state description of the voltage on the capacitor $v_{\mathrm{out,av}}(t)$ is in an accordance to (15) ($R_{\mathrm{L}} \to \infty$).

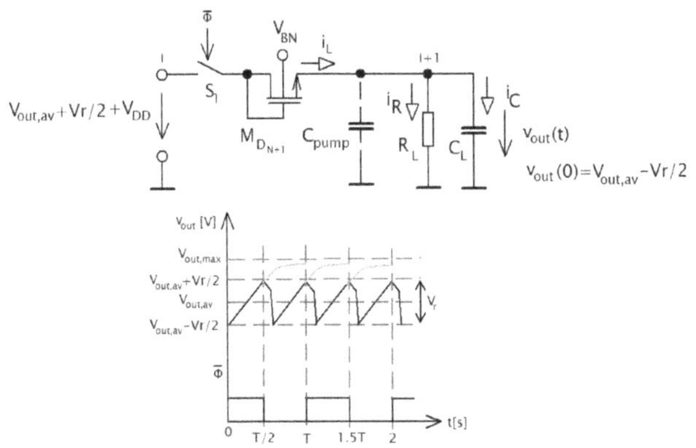

Fig. 11. The last stage of the charge pump and waveform of the output voltage in the steady state.

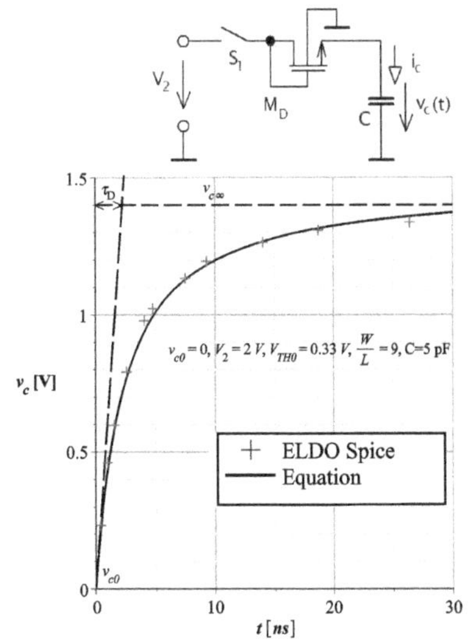

Fig. 12. Time response characteristics of the diode transistor.

The step response characteristics of the circuit in Fig. 12 after closed S at t=0, when the capacitor is charged from the initial value $v_{\mathrm{c}}(0_+) = v_{\mathrm{c}0}$ to the steady state $v_{\mathrm{c}\infty}$, is given by

$$v_{\mathrm{c}}(t) = \begin{cases} v_{\mathrm{c}0}, & \text{for } t \le 0, \\ \frac{W_{\mathrm{M_D}} \beta_{\mathrm{D}} v_{\mathrm{c}\infty} t (v_{\mathrm{c}\infty} - v_{\mathrm{c}0}) + C v_{\mathrm{c}0}}{W_{\mathrm{M_D}} \beta_{\mathrm{D}} t \left(v_{\mathrm{c}\infty} - v_{\mathrm{c}0}\right) + C}, & \text{for } t > 0. \end{cases} \tag{32}$$

The maximal output voltage value v_{c_∞} is equal to $V_{0\mathrm{F}}$ from (14) and β factor is calculated at the bias voltages in steady state,

$$\beta_{\mathrm{D}} = \frac{1}{2L} \frac{\mu_{\mathrm{eff}} c_{\mathrm{oxe}}}{A_{\mathrm{bulk}}} |_{V_{\mathrm{GS}} \approx V_{\mathrm{DD}}, V_{\mathrm{SB}} = V_{\mathrm{max}}}.$$

The increase of voltage α_{D} at time expressed as the multiples k of the τ_{D} parameter, $v_{\mathrm{c}}|_{t = k\tau_{\mathrm{D}}} = \alpha_{\mathrm{D}}(v_{\mathrm{c}_\infty} - v_{\mathrm{c}0})$, is listed in Tab. 4.

k	0.5	0.7	1	2	5	10	20	40	∞
α_{D}	0.32	0.39	0.48	0.63	0.77	0.85	0.91	0.97	1

Tab. 4. Relationship between parameters k and α_{D} calculated from (32).

The optimal width of the M_D transistor is determined from a condition, that the voltage v_{out} from Fig. 11 must achieve the maximal allowable ripple voltage V_{r} during T/2 at the desired average value of the output voltage. The maximal output voltage $V_{\mathrm{out,max}}$ is calculated from (32), into which the concrete values are substituted for input voltage V_2,

$$v_{\mathrm{c}_\infty} = V_{0\mathrm{F}}|_{V_2 = V_{\mathrm{out,av}} + V_{r_{\mathrm{max}}}/2 + V_{\mathrm{DD}}}. \tag{33}$$

Of course, the specified amplitude of the AC voltage value $v_{\mathrm{r}}(t)$ depends on both the external load R_{L}, C_{L} and on the equivalent internal pump impedance including R_{pump}, C_{pump}. Consequently, the following inequality must be true:

$$\left\{V_{\mathrm{out}}\left(W_{\mathrm{M_D}}\right) \ge v_{\mathrm{c}0} + \alpha_{\mathrm{D}}\left(v_{\mathrm{c}_\infty} - v_{\mathrm{c}0}\right)\right\}|_{t=T/2}. \tag{34}$$

Therefore,

$$W_{\mathrm{M_D}} \ge \frac{2\left(C_{\mathrm{L}} + C_{\mathrm{pump}}\right)}{T\beta_{\mathrm{D}}\left(v_{\mathrm{c}_\infty} - v_{\mathrm{c}0}\right)} \frac{\alpha_{\mathrm{D}}}{1 - \alpha_{\mathrm{D}}} \tag{35}$$

where pump capacitance may be neglected, provided $C_L \gg C_{pump}$ and $\alpha \geq \frac{V_{r_{max}}}{v_{C\infty} - v_{C0}}$ with the minimal value of the average voltage $V_{out,av}$. However, parameter α_D should be chosen, so that the load capacitor was charged by the large current all along of the active interval. Consequently, the transistor is fully switched on ($v_{gs} \gg v_{TH}$, strong inversion) and the load voltage is the approximately linear function of time. Results from Tab. 4 show that significant voltage change meets this assumption for α, which no exceeding the value about 0.7. Otherwise, the width quickly grows with $\alpha \rightarrow 1$ despite the improvement of dynamic properties. An example of the width calculation vs. α parameter is shown in Tab. 5.

Parameters
$L = 5\mu$m, $\beta_D = 135\,AV^{-2}m^{-1}$, $V_{DD} = 1$ V,
$V_{out,av} = 3.3$ V, $V_{r_{max}} = 50$ mV, $C_L = 20$ pF, $T = 100$ ns

$\alpha[-]$	0.1	0.2	0.4	0.6	0.8	0.9	1
$W_{M_D}[\mu m]$	5	11	30	68	182	411	∞

Tab. 5. Width of the M_{D_i} transistor vs. α parameter.

6. Experimental Part

The real circuits properties were simulated in the professional environment ELDO Spice, Design Architect-IC v2008.2_16.4. All assertions from the previous parts were be verified in the three-stages charge pump including the real models of all the components.

Parameter		Value
Temperature	ϑ	24° C
Number of stages	N	3
Supply voltage	V_{DD}	1.5 V
Main capacitance	C	40 pF
Load ressistance	R_L	200 kΩ
Load capacitance	C_L	20 pF
Threshold voltage of the	V_{THO_N}	0.35 V
hvt NMOS and PMOS at 0V	$\lvert V_{THO_P}\rvert$	0.33 V
Channel length of N(P)MOS	L	5μm
W/L ratio of the M_{S_i}	W_s/L_s	2
M_{P_i}	W_p/L_p	3
M_{N_i}	W_n/L_n	1
M_{D_i}	W_d/L_d	10

Tab. 6. Simulation parameters.

The HVT MOS transistors BSIM 4.2.0 were used that are available in the library MGC Design Kit. The various types of MOSFETs, M_{Si}, M_{N_i}, M_{Pi} and M_{D_i} are the same sized in each the pump stage. Simulation parameters (unless noticed otherwise) are specified in Tab. 6.

Firstly, the equation validity expressing the optimal point W_s/L_s (Fig. 13) will be tested via the comparison of the calculated functional values $V_c = v_{C_F} + v_{C_R}|_{t=T/2}$ from (16), (21) and the pump output voltage value $V_{out,av}$ dependending on the ratio W_s/L_s. The optimal width calculated from simplified (31) is listed in the last line. Setting of the voltages V_1, V_2 and source-bulk bias voltage of the N/PMOS for the calculation must first be resolved.

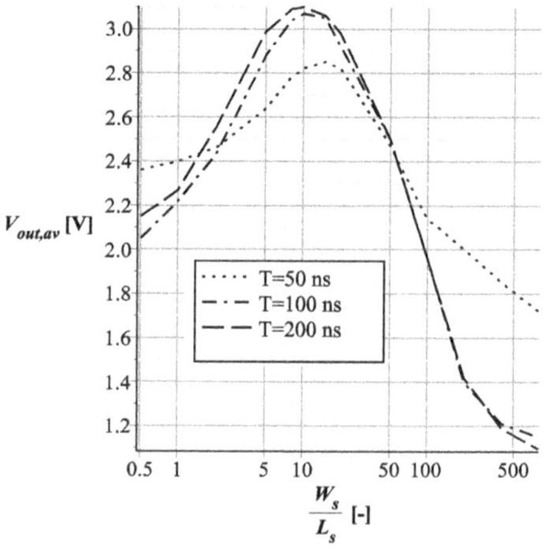

Fig. 13. Pump output voltage vs. the ratio W_s/L_s.

Starting from the fact, that the maximal output voltage value with a change of the circuit parameters (clock frequency, main capacitances, etc.) is achieved, just when the voltage gain of the first pump stage is maximal (it decreases with increasing the number of stages). In accordance to situation in Fig. 1, power supplies of the pump stage (Fig. 2) are $V_2 = V_{DD}$, $V_1 = 2V_{DD}$. The bulk of the NMOS is connected to ground ($V_{SB_N} = V_{DD}$) and bulk of the PMOS is selected so that the inverter switching point was the maximum (at the constant setting of the inverter transistors sizing). Using the definition V_{SP} [12], [15], then $V_{BS_P} = 0$ (in the last pump stage). As a consequence, the worst case of the V_{SP} voltage, labeled $V_{SP_{max}}$ in the N-stages pump is taken into account. Then, general formula of the $V_{SP_{max}}$ ($V_{B_P} = V_{B_N} = 0$) can be written as,

$$V_{SP_{max}} = V_{SP}|_{V_2=V_{DD}, V_1=2V_{DD}, V_{SB_N}=V_{DD}, V_{BS_P}=0}.$$

	$v_{C_\infty} = 2V_2 = 3$ V, $V_c(0) = V_{SP_{max}} = 2.41$ V					
	$T = 50$ ns		$T = 100$ ns		$T = 200$ ns	
$\frac{W_s}{L_s}$ [−]	$V_c[V]$	$V_{out,av}[V]$	$V_c[V]$	$V_{out,av}[V]$	$V_c[V]$	$V_{out,av}[V]$
0.5	2.39	2.36	2.42	2.05	2.49	2.15
1	2.42	2.4	2.48	2.22	2.59	2.27
2	2.49	2.46	2.60	2.43	2.75	2.55
5	2.64	2.64	2.81	2.88	2.99	2.99
8	2.76	2.78	2.94	3.04	3.06	3.09
10	2.81	2.82	2.99	3.07	**3.07**	**3.1**
15	2.92	2.85	3.05	3.05	**3.05**	**3.06**
20	**2.98**	**2.82**	**3.07**	**2.92**	3.04	2.98
50	**3.06**	**2.47**	**3.02**	**2.5**	3.01	2.5
100	3.02	2.14	3.01	1.96	3.01	1.96
200	3.01	2.0	3.0	1.4	3.0	1.42
400	3.0	1.85	3.0	1.21	3.0	1.19
800	3.0	1.72	3.0	1.15	3.0	1.097
$W_{MS_{opt}}[\mu m]$	43.8		21.9		10.9	

Tab. 7. Simulation results.

Example calculation of the switch transistor width for practical design will be shown in the following part. The charge pump parameters from Tab. 6 will be considered and: $T = 100$ ns, $V_{Bn} = 0$, $V_{Bp} = 0$. Channel length is same for all the transistors.

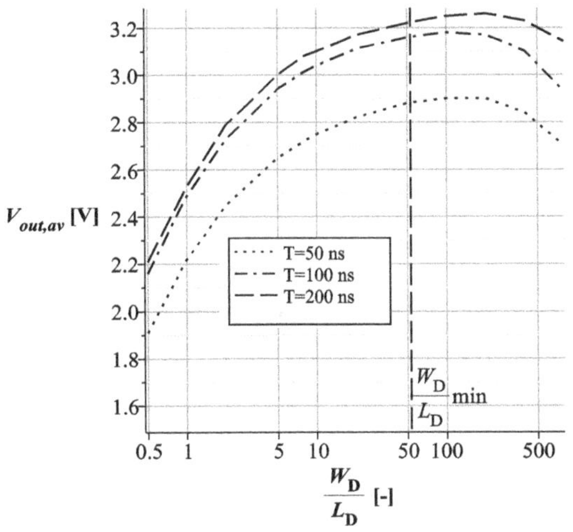

Fig. 14. Pump output voltage vs. the ratio W_D/L_D.

- the maximal inverter switching point [5] at appropriate bias voltage values $V_2 = V_{DD} = V_{SB_N} = 1.5\,V$, $V_1 = 2V_{DD} = 3\,V$ and $V_{BS_P} = 0$ is $V_{SP_{max}} = 2.41\,V$
- α factor is written as $\alpha = \frac{V_2}{V_{SP_{max}}} = \frac{1.5\,V}{2.41\,V} = 0.62\,[-]$,
- corresponding coefficient alpha deteremined based on the data from Tab. 3 is $k \approx 1\,[-]$,
- The drain current value in the specicific technology process and for unity width is $\hat{I}_{SR} = \hat{I}_{Dsat0}|_{V_{GS}=V_1-V_2=1.5\,V, V_{SB}=V_2=1.5\,V} \approx 34$ A/m, for $V_{IL} > V_2 + V_{DSsat_{M_S}}$,
- finally, the switch transistors width calculated from (31) is equal to

$$W_{M_S} \approx \frac{2kC(V_{SP} - V_2)}{T\hat{I}_{SR}} = \frac{2 \cdot 1 \cdot 40\,pF \cdot (2.41 - 1.5)\,V}{100\,ns \cdot 34\,A/m} \approx 21.9\,\mu m.$$

The results correspond with data from Tab. 6. They show that the pump output voltage is maximal (bold) if the time response characteristic of the pump stage at time $T/2$ does not exceed the value V_{max}, as it is shown in Fig. 9. The optimal width W_{M_s} must be less than the calculated value from (30), (31), otherwise, the pumping losses cause the discontinuous decrease of the output voltage due to the opening the feedback of the system (the condition 1 is not satisfied). It is a critical parameter from the view of the design process.

Conversely, the voltage gain is not changed in a wide ratio range W_D/L_D of the diode transistor, as it is shown in Fig. 14.

7. Conclusion

Guidelines on the design of the switch and "diode" transistors for the cross-coupled charge pump architecture without using long-time iteration process was presented in this paper. The symbolic description was used for calculation. The equivalent channel MOSFETs resistance was designed, so that the voltage gain and power efficiency were maximal. Analytical formulas are including the extreme values of circuit bias voltages to achieve satisfactory results in N-stage architecture, in which these types of the transistors have same dimensions in each of the stages. All the results were verified by the professional simulation software ELDO.

The switch and diode MOSFETs channel length was determined based on the critical electrical field in structure (breakdown voltage and maximal Drain-source voltage), see (2). The width of the switch and "diode" transistor were determined based on the analytical expression of the time response characteristics of the pump stage as an analog block. Switch transistors width, see (30), (31), was found using the criteria of the maximal difference between forward/charge and reverse/discharge current during the period of the clock signal. The ratio W_S/L_S, in which the sum of the time characteristics (reverse+forward) of the pump stage achieve the maximum value was compared with results fort three-stage pump, see Tab. 7. Compliance between both of the values is obvious. Simulation results show that switch resistance cannot be very small because increasing the ratio W/L, the reverse switch current decreases pump efficiency. This ratio can be less than one at the extreme values of input parameters (power supply, clock frequency). Exceeding the critical value of the reverse current, the pump output voltage is discontinuously decreased (see Fig. 13) because the basic condition (1) is not valid. This is an important practical result.

An estimation of the minimal width of the diode transistors W_D, see (35), is possible to determine from the requirements on the output load current and the output ripple voltage. Static and dynamic properties of the pump are quite stable over a wide range of the ratio W_D/L_D, see Fig. 14. Reverse current through the diode transistor is practically zero.

Acknowledgments

This work has been supported by the grant No. SGS17/183/OHK3/3T/13 of the CTU in Prague.

References

[1] PAN, F., SAMADDAR, T. *Charge Pump Circuit Design*. McGraw-Hill, 2006. ISBN: 978-007-1470-452

[2] NEW, L. F., BIN ABDUL, Z. A., LEONG, M. F. A low ripple CMOS charge pump for low-voltage application. In *Proceedings of the Intelligent and Advanced Systems (ICIAS)*. 2012. DOI: 10.1109/ICIAS.2012.6306120

[3] YIN, H., PENG, X., WANG, J., et al. Analysis and design of CMOS charge pump for EEPROM. In *Proceedings of the Solid-State and Integrated Circuit Technology (ICSICT)*. 2014. DOI: 10.1109/ICSICT.2014.7021464

[4] SHIAU, M. S., HSIEH, Z. H., HSIEH, C. C., et al. A novel static CTS charge pump with voltage level controller for DC-DC converters. In *Proceedings of the IEEE Conference on Electron Devices and Solid-State Circuits*. 2007. DOI: 10.1109/EDSSC.2007.4450167

[5] MAREK, J., HOSPODKA, J., SUBRT, O. Design aspects of the SC circuits and analysis of the cross-coupled charge pump. In *Proceedings of the International Conference on Applied Electronics (AE) 2016*. Pilsen (Czech Republic), 2016, p. 165–168. DOI: 10.1109/AE.2016.7577265

[6] SINGH, A., SINGH, T., PINDOO, I., et al. Transient response and dynamic power dissipation comparison of various Dickson charge pump configurations based on charge transfer switches. In *Proceedings of the 6th International Conference on Computing, Communication and Networking Technologies (ICCCNT)*. Denton, TX (USA), 2015, p. 1–6. DOI: 10.1109/ICCCNT.2015.7395219

[7] WANG, Y. R., YU, Z.G. A high-efficiency cross-coupled charge pump for flash memories. *IEEE Transaction on Circuits and Systems*, vol. 3, 2010. DOI: 10.1109/ICACC.2010.5486757

[8] KURTH, C. F., MOSCHYTZ, G. S. Nodal analysis of switched-capacitor networks. *IEEE Transaction on Circuits and Systems*, 1979, vol. 26, no. 2, p. 93–104. DOI: 10.1109/TCS.1979.1084613

[9] ANANDA MOHAN, P. V., RAMACHANDRAN, V., SWAMY, M. N. S. *Switched Capacitor Filters: Theory, Analysis and Design*. Prentice Hall PTR, 1995. ISBN: 0-13-879818-4

[10] C. HU, A. M. NIKENJAD, W. YANG, et al. *BSIM4.6.4 MOSFET Model: User's Manual*. UC Berkeley, 2009. Available at: www.device.eecs.berkeley.edu/bsim/

[11] STEFANOVIĆ, D., KAYL, M. *Structured Analog CMOS Design*. Dordrecht (Netherlands): Springer, 2008. ISBN: 9781402085727

[12] BAKER, R. *CMOS: Circuit Design, Layout and Simulation*. 3rd ed. Hoboken, NJ (USA): Wiley, 2010. ISBN: 9780470881323

[13] JAEGER, R., BLALOCK, T. *Microelectronic Circuit Design*. 3rd ed. Boston: Mcgraw-Hill, 2008. ISBN: 9780073309484

[14] PENG, X., ABSHIRE, P. Stochastic behavior of a CMOS inverter. *Electronics, Circuits and Systems*, 2007, p. 94–97. DOI: 10.1109/ICECS.2007.4510939

[15] TSIVIDIS, Y., MCANDREW, C. *Operation and Modeling of the MOS Transistor*. 3rd ed. New York: Oxford University Press, 2011. ISBN: 0195170156

About the Authors ...

Jan MAREK was born in Prague on April 8, 1990. He graduated from the Czech Technical University in Prague in 2014 and is currently studying towards Ph.D. at the Department of Circuit Theory. His research interests deal with the design of current analog circuits, as well as their optimization.

Jiří HOSPODKA was born in 1967. He received his Master's and Ph.D. degrees from the Czech Technical University in Prague in 1991 and 1995, respectively. Since 2007 he has been working as associate professor at the Department of Circuit Theory at the same university. Research interests: circuit theory, analog electronics, filter design, switched-capacitor, and switched current circuits.

Ondřej ŠUBRT was born in Hradec Králové on February 24, 1977. He works as analog design engineer with ASICentrum Prague, a company of the Swatch Group. At present, he has also been appointed an Ass. Prof. at the Faculty of Electrical Engineering, CTU Prague. His research interests being analog and mixed-signal integrated circuits design with emphasis to low-power low-voltage techniques and innovative design and verification methods of data converters.

Optimization of Adaptive Three-Mode GBN Scheme Control Parameters

Ranko VOJINOVIC [1], *Milos DAKOVIC* [2]

[1] Faculty of Information Technology, Bulevar Sv. Petra Cetinjskog 22, 81000, Podgorica, Montenegro
[2] University of Montenegro, Faculty of Electrical Engineering, Dz. Vasingtona bb, 81000, Podgorica, Montenegro

rankovoj@t-com.me, milos@ac.me

Abstract. *An adaptive three-mode system based on Go-Back-N (GBN) protocol is analyzed within this paper. An ideal mode selection procedure based on a-priori known packet error probability is defined. When packet error probability is unknown the system state transition is controlled by several system parameters. A procedure for optimal parameters selection is proposed and tested on a simulated system. The procedure is based on minimization of mean square deviation of the system throughput from the ideal one.*

Keywords

Go-Back-N, packet error probability, wireless communications, throughput, TS-ARQ scheme, optimization, control parameters

1. Introduction

One of the main disadvantages of "Go-Back-N" (GBN) protocol is the fact that system throughput significantly decreases when packet error rate (PER) increases. In order to solve this problem, several adaptive models of GBN protocol are proposed. An adaptive GBN scheme that dynamically adapts to varying channel conditions is proposed in [1] and further developed in [2–4]. In [5], the adaptive three-state system (TS-ARQ) that adapts the dynamic condition of the channel has been analyzed. In [6], the exact analysis of the throughput for the three-mode GBN (TM-GBN) system, based on retransmission cycles mechanism, is proposed. Throughput performances of various GBN adaptive schemes have been analyzed in [7–9].

The TM-GBN technique is proposed in order to improve system performances in the high noise case. Adaptation to the channel is done by classifying channel conditions into three states. The first state, denoted by L, is the low disturbances state. The moderate disturbances state is denoted by H, and intensive disturbances case is denoted by VH.

Formulated in this way, we get the three-state system. According to the current system state, the transmitter can apply the most appropriate procedure. In the L state, it follows standard GBN procedure. For the H state, the transmitter should switch to the multi-copy mode. Finally, for the VH state, the transmitter should continuously repeat packet until a positive acknowledgment is received.

The receiver provides corresponding acknowledgment for the each frame received. In the L mode, upon the receipt of α consecutive NAKs, the transmitter switches to the H mode, where the packets are emitted by n-copy GBN procedure. If the transmitter receives β consecutive ACKs in the H mode, the transmitter goes back to the L mode. If the transmitter receives γ consecutive NAKs in the H mode, then the transmitter will switch to the VH mode, in which it follows the procedure of continuous emitting of the same packet. The transmitter goes back to the H mode upon the receipt of δ consecutive positive acknowledgments. Channel state estimation and transmitter mode selection are based on the counting of positive and negative acknowledgments. These counters are compared with a given thresholds α, β, γ and δ in order to control the system and make decisions on the required system state transition. These parameters are essential for the adaptive system as the switching of the system state is controlled by them. In previous works, control parameters are obtained empirically. However, it is necessary to develop methods for determining these parameters in order to obtain optimal system. In this paper, we propose a technique for optimal control parameters selection.

The rest of this paper is organized as follows. In Section 2, the throughput analysis of general single-mode GBN mechanism and TM-GBN system are presented. In Section 3, optimization of control parameters is considered and discussed.

2. Adaptive GBN Protocol Analysis

Here we will give throughput analysis for single-mode adaptive and three-mode GBN protocol. The relations between the error probability and system throughput will be presented.

2.1 Single-Mode GBN Protocol Analysis

We will define the general model of GBN protocol considering the simplified scenario in which the receiver sends only the positive acknowledgments (ACK), and the transmitter starts the packets retransmission after a timeout. Expiration of the time-out interval is equivalent to the reception of the negative acknowledgments (NAK) from the receiver. The transmitting station forms the packet, marks them with consecutive numbers and stores the packets in the transmit buffer.

The transmitter sends the packets, without waiting for the ACK from the receiver. After the packet has been sent, transmitter activates the time control mechanism. If the positive acknowledgment is not received until the time-out, the ongoing transmission is interrupted. Then the unacknowledged packet and all the successive ones (even those with positive acknowledgment received) are being re-sent.

If the ACK is received for the packet in the transmit buffer with the lowest number, this packet is discarded from the buffer, and packet transmission continues with the next packet.

On the receiving side, upon the packet reception, the receiver performs error control. If the received packet is correctly received, the receiver sends positive acknowledgments for this packet.

One of the main communication system performance measures is system throughput. Throughput is defined as a ratio of a number of successfully received packets and a total number of transmitted packets.

The system throughput, in the single-copy GBN (SGBN) case, is

$$S_L = \frac{1 - P_e}{1 + (N-1)P_e} \tag{1}$$

where P_e is packet error probability and N is a standard value for GBN protocol.

In the n-copy GBN case (nGBN) transmitter sends n copies of each packet. The transmission is successful if a positive acknowledgment is received for at least one copy. The system throughput in this case is

$$S_H = \frac{1 - P_e^n}{n + (N-1)P_e^n}. \tag{2}$$

The continuous GBN procedure (CGBN) assumes that the same packet is transmitted until the positive acknowledgment is received. The system throughput is

$$S_V = \frac{1 - P_e}{1 + (N-1)(1-P_e)}. \tag{3}$$

Figure 1 presents throughput dependence of packet error probability for all considered single-mode protocols, with $N = 10$ and $n = 2$ Figure 2 illustrates throughput vs. P_e, for single-mode protocols, with $n = 2$ and $N = 30$.

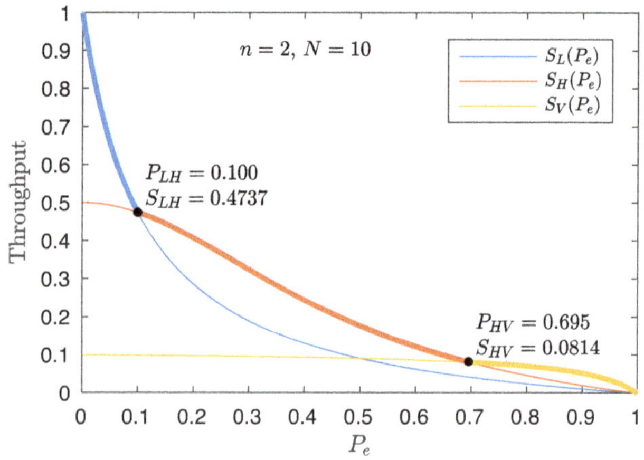

Fig. 1. System throughputs for $n = 2$ and $N = 10$. Probabilities P_{HV} and P_{LH} are obtained as intersection of the S_H curve with S_V and S_L curves.

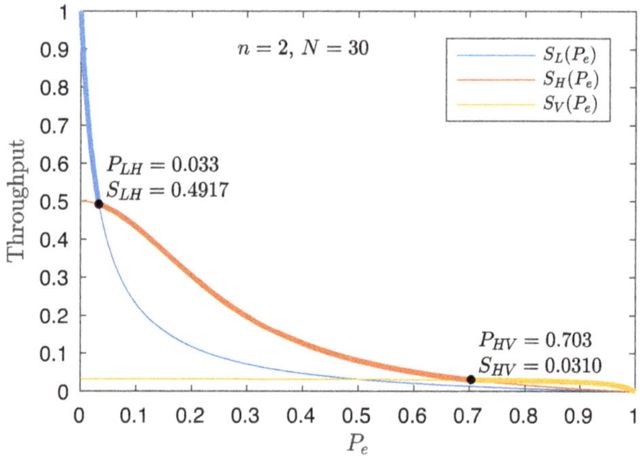

Fig. 2. System throughputs for $n = 2$, $N = 30$: Probabilities P_{HV} and P_{LH} are obtained as intersection of the S_H curve with S_V and S_L curves.

For each analyzed protocol we can identify range where it is the optimal protocol. Obviously for a very small probability of packet error SGBN protocol is optimal. The limit value of the packet error probability can be obtained by solving equation $S_L = S_H$ for unknown P_e. The smallest solution, denoted by P_{LH} is an upper limit when the SGBN protocol is optimal. P_{LH} can be obtained in a closed form for small n. For example:

$$P_{LH} = \frac{1}{N}, \qquad \text{for } n = 2, \tag{4}$$

$$P_{LH} = \sqrt{\frac{1}{4} + \frac{2}{N}} - \frac{1}{2}, \qquad \text{for } n = 3. \tag{5}$$

For large n value of P_{LH} can be obtained numerically.

For P_e higher than P_{LH} throughput obtained with nGBN protocol outperform SGBN case. In a similar way we can define upper limit for nGBN protocol by equating S_H and S_V and solving obtained equation for unknown limit P_{HV}. For $n = 2$ the solution is

$$P_{HV} = \frac{1 + \sqrt{8N^2 - 24N + 17}}{4N - 4}. \tag{6}$$

For higher values of n the solution can be found numerically. Now we have ranges for each protocol:

$$0 \leq P_e \leq P_{LH} \implies \text{SGBN protocol is optimal,}$$
$$P_{LH} < P_e \leq P_{HV} \implies \text{nGBN protocol is optimal,}$$
$$P_{HV} < P_e < 1 \implies \text{CGBN protocol is optimal.}$$

Limits P_{LH} and P_{HV} depend only on the system parameters N and n.

Presented results lead us to the conclusion that protocol choice highly depends on packet error probability. In the stationary channel case when P_e is a-priori known (or could be estimated) single protocol can be used. Adaptive systems [10] are able to track changes in the environment and to adjust system parameters in order to achieve optimal performances. For nonstationary channels, an adaptive procedure for transmitting protocol selection is implemented.

Having in mind that any adaptive system, where P_e is unknown, could not have better throughput than the system with optimally selected protocol, we can define optimal throughput S_{opt} as

$$S_{opt}(P_e) = \max\{S_L(P_e), S_H(P_e), S_V(P_e)\}. \tag{7}$$

Optimal throughput is presented in Figs. 1 and 2 with thick line. This case is an ideal one and can be used as a reference for adaptive system parameters optimization.

2.2 Three-Mode GBN Protocol Analysis

Let us review the basic foundations of the three-mode channel model, the concept which has been presented in details in [5].

TM-GBN model is given as scheme that uses three different methods of GBN protocols SGBN, nGBN, and CGBN. The transmitter sends the data packets to the receiver over the forward channel, and the receiver sends acknowledgments through the backward channel. In accordance with this model, transmitter's operating modes are different: in a state L - where the packet error probability is low, the transmitter follows basic SGBN procedure; in a state H - characterized by higher packet error probability, the transmitter operates in a n-copy mode; and in a state VH - where the packet error probability is very high, transmitter sends packet copies continuously until the first positive acknowledgment is received.

Switching from mode L to mode H, and from mode H to mode VH is managed by counting continuous NAKs for each packet and the switching from mode H to mode L, and from mode VH to mode H, by counting continuous ACKs for each packet. The receiver provides ACK for each received packet including the copies in the n-copy mode. The packets have been emitted in one of three possible ways. In the L mode, the standard GBN procedure is followed. Upon the reception of α consecutive NAKs, the transmitter switches to H mode, where the packets are emitted by the nGBN procedure. If the transmitter receives β consecutive ACKs in H mode, the transmitter returns to L mode. If the transmitter receives γ consecutive NAKs, while it is in H mode, the transmitter will switch to VH mode, in which it continuously emits the same packet. The transmitter returns to H mode upon the reception of δ consecutive ACKs.

Detailed analysis of TM-GBN system throughput is given in [6]. Here we will review final analysis results. The system throughput S_G is given by

$$\frac{1}{S_G} = \frac{C_L}{S_L} + \frac{C_H}{S_H} + \frac{C_V}{S_V} \tag{8}$$

where S_L, S_H and S_V are defined by (1), (2) and (3). According to [6], C_L, C_H and C_V are calculated as

$$C_L = \frac{(1 - P_L^\alpha)L}{L + H + V}, \tag{9}$$

$$C_H = \frac{(1 - P_H^\gamma)P_L^\alpha L + (1 - P_H^{\gamma+1})H}{L + H + V}, \tag{10}$$

$$C_V = \frac{P_L^\alpha P_H^\gamma L + P_H^{\gamma+1} H + V}{L + H + V} \tag{11}$$

with

$$L = \frac{(1 - P_H)^\beta}{P_L^\alpha}, \tag{12}$$

$$H = \frac{1 - (1 - P_H)^\beta}{P_H}, \tag{13}$$

$$V = \frac{1 - (1 - P_V)^{\delta-1}}{(1 - P_V)^{\delta-1}} \frac{P_H^\gamma}{P_V} \tag{14}$$

and

$$P_L = P_e,$$
$$P_H = P_e^n,$$
$$P_V = P_e.$$

It is obvious that the system throughput depends on packet error probability P_e and on system parameters N, n, α, β, γ and δ.

3. Parameters Optimization

In order to find optimal parameters, we should define optimization criterion. The optimization criterion should be based on the system throughput S_G. Since it depends on channel conditions (packet error rate P_e) averaging (integration) over all possible values of P_e should be used. The system optimization problem can be formulated as

$$\arg \min_{n,N,\alpha,\beta,\gamma,\delta} J(n, N, \alpha, \beta, \gamma, \delta). \tag{15}$$

Having in mind that in ideal case system throughput should be S_{opt} defined by (7), we can define cost function (optimization criterion) J as mean squared deviation of the system throughput S_G from the optimal throughput S_{opt}

$$J(n, N, \alpha, \beta, \gamma, \delta) = \int_0^1 \left(S_{opt}(P_e) - S_G(P_e)\right)^2 dP_e. \tag{16}$$

Note that integration limits in (16) could be reduced to $P_{\min}$,$P_{\max}$ if we expect that the packet error probability is within the range $P_{\min} \leq P_e \leq P_{\max}$.

Another approach could be that we optimize system for a given set of M error probabilities $\{P_1, P_2, \ldots, P_M\}$. In this case, the optimization criterion J could be defined as

$$J(n, N, \alpha, \beta, \gamma, \delta) = \sum_{k=1}^{M} (S_{\mathrm{opt}}(P_k) - S_{\mathrm{G}}(P_k))^2. \quad (17)$$

In all considered cases the function that should be minimized depends on integer parameters. Integer values of the parameters imply that classical minimization methods could not be directly applied.

When the minimization domain is finite, the direct search over all possible parameters combinations is a method that will surely lead us to the optimal solution. However, if the number of the minimization parameters is high, and when each parameter can achieve values from a large finite set, this approach could be inefficient. Of course, the optimization could be performed with other optimization methods as well [11], [12].

In the sequel, we will consider optimization with fixed N and n values. The ranges for remaining parameters are determined heuristically as

$$\begin{aligned} 1 &\leq \alpha \leq 5, \\ 1 &\leq \beta \leq 100, \\ 1 &\leq \gamma \leq 40, \\ 1 &\leq \delta \leq 30. \end{aligned}$$

In this case, the minimization implies calculation of $J(n, N, \alpha, \beta, \gamma, \delta)$ for a given N and n and for each combination of parameters α, β, γ, and δ. The total number of combinations is $5 \times 100 \times 40 \times 30 = 600,000$. The minimum of J is found and corresponding optimal parameters (within considered ranges) are obtained.

The optimization results for $N = 10$ and $n = 2$ are given in Fig. 3 (upper subplot). Obtained throughput is presented by thin black line, and theoretically optimal throughput with a thick gray line. We can see that for selected parameters actual throughput is very close to the ideal one. Deviation from the ideal throughput, used for criterion J calculation, is presented in the lower subplot. The same procedure is repeated for $N = 30$ and $n = 2$ case. The results are presented in Fig. 4.

From Figs. 3 and 4 we can conclude that adaptive system is very close to the ideal one. The highest deviation appears around $P_e = P_{\mathrm{HL}}$ i.e. when the system switches from standard GBN protocol to nGBN protocol and vice versa. Large values of β imply that when α consecutive errors are detected in the L mode (standard GBN protocol) and system switches to H mode (nGBN protocol) the system will go back to L mode very slowly (after β consecutive positive acknowledgments).

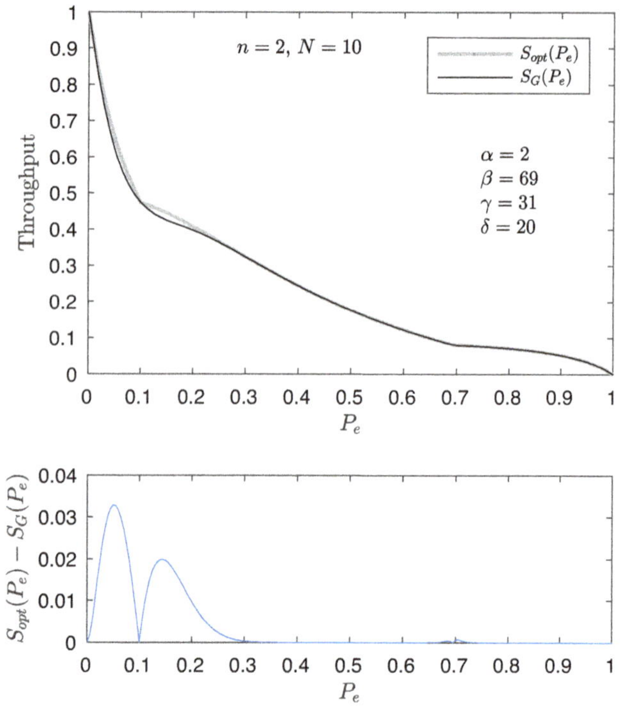

Fig. 3. Optimization of the system parameters for $n = 2$, $N = 10$. Throughput achieved through parameters optimization is presented with a thin black line. The thick gray line is theoretically optimal throughput for considered case. Deviation from the optimal throughput is presented in the lower subplot.

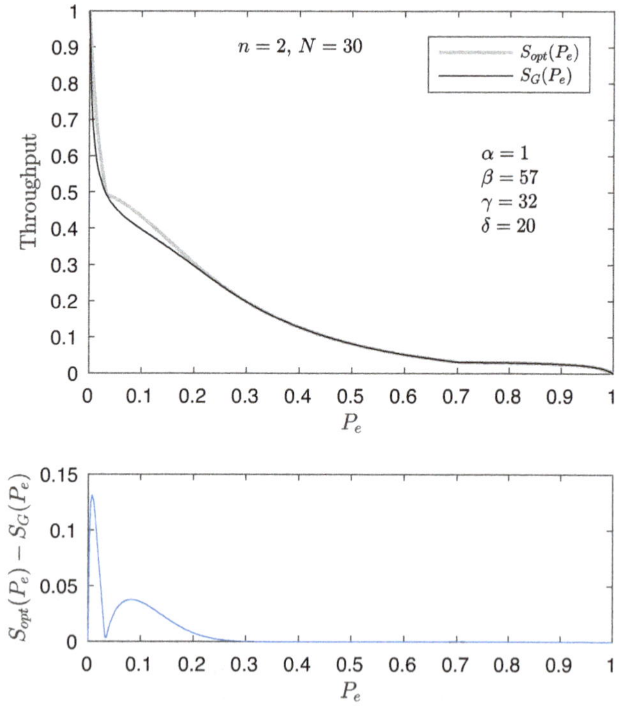

Fig. 4. Optimization of the system parameters for $n = 2$, $N = 30$. Throughput achieved through parameters optimization is presented with a thin black line. The thick gray line is theoretically optimal throughput for considered case. Deviation from the optimal throughput is presented in the lower subplot.

Note that the parameters optimization procedure should be performed once, so the high computational load caused by direct search procedure does not degrade system operating performances.

4. Conclusion

Three-mode GBN protocol is analyzed. The ideal case when we can select the optimal mode based on the a-priori known packet error probability is studied. This case is used in order to define optimization criterion. The procedure for adaptive system parameters optimization is derived. The optimization is performed and optimal system throughput is obtained for two system scenarios.

The presented results show that parameters choice in three-mode GBN system is very important. The optimization procedure is based on direct search within predefined parameter ranges. A large number of parameters combinations is not preferred, but when parameters take integer values this technique is highly reliable. It guarantees that our optimization procedure will not result in eventual local minimum and suboptimal parameters combination within considered ranges.

Acknowledgments

This research is partially supported by "New ICT Compressive Sensing Based Trends Applied to: Multimedia, Biomedicine and Communications (CS-ICT)" project (Montenegro Ministry of Science, Grant No. 01-1002).

References

[1] YAO, Y. D. An effective go-back-N ARQ scheme for variable-error-rate channels. *IEEE Transactions on Communications*, 1995, vol. 43, no. 1, p. 20–23. DOI: 10.1109/26.385946

[2] ANNAMALAI, A., BHARGAVA, V. K., LU. W. S. On adaptive go-back-N ARQ protocol for variable-error rate channels. *IEEE Transactions on Communications*, 1998, vol. 46, no. 11, p. 1405–1408. DOI: 10.1109/26.729379

[3] CHAKRABORTY, S. S., LIINAHARJA, M. Exact analysis of adaptive go-back-N ARQ scheme. *Electronics Letters*, 1999, vol. 35, no. 5, p. 379–380. DOI: 10.1049/el:19990165

[4] HUANG, S., LIU, J., YAN, Y., et al. Queuing analysis of co-operative GBN-ARQ in wireless networks with peers contending for a common helper. *AEU-International Journal of Electronics and Communications*, 2014, vol. 68, no. 5, p. 429–436. DOI: 10.1016/j.aeue.2013.11.006

[5] VOJINOVIĆ, R., PETROVIĆ, Z. A novel three-state ARQ scheme for variable error-rate channels. *AEU-International Journal of Electronics and Communications*, 2002, vol. 56, no. 6, p. 389–395. DOI: 10.1078/1434-8411-54100127

[6] VOJINOVIĆ, R., PETROVIĆ, G., PETROVIĆ, Z. The analysis of the adaptive three-mode ARQ GBN scheme using retransmission cycles mechanism. *AEU-International Journal of Electronics and Communications*, 2006, vol. 60, no. 2, p. 190–198. DOI: 10.1016/j.aeue.2005.03.002

[7] ALEXOVÁ, P., KOSUT, P., POLEC, J., et al. A comparison of selected GBN ARQ schemes for variable-error-rate channel using QAM. *Radioengineering*, 2002, vol. 11, no. 3, p. 43–47. ISSN: 1805-9600

[8] ZHOU, Y., LI, S., ZHOU, Y. Delay analysis of three ARQ protocols in Geom/G/1 Queue Model. *Journal of Theoretical and Applied Information Technology*, 2012, vol. 42, no. 2, p. 292–295.

[9] MAKKI, B., ERIKSSON, T. On the performance of MIMO-ARQ systems with channel state information at the receiver. *IEEE Transactions on Communications*, 2014, vol. 62, no. 5, p. 1588–1603. DOI: 10.1109/TCOMM.2014.033014.130223

[10] STANKOVIĆ, L. *Digital Signal Processing*. 1st ed. CreateSpace Amazon, 2015. ISBN: 978-1514179987

[11] KORTE, B., VYGEN, J. *Combinatorial Optimization*. Heidelberg: Springer, 2012. ISBN: 978-3540718437

[12] AHUJA, R. K., ERGUN, Ö., ORLIN, J. B., et al. A survey of very large-scale neighborhood search techniques. *Discrete Applied Mathematics*, 2002, vol. 123, no. 1:3, p. 75–102. DOI: 10.1016/S0166-218X(01)00338-9

About the Authors . . .

Ranko VOJINOVIĆ received the B.Sc. degree in Electrical Engineering from the University of Montenegro in 1982, and the M.Sc. and Ph.D. degrees from the University of Belgrade in 1992 and 2003, respectively. Since 1982 he has worked in a Department of Telecommunications in Montenegro Police Directorate. From 1996 to 2006, he was with the Maritime faculty at the University of Montenegro. In 2007 he joined the Faculty of Information Technologies of the University Mediterranean (Montenegro), where he is currently an Assistant Professor, teaching several courses in computer networks. His main research interests are in telecommunications and modeling and analysis of network protocols.

Miloš DAKOVIĆ was born in 1970 in Nikšić, Montenegro. He received a B.S. in 1996, an M.S. in 2001, and a Ph.D. in 2005, all in electrical engineering from the University of Montenegro. He is a professor at the University of Montenegro. His research interests are in signal processing, time-frequency signal analysis, compressive sensing, and radar signal processing. He is a member of the Time-Frequency Signal Analysis Group (www.tfsa.ac.me) at the University of Montenegro, where he is involved in several research projects.

10

Deceptive Jamming Method with Micro-Motion Property Against ISAR

Ning TAI [1], *Chao WANG* [1], *Liguo LIU* [2], *Weiwei WU* [1], *Naichang YUAN* [1]

[1] College of Electronic Science and Engineering, National University of Defense Technology, Changsha, Hunan, China
[2] College of Electronic Engineering, Naval University of Engineering, Wuhan, Hubei, China

358041578@qq.com, sywangc@163.com, poorthinker@sina.com, Shirleysp1130@hotmail.com, yuannaichang@hotmail.com

Abstract. *Airborne target's micro-motion such as rotation or vibration causes phase modulation, termed as micro-Doppler effect, into radar signals. The feature of micro-motion is one of the most obvious features for radar recognition in mid-course phase. In traditional works, it is assumed that the micro-motion of the scatterer is the same as the ballistic target. However, with the variation of the aspect angle of ISAR, the position of the scatterer changes. In this paper, the movement of a ballistic missile in mid-course is modeled and analyzed. A false target jamming method is proposed by combining the micro-motion modulation and the electromagnetic scattering modulation. Compared with the methods using ideal point models, our method is able to generate a vivid false target with structural information, micro-motion and variation of the scatterer's RCS. The micro-motion effect of the false target is presented through ISAR imaging and time-frequency analysis. The effectiveness and correctness of the algorithm is verified by simulation.*

Keywords

Micro-motion, false target, ISAR, jammer

1. Introduction

Inverse synthetic aperture radar (ISAR), a remote-sensing technique to provide a two-dimensional image of the interested target, has played a critical role in both military and civilian fields, such as target classification and recognition, ballistic missile defense [1], [2]. It enables the defenders with early warning, hostile target discrimination and military reconnaissance. Thus, great attention is paid to the researches on ISAR jamming and ant-jamming [3–5].

An ocean of works have been done to prevent a target from being detected by ISAR. Generally, the types of jamming methods can be classified to two categories: passive jamming and active jamming [6], [7]. Passive jamming produces jamming signals by the electromagnetic scattering from strong-reflecting objects, which do not change the properties of the surrounding environments. Due to the strong reflecting ability, the scattered signal of the reflector will induce bright spot on the image of ISAR and the protected target may be smeared or submerged. A passive jamming method based on rotating reflector to produce ghost target is proposed in [8], where a netlike jamming image is achieved through the reflector array. Xu proposes a method combing micro-motion modulation and phase-switched screen (PSS), which improves the jamming ability by enlarging the jamming strip both in range direction and azimuth direction [9].

Active jamming transmits an electromagnetic signal aiming at weakening the performance of the radar. By transmitting noise-like signals with strong power, barrage jamming results in severe degradation of the ISAR image, i.e. the signal-to-noise (SNR) ratio is weakened [10], [11]. The barrage jamming needs large jamming power since the transmitted signal is non-coherent with the matched filter. So deceptive jamming method which can be coherent or partial coherent with radar matched filter attracts more and more attentions. The jamming signal should contain the information as much the same as the real target echo and it makes the target's identification to be a hard task [12]. Through false target information modulation on the radar signals, sub-Nyquist sampling jamming signal can produce a train of false targets along the range direction [13]. With the idea of scatter-wave jamming, [14] deals with the jamming effect of sub-Nyquist jamming against ISAR where compressive sensing (CS) algorithm is applied to recover the image. It is further discussed in [15] that the situation when CS-based ISAR imaging algorithm is applied in fast-time domain to recover the high resolution range profile (HRRP).

Due to the relative movement between the radar and the target, the carrier frequency of returned radar signal will be shifted, which is known as Doppler phenomenon [16]. In addition to the bulk motion which causes a constant Doppler frequency shift, the target undergoes micro-motion dynamics, such as vibrating, rotating and rolling, which will also create time-vary frequencies [17]. The micro-motion effect is analyzed in detail by V.C. Chen in 2006 [18], [19]. The feature of micro-motion is academically acknowledged as one of the most important features in missile defense system. And this special characteristic is introduced in the algorithm for airborne target recognition [20] and ISAR imaging [21]. As the micro-motion will lead into additional Doppler frequency to the target echo, several researches are proposed to raise the image quality by eliminating the micro-Doppler frequency [22].

To maintain balance or interfered by external forces, airborne target, such as ballistic missile or airplane, always moves with micro-motion. Thus this phenomenon puts forward a challenge when we design a jamming method. The jamming signal should contain all the information as same as that of a real target, including micro-motion. The above false target jamming methods mainly focus on the reconstruction of the target's main body and neglect the micro-motion. The precession of a ballistic target is a typical kind of micro-motion and the caused Doppler frequency is considered as a sinusoid signal in the traditional works. These researches neglect two facts: one is that the scattering center will slide along the target with the change of radar aspect angle; the other is the changing in radar cross section (RCS). The target model made up of ideal scattering centers does not concentrate on the changing of RCS. And also certain scattering center may be invisible due to the shielding effect.

In this study, we propose a jamming method which contains the micro-motion property. The modulation effects of a ballistic target with precession both in the amplitude of the scatterer are modeled and analyzed. The modulation procedure of the jamming method is introduced in detail. The false target image, where the real target information is retained, may increase the burden of ISAR while recognizing the real one.

2. Trajectory of a Ballistic Target

Several coordinate systems are used extensively in Orbit mechanics. The coordinates used to analyze the trajectory of a ballistic target are Earth centered inertial (ECI) coordinate, Earth centered fixed (ECF) coordinate and East, North, Up (ENU) coordinate, as shown in Fig. 1.

It is quite easy to build the dynamic model of an airborne target in ECI coordinate $ox_Iy_Iz_I$. In many tracking applications, the ENU coordinate is far more intuitive and practical than ECF or ECI coordinate. The origin of ENU coordinate is set as the location of the jammer. For simplicity, the ECI coordinate coincides with ECF coordinate at time 0. In the following section, the location of the radar is computed in ECI coordinate and the computation of the distance between the jammer and the radar is conducted in ENU coordinate.

In ECI coordinate, the position vector is labeled as $p = [x_I, y_I, z_I]^{\mathrm{T}}$ and $v = \dot{p} = [\dot{x}_I, \dot{y}_I, \dot{z}_I]^{\mathrm{T}}$ is for the velocity vector. The status vector of a ballistic target can be defined as $X = (p, v)^{\mathrm{T}}$ and its non-linearity function is:

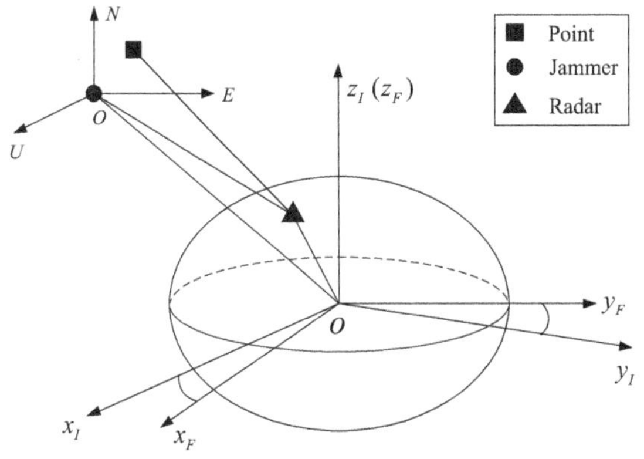

Fig. 1. Coordinates to analyze the target's movement.

$$\dot{X} = \begin{bmatrix} \dot{p} \\ \dot{v} \end{bmatrix} = \begin{bmatrix} v \\ a_{\mathrm{G}}(p) \end{bmatrix} \tag{1}$$

where $a_{\mathrm{G}}(p)$ is the Earth's gravity acceleration.

Considering that the flying time of the mid-course missile is quite long, we select ellipsoidal Earth model and the gravity acceleration can be [23].

$$a_{\mathrm{G}}(\boldsymbol{P}) = -\frac{\mu}{p^2}\{\boldsymbol{u}_p + \frac{3}{2}J_2(\frac{R_{\mathrm{e}}}{p})^2[(1-5(\boldsymbol{u}_p^{\mathrm{T}}\cdot\boldsymbol{u}_z)^2)\boldsymbol{u}_p + 2(\boldsymbol{u}_p^{\mathrm{T}}\cdot\boldsymbol{u}_z)\boldsymbol{u}_p]\} \tag{2}$$

where μ is the Earth gravity constant. P is the vector from Earth's center to the target's location, $p = \|\boldsymbol{P}\|$ is the length of vector, $\boldsymbol{u}_p = \boldsymbol{P}/p$ is the unit vector and $\boldsymbol{u}_z$ is the unit vector along the direction of oz_I. The corresponding dynamic model of the ballistic target is [24]:

$$\begin{bmatrix} \ddot{x}_I \\ \ddot{y}_I \\ \ddot{z}_I \end{bmatrix} = -\frac{\mu}{r_o}\begin{bmatrix} x_I + \frac{c_{\mathrm{e}}}{r_o^2}[1-5(\frac{z_{\mathrm{e}}}{r_o})^2]x_I \\ y_I + \frac{c_{\mathrm{e}}}{r_o^2}[1-5(\frac{z_{\mathrm{e}}}{r_o})^2]y_I \\ z_I + \frac{c_{\mathrm{e}}}{r_o^2}[3-5(\frac{z_{\mathrm{e}}}{r_o})^2]z_I \end{bmatrix} \tag{3}$$

where $c_{\mathrm{e}} = \frac{3}{2}J_2R_{\mathrm{e}}^2$ and J_2 is the second-order coefficient of zonal harmonics. R_{e} is the radius of equator. $r_o = \sqrt{x_I^2 + y_I^2 + z_I^2}$ stands for the distance from the target to the Earth's center.

We set the shut-down time of the ballistic target as the reference time, i.e. time 0. At time t, a scattering point of the target in ECI coordinate can be depicted with position vector $r_I = [x_I, y_I, z_I]^{\mathrm{T}}$ and velocity vector $v_I = \dot{r}_I = [\dot{x}_I, \dot{y}_I, \dot{z}_I]^{\mathrm{T}}$. These two vectors in ENU coordinate are: $r_{\mathrm{r}} = [x_r, y_r, z_r]^{\mathrm{T}}$ and $v_r = \dot{r}_{\mathrm{r}} = [\dot{x}_r, \dot{y}_r, \dot{z}_r]^{\mathrm{T}}$. The conversion from ECI coordinate to ENU coordinate is as below:

$$r_{\mathrm{r}} = C_I^r(t)\cdot r_I - \rho \tag{4}$$

where ρ is the vector from Earth's center to the location of the jammer in ENU coordinate

$$\rho = [0, -R_{\mathrm{N}}e^2\sin B\cos B, R_{\mathrm{N}} + H - R_{\mathrm{N}}e^2\sin^2 B]^{\mathrm{T}}, \tag{5}$$

$C_I^r(t)$ is the conversion matrix from ECI coordinate to ENU coordinate

$$C_I^r(t) = \begin{bmatrix} -\sin\varphi_r & \cos\varphi_r & 0 \\ -\cos\varphi_r\sin B & -\sin\varphi_r\sin B & \cos B \\ \cos\varphi_r\cos B & \sin\varphi_r\cos B & \sin B \end{bmatrix} \quad (6)$$

where ω_r is the angular velocity of Earth's rotating, $\varphi_r = L + \omega_r t$, L, B and H are the longitude, latitude and height of the jammer at time 0, e is the Earth's oblateness, $R_N = a/\sqrt{1 - e^2\sin^2 B}$ and a is the semi-major axis of the reference ellipsoid.

Equation (4) turns into (7) by taking the time derivative

$$\ddot{r}_r = \ddot{C}_I^r(t)\cdot r_I + 2\dot{C}_I^r(t)\cdot\dot{r}_I + C_I^r(t)\cdot\ddot{r}_I. \quad (7)$$

Equation (7) indicates that to build the dynamic equation in ENU coordinate, $\dot{r}_I$ and $\ddot{r}_I$ are needed. $\dot{r}_I$ and $\ddot{r}_I$ can be represented by the linear function which contains $\dot{r}_r$ and $\ddot{r}_r$ according to (4). The Runge-Kutta method can be used to solve the differential equation in (7) in ENU coordinate. Once we know the position of the jammer and the radar in real-time, the distance between them can be used to derive the modulation function which creates the false target jamming.

3. Micro-Motion of the False Target

3.1 Basic of Mirco-Motion

In traditional works, an airborne target with precession is always modeled as the isotropic scattering points, which differs from a real target. Thus when building up the model of a false target, we consider the micro-motion of the scatterer and its amplitude variation. The target's movements can be deposited into translational motion and micro-motion. Assume that the velocity of the false target is $\vec{v}$, the vector from the radar to the target's centroid o is expressed as below:

$$\vec{R}(t) = \vec{R}_0 + \vec{v}t \quad (8)$$

where $\vec{R}_0$ is the vector from the radar to the target at time 0. The point p on the target turns to p_1 after time t.

We need two coordinates named as local coordinate and velocity coordinate to analyze the motion of P. The details of these two coordinates will be introduced in Sec. 3.2. The location of p in local coordinate is deemed as $(\varepsilon_0, \eta_0, \mu_0)^{\mathrm{T}}$ and the corresponding location in velocity coordinate at time t is:

$$\vec{r}_t = M_t \cdot M_{\text{init}} \cdot (\varepsilon_0, \eta_0, \mu_0)^{\mathrm{T}} \quad (9)$$

where M_{init} stands for the initial Euler matrix and M_t is the rotating matrix. So the location of p in velocity coordinate is:

$$\vec{R}(t) = \vec{R}_0(t) + \vec{v}t + \vec{r}_t \quad (10)$$

Equation (9) and (10) indicate that M_{init} and M_t determine the target's moving properties. Next we will present the exact procedure to compute the micro-motion of the scatterer.

3.2 Target Model

The false ballistic target with precession is shown in Fig. 2. The velocity coordinate with unit vector $oXYZ$ is set up. The origin O is placed at the centroid of the false target and the direction of OY is the same as that of the false target's velocity. Another coordinate named local coordinate with unit vector $[e_\epsilon, e_\eta, e_\mu]^{\mathrm{T}}$ which moves with the target is also established. The $O\mu$ axis coincides with the symmetric axis of the target and points to the target's head. $O\epsilon$ and $O\eta$ reside in the plane which is normal to $O\mu$.

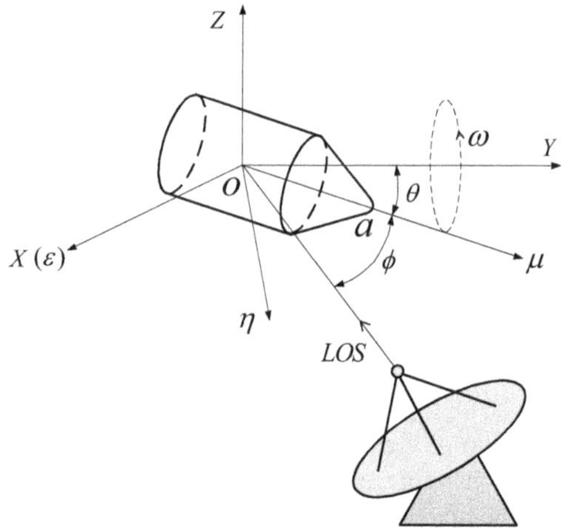

Fig. 2. False target with precession.

The target has coning motion along OY axis with angular velocity ω and spinning motion along its symmetric axis Oa. The included angle of Oa and OY is θ, which is the precession angle. Due to the symmetric structure, the spinning motion does not affect the position variation of scatterer. Thus only coning motion is considered in the following sections.

Firstly we analyze the distance from the radar to the jammer. Assume that at time 0, the distance from the radar to the origin of velocity coordinate is R_0, azimuth angle is φ_0 and pitch angle is θ_0. The position vector of the radar is

$$P_{b0} = \left[R_0\cos\varphi_0\cos\theta_0, R_0\sin\varphi_0\cos\theta_0, R_0\sin\theta_0\right]^{\mathrm{T}}. \quad (11)$$

Then radar position vector in ENU coordinate is

$$P_{R0} = C_b^r \cdot P_{b0} = [x_{R0}, y_{R0}, z_{R0}]^{\mathrm{T}} \quad (12)$$

where C_b^r is the conversion matrix from velocity coordinate to ENU coordinate

$$C_b^r = \begin{bmatrix} \cos\alpha & \sin\alpha\cos\beta & -\sin\alpha\sin\beta \\ -\sin\alpha & \cos\alpha\cos\beta & -\cos\alpha\sin\beta \\ 0 & \sin\beta & \cos\beta \end{bmatrix}. \quad (13)$$

If the velocity of the jammer is $V_J = [v_x, v_y, v_z]^{\mathrm{T}}$, then α and β in (13) can be computed by:

$$\begin{cases} \alpha = \sin^{-1}\left(\frac{v_x}{\sqrt{v_x^2+v_y^2}}\right), & -\frac{\pi}{2} \le \alpha \le \frac{\pi}{2}, \quad \text{if } v_y \ge 0, \\ \alpha = \pi - \sin^{-1}\left(\frac{v_x}{\sqrt{v_x^2+v_y^2}}\right), & \frac{\pi}{2} \le \alpha \le \frac{3\pi}{2}, \quad \text{if } v_y < 0, \\ \beta = \sin^{-1}\left(\frac{v_z}{\sqrt{v_x^2+v_y^2+v_z^2}}\right), & -\frac{\pi}{2} \le \beta \le \frac{\pi}{2}. \end{cases} \tag{14}$$

Further we derive the distance from the radar to the jammer:

$$R_J(t) = \sqrt{(x_J - x_{RO})^2 + (y_J - y_{RO})^2 + (z_J - z_{RO})^2} \tag{15}$$

where $[x_J, y_J, z_J]^T$ is the location of the jammer in ENU coordinate. The distance between the jammer and the radar is used to compute the modulation coefficient of the false target.

Next we present the micro-motion of the false ballistic target. For simplicity, the symmetric axis Oa resides in YOZ plane at time 0. The Euler angle of local coordinate is $(\alpha_c, \beta_c, \gamma_c) = (0, -\theta - \pi/2, 0)$. So the initial conversion matrix M_{init} from local coordinate to velocity coordinate can be derived

$$M_{\text{init}} = \begin{bmatrix} 1 & 0 & 0 \\ 0 & -\sin\theta & \cos\theta \\ 0 & -\cos\theta & -\sin\theta \end{bmatrix}. \tag{16}$$

Oa rotates along with OY axis with an angular velocity $\vec{\omega}$. According to Rodrigues transformation, the rotating matrix M_t at time t is

$$M_t = \begin{bmatrix} \cos\omega t & 0 & \sin\omega t \\ 0 & 1 & 0 \\ -\sin\omega t & 0 & \cos\omega t \end{bmatrix}. \tag{17}$$

Then Oa at time t can be expressed as:

$$Oa = [-\sin\theta\sin\omega t, \cos\theta, -\sin\theta\cos\omega t]^T. \tag{18}$$

The unit vector of light-of-sight (LOS) is

$$LOS_r = [\cos\varphi_r\cos\theta_r, \sin\varphi_r\cos\theta_r, \sin\theta_r]^T \tag{19}$$

where

$$\begin{cases} \varphi_r = \sin^{-1}\left(\frac{(y_{RO} - y_J)}{\sqrt{(x_{RO} - x_J)^2 + (y_{RO} - y_J)^2}}\right), \\ \theta_r = \sin^{-1}\left(\frac{(z_{RO} - z_J)}{\sqrt{(x_{RO} - x_J)^2 + (y_{RO} - y_J)^2 + (y_{RO} - y_J)^2}}\right). \end{cases} \tag{20}$$

The conversion matrix from ENU coordinate to velocity coordinate is $C_r^b = (C_b^r)^T$. Thus the LOS in velocity coordinate is:

$$LOS_b = C_r^b \cdot LOS_r = [m_x, m_y, m_z]^T. \tag{21}$$

So the included angle of Oa and LOS_b is

$$\phi = \cos^{-1}\big(-\sin\theta\sin\omega t \cdot m_x + \cos\theta \cdot m_y - \sin\theta\cos\omega t \cdot m_z\big). \tag{22}$$

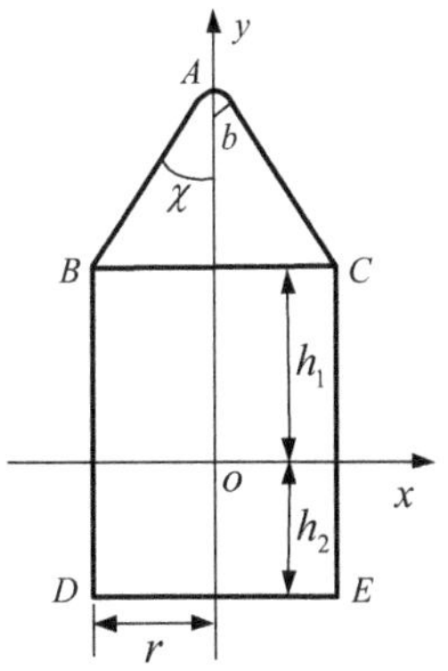

Fig. 3. Equivalent scattering points of the false target.

Assume that the plane determined by the Oa and LOS is labeled as $\prod_1$. In high-frequency region, the effective scattering points of the target is the intercross points of $\prod_1$ and the discontinuous area of the false target. Then the scatterers can be labeled as A, B, C, D and E, as shown in Fig. 3.

Firstly we analyze point A which is located at the head of target. According to the geometric structure, we know that:

$$\|OA\| = h_1 + \frac{r}{\tan\chi} + \left(1 - \frac{1}{\sin\chi}\right)b. \tag{23}$$

So the micro-motion of point A along LOS can be expressed as:

$$\begin{aligned} \|OA\|_{LOS} &= OA \cdot LOS \\ &= \left(h_1 + \frac{r}{\tan\chi} + \left(1 - \frac{1}{\sin\chi}\right)b\right)\cos\phi. \end{aligned} \tag{24}$$

From (24) we see that the micro-motion of A is determined by the aspect angle ϕ. According to (22), ϕ changes periodically due to the coning motion. Next let us analyze the micro-motion of point B and C. The normal vector of $\prod_1$ is:

$$\begin{aligned} n_1 &= Oa \times LOS \\ &= \begin{bmatrix} m_z \cdot \cos\theta + m_y \cdot \sin\theta\cos\omega t \\ -m_x \cdot \sin\theta\cos\omega t + m_z \cdot \sin\theta\sin\omega t \\ -m_y \cdot \sin\theta\sin\omega t - m_x \cdot \cos\theta \end{bmatrix} \end{aligned} \tag{25}$$

where $\times$ stands for cross product.

OB can be expressed as below [25]:

$$OB = k\,(n_1 \times Oa) + h_1 \cdot Oa \tag{26}$$

where k is unknown to us so far. Considering that n_1 is perpendicular to Oa and $n_1 \times Oa$ is perpendicular to Oa, we further derive

$$\|OB\|^2 = k^2\|\sin\phi\|^2 + \|h_1\|^2. \tag{27}$$

According to trigonometric function, the length of OB is

$$\|OB\|^2 = \|r\|^2 + \|h_1\|^2. \tag{28}$$

Combining (27) and (28), k can be obtained by

$$k = \pm \left| \frac{r}{\sin \phi} \right|. \tag{29}$$

Now we can obtain the projection of OB and OC on LOS

$$\|OB, C\|_{LOS} = (OB, C) \cdot LOS = \pm a \ \sin \phi + h_1 \cdot \cos \phi \tag{30}$$

where $'+'$ refers to OB and $'-'$ refers to OC.

In the same way, the projections of points D and E on the LOS are:

$$\|OD, E\|_{LOS} = \pm a \ \sin \phi - h_2 \cdot \cos \phi \tag{31}$$

where $'+'$ refers to OD and $'-'$ refers to OE.

According to micro-motion of each scatter and the jammer's distance in (15) we can obtain the range distance from the scatter to the radar, which can be used to build the radar echo.

3.3 Micro-Doppler Frequency

For simplicity, we assume that the radar locates in the YOZ plane and the included angle of OY axis and LOS is α. The unit vector of LOS in velocity coordinate can be depicted as $[0, \cos \alpha, -\sin \alpha]^{\mathrm{T}}$. So the included angle of Oa and LOS is : $\phi = \cos^{-1}(\cos \alpha \cos \theta + \sin \theta \sin \alpha \cos \omega t)$. Take point D as an example, its Doppler frequency can be computed by

$$\begin{aligned} f_{\mathrm{D}} &= \frac{2 f_{\mathrm{c}}}{c} \frac{d}{dt} (\|OD_{LOS}\|) \\ &= \frac{2 f_0 \omega \sin \alpha \sin \theta}{c} \cdot \sin \omega t \cdot \left(-h_2 + a \frac{\cos \phi}{\sqrt{1 - \cos^2 \phi}} \right). \end{aligned} \tag{32}$$

According to (32), the item $(a \cdot \cos \phi)/(\sqrt{1 - \cos^2 \phi})$ makes the Doppler frequency be not a sinusoid function since ϕ varies over time. When the relative position between radar and target changes, the Doppler frequency of D varies, too.

In the same way, the Doppler frequency of point A is

$$\begin{aligned} f_A = \left(h_1 + \frac{r}{\tan \chi} + \left(1 - \frac{1}{\sin \chi} \right) b \right) \cdot \\ \frac{2 f_{\mathrm{c}} \omega \sin \alpha \sin \theta}{c} \cdot \sin \omega t. \end{aligned} \tag{33}$$

From (33) we know the Doppler frequency of point A is fixed under the current LOS. Its Doppler frequency is related to the size of the false target and the precession angle θ. During the time of the ballistic flight, the Doppler frequency of point A or D will change because the LOS varies at different time.

4. Jamming Method

Suppose that ISAR transmits a liner frequency modulated (LFM) signal with pulse width T_{p}, carrier frequency f_{c} and range chirp k

$$s(t_r, t) = \mathrm{rect}\left(\frac{t_r}{T_{\mathrm{p}}}\right) \times \exp\left(\mathrm{j}2\pi\left(f_{\mathrm{c}} t + \frac{1}{2} k t_r^2\right)\right) \tag{34}$$

where t_r is fast time, t_a is slow time and $t = t_r + t_a$ is full time. rect $(\cdot)$ yields 1 when $|t_r/T_{\mathrm{p}}| \leq 1/2$ and otherwise is 0.

Assume that the distance from one certain point to the radar is marked as $R_i(t)$, thus the corresponding echo in baseband is:

$$\begin{aligned} s_t(t_r, t) = \mathrm{rect}\left(\frac{t_r - 2R_i(t)/c}{T_{\mathrm{p}}}\right) \times \exp\left(\mathrm{j}\pi k (t_r - \right. \\ \left. \frac{2R_i(t)}{c})^2\right) \times \exp\left(-\mathrm{j}\frac{4\pi}{\lambda} R_i(t)\right). \end{aligned} \tag{35}$$

The distance from the jammer to the radar is $R_j(t)$, so there is a range difference $\Delta R_i(t)$ between $R_i(t)$ and $R_j(t)$

$$R_i(t) = R_j(t) + \Delta R_i(t). \tag{36}$$

Substitute (36) into (35), we derive:

$$\begin{aligned} s_t(t_r, t) &= s_j(t_r, t) \otimes \delta\left(t_r - \frac{2\Delta R_i(t)}{c}\right) \times \exp\left(-\mathrm{j}\frac{4\pi}{\lambda} \Delta R_i(t)\right) \\ &= s_j(t_r, t) \otimes h(t_r, t) \end{aligned} \tag{37}$$

where $\delta(\cdot)$ is Dirac function and $\otimes$ means convolution. $s_j(t_r, t)$ is the radar signal intercepted by the jammer

$$\begin{aligned} s_j(t_r, t) = \mathrm{rect}\left(\frac{t_r - 2R_j(t)/c}{T_{\mathrm{p}}}\right) \times \exp\left(\mathrm{j}\pi k (t_r - \right. \\ \left. 2R_j(t)/c)^2\right) \times \exp\left(-\mathrm{j}\frac{4\pi}{\lambda} R_j(t)\right). \end{aligned} \tag{38}$$

From (37) we see that the radar echo of the scatterer i can be created by the time delay and a phase addition to $s_j(t_r, t)$. $h(t_r, t)$ is the modulation function to produce the corresponding radar echo and it is treated as a system response function. According to the property of Fourier transform, the modulation procedure on $s_j(t_r, t)$ can be implemented by fast Fourier transform (FFT)

$$S_T(f_{\mathrm{r}}, t) = S_{\mathrm{J}}(f_{\mathrm{r}}, t) \times H(f_{\mathrm{r}}, t) \tag{39}$$

where $S_T(f_{\mathrm{r}}, t)$, $S_{\mathrm{J}}(f_{\mathrm{r}}, t)$ and $H(f_{\mathrm{r}}, t)$ are the frequency expressions of the jamming signal, the radar signal, and the system response function, respectively

$$H(f_{\mathrm{r}}, t) = \exp\left(-\mathrm{j}2\pi f_{\mathrm{r}} \frac{2\Delta R_i(t)}{c}\right) \times \exp\left(-\mathrm{j}\frac{4\pi}{\lambda} \Delta R_i(t)\right). \tag{40}$$

Now we obtain the frequency expression of the jamming signal after deriving $H(f_{\mathrm{r}}, t)$. Equation (40) indicates that the jamming signal is derived from the multiplication between radar signal and the system response function. And the system response function is computed according to the current position of the jammer and the movement of the false target. Knowing that the modulation is based on the intercepted radar signal, the induced false target after ISAR imaging is around the jammer's location.

Equation (39) turns into (41) by multiplying the scattering coefficient σ and the inverse FFT

$$s_t(t_r,t) = \sigma \times \mathrm{IFFT}\left(S_\mathrm{J}(f_\mathrm{r},t) \times H(f_\mathrm{r},t)\right). \quad (41)$$

Considering the false target made up of N scatterers, the jamming signal is the superposition of all the echoes

$$s_t(t_r,t) = \sum_{i=1}^{N} \sigma_i \times \mathrm{IFFT}\left(S_\mathrm{J}(f_\mathrm{r},t) \times H_i(f_\mathrm{r},t)\right). \quad (42)$$

Equation (42) indicates that the modulation process, which includes FFT, multiplication and IFFT, is conducted in parallel and N times of modulation results need be summed up. Taking advantage of the linear property of FFT, the system response function of every scatterer can be summed up in advance. This work can be done when jammer is idle and the computation time is enough. The modified modulation process is

$$s_t(t_r,t) = \mathrm{IFFT}\left(S_\mathrm{J}(f_\mathrm{r},t) \times \left(\sum_{i=1}^{N} \sigma_i \times H_i(f_\mathrm{r},t)\right)\right). \quad (43)$$

To sum up, firstly the jammer computes the real-time distance between itself and the radar. Then according to the designed micro-motion of the false target, the jammer derives the system response function. At last the radar signal and the system response function is multiplied in frequency domain to deduce the frequency expression of the jamming signal.

5. Simulation Results

Assume that the location of ISAR is: latitude 25°, longitude 31° and height 100 m. The location of the jammer at time 0 is: latitude 30°, longitude 10° and height 100 km. The parameters of the false target is: $b = 0.08$ m, $h_1 = 1.6$ m, $h_2 = 0.7$ m, $a = 0.30$ m and $\chi = 15°$. The precession angle of the false target is set to $\theta = 4°$, angular velocity of precession is $\omega = 150°/\mathrm{s}$. The other parameters of the simulation is listed in Tab. 1.

The total flight time is about 180 s and it is sufficient to analyze the movements of the false target. The aspect angle changes periodically as revealed in Fig. 4. The period is 2.4 s since the angular velocity of the coning motion is 150°/s. The overall trend of the aspect angle changes due to the variation of the positions of the jammer and ISAR.

Parameter type	Value	Unit
Carrier frequency	10	GHz
Band width	1	GHz
Pulse width	0.5	μs
PRF	500	Hz
R_0	100	km
V_J	1000	m/s

Tab. 1. Simulation parameters.

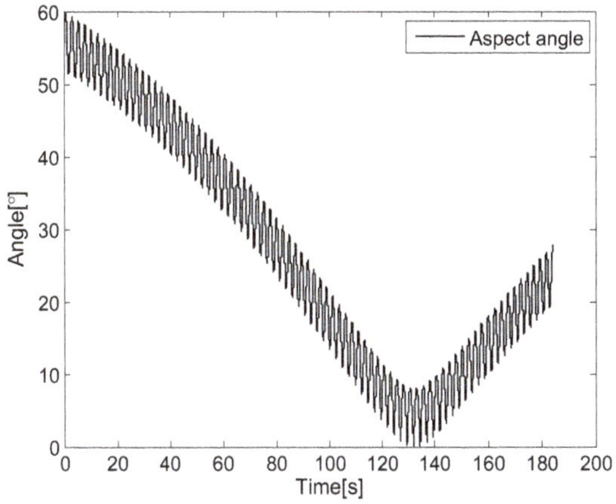

Fig. 4. Aspect angle of ISAR.

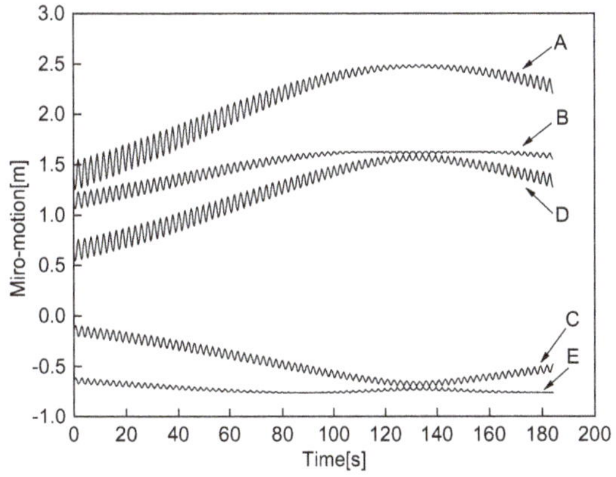

Fig. 5. Micro-motion of each scatterer.

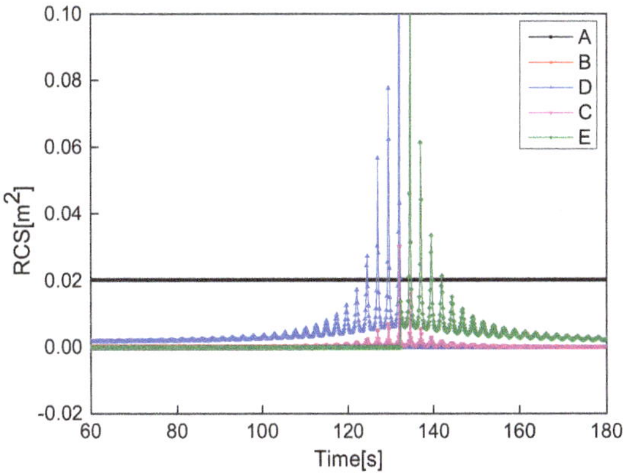

Fig. 6. RCS of each scattering point.

The micro-motion of each scattering point is demonstrated in Fig. 5. The amplitude of micro-motion changes with time and the overall trend is also related to the aspect angle. Take the movement of point A as an example. When the flight time is from 0 s to 120 s, the overall trend of the aspect angle decreases. So the amplitude of micro-motion of A increases in this time interval, which coincides with (24). The periodically change of the amplitude is caused by the coning motion of the false target.

The RCS of each scatterer is presented in Fig. 6, where we see that the RCS of A is fixed and the RCS of other scatterer is varying. Especially for point D and E, their RCS turns into 0 in some time due to the shielding effect. Because

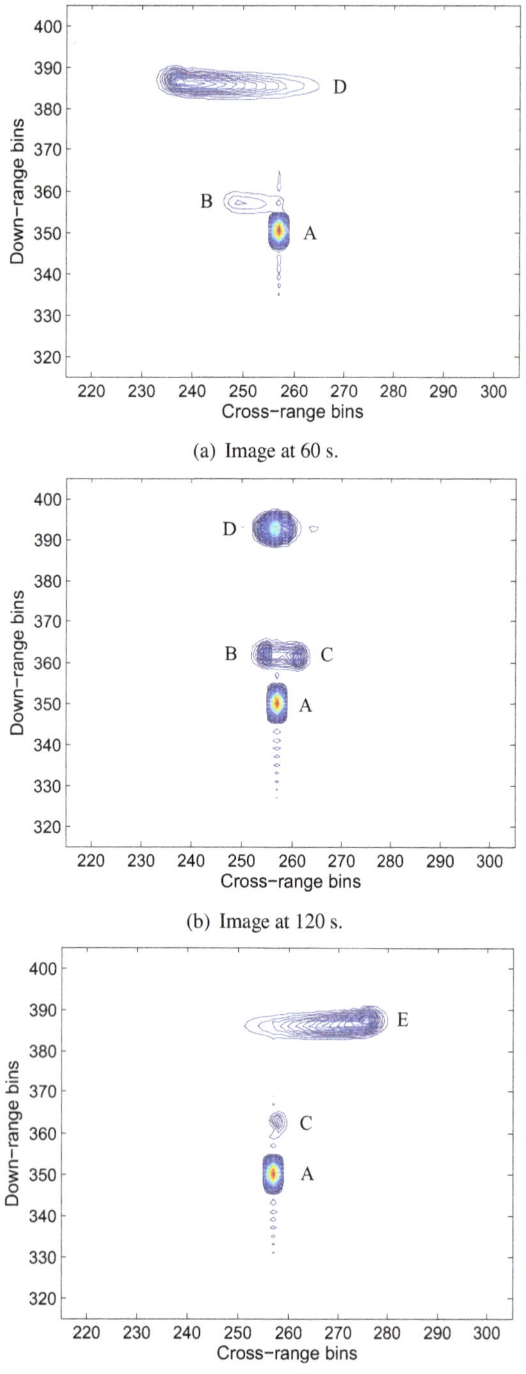

(a) Image at 60 s.

(b) Image at 120 s.

(c) Image at 180 s.

Fig. 7. ISAR image when $\theta = 4°$, $\omega = 150°/s$.

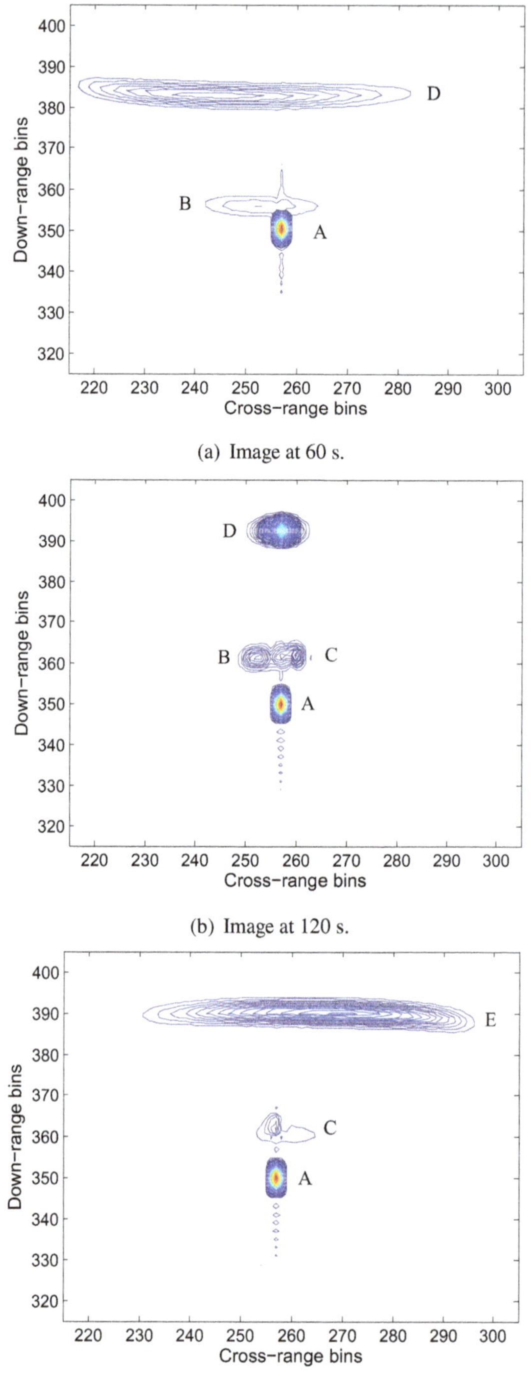

(a) Image at 60 s.

(b) Image at 120 s.

(c) Image at 180 s.

Fig. 8. ISAR image when $\theta = 10°$, $\omega = 120°/s$.

of the symmetric structure, the RCS of B is almost as same as that of C. The RCS of either B or C is small during the whole flight time except for the time interval from 110 s to 150 s. According to the value of RCS, the ISAR image of the false target should be in the form of 3 focused points.

Since the Doppler frequency of the false target is varying periodically, the imaging result will be smeared along the azimuth direction. Figure 7 demonstrates the ISAR image of the jamming signal when time is 60 s, 120 s and 180 s. There are 3 points to reflect the geometrical structure of the false target, which coincides with the RCS in Fig. 6. Only A, B and D are lighted by the radar's LOS at time 60 s, so the result in Fig. 7a shows a target with three obvious focused points. At 120 s, Fig. 7b shows four points because the RCS of A, B, C and D is relatively large at this time. Because the aspect angle of ISAR changes, the azimuth location of D moves obviously when compared with the location in Fig. 7a. At this time, B and C are barely visible because of the small RCS. The two

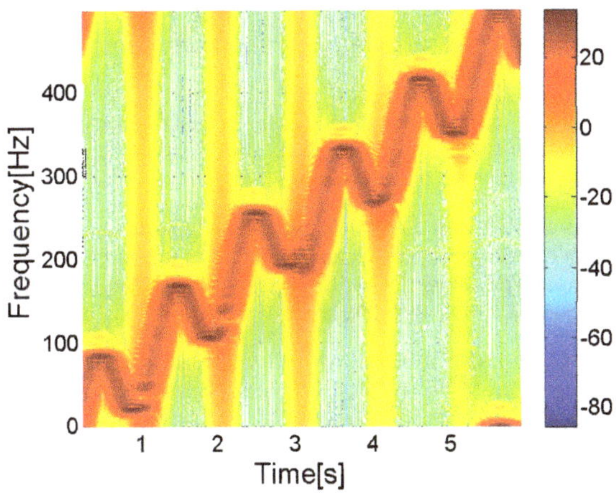

Fig. 9. STFT result of point *A*.

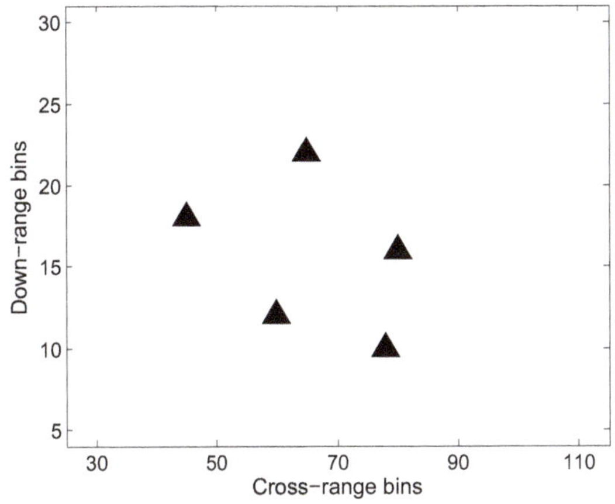

(a) Target model for DIS method.

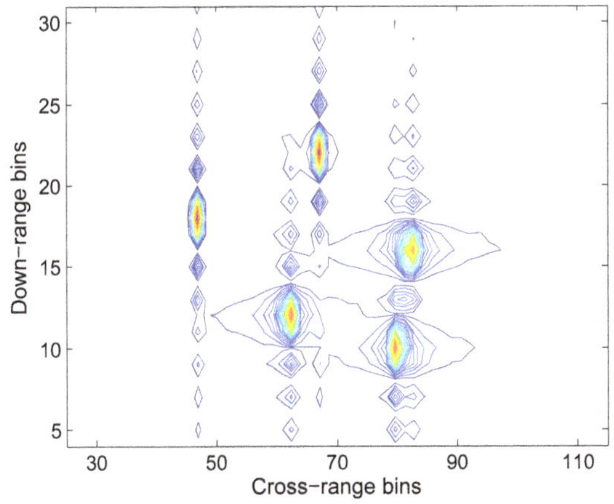

(b) ISAR image result.

Fig. 10. ISAR image of DIS method.

points are next to each other since the azimuth resolution of the ISAR image is not high enough. In Fig. 7c, E appears and D is invisible because of the shielding effect. The image of point D or E smeares along the azimuth direction, which is caused by its micro-Doppler frequency.

Another ISAR image of the jamming signal with different moving parameters is demonstrated in Fig. 8. According to (34), the Doppler frequency of a scattering point increases with a larger ω or θ. The point D or E in Fig. 8 smears heavily than that in Fig. 9 due to it's bigger Doppler frequency, which coincides with the previous analyses in Sec. 3.3.

The features of the false target can be revealed in the time, the frequency, or the joint time-frequency domains. To analyze the time-varying Doppler frequency caused by precession, the echo of the false target is analyzed by short time Fourier transform (STFT). The observation time is 6 s and the STFT result of point A when $\omega = 360°/\text{s}$ is shown in Fig. 9. From Fig. 9 we see that the overall trend of the frequency increases with time, which is caused by the translational movement of the false target. The frequency of A changes periodically due to the micro-motion and the period is 1 s, which coincides with the conning angular velocity.

The image result of DIS method is presented here as a comparison. The range-Doppler template is demonstrated in Fig. 10a, where 5 scattering points are used to imitate a ballistic target. The amplitude of the each point is set to 1 for all range bins. The range-Doppler template is processed to generate the jamming signal according to the procedure introduced in [26]. The corresponding ISAR image is presented in Fig. 10b, where we see that the scattering points are correctly resolved.

The jamming signal created by DIS method is capable of forming an ISAR image of the designed false target, but it fails to reflect the micro-motion information of a ballistic target, which does not coincide with the real target. If the template of the false target intends to reflect the RCS of each scattering point, the extraction procedure of the modulation parameters must be processed in real time, which is not an easy job for DIS method.

6. Conclusions

With the fast rate of the development in ISAR technology, the electronic counter-measurement against this high resolution imaging radar becomes a hot topic in recent decades. In this paper, a jamming method which creates a ballistic false target with precession is proposed. Through transformation in ECI coordinate, ENU coordinate and velocity coordinate, the dynamic movement of the false target is modeled and the micro-motion is analyzed. It is the first time that a deceptive jamming method with micro-motion property is proposed in the mid-course flight. The Doppler frequency of the scatterer is not a sinusoid curve. The introducing of the micro-motion reflects the moving property of a ballistic target and makes the false target more similar to a real one. The micro-motion modulation of the scattering points, together with the variation of RCS, demonstrates the scattering characteristic of the false target.

When this method is applied into the engineering implementation, we recommend that the jammer adopts quadrature channels to process the intermediate frequency radar signal. To reconstruct the echo of each scatterer, the method needs frequency modulation, which requires the phase of radar signal. Because the time-delay method is lack of precision. In our former design, the system clock is 150 MHz and it corresponds a delay precision of 1 meter, which is not proper to reflect the characteristic of a ballistic missile. After the phase modulation, the phase of the jamming signal which represents one scatterer is transformed into amplitude. Then all the scatterers' echoes are cumulative added to build the echo of the false target.

The novelties of this paper are the considerations in the micro-motion and the variation of RCS. The false target created by our method is with a constant trace and vivid micro-motion. Compared with DIS method, the proposed method is able to imitate the constant movement of the false target and reflects the micro-motion of each scatterer. The simulation result validates the correctness of the method and the ISAR image of the jamming signal presents a false target with micro-motion property.

Acknowledgments

This work was supported in part by the National Natural Science Foundation of China under Grant no. 61302017 and 61601492. Thanks to Dr. Xi Chen for analyzing the electromagnetic scattering property of a ballistic missile. Thanks to Dr. Hong Zhu for his brilliant suggestions.

References

[1] WANG, Y. X., HAO, L., CHEN, V. C. ISAR motion compensation via adaptive joint time-frequency technique. *IEEE Transaction on Aerospace and Electronic System*, 1998, vol. 34, no. 2, p. 670–677. DOI: 10.1109/7.670350

[2] LI, X., LIU, G. S., NI, G. L. Autofocusing of ISAR images based on entropy minimization. *IEEE Transaction on Aerospace and Electronic System*, 1999, vol. 35, no. 4, p. 1240–1252. DOI: 10.1109/7.805442

[3] BACHMANN, D. J., EVANS, R. J., MORAN, B. Game theoretic analysis of adaptive radar jamming. *IEEE Transaction on Aerospace and Electronic System*, 2011, vol. 47, no. 2, p. 1081–1100. DOI: 10.1109/TAES.2011.5751244

[4] ZHANG, J., LIU, N., ZHANG., L. R., et al. Active jamming suppression based on transmitting array designation for colocated multiple-input multiple-output radar. *IET Radar Sonar Navig.*, 2016, vol. 10, no. 3, p. 500–505. DOI: 10.1049/iet-rsn.2015.0215

[5] ALMSLMANY, A., WANG, C. Y., CAO, Q. S. Advanced deceptive jamming model based on DRFM sub-Nyquist sampling. In *Proceedings of the 13th International Bhurban Conference on Applied Sciences and Technology (IBCAST)*, Islamabad (Pakistan), 2016, p. 727–730. DOI: 10.1109/IBCAST.2016.7429963

[6] ZHANG, R. W., LI, Y. P., JIAO, Y. F. Cognitive radar waveform diversity for anti-passive false target jamming in an active radar seeker. In *Proceedings of the 15th International Conference on Instrumentation and Measurement, Computer, Communication and Control (IMCCC)*. Qinhuangdao (China), 2015, p. 1742–1745. DOI: 10.1109/IMCCC.2015.370

[7] LIU, Y. C., WANG, W., PAN, X. Y., et al. A frequency-domain three-stage algorithm for active deception jamming against synthetic aperture radar. *IET Radar Sonar Navigation*, 2014, vol. 8, no. 6, p. 639–646. DOI: 10.1049/iet-rsn.2013.0222

[8] BAI, X. R, SUN, G. C., ZHOU, F., et al. A novel ISAR jamming method based on rotating angular reflectors. *Chinese Journal of Radio Science*, 2008, vol .23, no. 5, p. 867–872. DOI: 10.13443/j.cjors.2008.05.019

[9] XU, L. T., FWNG, D. J., WANG, X. S. Improved synthetic aperture radar micro-Dopler jamming method based on phase-switched screen. *IET Radar Sonar Navigation*, 2015, vol. 10, no. 13, p. 525–534. DOI: 10.1109/JSEN.2015.2453163

[10] TAI, N., PAN, Y. J., YUAN, N. C. Quasi-coherent noise jamming to LFM radar based on pseudo-random sequence phase-modulation. *Radioengineering*, 2015, vol. 24, no. 4, p. 1013–1024. DOI: 10.13164/re.2015.1013

[11] GONG, S., WEI, X., LI, X., et al. Mathematic principle of active jamming against wideband LFM radar. *Journal of Systems Engineering and Electronics*, 2015, vol. 26, no. 1, p. 50–60. DOI: 10.1109/JSEE.2015.00008

[12] ZHAO, B., ZHOU, F., SHI, X. R., et al. Multiple targets deception jamming against ISAR using electromagnetic properties. *IEEE Sensors Journal*, 2015, vol. 15, no. 4, p. 2031–2038. DOI: 10.1109/JSEN.2014.2368985

[13] XU, S. K., LIU, J. H., FU, T. W., et al. Deception jamming method for ISAR based on sub-Nyquist sampling technology. In *Proceedings of the IEEE 10th International Conference on Signal Processing*. Beijing (China), 2010, p. 2023–2026. DOI: 10.1109/ICOSP.2010.5655854

[14] WANG, W., PAN, X. Y., LIU, Y. C., et al. Sub-Nyquist sampling jamming against ISAR with compressive sensing. *IEEE Sensors Journal*, 2014, vol. 14, no. 9, p. 3131–3136. DOI: 10.1109/JSEN.2014.2323978

[15] PAN, X. Y., WANG, W., WANG, G. Y. Sub-Nyquist sampling jamming against ISAR with CS-based HRRP reconstruction. *IEEE Sensors Journal*, 2016, vol. 16, no. 6, p. 1597–1602. DOI: 10.1109/JSEN.2015.2503419

[16] TAHMOUSH, D. Review of micro-Doppler signitures. *IET Radar Sonar Navigation*, 2015, vol. 9, no. 9, p. 1140–1146. DOI: 10.1049/iet-rsn.2015.0118

[17] ZHAO, G. H., FU, T. W., NIE, L. et al. Imaging and micro-Doppler analysis of vibrating target in multi-input multi-output synthetic aperture radar. *IET Radar Sonar Navigation*, 2015, vol. 9, no. 9, p. 1360–1365. DOI: 10.1049/iet-rsn.2014.0480

[18] CHEN, V. C., LI, F. Y., HO, S. S., et al. Micro-Doppler effect in radar: Phenomenon, model, and simulation study. *IEEE Transaction on Aerospace and Electronic System*, 2006, vol. 42, no. 1, p. 2–21. DOI: 10.1109/TAES.2006.1603402

[19] CHEN, V. C. Doppler signatures of radar backscattering from objects with micro-motions. *IET Signal Processing*, 2006, vol. 2, no. 3, p. 291–300. DOI: 10.1049/iet-spr:20070137

[20] COLEGROVE, S. B., DAVEY, S. J., CHEUNG, B. Separation of target rigid body and micro-Doppler effects in ISAR imaging. *IEEE Transaction on Aerospace and Electronic System*, 2006, vol. 42, no. 4, p. 1496–1506. DOI: 10.1109/TAES.2006.314590

[21] WANG, Y., LIN, Y. C., DAKOVIC, M., et al. ISAR imaging of non-uniformly rotating target via range-instantaneous-Doppler-derivatives algorithm. *IEEE Journal of Selected Topics in Applied Earth Observations and Remote Sensing*, 2014, vol. 7, no. 1, p. 167–176. DOI: 10.1109/JSTARS.2013.2257699

[22] STANKOVIC, L., THAYAPARAN, T., DAKOVIC, M., et al. Micro-Doppler removal in the radar imaging analysis. *IEEE Transaction on Aerospace and Electronic System*, 2013, vol. 49, no. 2, p. 1234–1250. DOI: 10.1109/TAES.2013.6494410

[23] COSTA, P. Adaptive model architecture and extended Kalman-Bucy filters. *IEEE Transactions on aerospace and electronics systems*, 1994, vol. 30, no. 2, p. 525–533. DOI: 10.1109/7.272275

[24] LI, X. R., JILKOV, V. P. Survey of maneuvering target tracking. Part II: Motion models of ballistic and space targets. *IEEE Transactions on Aerospace and Electronic Systems*, 2010, vol. 46, no. 1, p. 96–119. DOI: 10.1109/TAES.2010.5417150

[25] MA, L., LIU, J., WANG, T., et al. Micro-Doppler characteristic of sliding-type scattering center on rotationally symmertric target. *Science China Information Sciences*, 2011, vol. 54, no. 9, p. 1957–1967. DOI: 10.1007/s11432-011-4254-3

[26] PACE, P. E., FOUTS, D. J., EKESTORM, S., et al. Digital false-target image synthesiser for countering ISAR. *IEE Radar Sonar Navigation*, 2002, vol. 149, no. 5, p. 248–257. DOI: 10.1049/ip-rsn:20020635

About the Authors . . .

Ning TAI was born in 1989. He received his M.Sc. from National University of Defense Technology in 2013. His research interests include radar signal processing and radar system simulation.

Chao WANG was born in 1977. He received his M.Sc. and Ph.D. from National University of Defense Technology in 2002 and 2007, respectively. His research interests include radar signal processing and microwave circuit design.

Liguo LIU was born in 1983. He received his M.Sc. and Ph.D. from National University of Defense Technology in 2008 and 2013, respectively. His research interests include the novel invisible structure and target characteristic.

Weiwei WU was born in 1981. She received his M.Sc. and Ph.D. from National University of Defense Technology in 2008 and 2011, respectively. Her research interests include antenna design and microwave circuit design.

Naichang YUAN was born in 1965. He received his M.Sc. and Ph.D. from Electronic Science and Technology from University of Electronic Science and Technology of China in 1991 and 1994, respectively. His research interests include array signal processing, SAR imaging processing and signal processing in radar.

11

A Blind Adaptive Color Image Watermarking Scheme Based on Principal Component Analysis, Singular Value Decomposition and Human Visual System

Muhammad IMRAN, Bruce A. HARVEY

Department of Electrical & Computer Engineering, College of Engineering
Florida State University, 2525 Pottsdamer St, Tallahassee, FL 32310, USA
mi14@my.fsu.edu, bharvey@fsu.edu

Abstract. *A blind adaptive color image watermarking scheme based on principal component analysis, singular value decomposition, and human visual system is proposed. The use of principal component analysis to decorrelate the three color channels of host image, improves the perceptual quality of watermarked image. Whereas, human visual system and fuzzy inference system helped to improve both imperceptibility and robustness by selecting adaptive scaling factor, so that, areas more prone to noise can be added with more information as compared to less prone areas. To achieve security, location of watermark embedding is kept secret and used as key at the time of watermark extraction, whereas, for capacity both singular values and vectors are involved in watermark embedding process. As a result, four contradictory requirements; imperceptibility, robustness, security and capacity are achieved as suggested by results. Both subjective and objective methods are acquired to examine the performance of proposed schemes. For subjective analysis the watermarked images and watermarks extracted from attacked watermarked images are shown. For objective analysis of proposed scheme in terms of imperceptibility, peak signal to noise ratio, structural similarity index, visual information fidelity and normalized color difference are used. Whereas, for objective analysis in terms of robustness, normalized correlation, bit error rate, normalized hamming distance and global authentication rate are used. Security is checked by using different keys to extract the watermark. The proposed schemes are compared with state-of-the-art watermarking techniques and found better performance as suggested by results.*

Keywords

Image watermarking, principal component analysis, singular value decomposition, human visual system, imperceptibility, robustness

1. Introduction

The enormous usage and afford-ability of Internet across the world, has made it easy to access online literature in the form of pictures, audios, videos, books, etc. The data available online can be downloaded, and then can be redistributed its copies multiple times, without any distinguishable difference between original and copied material. This illegal distribution and copyright violation results in the form of millions of dollars loss [1]. To overcome this illicit distribution and copyright violation digital watermarking is proposed as a prominent solution [2–7].

Watermarking is the way of embedding some information (image, audio, strings) into another data (image, audio, video, pdf file). The embedded data can later be extracted to prove the ownership. If the data which is being protected by embedding information in it is image, then this type of watermarking is called image watermarking. The information which is being hidden is called watermark, and the image in which watermark is hidden is called host image, and the resultant image is known as watermarked image. In terms of information required at the time of watermark extraction, watermarking is divided into blind and non-blind [2]. In later case, original image is required (and may be key) at the time of watermark extraction, whereas in former watermarking type there is no need of original image for extraction of watermark [3]. Therefore, blind watermarking is considered to be more secure and convenient, hence in this paper blind watermarking schemes are proposed.

The watermarking scheme for gray scale [2], [4], [5] and for color images [1], [3], [6], [7] are available in literature. One additional challenge of color image watermarking is that the three channels R, G, and B are highly correlated [3], [8], hence modification of one channel affects other channels severally and as a result the quality of whole image is compromised. This adverse effect can be avoided if Principal Component Analysis (PCA) [9] is used. In the proposed scheme PCA is used to decorrelate the three channels. In most cases either the same amount of information is embedded or modification in host image is made, same is case with [3], [6]. Whereas in actual there are some regions in image which are more tolerable to noise as compared to others. Therefore embedding more information in areas which are more prone to noise and less data into regions which are more susceptible to alterations. In this way both imperceptibility and robustness can be achieved simultaneously. For this purpose Human Visual System (HVS) and Fuzzy Inference System (FIS) are used to find adaptive scaling factor so that the amount of information is embedded according to the acceptability of host image.

In addition to perceptual quality of watermarked image and robustness, an other concern is that neither original nor false watermark should be extracted intentionally or unintentionally as was the case with [10–12]. Where, with the help of fake keys (singular vectors) a watermark different than the embedded watermark is extracted, which completely destroys the purpose of copyright protection. Therefore in designing proposed scheme, special attention is given to security and it is ensured that neither original nor false watermark is extracted as suggest by results in Sec. 5.3. The proposed scheme is discussed in detail in the following sections.

2. Proposed Scheme

In this paper two watermarking schemes are proposed. In proposed scheme 1, HVS and FIS are used together to find the areas where more information can be embedded, and regions where less data should be concealed. In this way two contradictory requirements of watermarking techniques; imperceptibility and robustness can be achieved. The perceptual quality of watermarked image is also improved due to the use of PCA. The third and utmost important quality is the security of watermarking scheme, which is given special attention in designing both the techniques. To achieve security, a different way of choosing elements for modifications from U, S and V are selected. In this way not only security but robustness and capacity are also improved as suggested by results in Sec. 5.2 and 5.3. In proposed scheme 2, HVS and FIS are not employed to find adaptive scaling factors, hence proposed scheme 1 is general case of proposed scheme 2. Since the proposed schemes are mostly relying on SVD, therefore, techniques chosen for comparision are also based on SVD. Both schemes are compared with state-of-the-art techniques [3], [6], [13], [14] and found that proposed schemes outperform. As HVS and FIS are used in this paper, therefore, they are discussed in following sections.

2.1 Human Visual System

Given an image I of size $M \times N$, the luminance masking [15] M_L, is calculated as follows:

$$M_L(x, y) = \max\{f_1(bg(x, y), mg(x, y)), f_2(bg(x, y))\} \quad (1)$$

where

$$f_1(bg(x, y), mg(x, y)) = mg(x, y)\alpha\{bg(x, y)) + \beta(bg(x, y)\},$$

$$f_2(bg(x, y)) = \begin{cases} T_o\left(1 - \left(\frac{bg(x,y)}{127}\right)^{0.5}\right) + 3, & bg(x, y) \le 127, \\ \gamma(bg(x, y) - 127) + 3, & bg(x, y) > 127, \end{cases}$$

$$\alpha(bg(x, y)) = \begin{cases} 0.0001bg(x, y) + 0.115, \\ 1 \le x \le H, 1 \le y \le W, \end{cases}$$

$$\beta(bg(x, y)) = \lambda - 0.01bg(x, y),$$

f_1 is the spatial masking function, $bg(x, y)$ is the background luminance, $mg(x, y)$ is the maximum weighted average of luminance differences around the pixel at location (x, y), $W = M/2$ and $H = N/2$. f_2, α and β are the background luminance dependent functions. The value of some other parameters are: $T_0 = 17$, $\gamma = 3/128$, and $\lambda = 1/2$. In order to know about the selection of parameter's values readers may refer [15]. $mg(x, y)$ and $bg(x, y)$ are calculated as follows:

$$mg(x, y) = \max_{k=1,2,3,4} \{|\text{grad}_k(x, y)|\},$$

$$\text{grad}_k(x, y) = \frac{1}{16}\sum_{i=1}^{5}\sum_{j=1}^{5} I(x - 3 + i, y - 3 + j)G_k(i, j),$$

$$bg(x, y) = \frac{1}{32}\sum_{i=1}^{5}\sum_{j=1}^{5} I(x - 3 + i, y - 3 + j)B(i, j).$$

The values of G_1, G_2, G_3, G_4 and B are shown below:

$$G_1 = \begin{bmatrix} 0 & 0 & 0 & 0 & 0 \\ 1 & 3 & 8 & 3 & 1 \\ 0 & 0 & 0 & 0 & 0 \\ -1 & -3 & -8 & -3 & -1 \\ 0 & 0 & 0 & 0 & 0 \end{bmatrix}, G_2 = \begin{bmatrix} 0 & 0 & 1 & 0 & 0 \\ 0 & 8 & 3 & 0 & 0 \\ 1 & 3 & 0 & -3 & -1 \\ 0 & 0 & -3 & -8 & 0 \\ 0 & 0 & -1 & 0 & 0 \end{bmatrix},$$

$$G_3 = \begin{bmatrix} 0 & 0 & 1 & 0 & 0 \\ 0 & 0 & 3 & 8 & 0 \\ -1 & -3 & 0 & 3 & 1 \\ 0 & -8 & -3 & 0 & 0 \\ 0 & 0 & 1 & 0 & 0 \end{bmatrix}, G_4 = \begin{bmatrix} 0 & 1 & 0 & -1 & 0 \\ 0 & 3 & 0 & -3 & 0 \\ 0 & 8 & 0 & -8 & 0 \\ 0 & 3 & 0 & -3 & 0 \\ 0 & 1 & 0 & -1 & 0 \end{bmatrix},$$

$$B = \begin{bmatrix} 1 & 1 & 1 & 1 & 1 \\ 1 & 2 & 2 & 2 & 1 \\ 1 & 2 & 0 & 2 & 1 \\ 1 & 2 & 2 & 2 & 1 \\ 1 & 1 & 1 & 1 & 1 \end{bmatrix}.$$

Remark 1: Since, darkness is more tolerable (then brightness) to watermark [15], [16]. More alternation can be made into dark areas as compared to bright regions of image, in this way, imperceptibility and robustness can be achieved.

2.2 Fuzzy Inference System

Mamdani fuzzy inference system [17] is used to find the adaptive scaling factor α_{M_L} using luminance masking shown in (1). The membership functions for luminance masking is shown in Fig. 1.

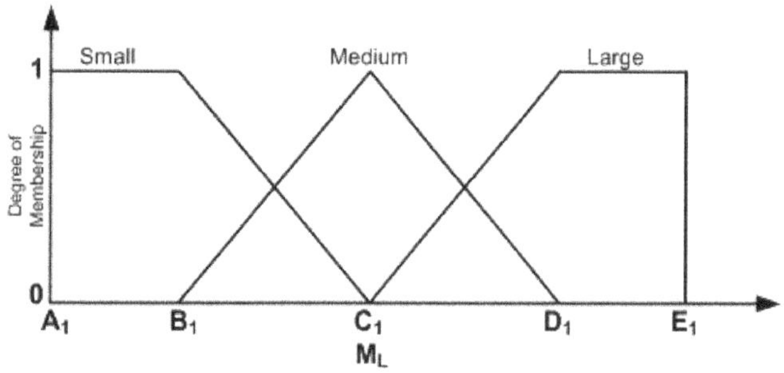

Fig. 1. Membership function for luminance masking.

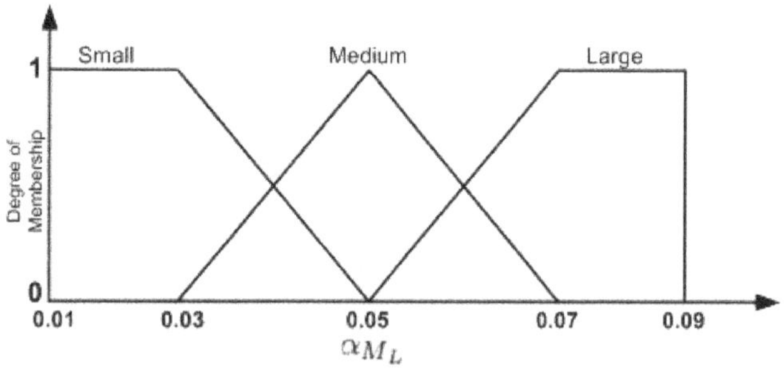

Fig. 2. Membership function for α_{M_L}.

Where the values A_1, B_1, C_1, D_1 and E_1 are calculated as:

$$A_1 = \min_{x=1}^{W} \min_{y=1}^{H} (M_L), \tag{2}$$

$$C_1 = \frac{1}{WH} \sum_{x=1}^{W} \sum_{y=1}^{H} M_L(x, y), \tag{3}$$

$$E_1 = \max_{x=1}^{W} \max_{y=1}^{H} M_L(x, y). \tag{4}$$

In order to chose B_1 and D_1, following condition must be satisfied

$$C_1 - B_1 = D_1 - C_1.$$

The membership function for adaptive scaling factor α_{M_L} is shown in Fig. 2.

The rules used for calculating α_{M_L} are as follows:

Ru^1 : IF M_L is large, THEN α_{M_L} is large,
Ru^2 : IF M_L is medium, THEN α_{M_L} is medium,
Ru^3 : IF M_L is small, THEN α_{M_L} is small.

The detailed procedure of watermark embedding and extraction are discussed in following sections.

3. Proposed Scheme 1

In proposed scheme 1, human visual system is used to find the areas of image which are more prone to alterations, and areas which are less prone to disturbances. The information of HVS is then used in FIS to find adaptive scaling factor for watermark embedding, so that the areas which are more prone to noise, can be embedded with more information as compared to areas which are less prone. In this way both; imperceptibility and robustness can be achieved as suggested by results in Sec. 5. The detailed procedure of watermark embedding and extraction are discussed in Sec. 3.1 and 3.2 respectively.

3.1 Watermark Embedding

1. Let the original image I is decomposed into three channels R, G, and B

$$R = \begin{bmatrix} r_{11} & r_{12} & \cdots & r_{1N} \\ r_{21} & r_{22} & \cdots & r_{2N} \\ \vdots & \vdots & \ddots & \vdots \\ r_{M1} & r_{M2} & \cdots & r_{MN} \end{bmatrix},$$

$$G = \begin{bmatrix} g_{11} & g_{12} & \cdots & g_{1N} \\ g_{21} & g_{22} & \cdots & g_{2N} \\ \vdots & \vdots & \ddots & \vdots \\ g_{M1} & g_{M2} & \cdots & g_{MN} \end{bmatrix},$$

$$B = \begin{bmatrix} b_{11} & b_{12} & \cdots & b_{1N} \\ b_{21} & b_{22} & \cdots & b_{2N} \\ \vdots & \vdots & \ddots & \vdots \\ b_{M1} & b_{M2} & \cdots & b_{MN} \end{bmatrix},$$

M and N defines the size of image.

2. Let a covariance matrix C be computed as

$$C = \frac{1}{MN}(AA^{\mathrm{T}}) = Q \wedge Q^{-1} \tag{5}$$

where

A =

$$\begin{bmatrix} r_{11} & \cdots & r_{1N} & r_{21} & \cdots & r_{2N} & \cdots & r_{M1} & \cdots r_{MN} \\ g_{11} & \cdots & g_{1N} & g_{21} & \cdots & g_{2N} & \cdots & g_{M1} & \cdots g_{MN} \\ b_{11} & \cdots & b_{1N} & b_{21} & \cdots & b_{2N} & \cdots & b_{M1} & \cdots b_{MN} \end{bmatrix},$$

$$Q = \begin{bmatrix} q_{11} & q_{12} & q_{13} \\ q_{21} & q_{32} & q_{33} \\ q_{31} & q_{32} & q_{33} \end{bmatrix}, \quad \wedge = \begin{bmatrix} \lambda_{11} & 0 & 0 \\ 0 & \lambda_{22} & 0 \\ 0 & 0 & \lambda_{33} \end{bmatrix},$$

$\lambda_{11} \geq \lambda_{22} \geq \lambda_{33}$ are eigenvalues in descending order.

3. The principal components [18] of covariance matrix C are calculated as

$$P = \begin{bmatrix} P_r \\ P_g \\ P_r \end{bmatrix} = Q^{\mathrm{T}} A =$$

$$= \begin{bmatrix} p_{r_{11}} & \cdots & p_{r_{1N}} & p_{r_{21}} & \cdots \\ p_{g_{11}} & \cdots & p_{g_{1N}} & p_{g_{21}} & \cdots \\ p_{b_{11}} & \cdots & p_{b_{1N}} & p_{b_{21}} & \cdots \end{bmatrix.}$$

$$\begin{matrix} \cdots & p_{r_{2N}} & \cdots & p_{r_{M1}} & \cdots & p_{r_{MN}} \\ \cdots & p_{g_{2N}} & \cdots & p_{g_{M1}} & \cdots & p_{g_{MN}} \\ \cdots & p_{b_{2N}} & \cdots & p_{b_{M1}} & \cdots & p_{b_{MN}} \end{matrix} \Bigg] .$$

Remark 2: Since R, G and B components are highly correlated [3], [7] and [8], modification in one channel causes alteration in other two channels, as a result imperceptibility is affected severely. However, this can be avoided if PCA [9] is used to decorrelate these R, G and B components.

4. Let matrix $\mathbf{P_{rn}}$ is formed from row vector $\mathbf{P_r}$ of matrix $\mathbf{P}$ as shown below

$$P_{rn} = \begin{bmatrix} p_{r11} & p_{r12} & \cdots & p_{r1N} \\ p_{r21} & p_{r22} & \cdots & p_{r2N} \\ \vdots & \vdots & \ddots & \vdots \\ p_{rM1} & p_{rM2} & \cdots & p_{rMN} \end{bmatrix}.$$

Remark 3: Since $\mathbf{P_{rn}}$ is composed of components from $\mathbf{P_r}$ and contains most of the information [9], therefore it is chosen for watermark embedding.

5. Let $\mathbf{P_{rn}}$ is divided into non-overlapping blocks $\mathbb{A}_1$, $\mathbb{A}_2, \ldots \mathbb{A}_{\mathbf{pq}}$. Where the size of each block is 4×4, $p \leq M/4$ and $q \leq N/4$.

6. Let each block is decomposed using SVD as follows

$$\mathbb{A}_i = U_i S_i V_i^{\mathrm{T}}, \quad i = 1, 2, \ldots, pq \tag{6}$$

where S is a diagonal matrix containing singular values in descending order, whereas, U and V represent the left and right singular vectors respectively.
Remark 4: Small perturbation in image does not cause large variation in singular values and vice versa [5], [7].
Remark 5: Singular values contain intrinsic properties of image, whereas geometric information is maintained by corresponding singular vectors [5], [7].

7. Compute the covariance matrix of each block $\mathbb{A}_i$. Select the column number with lowest covariance (variance) value, and then location of two values with lowest covariance values within that selected column from covariance matrix of each block. In this way three numbers are selected for each block, and those selected locations will later be used as key at the time of watermark extraction. For instance (f, e and g) are selected, where, 'f' represents the column number with lowest covariance value, 'e' and 'g' denoting the locations of two values with lowest covariance values from selected columns. These three numbers are used to select two numbers from U and V for modifications. For example in case of (f, e, and g), tow numbers at location (e, f) and (g, f) from U, similarly two values from location (f, e) and (f, g) from V are selected for watermark embedding. In addition to that singular values either at location (e, e) or (g, g) based on embedding bits are modified as shown in (7)–(10). It is shown in result in Sec. 5 that the procedure described above for choosing two numbers for modifications gives better results in terms of imperceptibility.

8. Let the luminance masking M_{L_i} for each block $\mathbb{A}_i$, where $i = 1, 2, 3, \ldots, pq$ is calculated using (1).

9. The adaptive scaling factor $\alpha_{ML_i}(g, f)$, $\alpha_{ML_i}(e, f)$, $\alpha_{ML_i}(f, g)$ and $\alpha_{ML_i}(f, e)$ of each block $\mathbb{A}_i$ for values of selected positions are calculated by utilizing luminance masking values M_{L_i} of block 'i'.

10. Given a watermark W of size $M/4 \times N/4$, the watermarking bits W_i are embedded as described below in (7)–(10).

Case 1: If embedding bit is 1 i.e. $(W_i = 1)$.

$$\overbrace{\begin{aligned} u_{wi(e,f)} &= \mathrm{sgn}\big(u_{i(e,f)}\big) \times \big(\bar{U}_i + (\beta\alpha_{ML_i}(e,f))/2\big) \\ u_{wi(g,f)} &= \mathrm{sgn}\big(u_{i(g,f)}\big) \times \big(\bar{U}_i - (\beta\alpha_{ML_i}(g,f))/2\big) \\ v_{wi(f,e)} &= \mathrm{sgn}\big(v_{i(f,e)}\big) \times \big(\bar{V}_i + (\beta\alpha_{ML_i}(f,e))/2\big) \\ v_{wi(f,g)} &= \mathrm{sgn}\big(v_{i(f,g)}\big) \times \big(\bar{V}_i - (\beta\alpha_{ML_i}(f,g))/2\big) \\ s_{wi(g,g)} &= 3 \times s_{i(e,e)} \end{aligned}}^{\text{if } e < g} \tag{7}$$

$$\overbrace{\begin{aligned} u_{wi(g,f)} &= \mathrm{sgn}\big(u_{i(g,f)}\big) \times \big(\bar{U}_i + (\beta\alpha_{ML_i}(g,f))/2\big) \\ u_{wi(e,f)} &= \mathrm{sgn}\big(u_{i(e,f)}\big) \times \big(\bar{U}_i - (\beta\alpha_{ML_i}(e,f))/2\big) \\ v_{wi(f,g)} &= \mathrm{sgn}\big(v_{i(f,g)}\big) \times \big(\bar{V}_i + (\beta\alpha_{ML_i}(f,g))/2\big) \\ v_{wi(f,e)} &= \mathrm{sgn}\big(v_{i(f,e)}\big) \times \big(\bar{V}_i - (\beta\alpha_{ML_i}(f,e))/2\big) \\ s_{wi(e,e)} &= 3 \times s_{i(g,g)} \end{aligned}}^{\text{If } e > g} \tag{8}$$

Case 2: If embedding bit is 0 i.e. $(W_i = 0)$.

$$\overbrace{\begin{aligned} u_{wi(e,f)} &= \mathrm{sgn}\big(u_{i(e,f)}\big) \times \big(\bar{U}_i - (\beta\alpha_{ML_i}(e,f))/2\big) \\ u_{wi(g,f)} &= \mathrm{sgn}\big(u_{i(g,f)}\big) \times \big(\bar{U}_i + (\beta\alpha_{ML_i}(g,f))/2\big) \\ v_{wi(f,e)} &= \mathrm{sgn}\big(v_{i(f,e)}\big) \times \big(\bar{V}_i - (\beta\alpha_{ML_i}(f,e))/2\big) \\ v_{wi(f,g)} &= \mathrm{sgn}\big(v_{i(f,g)}\big) \times \big(\bar{V}_i + (\beta\alpha_{ML_i}(f,g))/2\big) \\ s_{wi(g,g)} &= 3 \times s_{i(e,e)} \end{aligned}}^{\text{If } e < g} \tag{9}$$

$$\overbrace{\begin{aligned} u_{wi(g,f)} &= \mathrm{sgn}\big(u_{i(g,f)}\big) \times \big(\bar{U}_i - (\beta\alpha_{ML_i}(g,f))/2\big) \\ u_{wi(e,f)} &= \mathrm{sgn}\big(u_{i(e,f)}\big) \times \big(\bar{U}_i + (\beta\alpha_{ML_i}(e,f))//2\big) \\ v_{wi(f,g)} &= \mathrm{sgn}\big(v_{i(f,g)}\big) \times \big(\bar{V}_i - (\beta\alpha_{ML_i}(f,g))//2\big) \\ v_{wi(f,e)} &= \mathrm{sgn}\big(v_{i(f,e)}\big) \times \big(\bar{V}_i + (\beta\alpha_{ML_i}(f,e))//2\big) \\ s_{wi(e,e)} &= 3 \times s_{i(g,g)} \end{aligned}}^{\text{If } e > g} \tag{10}$$

where

$$\mathrm{sgn}(x) = \begin{cases} -1, & \text{if} x < 0, \\ 0, & \text{if} x = 0, \\ 1, & \text{if} x > 0, \end{cases}$$

$$\bar{U}_i = \frac{|u_{i(e,f)} + u_{i(g,f)}|}{2}, \quad \bar{V}_i = \frac{|v_{i(f,e)} + v_{i(f,g)}|}{2}.$$

W_i represents the watermark bit, where $1 \leq i \leq MN/16$. In (7)–(10), β represents the threshold defining the amount of change in values, whereas, $u_{wi(e,f)}$, $u_{wi(g,f)}$, $v_{wi(f,e)}$, $v_{wi(f,g)}$, $s_{wi(e,e)}$, and $s_{wi(g,g)}$ represents the modified (watermark added) values at locations (e, f), (g, f), (f, e), (f, g), (e, e) and (g, g) respectively for block i. For (7) and (9) the key would be

$\mathbb{K}_i = \{f, e, g\}$ whereas in case of (8) and (10) key would be $\mathbb{K}_i = \{f, g, e\}$ for block i.

Remark 6: Modification in columns of U as compared to rows of U are less visible, likewise, in V the modification in rows are less visible as compared to changes in columns in V [4].

11. Let the modified blocks are formed as follows

$$\mathbb{A}_{wi} = U_{wi} S_{wi} V_{wi}^{\mathrm{T}}, \qquad i = 1, 2, \ldots pq, \tag{11}$$

where, the subscript 'w' is representing the modified (or watermark added) blocks.

12. The modified blocks $\mathbb{A}_{w1}, \mathbb{A}_{w2}, \ldots, \mathbb{A}_{wpq}$ are combined to form modified first principal component P_{rnw}, where

$$P_{rnw} = \begin{bmatrix} P_{rw_{11}} & P_{rw_{12}} & \cdots & P_{rw_{1N}} \\ P_{rw_{21}} & P_{rw_{22}} & \cdots & P_{rw_{2N}} \\ \vdots & \vdots & \ddots & \vdots \\ P_{rw_{M1}} & P_{rw_{M2}} & \cdots & P_{rw_{MN}} \end{bmatrix}.$$

13. The modified principal components are obtained as

$$P_w = \begin{bmatrix} P_{rw} \\ P_g \\ P_b \end{bmatrix}$$

where $\mathbf{P_{rw}}$ is a row vector, obtained from $\mathbf{P_{rnw}}$.

14. The matrix A_w is obtained as

$$A_w = Q P_w =$$

$$= \begin{bmatrix} rw_{11} & \ldots & rw_{1N} & rw_{21} & \ldots \\ g_{11} & \ldots & g_{1N} & g_{21} & \ldots \\ b_{11} & \ldots & b_{1N} & b_{21} & \ldots \end{bmatrix}$$

$$\begin{matrix} \ldots & rw_{2N} & \ldots & rw_{M1} & \ldots rw_{MN} \\ \ldots & g_{2N} & \ldots & g_{M1} & \ldots g_{MN} \\ \ldots & b_{2N} & \ldots & b_{M1} & \ldots b_{MN} \end{matrix} \Bigg].$$

15. Finally, R_W, G and B channels are combined to form the watermarked image I_w, where

$$R_W = \begin{bmatrix} rw_{11} & rw_{11} & \cdots & rw_{1N} \\ rw_{21} & rw_{22} & \cdots & rw_{2N} \\ \vdots & \vdots & \ddots & \vdots \\ rw_{M1} & rw_{M2} & \cdots & rw_{MN} \end{bmatrix}.$$

3.2 Watermark Extraction

1. Let received watermarked image $\hat{I}_w$ be decomposed into its components $\hat{R}_w$, $\hat{G}$, and $\hat{B}$, where

$$\hat{R}_w = \begin{bmatrix} r\hat{w}_{11} & r\hat{w}_{12} & \cdots & r\hat{w}_{1N} \\ r\hat{w}_{21} & r\hat{w}_{22} & \cdots & r\hat{w}_{2N} \\ \vdots & \vdots & \ddots & \vdots \\ r\hat{w}_{M1} & r\hat{w}_{M2} & \cdots & r\hat{w}_{MN} \end{bmatrix},$$

$$\hat{G} = \begin{bmatrix} \hat{g}_{11} & \hat{g}_{12} & \cdots & \hat{g}_{1N} \\ \hat{g}_{21} & \hat{g}_{22} & \cdots & \hat{g}_{2N} \\ \vdots & \vdots & \ddots & \vdots \\ \hat{g}_{M1} & \hat{g}_{M2} & \cdots & \hat{g}_{MN} \end{bmatrix},$$

$$\hat{B} = \begin{bmatrix} \hat{b}_{11} & \hat{b}_{12} & \cdots & \hat{b}_{1N} \\ \hat{b}_{21} & \hat{b}_{22} & \cdots & \hat{b}_{2N} \\ \vdots & \vdots & \ddots & \vdots \\ \hat{b}_{M1} & \hat{b}_{M2} & \cdots & \hat{b}_{MN} \end{bmatrix}.$$

2. Let a covariance matrix $\hat{C}$ be computed as

$$\hat{C} = \frac{1}{MN}(\hat{A}\hat{A}^{\mathrm{T}}) = \hat{Q}\hat{\wedge}\hat{Q}^{-1} \tag{12}$$

where

$$\hat{A} = \begin{bmatrix} r\hat{w}_{11} & \ldots & r\hat{w}_{1N} & r\hat{w}_{21} & \ldots \\ \hat{g}_{11} & \ldots & \hat{g}_{1N} & \hat{g}_{21} & \ldots \\ \hat{b}_{11} & \ldots & \hat{b}_{1N} & \hat{b}_{21} & \ldots \end{bmatrix}$$

$$\begin{matrix} \ldots & r\hat{w}_{2N} & \ldots & r\hat{w}_{M1} & \ldots r\hat{w}_{MN} \\ \ldots & \hat{g}_{2N} & \ldots & \hat{g}_{M1} & \ldots \hat{g}_{MN} \\ \ldots & \hat{b}_{2N} & \ldots & \hat{b}_{M1} & \ldots \hat{b}_{MN} \end{matrix} \Bigg].$$

$$\hat{Q} = \begin{bmatrix} \hat{q}_{11} & \hat{q}_{12} & \hat{q}_{13} \\ \hat{q}_{21} & \hat{q}_{32} & \hat{q}_{33} \\ \hat{q}_{31} & \hat{q}_{32} & \hat{q}_{33} \end{bmatrix}, \qquad \hat{\wedge} = \begin{bmatrix} \hat{\lambda}_{11} & 0 & 0 \\ 0 & \hat{\lambda}_{22} & 0 \\ 0 & 0 & \hat{\lambda}_{33} \end{bmatrix},$$

$\hat{\lambda}_{11} \geq \hat{\lambda}_{22} \geq \hat{\lambda}_{33}$ are eigenvalues in descending order.

3. The principal components of covariance matrix $\hat{C}$ are calculated as

$$\hat{P}_w = \begin{bmatrix} \hat{P}_{rw} \\ \hat{P}_g \\ \hat{P}_b \end{bmatrix} = \hat{Q}^{\mathrm{T}}\hat{A} =$$

$$= \begin{bmatrix} \hat{P}_{rw_{11}} & \ldots & \hat{P}_{rw_{1N}} & \hat{P}_{rw_{21}} & \ldots \\ \hat{P}_{g_{11}} & \ldots & \hat{P}_{g_{1N}} & \hat{P}_{g_{21}} & \ldots \\ \hat{P}_{b_{11}} & \ldots & \hat{P}_{b_{1N}} & \hat{P}_{b_{21}} & \ldots \end{bmatrix}$$

$$\begin{matrix} \ldots & \hat{P}_{rw_{2N}} & \ldots & \hat{P}_{rw_{M1}} & \ldots & \hat{P}_{rw_{MN}} \\ \ldots & \hat{P}_{g_{2N}} & \ldots & \hat{P}_{g_{M1}} & \ldots & \hat{P}_{g_{MN}} \\ \ldots & \hat{P}_{b_{2N}} & \ldots & \hat{P}_{b_{M1}} & \ldots & \hat{P}_{b_{MN}} \end{matrix} \Bigg].$$

4. Let matrix $\hat{\mathbf{P}}_{\mathbf{rnw}}$ is obtained by converting row vector $\mathbf{P_{rw}}$ into a matrix of size $M \times N$ as shown below

$$\hat{P}_{rnw} = \begin{bmatrix} \hat{P}_{rw_{11}} & \hat{P}_{rw_{12}} & \cdots & \hat{P}_{rw_{1N}} \\ \hat{P}_{rw_{21}} & \hat{P}_{rw_{22}} & \cdots & \hat{P}_{rw_{2N}} \\ \vdots & \vdots & \ddots & \vdots \\ \hat{P}_{rw_{M1}} & \hat{P}_{rw_{M2}} & \cdots & \hat{P}_{rw_{MN}} \end{bmatrix}.$$

5. The $\hat{\mathbf{P}}_{\mathbf{rnw}}$ is divided into non-overlapping blocks $\hat{\mathbb{A}}_{w1}, \hat{\mathbb{A}}_{w2}, \ldots \hat{\mathbb{A}}_{wpq}$, each of size 4×4. Using SVD, each block is decomposed as follows:

$$\hat{A}_{wi} = \hat{U}_{wi}\hat{S}_{wi}\hat{V}_{wi}^{\mathrm{T}}, \quad i = 1, 2, \ldots, pq. \tag{13}$$

6. Based on keys: $\mathbb{K}_i = \{k_1, k_2, k_3\}$, the watermarking bits are extracted as follows:

$$\xi = \begin{cases} 1, & \text{if} \quad \hat{u}_{wi(k_2,K_1)} \leq \hat{u}_{wi(k_3,K_1)}, \\ 0, & \text{otherwise.} \end{cases} \tag{14}$$

$$\zeta = \begin{cases} 1, & \text{if} \quad \hat{v}_{wi(k_1,K_2)} \leq \hat{v}_{wi(k_1,K_3)}, \\ 0, & \text{otherwise.} \end{cases} \tag{15}$$

$$\psi = \begin{cases} 1, & \text{if} \quad \hat{s}_{wi(k_2,K_2)} \leq \hat{s}_{wi(k_3,K_3)}, \\ 0, & \text{otherwise.} \end{cases} \tag{16}$$

Once three values ξ, ζ, and ψ are calculated for each block, then watermarking bits are extracted as follows.

$$\hat{W}_i = \begin{cases} \psi, & \text{if} \quad (\xi = \psi) \vee (\zeta = \psi), \\ \vartheta, & \text{otherwise} \end{cases} \tag{17}$$

where

$$\vartheta = \text{Mode}\{\xi,\ \psi,\ \zeta\}.$$

4. Proposed Scheme 2

The only difference in proposed scheme 1 and 2 is that, in scheme 2 neither HVS nor FIS used to find adaptive factor α_{M_L} for watermark embedding. Instead only constant scaling factor β having values from 0.1 to 0.9 with a step size of 0.2 is used. The watermark embedding and extraction procedure are described in following sections.

4.1 Watermark Embedding

Eliminating step 8 and step 9, and the terms $\alpha_{ML_i}(g, f)$, $\alpha_{ML_i}(e, f)$, $\alpha_{ML_i}(f, g)$ and $\alpha_{ML_i}(f, e)$ from (7)–(10) of Sec. 3.1 will reduce to watermark embedding procedure of proposed scheme 2.

4.2 Watermark Extraction

The watermark extraction procedure of proposed scheme 2 is same as of proposed scheme 1 which is described in Sec. 3.2.

5. Experimental Results

In order to test the performance of proposed schemes a number of experiments were performed. For this purpose six different images shown in Fig. 3, each of size 512 × 512 were used as host images. Whereas for watermark a binary image shown in Fig. 4(a) of size 64 × 64 was used. The performance of proposed schemes is evaluated in terms of imperceptibility, robustness, security and capacity, which are discussed in following sections.

5.1 Imperceptibility

To end user the original and watermarked image should look similar, in other words, there should be no visible difference between original and watermarked images [3], this is referred as imperceptibility. For qualitative analysis watermarked images are shown in Fig. 3, whereas, for quantitative evaluation Peak-Signal-to-Noise-Ration (PSNR) [7] shown in (18) is used to measure imperceptibility. The PSNR for proposed scheme 1 for different images and for distinct constant scaling factor is shown in Tab. 1. The use of constant scaling factor here is only for reference, otherwise, proposed scheme 1 performs well without constant scaling factor. The watermarked images are shown for proposed scheme 1, however, the watermarked images obtained from proposed scheme 2 also have good perceptual quality.

$$PSNR(\text{dB}) = 10\log_{10}\left(\frac{G^2}{H}\right) \tag{18}$$

where

$$G = \max\{I(m,n) : 1 \leq m \leq M, 1 \leq n \leq N\},$$

$$H = \frac{1}{M \times N}\sum_{m=1}^{M}\sum_{n=1}^{N}(I(m,n) - I_w(m,n))^2.$$

Images	Constant Scaling Factor (β)				
	0.1	0.3	0.5	0.7	0.9
Lena	57.2311	56.5784	55.9255	55.2696	54.6518
Baboon	46.7905	46.5187	46.2263	45.9465	45.6602
Autumn	55.6855	53.8721	51.8756	50.1673	48.5929
Airplane	55.0401	54.4737	53.9122	53.3167	52.7232
Peppers	59.5511	58.4239	57.2720	56.1712	55.0926
Crane	54.5905	54.1370	53.6779	53.2249	52.7693

Tab. 1. PSNR for different values of scaling factors (Proposed scheme 1).

The proposed schemes are compared with state-of-the-art schemes in terms of PSNR as shown in Tab. 2. It is clearly visible that both proposed schemes performs well with the compared techniques in terms of PSNR. However, the proposed scheme 1 outperforms proposed scheme 2 in terms of imperceptibility, and that is due to the use of HVS and FIS to find adaptive scaling factors.

Images	For Constant Scaling Factor $\beta = 0.5$					
	Proposed		Presented in			
	1	2	[6]	[3]	[13]	[14]
Lena	55.9255	42.0455	38.7070	36.1329	38.9620	34.3263
Baboon	46.2263	37.3587	30.2003	27.5918	33.0635	27.3159
Autumn	51.8756	41.7374	37.9715	35.0159	36.0157	33.6153
Airplane	53.9122	39.9456	38.8983	35.4927	37.3391	34.7828
Peppers	57.2720	39.6113	38.6906	34.2894	38.8031	33.2607
Crane	53.6779	45.7264	44.9576	38.5728	40.0731	37.0299

Tab. 2. PSNR for different techniques.

Generally the quality of watermarked images can either be evaluated qualitatively or qualitatively. For qualitative evaluation of proposed scheme both original and watermarked images are shown in Fig. 3. For quantitative measurement, PSNR shown in (18) is used, the results and comparision are shown in Tab. 1 and Tab. 2 respectively. However, PSNR is not only the metric to measure the quality of watermarked images, Structural Similarity Index (SSIM), Visual

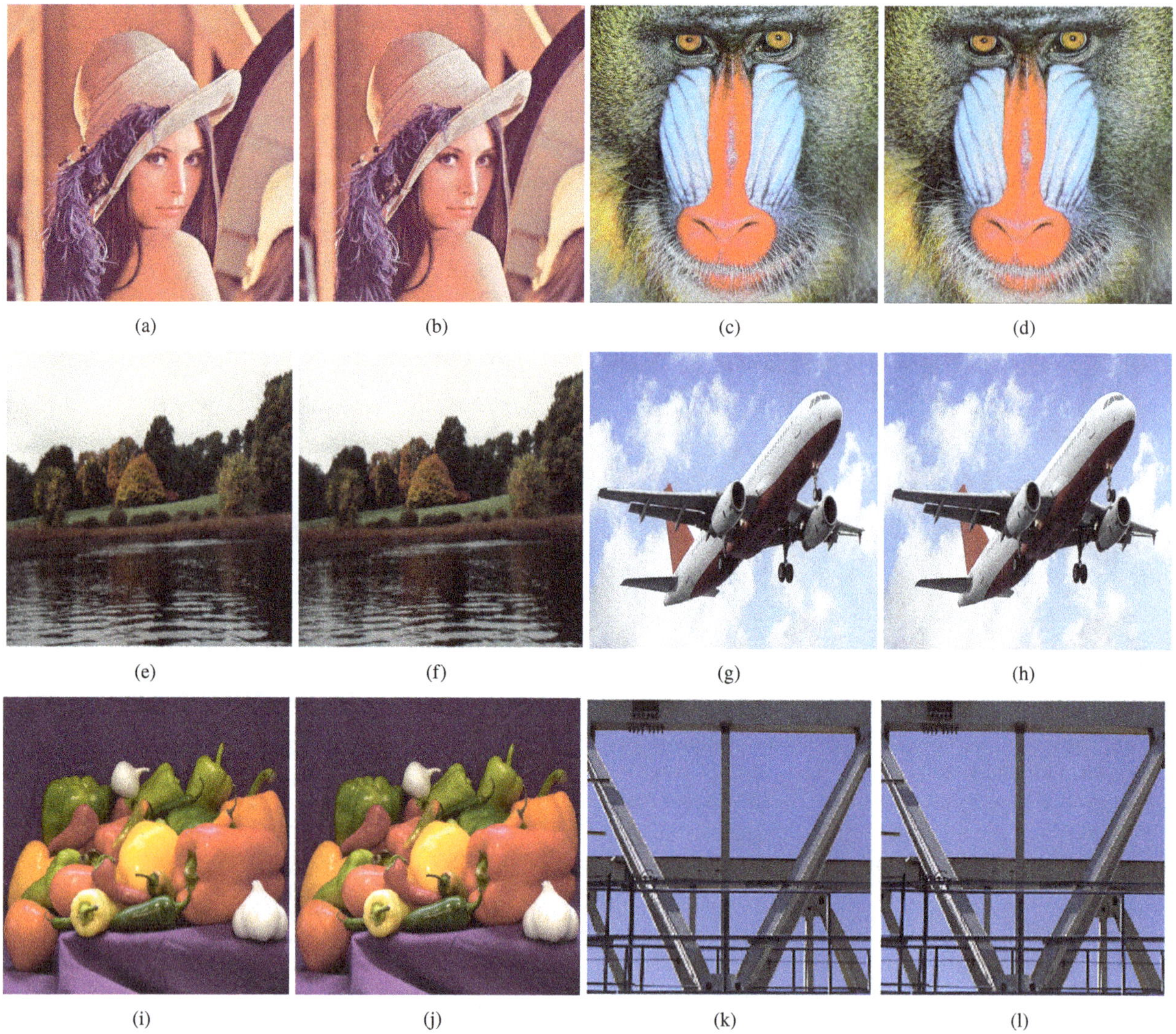

Fig. 3. Host and watermarked images (512 × 512) for proposed scheme 1 (β = 0.5) (a). Lena (original), (b). Lena (watermarked) (c). Baboon (original) (d). Baboon (watermarked) (e). Autumn (original) (f). Autumn (watermarked) (g). Airplane (original) (h). Airplane (watermarked) (i). Peppers (original) (j). Peppers (watermarked) (k). Crane (original) (l). Crane (watermarked).

Information Fidelity (VIF) and Normalized Color Difference (NCD) also used to examine the quality of watermarked images, quantitatively. Therefore, in this paper in addition to PSNR, SSIM, VIF and NCD are also utilized for evaluation [3]. SSIM [19] utilizes luminance ($\mathcal{L}$), contrast (C) and structural information ($\mathcal{S}$) to find the distortion introduced in watermarked images. Given original I and watermarked I_w images, the SSIM can be calculated using (19).

$$SSIM(I, Iw) = \mathcal{L}(I, I_w)^{\gamma_{\mathcal{L}}} \times \mathcal{S}(I, I_w)^{\gamma_{\mathcal{S}}} \times C(I, I_w)^{\gamma_C} \quad (19)$$

where $\gamma_{\mathcal{L}} > 0$, $\gamma_{\mathcal{L}} > 0$ and $\gamma_{\mathcal{L}} > 0$, are constant used to describe the dependency of each component. For equal contribution of $\mathcal{L}$, $\mathcal{S}$ and C in the calculation of SSIM, $\gamma_{\mathcal{L}} = \gamma_{\mathcal{S}} = \gamma_C = 1$ are set equal to 1. The $\mathcal{L}$, C and $\mathcal{S}$ defined in (19) are calculated as follows.

$$\mathcal{L}(I, I_w) = \frac{2\mu_I \mu_{I_w} + C_1}{\mu_I^2 + \mu_{I_w}^2 + C_1}, \quad C(I, I_w) = \frac{2\rho_I \rho_{I_w} + C_2}{\rho_I^2 + \rho_{I_w}^2 + C_2},$$
$$\mathcal{S}(I, I_w) = \frac{\rho_{II_w} + C_3}{\rho_I^2 \rho_{I_w}^2 + C_3} \quad (20)$$

where μ_I, μ_{I_w}, ρ_I and ρ_{I_w} denotes the mean and covariance of host and watermarked images respectively. C_1, C_2 and C_3 are small constants used to avoid the situations where the sum of means or covariances can be zero. Using (20) and setting $C_3 = C_2/2$ in (19), will result in the form of equation shown below

$$SSIM = \frac{(2\mu_I \mu_{I_w} + C_1) \times (2\rho_I \rho_{I_w} + C_2)}{(2\mu_I^2 + \mu_{I_w}^2 + C_1) \times (2\rho_I^2 + \rho_{I_w}^2 + C_2)}. \quad (21)$$

Images	For Constant Scaling Factor $\beta = 0.5$					
	Proposed		Presented in			
	1	2	[6]	[3]	[13]	[14]
Lena	0.9580	0.9088	0.8366	0.7810	0.8422	0.6881
Baboon	0.9118	0.8796	0.7911	0.7228	0.8661	0.5972
Autumn	0.9439	0.9279	0.8442	0.7785	0.8007	0.7266
Airplane	0.9428	0.8921	0.8687	0.7926	0.8338	0.7383
Peppers	0.9653	0.9130	0.8918	0.7904	0.8944	0.6775
Crane	0.9259	0.9033	0.8574	0.7356	0.7642	0.6797

Tab. 3. SSIM for different techniques.

The comparision of proposed schemes with existing techniques in terms of SSIM shown in Tab. 3, clearly demonstrates the improvement of proposed schemes over present watermarking techniques.

VIF [20] introduced in 2006 is also used to asses the quality of images. In this paper VIF shown in (22), is also used to examine the quality of watermarked images with respect to original host images

$$VIF = \frac{\sum_{j\in\text{channels}} \mathbb{I}(\vec{C}^{N,j}, \vec{I}^{N,j}|s^{N,j})}{\sum_{j\in\text{channels}} \mathbb{I}(\vec{C}^{N,j}, \vec{I_w}^{N,j}|s^{N,j})} \tag{22}$$

where $\mathbb{I}(\vec{C}^{N,j}, \vec{I}^{N,j}|s)$ and $\mathbb{I}(\vec{C}^{N,j}, \vec{I}^{N,j}|s)$ represent the information that a brain can extract from images using HVS of original and watermarked images respectively, $\vec{C}^{N,j}$ represents the N elements of random field C_j for $j-th$ sub-band. For detailed description of parameters and calculation of information $\mathbb{I}$ from host and watermarked images, [20] can be referred.

The comparison in terms of VIF of proposed schemes with that of state-of-the-art techniques is shown in Tab. 4. The better visual quality of watermarked images obtained from proposed schemes is clearly visible from the results shown in Tab. 4.

Finally, NCD [3] shown in (23), is also used to evaluate the quality of watermarked images.

$$NCD = \frac{\sum_{i=1}^{M}\sum_{j=1}^{N}\sqrt{(\mathcal{L}(i,j)-\mathcal{L}_w(i,j))^2+(a(i,j)-a_w(i,j))^2+(b(i,j)-b_w(i,j))^2}}{\sum_{i=1}^{M}\sum_{j=1}^{N}\sqrt{(\mathcal{L}(i,j))^2+(a(i,j))^2+(b(i,j))^2}} \tag{23}$$

where $\mathcal{L}$ represents the luminance, a and b denote the chrominance. It should be noted that in order to calculate NCD, the RGB color model must be converted to Lab color space. The performance of proposed scheme is calculated in terms of NCD and compared with existing schemes as shown in Tab. 5.

Images	For Constant Scaling Factor $\beta = 0.5$					
	Proposed		Presented in			
	1	2	[6]	[3]	[13]	[14]
Lena	0.8959	0.8423	0.7954	0.7425	0.8006	0.7053
Baboon	0.8584	0.8154	0.7592	0.6936	0.7811	0.6866
Autumn	0.8764	0.8282	0.7535	0.6948	0.7146	0.6670
Airplane	0.8886	0.8343	0.8124	0.7413	0.7799	0.7265
Peppers	0.9138	0.8779	0.8575	0.7600	0.8600	0.7372
Crane	0.8805	0.8548	0.8404	0.7210	0.7490	0.6921

Tab. 4. VIF for different techniques.

Images	For Constant Scaling Factor $\beta = 0.5$					
	Proposed		Presented in			
	1	2	[6]	[3]	[13]	[14]
Lena	0.0233	0.0408	0.0784	0.1516	0.0619	0.1164
Baboon	0.0716	0.1295	0.2342	0.4482	0.1889	0.3450
Autumn	0.0579	0.1045	0.1996	0.3837	0.2110	0.4079
Airplane	0.0209	0.0364	0.0718	0.1373	0.0671	0.1296
Peppers	0.0448	0.0758	0.1498	0.2826	0.0972	0.1805
Crane	0.0268	0.0496	0.0984	0.1828	0.0871	0.1676

Tab. 5. NCD for different techniques.

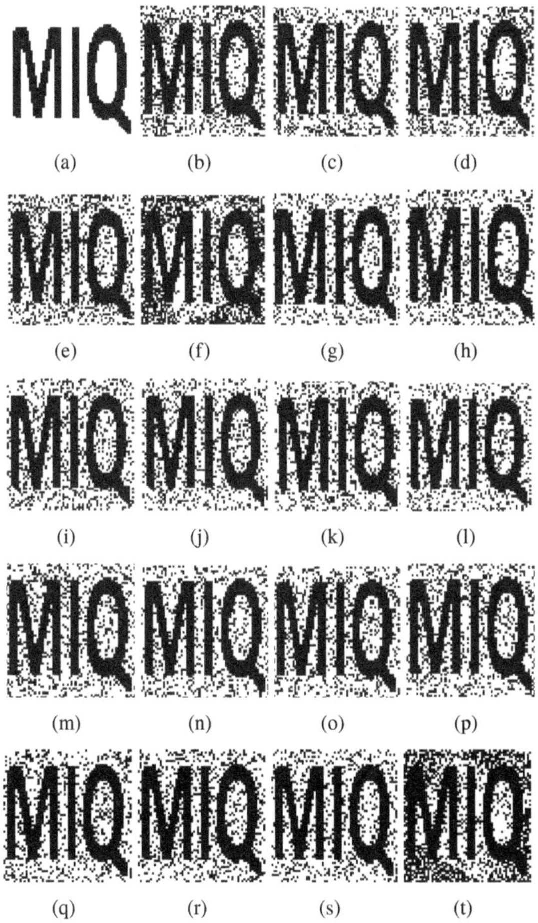

Fig. 4. Extracted watermarks from attacked watermarked images (proposed scheme 1 ($\beta = 0.5$)): (a) Original watermark (b) RO, (c) TR, (d) XSH, (e) YSH, (f) AFT, (g). SC, (h) CR, (i). GN, (j) S&P (k) SN, (l) MB, (m) SB, (n) AF, (o) HE, (p) JQ (q) JQ+AF, (r) JQ+GN, (s) JQ+SC (t) RO+SC.

5.2 Robustness

Robustness refers to the ability of watermarking scheme to withstand against intentional or unintentional attacks, that may be applied on watermarked images either to remove or to destroy the hidden information [7], [21]. Robustness is measured using normalized correlation [6] shown in (24). In order to check the robustness of proposed schemes different attacks like rotation (RO), translation (TR), x-shearing (XSH), y-shearing (YSH), scaling (SC), cropping (CR), affine transformation (AFT), Gaussian noise (GN), salt & pepper noise

Attacks		Constant Scaling Factor (β)				
Attack	**Parameters**	**0.1**	**0.3**	**0.5**	**0.7**	**0.9**
Rotation	$\theta = 45$	0.8031	0.7283	0.8229	0.8211	0.7286
	$\theta = 60$	0.8258	0.8279	0.8103	0.8098	0.8112
	$\theta = 125$	0.8272	0.8064	0.8068	0.8064	0.8066
Translation	displacement 60%	0.8142	0.8135	0.8135	0.8596	0.8117
	displacement 120%	0.8089	0.7444	0.7444	0.8084	0.7441
X-Shearing	Shearing Factor=0.4	0.7998	0.7987	0.8466	0.7982	0.8466
	Shearing Factor=-0.5	0.7989	0.7996	0.8466	0.8488	0.7980
Y-Shearing	Shearing Factor=-0.4	0.8488	0.8322	0.7873	0.7899	0.8455
	Shearing Factor=0.5	0.7715	0.8413	0.8418	0.7722	0.7717
Affine Transformation	Transform Factor=0.5	0.7563	0.7551	0.7558	0.8188	0.7556
Scaling	3times	0.8712	0.8695	0.8693	0.8682	0.8693
	0.5times	0.8792	0.8800	0.8796	0.8787	0.8783
Cropping	10%	0.8676	0.8682	0.8678	0.8676	0.8674
	25%	0.8646	0.8646	0.8648	0.8641	0.8646
Gaussian Noise	$\mu = 0.4, \sigma^2 = .01$	0.8654	0.8635	0.8659	0.8678	0.8643
	$\mu = 0.5, \sigma^2 = 0.5$	0.8641	0.8563	0.8587	0.8663	0.8700
Salt & Pepper Noise	Density=0.1	0.8652	0.8633	0.8704	0.8654	0.8691
	Density=0.5	0.8609	0.8650	0.8704	0.8652	0.8609
Speckle Noise	Density=0.1	0.8725	0.8689	0.8667	0.8689	0.8732
	Density=0.5	0.8695	0.8738	0.8585	0.8567	0.8598
Blurring	Motion Blurring	0.8678	0.8687	0.8689	0.8700	0.8691
	Simple Blurring	0.8719	0.8721	0.8715	0.8706	0.8704
Average Filtering	5×5	0.8836	0.8834	0.8834	0.8830	0.8836
	7×7	0.8757	0.8747	0.8730	0.8727	0.8738
Histogram Equalization		0.8661	0.8667	0.8665	0.8665	0.8663
JPEG Compression	Quality Factor=50	0.8611	0.8624	0.8617	0.8617	0.8600
Combined Watermarking Attacks	JPEG Comp+Filtering(5×5)	0.8745	0.8753	0.8762	0.8751	0.8747
	JPEG Comp + Gaussian Noise (0.5)	0.8643	0.8730	0.8680	0.8723	0.8736
	JPEG Comp + Scaling (Half)	0.8811	0.8813	0.8804	0.8811	0.8800
	Translation+Shearing	0.8429	0.8493	0.8521	0.8623	0.8670
	Rotation(45°)+ scaling	0.8133	0.8084	0.7479	0.7471	0.8270

Tab. 6. NC For Different Attacks (Proposed Scheme 1).

(S&P), speckle noise (SN), motion blurring (MB), simple blurring (SB), average filtering (AF), histogram equalization (HE) and JPEG compression (JQ) were applied on watermarked image. In addition to conventional watermarking attacks, combined attacks are formed by combining two or more conventional attacks. For instance, watermark is tried to be extracted form a watermarked images that has been subjected to JPEG compression and filtering attack. In this way five additional attacks – JPEG compression plus filtering (JQ+AF), JPEG compression plus Gaussian noise (JQ+GN), JPEG compression plus scaling (JQ+SC), translation plus x and y shearing (TR+SH), and rotation plus scaling (RO+SC) – are also used to check robustness of proposed schemes

$$NC = \frac{\sum_{p=1}^{P}\sum_{q=1}^{Q}\left(W(p,q) \times \hat{W}(p,q)\right)}{\sqrt{\sum_{p=1}^{P}\sum_{q=1}^{Q} W^2(p,q)} \times \sqrt{\sum_{p=1}^{P}\sum_{q=1}^{Q} \hat{W}^2(p,q)}}. \quad (24)$$

For qualitative assessment of extracted watermarks after applying above mentioned attacks on watermarked image, Fig. 7 can be referred. It can be seen that in all cases the extracted watermarks are clearly visible hence can be used to prove ownership. The performance of proposed scheme 1 in terms of robustness for different scaling factors is shown in Tab. 6. In general with increasing scaling factor the robustness is increased [3], whereas, the change in NC values shown in Tab. 6, is random and that is due to the adaptive scaling factor (α_{ML}).

The comparison of proposed schemes in terms of NC with existing technique [3], [6], [13] and [14] is shown in Tab. 7.

Robustness of watermarking techniques are also measured using Bit Error Rate (BER) [3], shown in (25). The comparision of proposed schemes in terms of BER with [3], [6], [13] and [14] is shown in Tab. 8. From Tab. 7 and Tab. 8, it is clear that the proposed schemes give better results in terms of robustness as well.

$$BER = \frac{\text{Number of wrong bits extracted}}{\text{Total number of bits embedded}}. \quad (25)$$

The extracted watermarks shown in Fig. 4 are for proposed scheme 1, however, the quality of watermarks obtained using proposed scheme 2 is also good. This can be seen from Tab. 7 that there is not significant difference between NC

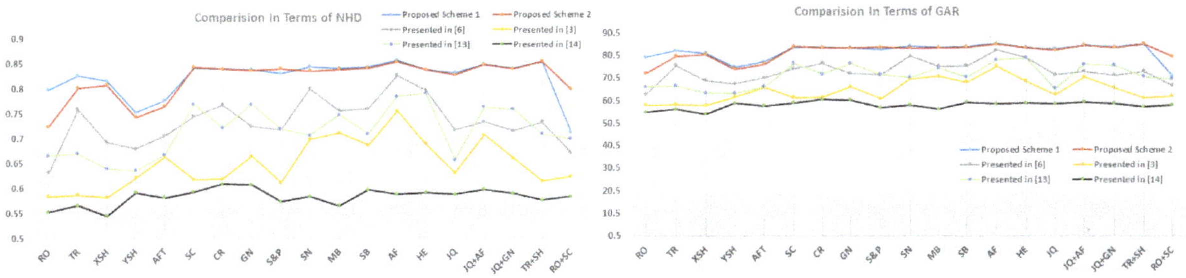

(a) (b)

Fig. 5. Comparison of proposed schemes with existing techniques (a). Comparison in terms of NHD (b). Comparison in terms of GAR.

Attacks	For Constant Scaling Factor $\beta = 0.5$					
	Proposed Schemes		Presented in			
	1	2	[6]	[3]	[13]	[14]
RO	0.8229	0.8036	0.5894	0.5377	0.6274	0.5218
TR	0.8135	0.8529	0.5946	0.5182	0.6056	0.5593
XSH	0.8466	0.8384	0.6148	0.5002	0.6633	0.5442
YSH	0.8418	0.8453	0.6582	0.4788	0.6779	0.6278
AFT	0.7558	0.8060	0.6376	0.4885	0.7029	0.5385
SC	0.8796	0.8672	0.7664	0.6356	0.7909	0.5386
CR	0.8648	0.8570	0.6825	0.6322	0.7370	0.6216
GN	0.8659	0.8583	0.7438	0.5797	0.7896	0.5823
S&P	0.8704	0.8656	0.7406	0.5270	0.7420	0.5436
SN	0.8667	0.8716	0.7304	0.6247	0.7375	0.5565
MB	0.8689	0.8716	0.7859	0.6355	0.7772	0.5873
SB	0.8715	0.8665	0.7824	0.6555	0.7312	0.5527
AF	0.8834	0.8073	0.7811	0.6577	0.7417	0.5562
HE	0.8665	0.7929	0.7539	0.5576	0.7493	0.5601
JQ	0.8617	0.8639	0.7488	0.5538	0.6861	0.5621
JQ+AF	0.8762	0.8753	0.7568	0.5234	0.7882	0.5343
JQ+GN	0.8680	0.8643	0.7375	0.5769	0.7822	0.5870
JQ+SC	0.8804	0.8813	0.7554	0.5305	0.7316	0.5425
TR+SH	0.8521	0.8475	0.7058	0.5028	0.7238	0.5123
RO+SC	0.7479	0.8308	0.6973	0.5028	0.7276	0.5348

Tab. 7. NC for different techniques.

values of both proposed schemes. The recognizable watermarks can be extracted as long as the distorted watermarked image are visually identifiable.

Besides NC and BER, there are other ways to calculate the credibility of extracted watermarks quantitatively. For instance, Normalized Hamming Distance (NHD) [6] shown in 26, where, w and $\hat{w}$ represent embedded and extracted watermark respectively, can be used to calculate the similarity (or difference) between embedded and extracted watermarks.

$$H_{\mathrm{D}} = \frac{\sum_{1=1}^{m} \sum_{j=1}^{n} h_{w\hat{w}}(i,j)}{m \times n} \tag{26}$$

where

$$h_{w\hat{w}}(i,j) = \begin{cases} 1, & \text{if } \ w(i,j) = \hat{w}(i,j), \\ 0, & \text{otherwise.} \end{cases}$$

Attacks	For Constant Scaling Factor $\beta = 0.5$					
	Proposed Schemes		Presented in			
	1	2	[6]	[3]	[13]	[14]
RO	0.2107	0.2256	0.2360	0.3679	0.3153	0.3791
TR	0.1792	0.1736	0.1907	0.2933	0.2510	0.2717
XSH	0.1848	0.1892	0.2268	0.2951	0.2225	0.2712
YSH	0.1902	0.1819	0.2965	0.2984	0.2108	0.2276
AFT	0.2798	0.2231	0.2577	0.3570	0.2481	0.3239
SC	0.1477	0.1580	0.1939	0.2738	0.2200	0.3231
CR	0.1611	0.1692	0.1920	0.2799	0.2401	0.2847
GN	0.1672	0.1677	0.1829	0.2801	0.2056	0.2788
S&P	0.1582	0.1597	0.1454	0.2798	0.1987	0.2713
SN	0.1660	0.1531	0.1825	0.2857	0.2420	0.3207
MB	0.1599	0.1531	0.1833	0.2769	0.2264	0.2996
SB	0.1570	0.1587	0.1761	0.2758	0.2472	0.3271
AF	0.1433	0.1477	0.1652	0.2601	0.2306	0.3076
HE	0.1626	0.2517	0.1130	0.2769	0.2061	0.2757
JQ	0.1680	0.1616	0.2064	0.2760	0.2228	0.2719
JQ+AF	0.1516	0.1526	0.1985	0.2864	0.2318	0.2945
JQ+GN	0.1609	0.1650	0.2053	0.2356	0.2514	0.2718
JQ+SC	0.1467	0.1458	0.1856	0.2281	0.2346	0.2815
TR+SH	0.1656	0.1715	0.1983	0.2459	0.2287	0.2568
RO+SC	0.1876	0.2021	0.2598	0.2983	0.2795	0.3143

Tab. 8. BER for different techniques.

Similarly, Global Authentication Rate (GAR) [22] shown in (27) is also used to check the quality of extracted watermark. It should be noted that GAR and NHD are almost same just different form of representations

$$\varrho^{\mathrm{GAR}} = \left(1 - \frac{1}{m \times n} \sum_{i=1}^{m} \sum_{j=1}^{n} \left(w(i,j) \oplus \hat{w}(i,j)\right)\right) \times 100\,\%. \tag{27}$$

The performance of proposed schemes is examined using both NHD and GAR in terms of robustness and also compared with other watermark techniques.

The comparision of proposed and existing watermarking schemes in terms of NHD is shown in Fig. 5(a) and in terms of GAR in Fig. 5(b). Both figures clearly demonstrate the improvement of proposed scheme over existing watermarking techniques.

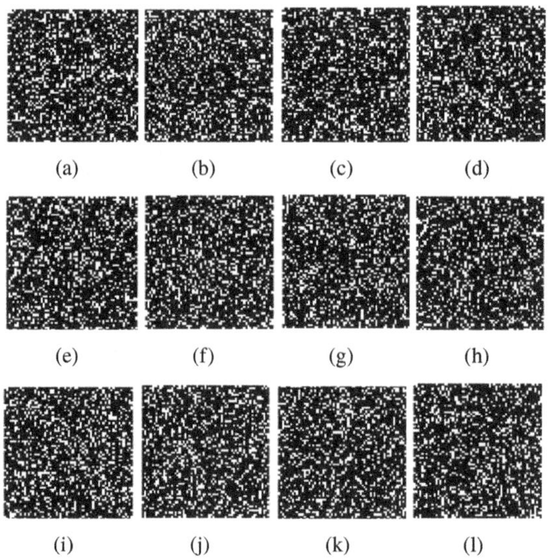

Fig. 6. Extracted watermarks with fake keys (proposed scheme 1).

5.3 Security

Security refers to the resistance against false positive or true negative extraction of watermark [5]. In designing the proposed schemes security is given high importance and it was ensured to nullify the chances of false positive or true negative detection or extraction of hidden information. To test the security ample number of fake keys were used to extract the watermark neither true nor false watermark was extracted. In Fig. 6 the extracted watermarks with false keys are shown. It is evident that no watermark was extracted.

5.4 Capacity

Capacity refers to the amount of data that can be embedded into the host image without degrading the quality of watermarked image. The capacity of proposed schemes is ($64 \times 64 \times 3 = 12288$), since watermark of size 64×64, and all three components (U, S and V) of SVD decomposition are used. This is considered to be good capacity for a watermarking scheme. The capacities of proposed schemes are 3 times more than the capacities of [3] and [6].

6. Conclusion

In this paper, two watermarking schemes are proposed. In order to achieve four conflicting requirements for a good watermarking scheme, HVS, FIS, PCA and SVD are used together. The perceptual quality is improved by using PCA to decorrelate the three channels of color image, afterwards HVS and FIS are used to find adaptive scaling factor so that the amount of information embedded is subject to the acceptability of host image. For instance, the areas which are less prone to modifications are modified to lesser extent as compared to areas which are open to changes. In this way imperceptibility is further improved, and can be seen from results that the imperceptibility of proposed scheme 1 is much better than proposed scheme 2 (HVS and FIS are not employed in proposed scheme 2, otherwise it is same as proposed scheme 1). To achieve robustness, SVD is used, as changes in singular values does not change the image and vice versa. Whereas to obtain security, based on correlation certain elements from SVD components are selected for modification, and then those locations served as key at the time of watermark extraction. In this way not only security and capacity are achieved but this method also helped to improve the imperceptibility as suggested by results. The proposed schemes are compared with state-of-the-art watermarking techniques and obtained better results.

References

[1] TSUI, K. T., ZHANG, P. X., ANDROUTSOS, D. Color image watermarking using multidimensional Fourier transforms. *IEEE Transactions on Information Forensics and Security*, 2008, vol. 3, no. 1, p. 1556–6013. DOI: 10.1109/TIFS.2007.916275

[2] CHOU, H. C., LIU, C. K. A perceptually tuned watermarking scheme for color images. *IEEE Trans. on Image Processing*, 2010, vol. 19, no. 11, p. 2966–2982. DOI: 10.1109/TIP.2010.2052261

[3] CEDILLO-HERNANDEZ, M. CEDILLO-HERNANDEZ, A. GARCIA-UGALDE, F., et. al. Copyright protection of color imaging using robust-encoded watermarking. *Radioengineering*, 2015, vol. 24, no. 1, p. 240–251. DOI: 10.13164/re.2015.0240

[4] SU, Q., NIU, Y., ZOU, H., et el. A blind dual color image watermarking based on singular value decomposition. *International Journal of Applied Mathematics and Computation*, 2013, vol. 219, no. 16, p. 8455–8466. DOI: 10.1016/j.amc.2013.03.013

[5] MAKBOOL, N. M., KHOO, E. B. A new robust and secure digital image watermarking scheme based on the integer wavelet transform and singular value decomposition. *International Journal of Digital Signal Processing*, 2014, vol. 33, p. 134–147. DOI: 10.1016/j.dsp.2014.06.012

[6] PRATHAP, I., NATARAJAN, V., ANITHA, R. Hybrid robust watermarking for color images. *International Journal of Computers & Electrical Engineering*, 2014, vol. 14, no. 3, p. 920–930. DOI: 10.1016/j.compeleceng.2014.01.006

[7] GHAFOOR, A., IMRAN, M. A non-blind color image watermarking scheme resistant against geometric attacks. *Radioengineering*, 2012, vol. 21, no. 4, p. 1246–1251. ISSN: 1805-9600

[8] BAISA, L., GUNJAL, MALI, N. S. Comparative performance analysis of DWT-SVD based color image watermarking technique in YUV, RGB and YIQ color spaces. *International Journal of Computer Theory and Engineering*, 2011, vol. 3, no. 6, p. 714–717.

[9] SUN, X., BO, S. A blind digital watermarking for color medical images based on PCA. In *Proceedings of the IEEE International Conference on Wireless Communications, Networking, and Information Security (WCNIS)*. Beijing (China), 2010, p. 421–427. DOI: 10.1109/WCINS.2010.5541812

[10] LIU, R., TAN, T. An SVD based watermarking scheme for protecting rightful information. *IEEE Transactions on Multimedia*, 2002, vol. 4, no. 1, p. 121–128. DOI: 10.1109/6046.985560

[11] LAI, C. C., TAI, C. C. Digital image watermarking using discrete wavelet transform and singular value decomposition. *IEEE Transactions on Instrumentation and Measurement*, 2010, vol. 59, no. 11, p. 3060–3063. DOI: 10.1109/TIM.2010.2066770

[12] LAI, C. C. A digital watermarking scheme based on singular value decomposition and tiny genetic algorithm. *International Journal of Digital Signal Processing*, 2011, vol. 21, no. 4, p. 522–527. DOI: 10.1016/j.dsp.2011.01.017

[13] FAZLI, S., MOEINIi, M. A robust image watermarking method based on DWT, DCT, and SVD using a new technique for correction of main geometric attacks. *International Journal of Optik*, 2016, vol. 127, no. 2, p. 964–972. DOI: 10.1016/j.ijleo.2015.09.205

[14] ROY, A., MAITI, K. A., GHOSH, K. A perception based color image adaptive watermarking scheme in YCbCr space. In *Proceedings of the IEEE International Conference on Signal Processing and Integrated Networks*. 2015, p. 537–543. DOI: 10.1109/SPIN.2015.7095399

[15] QI, H., ZHENG, D., ZHAO, J. Human visual system based adaptive digital image watermarking. *International Journal of Signal Processing*, 2008, vol. 88, no. 1, p. 174–188. DOI: 10.1016/j.sigpro.2007.07.020

[16] GOU, J. M., PRASETYO, H. A novel gray-scale image watermarking using hybrid fuzzy-BPN architecture. *International Journal of Visual Communication and Image Representation*, 2014, vol. 25, no. 5, p. 1149–1163. DOI: 10.1016/j.eij.2015.01.002

[17] WANG, X., L. *A Course in Fuzzy Systems and Control*. 1st ed., USA: Prentice Hall, 1997. ISBN: 978-0135408827

[18] JOLLIFFE, T. I. *Principal Component Analysis*. 2nd ed., USA: Springer, 2002. ISBN: 978-0-387-22440-4

[19] WANG, Z. BOVIK, C. A., SHEIKH, R. H., et al. Image quality assessment: from error visibility to structural similarity. *IEEE Transactions on Image Processing*, 2004, vol. 13, no. 4, p. 600–612. DOI: 10.1109/TIP.2003.819861

[20] SHEIKH, R. H., BOVIK, C. A. T. S. Image information and visual quality. *IEEE Transactions on Image Processing*, 2006, vol. 15, no. 2, p. 430–444. DOI: 10.1109/TIP.2005.859378

[21] LEVICKY. D., FORIS, P. Human visual system models in digital image watermarking. *Radioengineering*, 2004, vol. 13, no. 4, p. 28–43. ISSN: 1805-9600

[22] LIE, N. W., LIN, S. G., CHENG, L. S. Dual protection of JPEG images based on informed embedding and two-stage watermark extraction techniques. *IEEE Transactions on Information Forensics and Security*, 2006, vol. 1, no. 3, p. 330–341. DOI: 10.1109/TIFS.2006.879297

About the Authors . . .

Muhammad IMRAN received the B.Eng. degree in electronic engineering from Mehran University of Engineering and Technology (MUET), Jamshoro, Pakistan, in 2007 and the M. Sci. degree in electrical engineering from National University of Sciences and Technology (NUST), Islamabad, Pakistan, in 2012. He is currently pursuing the PhD degree in electrical engineering at the Florida State University. He is a research assistant in Image Processing and Communication Research Laboratory. His research interests are in digital image processing, digital image watermarking and steganography.

Bruce A. HARVEY received the B.E.E. (Co-op, Highest Honor) from Auburn University in 1984, the M.S.E.E. from the University of Alabama in Huntsville in 1987 and the Ph.D. in Electrical Engineering from Georgia Institute of Technology in 1991. He was a Research Engineer I at the Georgia Tech Research Institute (GTRI) from 1984-1986 and a Lead Engineer at Phase IV Systems, Inc. from 1986-1988. From 1991-1997 he was a Research Engineer in the Communications Division of GTRI. In 1997 he joined the Department of Electrical and Computer Engineering at the FAMU-FSU College of Engineering in Tallahassee, Florida, and currently holds the rank of Associate Professor. His current fields of interest include lightning surge suppression, wireless communication, error control coding, wireless networks, modulation techniques, digital image processing, digital image watermarking, and modeling and analysis. Dr. Harvey is a member of the IEEE Communications and Education Societies.

12

On the Performance of a Wireless Powered Communication System Using a Helping Relay

Tan N. NGUYEN [1], *Phuong T. TRAN* [1], *Hoang-Sy NGUYEN* [1,2], *Dinh-Thuan DO* [3], *Miroslav VOZNAK* [1,2]

[1] Wireless Communications Research Group, Faculty of Electrical and Electronics Engineering, Ton Duc Thang University, No. 19 Nguyen Huu Tho Street, Tan Phong Ward, District 7, Ho Chi Minh City, Vietnam
[2] VSB Technical University of Ostrava, 17. listopadu 15/2172, 708 33 Ostrava - Poruba, Czech Republic
[3] Faculty of Electronics Technology, Industrial University of Ho Chi Minh City, Vietnam

{nguyennhattan, tranthanhphuong, nguyenhoangsy}@tdt.edu.vn, dodinhthuan@iuh.edu.vn, miroslav.voznak@vsb.cz

Abstract. *This paper studies the outage performance and system throughput of a bidirectional wireless information and power transfer system with a helping relay. The relay helps forward wireless power from the access point (AP) to the user, and also the information from the user to the AP in the reverse direction. We assume that the relay uses time switching based energy harvesting protocol. The analytical results provide theoretical insights into the effect of various system parameters, such as time switching factor, source transmission rate, transmitting-power-to-noise ratio to system performance for both amplify-and-forward and decode-and-forward relaying protocols. The optimal time switching ratio is determined in each case to maximize the information throughput from the user to the AP subject to the energy harvesting and consumption balance constraints at both the relay and the user. All of the above analyses are confirmed by Monte-Carlo simulation.*

Keywords

Amplify-and-forward, bidirectional relay, decode-and-forward, wireless powered communications, time-switching

1. Introduction

Recently, radio frequency (RF) signal based wireless energy transfer (WET) has emerged as a perpetual and cost-effective solution to power wireless devices, such as mobile sensors, electronic tags, etc. [1]. While numerous works have focused on WET systems to optimize the energy harvesting process at energy receivers [2], the authors of this paper are more interested in another line of WET research, where WET could be integrated with wireless communication by exploiting the dual use of RF signals. Especially, we focus on wireless powered communication (WPC) [3], where the energy for wireless communication at the device is obtained via the WET technology. This advanced technology has been deployed and investigated in various wireless system models, including cellular networks [4], relay systems [5], [6], cognitive radio networks [7], [8].

In last decade, wireless sensor networks have been more and more attracted by research community, due to their ability to carry out different kinds of tasks, from traffic monitoring, agriculture monitoring, to smart home and health-care applications. For these networks, network life time is a critical aspect to the success of the system. By some previous research, battery life time is the bottleneck in determining the life time of the whole system. In [9], three wireless powered sensor network models for infrastructure monitoring application have been proposed and their performances have been investigated. Another wireless powered sensor network model was presented in [10], in which a number of sensor nodes send common information to a far apart information access point via distributed beamforming, by using the wireless energy transferred from a set of nearby multi-antenna energy transmitters.

Both of the works in [9] and [10] only introduce normal sensor networks without the helping of relay nodes. Furthermore, the RF energy transmitters in those works are independent of the information transfer process. This would increase the cost of implementation of these models in practice. In [11], the authors have tried to overcome this drawback by considering a new WPC system, where a wireless user communicates with an access point (AP) assisted by a bidirectional relay. The user and the relay are both powered by the RF energy from the AP. Here, the role of the relay is to forward the energy from the AP to the user, as well as to forward information from the user to the AP. However, the authors in [11] only considered the case that channel gains are constant, and estimate the maximum achievable throughput of the system. Because of this limitation, there is a large difficulty to apply this result to practical sensor networks. In addition, the work in [11] only considers amplify-and-forward as the relaying strategy.

Continuing to the work of [11], in this paper we provide a rigorous analysis on the same wireless powered sensor network model. We apply a Rayleigh distribution model for the channel gains between nodes, including the AP, relay node and the wireless user. For information transfer, both amplify-and-forward (AF) and decode-and-forward (DF) relaying protocols are investigated. Regarding to the energy harvesting protocol, we focus on time switching (TS) strategy at the relay. The outage probability and the average throughput of the system are derived mathematically. The optimal time switching factor to maximize the system throughput is obtained via numerical algorithm. To verify the analysis mentioned above, Monte-Carlo simulations are also conducted and the results are reported in this paper, too.

The rest of this paper is organized as follows. The next section introduces the system model that we are going to analyze. The detailed performance analysis is provided in Sec. 3. The numerical results to support the analysis are given in Sec. 4. Finally, Sec. 5 concludes the paper.

2. System Model

We consider a wireless powered system as illustrated in Fig. 1, where a mobile user is intended to send information to the AP with the assistance of a relay R. Assume that both the user and relay R have no other energy supply but solely the energy harvested from the AP. Furthermore, we assume that direct connection between the AP and the user is so weak, hence, the only available communication path as well as power transfer path is via the relay R. The relay serves the dual roles of both energy relaying from the AP to the user and information forwarding from the user to the AP [11]. To initialize the communication process, a sufficient amount of initial energy is stored in the battery to conduct the first transmission block before energy harvesting, as in [12]. After that, the energy consumed by the user/relay is kept lower than or at most equal to the harvested energy amount during each block, thus no further manual battery replacement/recharging is needed.

All nodes are assumed to operate in half-duplex mode, and either amplify-and-forward (AF) or decode-and-forward (DF) relaying strategy can be used at the relay for information transferring. Regarding to the channel model, we consider the case that perfect channel state information (CSI)

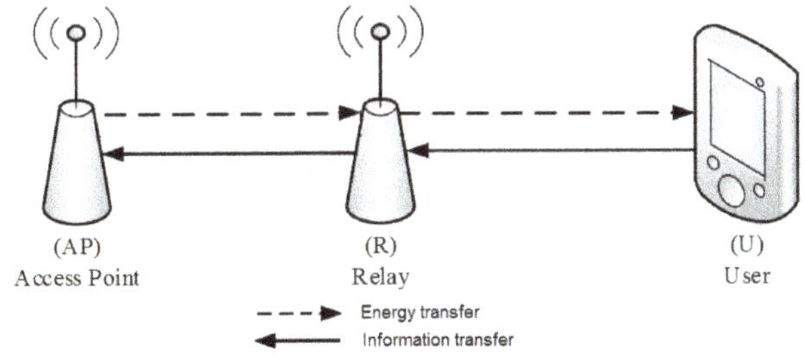

Fig. 1. System model.

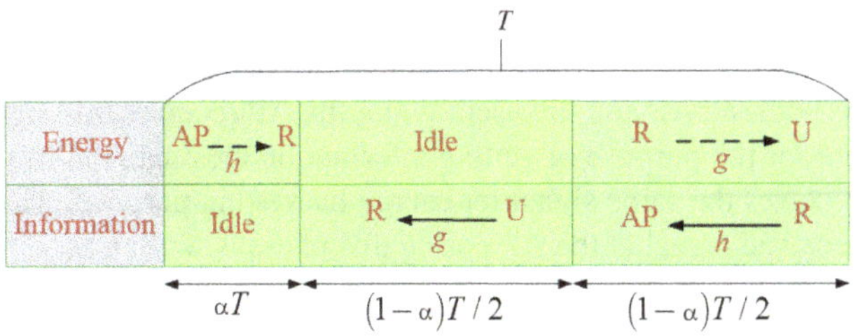

Fig. 2. Time switching based protocol.

is available at the relay and the AP. Let h and g denote the channels from the AP to the relay and from the user to the relay, respectively. In addition, we assume for simplicity that these channels are reciprocal. Different from the work in [11], all channels here experience Rayleigh fading and keep constant during each transmission block so that they can be considered as slow fading. As a result, $|h|^2$ and $|g|^2$ are an exponential random variables with parameters λ_h and λ_g, respectively.

For energy harvesting, we employ the time switching relaying (TSR) protocol, which is more convenient to implement in practice. As shown in Fig. 2, the total symbol duration T is divided into three intervals with the lengths of αT, $(1-\alpha)T/2$, and $(1-\alpha)T/2$, respectively, where $0 < \alpha < 1$ denotes the time-switching ratio. The first interval corresponds to the energy harvesting phase at the relay R, in which the AP wirelessly sends its energy to R with power P_{ap}. Then, the total energy harvested at R during each block is given by $E_{\text{r}} = \eta P_{\text{ap}}.|h|^2.\alpha T$, where $0 \leq \eta \leq 1$ is the energy conversion efficiency. The second phase of duration $(1-\alpha)T/2$ corresponds to the information transmission from the user to the relay. In the third phase of the transmission block, R forwards an amplified or decoded signal to the AP and also forwards energy to the user. We assume that the circuit power consumption is negligible as compared to the radiation power, which is reasonable for low-power devices such as sensor nodes.

3. Performance Analysis

In this section, the throughput and outage performance of the proposed system are analyzed mathematically. The impact of time-switching factor on system performance is investigated. We consider both AF and DF protocols in our analysis.

3.1 Amplify-and-Forward Protocol

Let x_{u} denote the transmitted signal from the user during the second phase and P_{u} denote the power of this signal. The received signal at R during this phase is espressed as

$$y_{\text{r}} = \sqrt{P_{\text{u}}} g x_{\text{u}} + n_{\text{r}} \tag{1}$$

where $n_{\text{r}} \sim N(0, N_0)$ denotes the Gaussian distributed noise at the relay R.

During the third phase of transmission, the relay amplifies the received signal from mobile user and forwards it to both the AP and the user. While the AP receives this signal for the purpose of getting information message, the user receives the same signal for energy harvesting purpose. The received signal at the AP during this phase is written as

$$y_d = \sqrt{P_r} h x_r + n_d \quad (2)$$

where x_r is the signal transmitted by the relay, which has the power of P_r, and n_d is the zero-mean Gaussian noise at the AP with variance N_0. Because the transmit power of the relay comes from the energy supplied by the AP in the first phase, we must have [11]

$$P_r = \frac{E_r}{(1-\alpha)T/2} = \frac{2\alpha\eta P_{ap}|h|^2}{1-\alpha} = k\eta P_{ap}|h|^2 \quad (3)$$

where $k = \frac{2\alpha}{1-\alpha}$.

The signal transmitted by the relay is an amplified version of y_r:

$$x_r = \sqrt{\beta} y_r. \quad (4)$$

According to energy conservation law, the energy consumed by the relay cannot exceed its available energy, which yields [11]

$$\beta = \frac{1}{P_u|g|^2 + N_0}. \quad (5)$$

Now, we can substitute (1), (4), and (5) into (2) and get

$$\begin{aligned} y_d &= \sqrt{P_r} h \sqrt{\beta}(\sqrt{P_u} g x_u + n_r) + n_d \\ &= \underbrace{\sqrt{P_r} h \sqrt{\beta}\sqrt{P_u} g x_u}_{\text{Signal part}} + \underbrace{\sqrt{P_r} h \sqrt{\beta} n_r + n_d}_{\text{Noise part}}. \end{aligned} \quad (6)$$

From (6), the signal-to-noise ratio at the AP can be computed by

$$SNR = \frac{P_r\beta P_u|h|^2|g|^2}{P_r\beta|h|^2 N_0 + N_0} = \frac{P_u|h|^2|g|^2}{N_0\left(|h|^2 + \frac{P_u|g|^2+N_0}{k\eta P_{ap}|h|^2}\right)}. \quad (7)$$

Let's move on to determine P_u. We know that the received signal at the mobile user during the third transmission phase is $y_u = g x_r = g\sqrt{\beta} y_r$. Hence, the energy harvested during this phase can be determined by

$$\begin{aligned} E_u &= \eta|g|^2\beta \mathrm{E}\{|y_r|^2\}(1-\alpha)T/2 \\ &= \eta|g|^2 E_r = \eta^2 P_{ap}|h|^2|g|^2\alpha T. \end{aligned} \quad (8)$$

So, the transmit power of the mobile user during the second phase is expressed as

$$P_u = \frac{E_u}{(1-\alpha)T/2} = k\eta^2 P_{ap}|h|^2|g|^2. \quad (9)$$

By substituting (9) into (7) and doing some algebra, we obtain the overall SNR for AF protocol:

$$SNR_{AF} = \frac{k\eta^2 P_{ap}|h|^4|g|^4}{|h|^2 N_0 + \eta N_0|g|^4 + \frac{N_0^2}{k\eta P_{ap}|h|^2}}. \quad (10)$$

Assume that the source transmits at a constant rate R, then $\gamma = 2^R - 1$ is the lower threshold for SNR. Here, the outage probability P_{out} and the average throughput of the system can be evaluated by [6]

$$P_{out,AF} = \Pr\{SNR < \gamma\}, \quad (11)$$

$$R_{AF} = (1 - P_{out,AF})R\frac{1-\alpha}{2}. \quad (12)$$

The main contribution of this paper is to derive the closed-form expression of the outage probability and average throughput of the system of interest, as well as to figure out the optimal time-switching factor for energy harvesting. The results for AF protocol are formally stated in the following theorems. Theorem 1 provides the exact integral forms for the outage probability and throughput of the proposed system with AF protocol. In Theorem 2, closed-form approximations of the outage probability and throughput in terms of Meijer function are derived for high source-power-to-noise-ratio regime.

Theorem 1 (AF Protocol) *For the AF protocol, the outage probability and the average throughput of the proposed system can be expressed as*

$$P_{out,AF} = 1 - \lambda_h \int_{\sqrt{\frac{\gamma}{\delta}}}^{\infty} e^{-\lambda_h x - \lambda_g \sqrt{\frac{\gamma\delta x^2+\gamma}{\eta\delta^2 x^3 - \eta\gamma\delta x}}} dx \quad (13)$$

and

$$R_{AF} = \frac{R\lambda_h(1-\alpha)}{2} \int_{\sqrt{\frac{\gamma}{\delta}}}^{\infty} e^{-\lambda_h x - \lambda_g \sqrt{\frac{\gamma\delta x^2+\gamma}{\eta\delta^2 x^3 - \eta\gamma\delta x}}} dx \quad (14)$$

where $\delta = k\eta\frac{P_{ap}}{N_0}$.

Proof 1 *See Appendix A.*

Theorem 2 (AF Protocol - Closed-form approximation) *At high* P_{ap}/N_0 *regime, the outage probability and average throughput of the proposed system with AF protocol can be respectively approximated to*

$$P_{out,AF} \approx 1 - \frac{e^{-\lambda_h\sqrt{\frac{\gamma}{\delta}}}}{\sqrt{\pi}}.G_{0,3}^{3,0}\left(\frac{\lambda_g^2\gamma\lambda_h}{4\eta\delta}\middle| 0, \frac{1}{2}, 1\right) \quad (15)$$

and

$$R_{AF} \approx \frac{R(1-\alpha)e^{-\lambda_h\sqrt{\frac{\gamma}{\delta}}}}{2\sqrt{\pi}}.G_{0,3}^{3,0}\left(\frac{\lambda_g^2\gamma\lambda_h}{4\eta\delta}\middle| 0, \frac{1}{2}, 1\right) \quad (16)$$

where $G_{p,q}^{m,n}(\cdot|\cdots)$ *is the Meijer function (Sec. 9.3 of [13]).*

Proof 2 *See Appendix B.*

3.2 Decode-and-Forward Protocol

For DF relaying protocol, the data communication is divided into two separating hops, which do not depend on each other. Hence, the outage occurs if and only if either the source-relay path or the relay-destination path fails to satisfy the corresponding SNR constraint. Different from the AF protocol, the message transmitted by the relay during the third transmission phase is the decoded message $\hat{x}_r$, instead of x_r, and the transmit power of the relay in this phase is the same as the one given in (3). Hence, the energy harvested by the mobile user during the same transmission phase is $E_u = \eta|g|^2 P_r \frac{(1-\alpha)T}{2} = \eta^2 P_{ap}|h|^2|g|^2\alpha T$. As a result, the transmit power of the mobile user in the second phase is the same as in (9).

According to the equations (1) and (2), the SNR values at the relay R and the AP are respectively determined by

$$SNR_R = \frac{P_u|g|^2}{N_0} = \delta\eta|h|^2|g|^4 \quad (17)$$

and

$$SNR_{AP} = \frac{P_r|h|^2}{N_0} = \delta|h|^4. \quad (18)$$

The outage probability of the system can be written as

$$\begin{aligned} P_{out,DF} &= \Pr(\min\{SNR_R, SNR_{AP}\} < \gamma) \\ &= 1 - \Pr(SNR_R \geq \gamma, SNR_{AP} \geq \gamma). \end{aligned} \quad (19)$$

Now we can claim the following theorem on the outage probability and the average throughput of the system of interest.

Theorem 3 (DF Protocol) *For the DF protocol, the outage probability and the average throughput of the proposed system can be expressed as*

$$P_{out,DF} = 1 - \Gamma\left(1, \lambda_h x_0; \lambda_g y_0\sqrt{x_0\lambda_h}, \frac{1}{2}\right) \quad (20)$$

and

$$R_{DF} = \frac{R(1-\alpha)}{2}\Gamma\left(1, \lambda_h x_0; \lambda_g y_0\sqrt{x_0\lambda_h}, \frac{1}{2}\right) \quad (21)$$

where $\Gamma(\alpha, x; b, \beta) \triangleq \int_x^\infty t^{\alpha-1}e^{-t-bt^{-\beta}}dt$ *is the extended incomplete gamma function, which is defined in [14], and* x_0, y_0 *are defined by*

$$x_0 = \sqrt{\frac{\gamma N_0}{k\eta P_{ap}}}, \quad (22)$$

$$y_0 = \sqrt{\frac{\gamma N_0}{k\eta^2 P_{ap} x_0}}. \quad (23)$$

Proof 3 *See Appendix C.*

3.3 Optimal Time-Switching

To find the optimal time-switching factors that give the best performance in terms of outage probability or average throughput, we solve the equations $\frac{dP_{out}(\alpha)}{d\alpha} = 0$ and $\frac{dR(\alpha)}{d\alpha} = 0$, respectively, where $P_{out}(\alpha)$ and $R(\alpha)$ are outage probability and throughput functions with respect to the time-switching factor.

By investigating the outage probability functions with respect to α for both AF and DF, we can easily see that these are non-increasing functions. That means, the best outage performance is obtained when we exploit energy harvesting at full-scale. However, we should keep in mind that this outage performance only based on the comparison of power between signal and noise. It ignores other factors of communication process. In practice, we cannot set α to 1 because it means that no communication data is transferred.

Hence, the average throughput should be a more reasonable performance factor to be optimized. By plotting the throughput functions for AF and DF protocols versus α, we learn that these functions are concave functions, which have a unique maxima on the interval [0, 1]. The optimal factor α^* can be found numerically by some iterative methods, for instance, Golden section search method [15].

4. Numerical Results and Discussion

In this section, we conduct Monte Carlo simulation to verify the analysis developed in the previous section. For simplicity, in our simulation model, we assume that the source-relay and relay-destination distances are both normalized to unit value. Other simulation parameters are listed in Tab. 1.

Symbol	Name	Values
R	Source rate	3 bps/Hz
γ	SNR threshold	7
η	Energy harvesting efficiency	0.9
λ_h	Mean of $\|h\|^2$	0.5
λ_g	Mean of $\|g\|^2$	0.5
P_{ap}/N_0	Source (AP) Power to Noise Ratio	0-20 dB

Tab. 1. Simulation parameters.

4.1 Amplify-and-Forward Protocol

In Figures 3 and 4, the achievable throughput and outage probability of the system with AF protocol are plotted against P_{ap}/N_0 ratio with the data rate set to be 3 bps. The time-switching factor α is chosen to be 0.3 and 0.7. It's can be observed that the outage probability is a decreasing function with respect to P_{ap}/N_0, while the throughput grows with P_{ap}/N_0. In addition, the simulation and the analysis curves are overlapping. The approximate outage probability and throughput are also plotted in these figures. They are close to the exact curves, especially when P_{ap}/N_0 is large. This confirms the correctness of our analysis in the previous section.

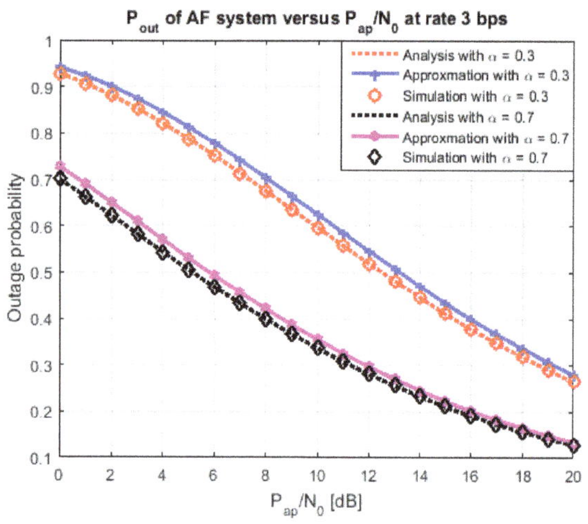

Fig. 3. Outage probability versus source power to noise ratio for AF protocol.

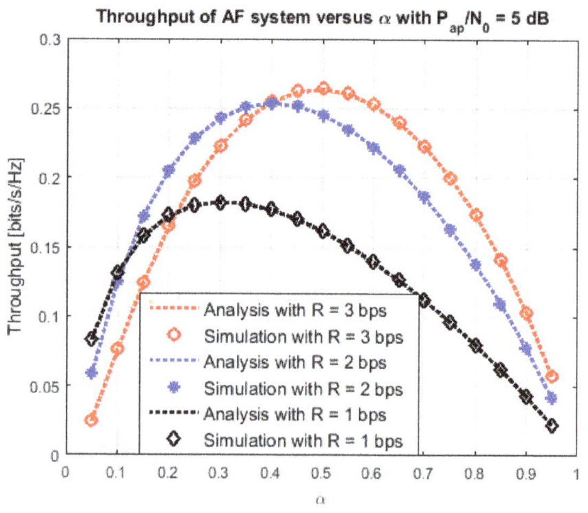

Fig. 6. Throughput versus time-switching factor for AF protocol.

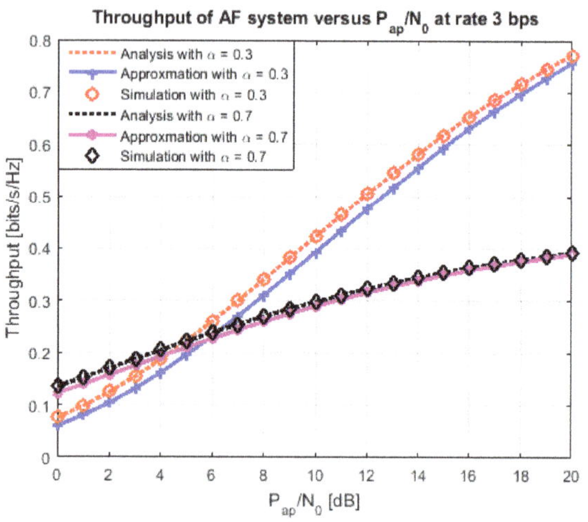

Fig. 4. Throughput versus source power to noise ratio for AF protocol.

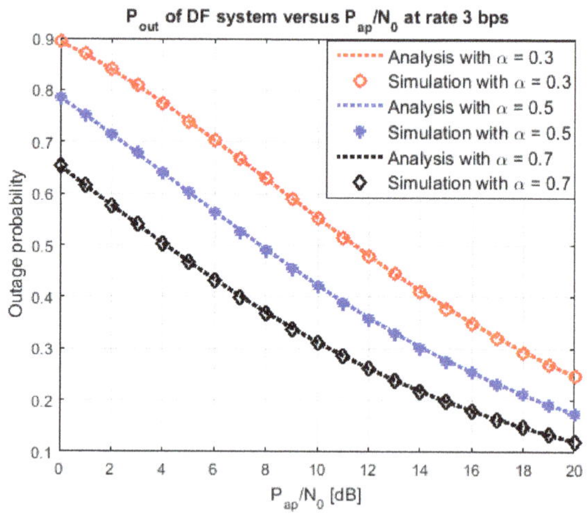

Fig. 7. Outage probability versus source power to noise ratio for DF protocol.

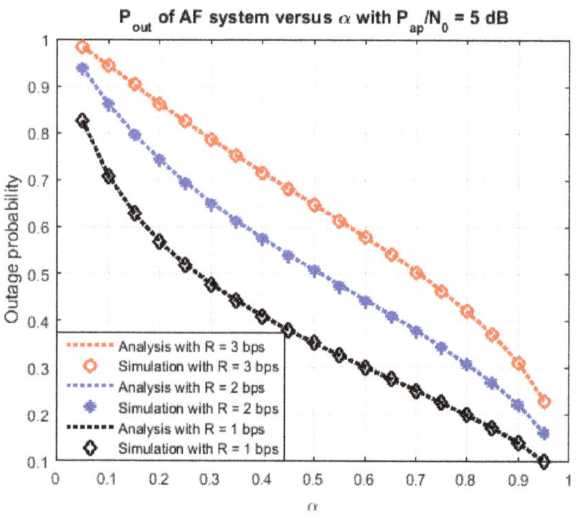

Fig. 5. Outage probability versus time-switching factor for AF protocol.

The impact of time switching factor on system performance with AF protocol is illustrated in Fig. 5 and 6. In this experiment, P_{ap}/N_0 is set to 5 dB, and the rate can be varied at 3 bps, 2 bps, and 1 bps. It can be observed that the outage probability is reduced when we increase the value of α. On the other hand, the simulation result shows that there exists a unique time switching factor at which the average throughput is maximized. Indeed, this optimal factor can be found iteratively using numerical methods.

4.2 Decode-and-Forward Protocol

For decode-and-forward protocol, we also have similar results about the impact of various parameters, such as P_{ap}/N_0 and α on the average throughput and the outage probability of the system. Specifically, Fig. 7 and 8 respectively plot the outage probability and throughput against P_{ap}/N_0, while Fig. 9 and 10 show the dependence of these performance characteristics on time-switching factor α.

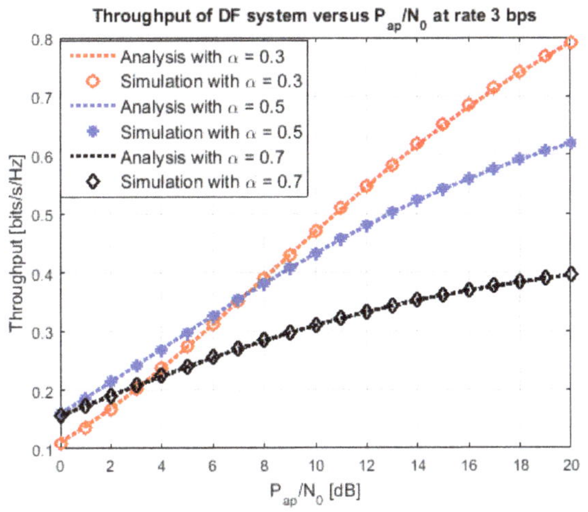

Fig. 8. Throughput versus source power to noise ratio for DF protocol.

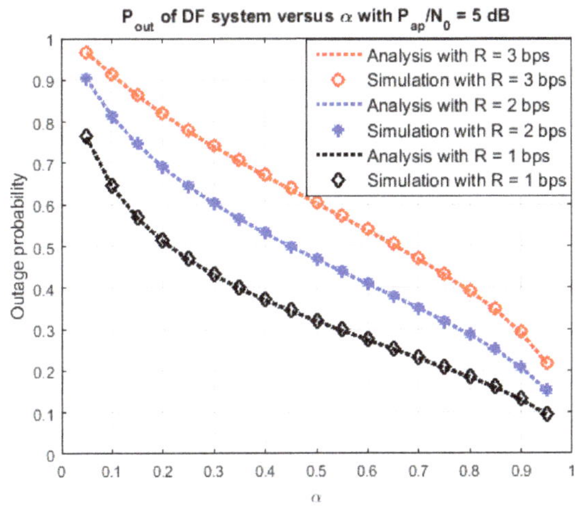

Fig. 9. Outage probability versus time-switching factor for DF protocol.

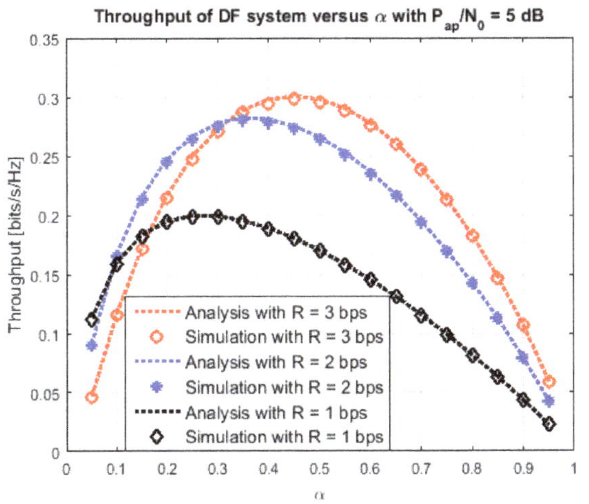

Fig. 10. Throughput versus time-switching factor for DF protocol.

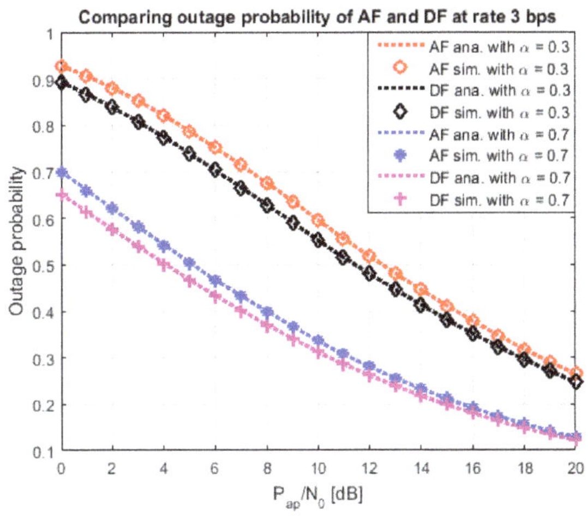

Fig. 11. Outage probability of AF and DF protocol at rate 3 bps.

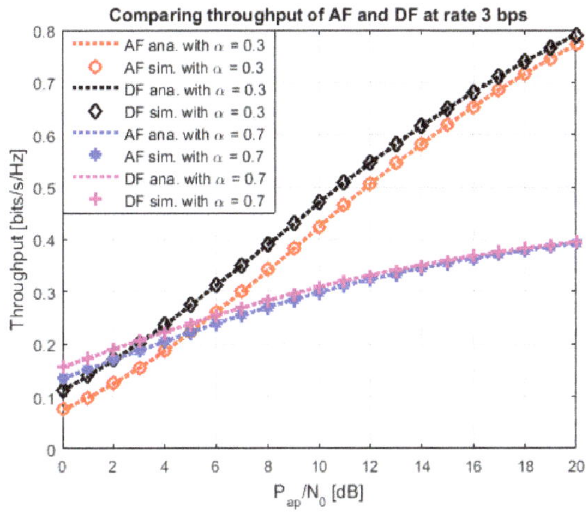

Fig. 12. Throughput of AF and DF protocol at rate 3 bps.

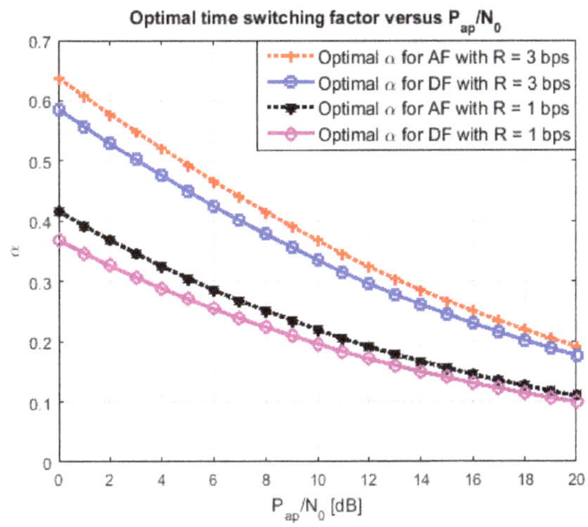

Fig. 13. Optimal time-switching factor versus P_{ap}/N_0.

4.3 Comparison Between DF and AF

Figures 11 and 12 compare the performance of two protocols that are considered in this paper. The results show that the DF protocol is slightly better than AF protocol in terms of both outage probability and throughput, because the noise at relay is eliminated in DF protocol, while it's accumulated and amplified in AF protocol.

4.4 Optimal Time-Switching Factor

Finally, the optimal values of α at different values of source-power-to-noise-ratio for both AF and DF protocols are shown in Fig. 13. We can see that the α value that optimizes the throughput has tendency to decrease when P_{ap}/N_0 increases.

5. Conclusions

In this paper, we investigate the performance of a new WPC system with a bidirectional information/energy forwarding relay in the Rayleigh fading environment. Two relaying protocols based on AF and DF strategies at the relay are considered in our work. For practical orientation, we employ the time switching protocol for energy harvesting. The exact-forms of outage probability as well as the average throughput of the proposed system are derived rigorously. Numerical results are provided to verify our analysis. The results show that the outage probability decreases as the time switching factor increases, while there is a unique value of time switching factor such that the throughput is maximized. For comparison between relaying protocols, the DF protocol is slightly better than its counterpart.

While the motivation of this paper comes from the energy problem in wireless sensor networks, the analysis obtained in this work does not limit to wireless sensor networks themselves, but can be applied for a wide range of wireless applications that employ the relay-node idea. Due to this reason, some specific issues related to sensor networks have not been considered in this paper. For example, the energy required by sensor nodes when collecting data or making measurement should be taken into account. In that case, the energy source not only comes from the information source node but also from other available nodes. This harvested energy can be modeled as a randomly varying variable. That should be our further work in this topic. In addition, we can take into account other factors such as CSI error and hardware impairment.

References

[1] BI, S., HO, C. K., ZHANG, R. Wireless powered communication: opportunities and challenges. *IEEE Communications Magazine*, 2015, vol. 53, no. 4, p. 117–125. DOI: 10.1109/MCOM.2015.7081084

[2] ZENG, Y. ZHANG, R. Optimized training design for wireless energy transfer. *IEEE Transactions on Communications*, 2015, vol. 63, no. 2, p. 536–550. DOI: 10.1109/TCOMM.2014.2385077

[3] BI, S., ZENG, Y., ZHANG, R. Wireless powered communication networks: An overview. *IEEE Wireless Communications*, 2016, vol. 23, no. 2, p. 10–18. DOI: 10.1109/MWC.2016.7462480

[4] HUANG, K. LAU, V. K. N. Enabling wireless power transfer in cellular networks: Architecture, modeling and deployment. *IEEE Transactions on Wireless Communications*, 2014, vol. 13, no. 2, p. 902–912. DOI: 10.1109/TWC.2013.122313.130727

[5] NASIR, A. A., ZHOU, X., DURRANI, S., et al. Relaying protocols for wireless energy harvesting and information processing. *IEEE Transactions on Wireless Communications*, 2013, vol. 12, no. 7, p. 3622–3636. DOI: 10.1109/TWC.2013.062413.122042

[6] NGUYEN, T. N., DO, D.-T., TRAN, P. T., et al. Time switching for wireless communications with full-duplex relaying in imperfect CSI condition. *KSII Transactions on Internet and Information Systems*, 2016, vol. 10, no. 9, p. 4223–4239. DOI: 10.3837/tiis.2016.09.011

[7] DUY, T. T. KONG, H. Y. Outage analysis of cognitive spectrum sharing for two-way relaying schemes with opportunistic relay selection over i.n.i.d. rayleigh fading channels. *IEICE Transactions on Communications*, 2013, vol. E96.B, no. 1, p. 348–351. DOI: 10.1587/transcom.E96.B.348

[8] SON, P. N. KONG, H. Y. Exact outage analysis of energy harvesting underlay cooperative cognitive networks. *IEICE Transactions on Communications*, 2015, vol. E98.B, no. 4, p. 661–672. DOI: 10.1587/transcom.E98.B.661

[9] ZHANG, Y., PFLUG, H., VISSER, H. J., et al. Wirelessly powered energy autonomous sensor networks. In *2014 IEEE Wireless Communications and Networking Conference (WCNC)*. 2014, p. 2444–2449. DOI: 10.1109/WCNC.2014.6952732

[10] XU, J., ZHONG, Z., AI, B. Wireless powered sensor networks: Collaborative energy beamforming considering sensing and circuit power consumption. *IEEE Wireless Communications Letters*, 2016, vol. 5, no. 4, p. 344–347. DOI: 10.1109/LWC.2016.2558503

[11] ZENG, Y., CHEN, H., ZHANG, R. Bidirectional wireless information and power transfer with a helping relay. *IEEE Communications Letters*, 2016, vol. 20, no. 5, p. 862–865. DOI: 10.1109/LCOMM.2016.2549515

[12] GURAKAN, B., OZEL, O., YANG, J., et al. Energy cooperation in energy harvesting communications. *IEEE Transactions on Communications*, 2013, vol. 61, no. 12, p. 4884–4898. DOI: 10.1109/TCOMM.2013.110113.130184

[13] ZWILLINGER, D., MOLL, V., GRADSHTEYN, I., et al. *Table of Integrals, Series, and Products*. 8th ed. Boston : Academic Press, 2015. ISBN: 978–0–12–384933–5

[14] CHAUDHRY, M. A. ZUBAIR, S. M. Extended incomplete gamma functions with applications. *Journal of Mathematical Analysis and Applications*, 2002, vol. 274, no. 2, p. 725 – 745. DOI: http://dx.doi.org/10.1016/S0022-247X(02)00354-2

[15] CHONG, E. K. P. ZAK, S. H., *An Introduction to Optimization*. 4th ed. John Wiley & Sons, 2013. ISBN: 978–1–118–27901–4

About the Authors . . .

Tan N. NGUYEN was born in 1986 in Nha Trang City, Vietnam. He received B.S. and M.S. degrees in Electronics and Telecommunications Engineering from Ho Chi Minh University of Natural Sciences, Ho Chi Minh City, Vietnam in 2008 and 2012, respectively. In 2013, he joined the Faculty of Electrical and Electronics Engineering of Ton Duc Thang University, Vietnam as a lecturer. He is currently pursuing his Ph.D. degree in Electrical Engineering at VSB Technical University of Ostrava, Czech Republic. His major interests are cooperative communications, cognitive radio, and physical layer security.

Phuong T. TRAN (corresponding author) was born in 1979 in Ho Chi Minh City, Vietnam. He received B.Eng. and M.Eng degrees in Electrical Engineering from Ho Chi Minh University of Technology, Ho Chi Minh City, Vietnam in

2002 and 2005, respectively. In 2007, he became a Vietnam Education Foundation Fellow at Purdue University, U.S.A., where he received his Ph.D. degree in Electrical and Computer Engineering in 2013. In 2013, he joined the Faculty of Electrical and Electronics Engineering of Ton Duc Thang University, Vietnam and served as the Vice Dean of Faculty since October 2014. His major interests are in the area of wireless communications and network information theory.

Hoang-Sy NGUYEN was born in Binh Duong province, Vietnam. He received the B.S. and MS.c degree from the Department of Computer Science from Ho Chi Minh City University of Information Technology (UIT-HCMC), Vietnam in 2007, 2013, respectively. He is currently pursuing the Ph.D. in School of Electrical Engineering and Computer Science, Technical University of Ostrava, Czech Republic. His research interests include energy efficient wireless communications, 5G wireless, network security, low-power networks, cloud and distributed networks.

Dinh-Thuan DO received the B.S. degree, M. Eng. degree, and Ph.D. degree from Vietnam National University (VNU-HCMC) in 2003, 2007, and 2013 respectively, all in Communications Engineering. He was a visiting Ph.D. student with Communications Engineering Institute, National Tsing Hua University, Taiwan from 2009 to 2010. Prior to joining Ton Duc Thang University in 2010, he was senior engineer at the VinaPhone Mobile Network from 2003 to 2009. Dr. Thuan was recipient of Golden Globe Award from Vietnam Ministry of Science and Technology in 2015. His research interest includes signal processing in wireless communications network, mmWave, device-to-device networks, cooperative communications, full-duplex transmission and energy harvesting.

Miroslav VOZNAK born in 1971 is an associate professor with the Department of Telecommunications, Technical University of Ostrava, Czech Republic and foreign professor with Ton Duc Thang University in Ho Chi Minh City, Vietnam. He received his Ph.D. degree in telecommunications in 2002 at the Technical University of Ostrava. He is a senior researcher in the Supercomputing center IT4Innovations in Ostrava, Czech Republic, a member of editorial boards of several journals and boards of conferences. Topics of his research interests are IP telephony, wireless networks, speech quality and network security.

Appendix A: Proof of Theorem 1

Let's denote $X = |h|^2$, $Y = |g|^2$, and $\delta = k\eta\frac{P_{\text{ap}}}{N_0}$. Note that X and Y are exponential random variables with the parameters λ_h and λ_g, respectively. The end-to-end SNR of the system can be rewritten as

$$SNR_{\text{AF}} = \frac{\eta\delta X^2Y^2}{X+\eta Y^2+\frac{1}{\delta X}} = \frac{\eta\delta^2X^3Y^2}{\delta X^2+\eta\delta XY^2+1}. \tag{A.1}$$

By substituting (A.1) to (11), we obtain

$$\begin{aligned} P_{\text{out,AF}} &= \Pr\left(\frac{X^3Y^2}{\delta X^2+\eta\delta XY^2+1} < \frac{\gamma}{\eta\delta^2}\right) \\ &= \Pr\left\{Y^2\left(X^3-\frac{\gamma X}{\delta}\right) < \left(\frac{\gamma X^2}{\eta\delta}+\frac{\gamma}{\eta\delta^2}\right)\right\} \\ &= \Pr\left(X^3 < \frac{\gamma X}{\delta}\right) \\ &+ \Pr\left\{Y^2 < \frac{\frac{\gamma X^2}{\eta\delta}+\frac{\gamma}{\eta\delta^2}}{X\left(X^2-\frac{\gamma}{\delta}\right)} \text{ and } X^3 \geqslant \frac{\gamma X}{\delta}\right\} \end{aligned} \tag{A.2}$$

The first term of (A.2) can be determined easily as

$$\Pr\left(X^3 < \frac{\gamma X}{b}\right) = \Pr\left(X < \sqrt{\frac{\gamma}{\delta}}\right) = 1-\mathrm{e}^{-\lambda_h\sqrt{\frac{\gamma}{\delta}}} \tag{A.3}$$

We proceed to evaluate the second term of (A.2). Let I denote this term, then we can write

$$\begin{aligned} I &= \Pr\left\{Y < \sqrt{\frac{\gamma(\delta X^2+1)}{\eta\delta X\left(\delta X^2-\gamma\right)}} \text{ and } X \geqslant \sqrt{\frac{\gamma}{\delta}}\right\} \\ &= \int_{\sqrt{\frac{\gamma}{\delta}}}^{\infty} f_X(x)\mathrm{d}x \int_0^{g(x)} f_Y(y)\mathrm{d}y \\ &= \int_{\sqrt{\frac{\gamma}{\delta}}}^{\infty} \lambda_h\mathrm{e}^{-\lambda_h x}\left(1-\mathrm{e}^{-\lambda_g g(x)}\right)\mathrm{d}x \\ &= \mathrm{e}^{-\lambda_h\sqrt{\frac{\gamma}{\delta}}} - \int_{\sqrt{\frac{\gamma}{\delta}}}^{\infty} \lambda_h\mathrm{e}^{-\lambda_h x-\lambda_g g(x)}\mathrm{d}x \end{aligned} \tag{A.4}$$

where $g(x) = \sqrt{\frac{\gamma(\delta x^2+1)}{\eta\delta x(\delta x^2-\gamma)}}$; $f_X(x)$ and $f_Y(y)$ are the pdf of X and Y, respectively.

Now, (13) can be obtained by substituting (A.3) and (A.4) into (A.2). Then, we substitute (13) into (12) to get (14).

Appendix B: Proof of Theorem 2

Obviously, $\frac{1}{\delta} \to 0$ as $\frac{P_{\text{ap}}}{N_0} \to \infty$. Hence, at high values of $\frac{P_{\text{ap}}}{N_0}$, we can use the following approximation:

$$\frac{X+\eta Y^2+\frac{1}{\delta X}}{X+\eta Y^2+\sqrt{\frac{\gamma}{\delta}}} \approx 1.$$

Now, by multiplying this term to the expression of SNR_{AF} in (A.1), we obtain

$$SNR_{\text{AF}} = \frac{\eta\delta X^2Y^2}{X+\eta Y^2+\sqrt{\frac{\gamma}{\delta}}} \tag{B.1}$$

Using the same procedure as in Appendix A, we have

$$P_{\text{out,AF}} = \Pr\left(\frac{\eta\delta X^2Y^2}{X+\eta Y^2+\sqrt{\frac{\gamma}{\delta}}} < \gamma\right)$$
$$= \Pr\left\{\eta\delta Y^2\left(X^2-\frac{\gamma}{\delta}\right)\right\} < \gamma\left(X+\sqrt{\frac{\gamma}{\delta}}\right)$$
$$= \Pr\left(X^2 < \frac{\gamma}{\delta}\right)$$
$$+ \Pr\left\{Y^2 < \frac{\gamma\left(X+\sqrt{\frac{\gamma}{\delta}}\right)}{\eta\delta\left(X^2-\frac{\gamma}{\delta}\right)} \text{ and } X^2 \geqslant \frac{\gamma}{\delta}\right\}$$
$$= 1-\mathrm{e}^{-\lambda_h\sqrt{\frac{\gamma}{\delta}}}$$
$$- \Pr\left\{Y^2 < \frac{\gamma}{\eta\delta\left(X-\sqrt{\frac{\gamma}{\delta}}\right)} \text{ and } X^2 \geqslant \frac{\gamma}{\delta}\right\} \quad \text{(B.2)}$$

Again, we use the same argument as in the proof of Theorem 1 to rewrite the last term of (B.2) as

$$I = \Pr\left\{Y < \sqrt{\frac{\gamma}{\eta\delta\left(X-\sqrt{\frac{\gamma}{\delta}}\right)}} \text{ and } X \geqslant \sqrt{\frac{\gamma}{\delta}}\right\}$$
$$= \mathrm{e}^{-\lambda_h\sqrt{\frac{\gamma}{\delta}}} - \int_{\sqrt{\frac{\gamma}{\delta}}}^{\infty} \lambda_h \mathrm{e}^{-\lambda_h x-\lambda_g h(x)}\mathrm{d}x \quad \text{(B.3)}$$

where $h(x) = \sqrt{\frac{\gamma}{\eta\delta\left(x-\sqrt{\frac{\gamma}{\delta}}\right)}}$; $f_X(x)$ and $f_Y(y)$ are the pdf of X and Y, respectively.

Then, (B.2) can be rewritten as

$$P_{\text{out,AF}} = 1-\int_{\sqrt{\frac{\gamma}{\delta}}}^{\infty} \lambda_h \mathrm{e}^{-\lambda_h x-\lambda_g\sqrt{\frac{\gamma}{\eta\delta\left(x-\sqrt{\frac{\gamma}{\delta}}\right)}}}\mathrm{d}x. \quad \text{(B.4)}$$

Finally, by changing variable $u = \lambda_h\left(x-\sqrt{\frac{\gamma}{\delta}}\right)$, we obtain

$$P_{\text{out,AF}} = 1-\int_0^{\infty} \mathrm{e}^{-u-\lambda_h\sqrt{\frac{\gamma}{\delta}}-\lambda_g\sqrt{\frac{\gamma\lambda_h}{\eta\delta u}}}\mathrm{d}u$$
$$= 1-\mathrm{e}^{-\lambda_h\sqrt{\frac{\gamma}{\delta}}}\Gamma\left(1,0;\lambda_g\sqrt{\frac{\gamma\lambda_h}{\eta\delta}},\frac{1}{2}\right)$$
$$= 1-\frac{\mathrm{e}^{-\lambda_h\sqrt{\frac{\gamma}{\delta}}}}{\sqrt{\pi}}G_{0,3}^{3,0}\left(\frac{\lambda_g^2\gamma\lambda_h}{4\eta\delta}\middle| 0,\frac{1}{2},1\right) \quad \text{(B.5)}$$

where $\Gamma(\alpha,x;b,\beta)$ is the extended incomplete Gamma function [14] and the last equality comes from the Corollary (3.20) of [14]. From (B.5), we can easily get the throughput formula (16).

Appendix C: Proof of Theorem 3

Again, let X denote $|h|^2$ and Y denote $|g|^2$. The SNR values in (17) and (18) now become

$$SNR_{\text{R}} = \delta\eta|h|^2|g|^4 = \delta\eta XY^2$$
$$SNR_{\text{AP}} = \delta|h|^4 = \delta X^2$$

where $\delta = k\eta\frac{P_{\text{ap}}}{N_0}$. Also, from (21) and (23), we have $x_0 = \sqrt{\frac{\gamma}{\delta}}$ and $y_0 = \sqrt{\frac{\gamma}{\delta\eta x_0}}$.

From (19), the outage probability can be rewritten as

$$P_{\text{out,DF}} = 1-\Pr\{\delta\eta XY^2 \geq \gamma, \delta X^2 \geq \gamma\}$$
$$= 1-\Pr\{Y \geq \sqrt{\frac{\gamma}{\eta\delta X}}, X \geq x_0\}$$
$$= 1-\Pr\{Y \geq y_0\sqrt{\frac{x_0}{X}}, X \geq x_0\}$$
$$= 1-\Pr\{X \geq x_0, y_0\sqrt{\frac{x_0}{X}} \leq Y < y_0\}$$
$$-\Pr\{X \geq x_0, Y \geq y_0\}$$
$$= 1-\mathrm{e}^{-\lambda_h x_0-\lambda_g y_0}$$
$$-\int_{x_0}^{\infty} f_X(x)\left(\int_{y_0\sqrt{\frac{x_0}{x}}}^{y_0} f_Y(y)\mathrm{d}y\right)\mathrm{d}x. \quad \text{(C.1)}$$

Denote the last term in (C.1) as I_1, then we have

$$I_1 = -\int_{x_0}^{\infty} \lambda_h\mathrm{e}^{-\lambda_h x}\left(\int_{y_0\sqrt{\frac{x_0}{x}}}^{y_0} \lambda_g\mathrm{e}^{-\lambda_g y}\mathrm{d}y\right)\mathrm{d}x$$
$$= -\int_{x_0}^{\infty} \lambda_h\mathrm{e}^{-\lambda_h x}\left(\mathrm{e}^{-\lambda_g y_0\sqrt{\frac{x_0}{x}}}-\mathrm{e}^{-\lambda_g y_0}\right)\mathrm{d}x$$
$$= -\int_{x_0}^{\infty} \lambda_h\mathrm{e}^{-\lambda_h x}\mathrm{e}^{-\lambda_g y_0\sqrt{\frac{x_0}{x}}}\mathrm{d}x+\mathrm{e}^{-\lambda_g y_0-\lambda_h x_0}$$
$$= -\int_{\lambda_h x_0}^{\infty} \mathrm{e}^{-u}.\mathrm{e}^{-\lambda_g y_0\sqrt{\frac{\lambda_h x_0}{u}}}\mathrm{d}u+\mathrm{e}^{-\lambda_g y_0-\lambda_h x_0} \quad \text{(C.2)}$$

By substituting this into (C.1) and using the definition of extended incomplete gamma function (formula (1.9) in [14]) we obtain (20). Finally, (21) is obtained by including (20) in the definitive formula of average throughput.

Wireless Powered Relaying Networks Under Imperfect Channel State Information: System Performance and Optimal Policy for Instantaneous Rate

Dinh-Thuan DO [1], *Hoang-Sy NGUYEN* [2], *Miroslav VOZNAK* [2,3], *Thanh-Sang NGUYEN* [4]

[1] Faculty of Electronics Technology, Industrial University of Ho Chi Minh City, 12 Nguyen Van Bao St., Go Vap Dist., Ho Chi Minh City, Vietnam

[2] Wireless Communications Research Group, Faculty of Electrical and Electronics Engineering, Ton Duc Thang University, 19 Nguyen Huu Tho St., Tan Phong Ward, Dist. 7, Ho Chi Minh City, Vietnam

[3] Faculty of Electrical Engineering and Computer Science, Technical University of Ostrava, 17. listopadu 2172/15, 708 33 Ostrava-Poruba, Czech Republic

[4] Binh Duong University, Binh Duong Province, Vietnam

Corresponding author: Hoang-Sy Nguyen

dodinhthuan@iuh.edu.vn, nguyenhoangsy@tdt.edu.vn, miroslav.voznak@vsb.cz, ntsang@bdu.edu.vn

Abstract. *In this investigation, we consider wireless powered relaying systems, where energy is scavenged by a relay via radio frequency (RF) signals. We explore hybrid time switching-based and power splitting-based relaying protocol (HTPSR) and compare performance of Amplify-and-Forward (AF) with Decode-and-Forward (DF) scheme under imperfect channel state information (CSI). Most importantly, the instantaneous rate, achievable bit error rate (BER) are determined in the closed-form expressions under the impact of imperfect CSI. Through numerical analysis, we evaluate system insights via different parameters such as power splitting (PS) and time switching (TS) ratio of the considered HTPSR which affect outage performance and BER. It is noted that DF relaying networks outperform AF relaying networks. Besides that, the numerical results are given to prove the optimization problems of PS and TS ratio to obtain optimal instantaneous rate.*

Keywords

Amplify-and-forward, decode-and-forward, throughput, channel state information, outage probability, cooperative network, bit error rate, energy harvesting

1. Introduction

A considerable number of different architectures and protocols related to radio frequency (RF) power transfer in wireless powered communication networks have recently been introduced in many studies. In such systems, the relay node receives energy from the source node and then utilizes the scavenged energy to transfer information to the destination node [1–5]. Depending on time switching-based relaying (TSR) and power splitting-based relaying (PSR) which are the two primary mechanisms, in which data and energy are transmitted between the source and the relay node [2]. In particular, there was a new protocol proposed by Do in [3] for energy harvesting (EH) at the mobile relay node in wireless communications, namely energy harvesting cooperative networks (EHCN). Besides that, according to several studies, they depict that lower transmission rate is a challenge of wireless energy transfer, since data processing requires less energy. There are also many applications of energy harvesting in wireless sensor networks, including heterogeneous networks, small-cell networks, etc. For instance, the deployment of energy harvesting models was accomplished in cellular networks [6] while full-duplex (FD) transmission systems underwent the use of radiated power [7]. A hybrid AF and DF scheme associated with network coding for Two Way Relay Networks (TWRN) was taken into account in [8].

In terms of perfect and imperfect channel state information (CSI) addressed in [9–13]. In particular, a multiple-input single-output (MISO) system was studied [9] while the authors in [10] considered a transmit power allocation issue for a hybrid EH single relay network with channel and energy state uncertainties with the aim to optimizing system throughput over a limited number of transmission intervals, in which sub-optimal online, optimal online and optimal offline allocation schemes were put forward. In [11], two-way full-duplex (TWFD) relaying with a residual loop interference (LI) was studied, where data is exchanged between users by the assistance of a FD relay. In [12], under the impact of imperfect CSI, the cognitive relay network performance was investigated. In an interference-limited environment, FD cooperative networks were proposed in [13]. Furthermore, fault-tolerant schemes were analyzed in the presence of imperfect CSI [14]. In order to optimize throughput in energy-aware cooperative networks, the optimal time switching and power splitting fraction in the proposed TPSR protocol were found. Additionally, the work in [16] focused on optimizing throughput in wireless powered communication networks. In addition, energy harvesting is developed in multi-antenna systems, i.e. some studies on Multiple-Input Multiple-Output (MIMO) were conducted [17–19]. For example, in [18], two-hop MIMO AF relay communication systems with simultaneous WIPT at the multi-antenna EH relay were considered.

Meanwhile, it is shown that in wireless cognitive radio networks, the authors in [20] addressed the joint problem of optimization over relay selection, subcarrier assignment, power splitting ratio determination in scenario of imperfect channel state information (CSI) conditions. Such resource allocation problem is required to maximize total throughput in the secondary network in terms of guaranteeing quality-of-service (QoS) requirements of the primary network. Regarding imperfect channel estimations as in [21], an adaptive power allocation and splitting (APAS) scheme was proposed to obtain near-optimal performances for both energy and data transmission over a single RF. While considering the impact of imperfect CSI in amplify-and-forward (AF) full-duplex relay network (FDRN), the optimal energy time switching coefficients are calculated through the numerical search method as in [22].

Motivated by [10], [20], [21] and [22], the optimal time and power splitting ratio of the energy harvesting protocol for instantaneous rate is not considered, so we consider an optimal policy to improve energy efficiency for energy harvesting. In the proposed TPSR protocol [3], the harvested energy highly depends on the average channel gain, but the existence of channel estimation error is the key parameter which needs to be tackled for performance evaluation. Besides that, the impact of the correlation between the actual CSI and its estimation value should be considered.

The major contributions of this paper are summarized as follows:

- We consider hybrid time switching-based and power splitting-based relaying protocol (HTPSR) considering power splitting and time switching fraction for EH efficiency of two-hop relaying networks under the impact of imperfect CSI.
- We derive expressions of the instantaneous transmission mode and delay-limited transmission mode in both AF and DF protocols.
- The performance of BER is evaluated by the outage probability and signal modulation techniques.
- The power splitting and time switching fraction in HTPSR are calculated by closed-form expressions for DF and numerical methods for AF.
- The impact of estimation channel errors on the performance is evaluated by throughput analysis. The system performance declines as there is a rise in the channel estimation error. In particular, the impact of estimation channel errors in throughput is trivial when P_S is low. Meanwhile, the performance gap between perfect and imperfect CSI in outage probability can be clearly seen when approximate $\alpha = 0.9$, $\beta = 0.7$.

The remainder of the paper is organized as follows. The system considering channel estimation errors is modeled in Sec. 2. Meanwhile, in Sec. 3, we derive expressions of throughput, BER and optimization problems for EH time and power fraction in both AF and DF relaying schemes. Section 4 provides the numerical results. Finally, Sec. 5 draws a conclusion for the paper.

2. System Model

In this system, we consider a relaying network, in which the source node (S) forwards signal to the destination node (D) via the immediate node (R). We denote $\widetilde{h}_1$ and $\widetilde{h}_2$ as first hop S-R and second hop R-D, respectively.

In each hop, channel state information (CSI) knowledge is required by the relay for self-information removal and signal detection. Unfortunately, channel estimation errors (CEE) always exist which affect negatively the system performance and energy harvesting efficiency. As illustrated in Fig. 1, l_1 and l_2 denote as the distances between $(S) \to (R)$ and $(R) \to (D)$, respectively. All channels are assumed to be Rayleigh block fading, i.e., in which they are independent and identically distributed from one slot to another. In this system herein, the fading channel is considered as the sum of the channel estimation (CE) and the CEE, in which the fading channel is distributed by $\widetilde{h}_1 \sim CN\left(0, \sigma^2_{\widetilde{h}_1}\right)$, $\widetilde{h}_2 \sim CN\left(0, \sigma^2_{\widetilde{h}_2}\right)$ denotes zero mean circularly symmetric complex Gaussian (CSCG) random variable.

As illustrated in Tab. 1, T is the block time, in which the (D) node receives a certain block of information from the (S) node. The first time slot is designed for EH and information transmission (IT) in the first hop $(S) \to (R)$ during αT while the second time slot is responsible for IT equivalent to the second hop $(R) \to (D)$ and accounts $(1 - \alpha)T$. Furthermore, while the signal is forwarded from (S) to (D), the relay uses the entire received energy not only via energy circuit but

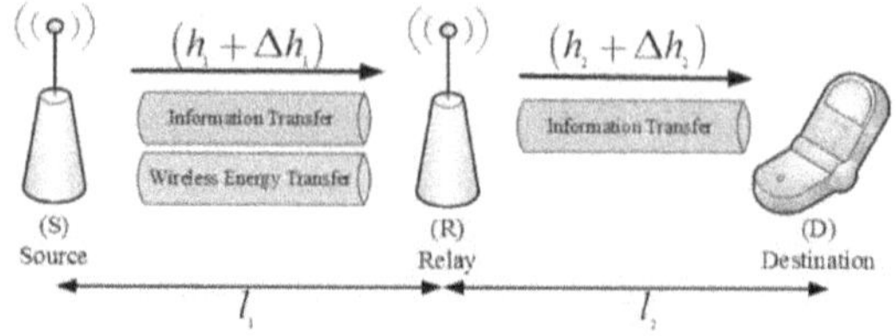

Fig. 1. The system model consists of a source, a relay and a destination node which are denoted by (S), (R) and (D), respectively.

Symbol	Description
αT	Percentage of the time switching-based IT from (S) to (R)
$(1-\alpha)T$	Percentage of the time switching-based IT from (R) to (D)
βP_S	Percentage of the power splitting-based EH at (R)
$(1-\beta)P_S$	Percentage of the power splitting-based IT from (S) to (R)
P_S	Power transmitted from (S) to (R)
P_R	Power received from (S) at (R)
T	Block time of transmission from (S) to (D)

Tab. 1. Summary of energy harvesting HTPSR protocol for relay.

also the information processing phase. In particular, there are two separate circuit components for EH, IT transmitter and different parts of the transmitted signal power: βP_S is used to transmit the amount of EH to the relay while IT from the source to the relay node accounts for $(1-\beta)P_S$, where P_S is the transmitted source power. In terms of HTPSR protocol, α denotes time switching fraction while β stands for power splitting fraction. It is noted that $\alpha \in (0,1)$, $\beta \in (0,1)$.

In the first link, $S \to R$, the calculation of the fading channel $\widetilde{h}_1$ can be expressed by [10]

$$\widetilde{h}_1 = h_1 + \Delta h_1 \tag{1}$$

and similarly in the second link, $R \to D$ the fading channel $\widetilde{h}_2$ can be expressed by

$$\widetilde{h}_2 = h_2 + \Delta h_2 \tag{2}$$

where h_1, h_2 and Δh_1, Δh_2 are CE and CEE, respectively, CSCG random variables are denoted by $h_1 \sim CN\left(0, \sigma^2_{h_1}\right)$, $h_2 \sim CN\left(0, \sigma^2_{h_2}\right)$, and $\Delta h_1 \sim CN\left(0, \sigma^2_{\Delta h_1}\right)$, $\Delta h_2 \sim CN\left(0, \sigma^2_{\Delta h_2}\right)$, respectively, with $\sigma^2_{\Delta h_1} = \sigma^2_{\widetilde{h}_1} - \sigma^2_{h_1}$, and $\sigma^2_{\Delta h_2} = \sigma^2_{\widetilde{h}_2} - \sigma^2_{h_2}$.

In the considered HTPSR protocol, the harvested energy depends on power splitting coefficient and time switching coefficient and hence it is expressed by [15]

$$E^{\mathrm{HTPSR}}_{\widetilde{h}_1} = \eta P_S \left(|h_1|^2 + \sigma^2_{\Delta h_1}\right) l_1^{-m} \alpha \beta T \tag{3}$$

where the energy conversion efficiency is denoted by η , which relies on the rectification process and the energy harvesting circuitry, $\eta \in (0,1)$ and m stands for the path loss exponent relies on the transmission medium.

At the relay node, the received power P_R is presented according to the communication between the relay and the destination node during the time slot, $(1-\alpha)T$

$$\begin{aligned} P_R &= \frac{E^{\mathrm{HTPSR}}_{\widetilde{h}_1}}{(1-\alpha)T} = \frac{\eta\alpha\beta P_S\left(|h_1|^2+\sigma^2_{\Delta h_1}\right)l_1^{-m}}{(1-\alpha)} \\ &= \varphi P_S \left(|h_1|^2 + \sigma^2_{\Delta h_1}\right) l_1^{-m} \end{aligned} \tag{4}$$

where $\varphi = \eta\alpha\beta(1-\alpha)^{-1}$.

In AF and DF relaying networks, the sampled signal at the relay in the first phase is depicted as

$$y_R(k) = \sqrt{l_1^{-m}(1-\beta)}\,(h_1 + \Delta h_1)\, x_S(k) + n_R \tag{5}$$

where data symbol is denoted by $x_S(k)$ from the source at time slot k $(k = 1, 2, ..., N)$, and it satisfies $E\left\{|x_S(k)|^2\right\} = P_S$ with the additive white Gaussian noise (AWGN) denoted by n_R is zero-mean and noise variance, σ^2_R.

In terms of AF protocol, after being amplified at the relay node, the received signal is forwarded to the destination node. In particular, the received signal is processed by the amplification factor denoted by $\mathcal{G}$ which is expressed by [5]

$$\begin{aligned} \mathcal{G}^2 &= 1/\left(l_1^{-m}(1-\beta)P_S\left(|h_1|^2 + \sigma^2_{\Delta h_1}\right) + \sigma^2_R\right) \\ &\approx 1/\left(l_1^{-m}(1-\beta)P_S\left(|h_1|^2 + \sigma^2_{\Delta h_1}\right)\right) \end{aligned}, \tag{6}$$

here, an approximation of amplify factor can be obtained due to the trivial value of AWGN when there is a significant increase in SNR. Consequently, $\mathcal{G}$ relies solely on the instantaneous CSI. Regarding DF protocol, the received signal is decoded at the relay before being regenerated. Therefore, the received signal at the (R) node [5] for both AF and DF protocol can be expressed, respectively as

$$x_R(k) = \mathcal{G} y_R(k) \tag{7}$$

and for DF case

$$x_R(k) = \frac{1}{P_S} x_S(k). \tag{8}$$

Next, the received signal at the (D) node can be calculated as

$$y_D(k) = \sqrt{l_2^{-m} P_R}(h_2 + \Delta h_2)x_R(k) + n_D \tag{9}$$

where n_D is denoted as AWGN at the (D) node with zero-mean and variance of σ^2_D.

3. Performance Analysis

In this section, the instantaneous rate and throughput performance for half duplex relaying networks using RF energy harvesting are investigated under the impact of imperfect CSI. In addition, the comparison of both AF and DF relaying protocols with the imperfect CSI is presented. In order to find detailed parameters for the design, CSI impairments are calculated to satisfy the acceptable outage performance.

3.1 SNR Calculation

In this subsection, we formulate instantaneous rate for AF and DF relaying protocols.

3.1.1 AF Based Relaying

At the destination node, we substitute the values of (5) and (7) into (9). Thus, the signal, $y_D(k)$ can be computed as (for simplicity we omit index time instant (k))

$$\begin{aligned} y_D(k) &= \sqrt{(1-\beta) P_R l_2^{-m} \mathcal{G}} x_S(k) h_1 h_2 \\ &+ \sqrt{(1-\beta) P_R l_2^{-m} \mathcal{G}} x_S(k) (h_2 \Delta h_1 + h_1 \Delta h_2 + \Delta h_1 \Delta h_2) \\ &+ \sqrt{(1-\beta) P_R l_2^{-m} \mathcal{G}} (h_2 + \Delta h_2) n_R + n_D. \end{aligned} \quad (10)$$

Based on (10), the end-to-end SNR at the (D) node can be computed by

$$\gamma_{AF} = \frac{|h_1|^2 |h_2|^2}{|h_2|^2 \mathcal{W}_1 + |h_1|^2 \mathcal{W}_2 + \mathcal{W}_3} \quad (11)$$

where $\mathcal{W}_1 = \sigma_{\Delta h_1}^2 + \frac{\sigma_R^2}{(1-\beta) l_1^{-m} P_S}$, $\mathcal{W}_2 = \sigma_{\Delta h_2}^2$, and $\mathcal{W}_3 = \sigma_{\Delta h_2}^2 \sigma_{\Delta h_1}^2 + \frac{\sigma_{\Delta h_2}^2 \sigma_R^2}{(1-\beta) l_1^{-m} P_S} + \frac{l_1^m l_2^m \sigma_D^2}{\varphi P_S}$.

3.1.2 DF Based Relaying

From (5) at the (R) node and based on (10) at the (D) node. The received SNRs at (R) and (D) in terms of DF protocol are calculated, respectively as

$$\gamma_R = \frac{(1-\beta) P_S |h_1|^2}{(1-\beta) P_S \sigma_{\Delta h_1}^2 + l_1^m \sigma_R^2}, \quad (12a)$$

$$\gamma_D = \frac{|h_1|^2 |h_2|^2}{\left(|h_2|^2 \mathcal{Z}_1 + |h_1|^2 \mathcal{Z}_2 + \mathcal{Z}_3\right)} \quad (12b)$$

where $\mathcal{Z}_1 = \sigma_{\Delta h_1}^2$, $\mathcal{Z}_2 = \sigma_{\Delta h_2}^2$, and $\mathcal{Z}_3 = \sigma_{\Delta h_1}^2 \sigma_{\Delta h_2}^2 + \frac{l_1^m l_2^m}{\varphi P_S} \sigma_D^2$.

Therefore, the calculation of end-to-end SNR, γ_{DF} can be given by

$$\gamma_{DF} = \min(\gamma_R, \gamma_D), \quad (13)$$

in which γ_R, γ_D follows from (12a) and (12b).

3.2 Optimization Problems of Instantaneous Rate

In this section, we depict the optimization problems under the power splitting ratio and time switching ratio for both AF and DF protocol. Accordingly, the data rates achieved of AF and DF protocol can be given by

$$R_{i \in \{AF,DF\}} = \log_2(1 + \gamma_i). \quad (14)$$

3.2.1 Case AF:

We mathematically formulate the optimization problem (OPT) as

$$\max_{\alpha,\beta} R_{i \in \{AF,DF\}} \quad \text{subject to} \quad \alpha,\beta \in (0,1) \quad (15)$$

Due to the fact that the logarithmic function is a monotonically increasing function of its arguments, the OPT is equivalent to follow

$$\max_{\alpha,\beta} (\gamma_{AF}) \quad \text{subject to} \quad \alpha,\beta \in (0,1) \quad (16)$$

However, due to the complexity of the aforementioned expressions, using a closed-from solution is impossible. The optimal instantaneous rate is biconvex function of α and β is numerically evaluated by taking advantage of the Golden Selection Search [25] algorithm which is similar to *Algorithm 1: Optimal solution to finding the optimal α_{opt} and β_{opt}*

Define: $f(u, v)$ is a strictly unimodal function on the boundaries of the interval $[a, b]$. Set x_1 and x_2 as two test points for the argument, in which k is the number of loops. λ is a golden proportion coefficient, around $\lambda = \frac{-1+\sqrt{5}}{2}$ and an absolute tolerance of $\phi = 1e-5$.
Set f_{max}, g_{max} is zero.
Step 1:
for $i := a$ to b **do**
replace u by i of $f(u, v)$,
then optimization of $f(i, v)$ subject to v.
Step 2:
while $|a - b| > \phi$ **do**
re-compute values $x_1 := b - (b-a)/\lambda$ and $x_2 := a + (b-a)/\lambda$ with $x_1 < x_2$.
Find $f(i, x_1)$ and $f(i, x_2)$.
Step 3:
if $f(i, x_1) < f(i, x_2)$ **then**
a new set of boundaries $[x_1, b]$,
update $g_{max} := f(i, x_2)$ and $\beta_{opt} := x_2$.
else
a new set of boundaries $[a, x_2]$,
update $g_{max} := f(i, x_1)$ and $\beta_{opt} := x_1$.
end if
end while
Step 4: Choose maximum point
if $g_{max} > f_{max}$ **then**
$f_{max} := g_{max}$ and $\alpha_{opt} := i$
Next $i := i + b/k$ and go back Step 1.
end if
end for

Regarding the optimization of α and β, while the passive variables are fixed, optimization only occurs with active variables. Consequently, looking for the partial optimum based on Algorithm 1 is suitable solution in this manner.

3.2.2 Case DF:

Based on (12a), (12b) the received SNRs can be rewritten respectively

$$\gamma_R = \frac{1}{\omega_1 + \frac{\omega_2}{(1-\beta)}}, \tag{17a}$$

$$\gamma_D = \frac{1}{\omega_3 + \frac{(1-\alpha)}{\alpha\beta}\omega_4} \tag{17b}$$

where $\omega_1 = \frac{\sigma^2_{\Delta h_1}}{|h_1|^2}$, $\omega_2 = \frac{l_1^m \sigma_R^2}{P_S |h_1|^2}$, $\omega_4 = \frac{l_1^m l_2^m \sigma_D^2}{2\eta P_S |h_1|^2 |h_2|^2}$ and $\omega_3 = \frac{\sigma^2_{\Delta h_1}}{|h_1|^2} + \frac{\sigma^2_{\Delta h_2}}{|h_2|^2} + \frac{\sigma^2_{\Delta h_1}\sigma^2_{\Delta h_2}}{|h_1|^2|h_2|^2}$.

The optimal α_{opt}, β_{opt} could be obtained by solving the following optimization

$$\max R_{DF} = \arg\max \gamma_{DF}(\alpha, \beta) \tag{18}$$

where subject to $0 < \alpha < \alpha_{\text{opt}} < 1, 0 < \beta < \beta_{\text{opt}} < 1$.

The above optimization could solved analytically (when $\gamma_R = \gamma_D$), and we have the following key result:

$$\alpha\left[\left(\omega_1 + \frac{\omega_2}{(1-\beta)} - \omega_3\right)\beta + \omega_4\right] = \omega_4. \tag{19}$$

Thus, β is fixed, the optimal α_{opt} is calculated by

$$\alpha_{\text{opt}} = \frac{\omega_4}{\left[\left(\omega_1 + \frac{\omega_2}{(1-\beta)} - \omega_3\right)\beta + \omega_4\right]}, \tag{20}$$

or α is fixed , the optimal β_{opt} is given by

$$\beta_{\text{opt}} = \frac{-b + \sqrt{b^2 - 4ac}}{2a} \tag{21}$$

where $b = (\alpha\omega_1 + \alpha\omega_2 - \alpha\omega_3 + (1-\alpha)\omega_4)$, $c = (\alpha - 1)\omega_4$, and $a = \alpha(\omega_3 - \omega_1)$.

As a result, it seems appropriate to find the partial optimum.

3.3 BER Analysis

In this section, to obtain BER calculation we first find outage probability in two cases of AF and DF protocols.

3.3.1 Outage Probability in AF

In the delay-limited transmission mode, the throughput is specified by determining the outage probability, OP, with a fixed source transmission rate, R_0 (bps/Hz), and the threshold value of SNR for detecting information precisely at the destination is $\gamma_{th} = 2^{R_0} - 1$. In that way, OP is given by

$$OP_{AF} = \Pr(\gamma_{AF} < \gamma_{th}), \tag{22}$$

in which Pr(.) denotes the probability function.

The analytical expression for OP_{AF} is determined in the following Proposition 1.

Proposition 1: At the (D) node, the OP for the HTPSR AF protocol is computed by

$$OP_{AF} \approx 1 - (\mathcal{A}_{AF})^{-1}\mathcal{B}_{AF} \times K_1(\mathcal{B}_{AF}) \tag{23}$$

where $K_1(\cdot)$ is the first order Bessel function of the second kind in [23], $\mathcal{A}_{AF} = \exp\left(\gamma_{th}\left(\frac{\mathcal{W}_1}{\sigma^2_{h_2}} + \frac{\mathcal{W}_2}{\sigma^2_{h_1}}\right)\right)$ and $\mathcal{B}_{AF} = 2\sqrt{\gamma_{th}(\mathcal{W}_3 + \gamma_{th}\mathcal{W}_1\mathcal{W}_2)\left(\sigma^2_{h_1}\sigma^2_{h_2}\right)^{-1}}$. The channel gain of the exponential random variables $|h_1|^2$ and $|h_2|^2$ are characterized $\sigma^2_{h_1}$ and $\sigma^2_{h_2}$, respectively.

Proof:

The general SNR at the (D) node of the imperfect of CSI for the considered protocol is depicted as

$$Y = \frac{X_1 X_2}{\mathcal{W}_1 X_2 + \mathcal{W}_2 X_1 + \mathcal{W}_3} \tag{24}$$

where $X_1 = |h_1|^2$ and $X_2 = |h_2|^2$ with means $\sigma^2_{X_1}$, $\sigma^2_{X_2}$, respectively.

We will first derive the cumulative distribution function (CDF), $F_Y(x)$ of x, which is the exponential random variables (RVs). In addition, we derive the probability density function (PDF) of RV X_1 is $f_{X_1}(x) \triangleq \frac{1}{\sigma^2_{X_1}}\exp\left(-\frac{x}{\sigma^2_{X_1}}\right)$.

We apply the formula to guarantee the last equality, $\int_0^\infty e^{-\frac{\beta}{4x} - yx}dx = \sqrt{\frac{\beta}{y}}K_1\left(\sqrt{\beta y}\right)$, in ([23], 3.324.1), and $F_Y(x) = \Pr(Y < x)$, which is described by

$$F_Y(x) = \int_0^{z=x.\mathcal{W}_2} f_{|X_1|}(z)\,dz + \int_{z=x.\mathcal{W}_2}^{\infty} f_{|X_1|}(z)\Pr\left(1 - \exp\left(-\frac{x(\mathcal{W}_2 z + \mathcal{W}_3)}{(\mathcal{W}_3 - x\mathcal{W}_1)\sigma^2_{X_2}}\right)\right)dz, \tag{25a}$$

$$F_Y(x) \approx 1 - \frac{1}{\sigma^2_{X_1}} \times \int_{y=x.\mathcal{W}_2}^{\infty} \exp\left(-\left(\frac{y}{\sigma^2_{X_1}} + \frac{x(\mathcal{W}_3 + x\mathcal{W}_2)}{(y - x\mathcal{W}_1)\sigma^2_{X_2}}\right)\right)dy, \tag{25b}$$

$$F_Y(x) \approx 1 - (\mathcal{A})^{-1}\mathcal{B} \times K_1(\mathcal{B}) \tag{25c}$$

where $\mathcal{A} = \exp\left(x\left(\frac{\mathcal{W}_1}{\sigma^2_{X_2}} + \frac{\mathcal{W}_2}{\sigma^2_{X_1}}\right)\right)$, $\mathcal{B} = 2\sqrt{\frac{x(\mathcal{W}_3 + x\mathcal{W}_1\mathcal{W}_2)}{\sigma^2_{X_1}\sigma^2_{X_2}}}$.

This ends the proof for Proposition 1.

3.3.2 Outage Probability in DF

In this subsection, the closed-form expressions of outage probability in HD DF protocol will be obtained. Besides that, a pre-set threshold at R_0 is represented by γ_{th}. Thus, OP is expressed by

$$OP_{DF} = \Pr(\min\{\gamma_{DF}\} < \gamma_{th}). \tag{26}$$

Proposition 2: The outage probability at the (D) node for DF protocol is given by

$$\mathrm{OP_{DF}}(\gamma_{\mathrm{th}}) \approx 1-\exp\left(-\frac{\psi\gamma_{\mathrm{th}}}{\sigma^2_{\mathrm{h_1}}}\right)\times(\mathcal{A}_{\mathrm{DF}})^{-1}\mathcal{B}_{\mathrm{DF}}\times K_1\left(\mathcal{B}_{\mathrm{DF}}\right) \tag{27}$$

where $\psi = \sigma^2_{\Delta h_1} + \frac{l_1^m \sigma_R^2}{(1-\beta)P_S}$, $\mathcal{A}^{\mathrm{DF}} = \exp\left(\gamma_{\mathrm{th}}\left(\frac{Z_1}{\sigma^2_{h_2}} + \frac{Z_2}{\sigma^2_{h_1}}\right)\right)$ and $\mathcal{B}_{\mathrm{DF}} = 2\sqrt{\frac{\gamma_{\mathrm{th}}\left(\sigma^2_{h_1} Z_3+\gamma_{\mathrm{th}} Z_1 Z_2\right)}{\sigma^2_{h_1}\sigma^2_{h_2}}}$.

Proof:

Similarly, according to the expression of OP at (D) γ_D (in (12b) for DF protocol, as introduced in Proof of Proposition 1 in (24)), we have

$$F_{\gamma_D}(\gamma_{\mathrm{th}}) \approx 1-(\mathcal{A}_{\mathrm{DF}})^{-1}\mathcal{B}_{\mathrm{DF}}\times K_1\left(\mathcal{B}_{\mathrm{DF}}\right) \tag{28}$$

where $\gamma_{\mathrm{th}} > 0$, and $\mathcal{A}_{\mathrm{DF}}$, $\mathcal{B}_{\mathrm{DF}}$ in (27).

The imperfect CSI for the DF protocol, the OP at the (R) node (in (12a) is calculated as

$$F_{\gamma_R}(\gamma_{\mathrm{th}}) = 1-\exp\left(-\frac{\psi\gamma_{\mathrm{th}}}{\sigma^2_{h_1}}\right) \tag{29}$$

where the PDF $f_{\gamma_R}(\gamma_{\mathrm{th}})$ of γ_R is presented by $f_{\gamma_R}(\gamma_{\mathrm{th}}) = \frac{\psi}{\sigma^2_{h_1}}\exp\left(-\frac{\psi\gamma_{\mathrm{th}}}{\sigma^2_{h_1}}\right)$, and ψ can be seen in (27). Hence, the CDF $\gamma_{\mathrm{DF}} = \min\{\gamma_R, \gamma_D\}$ can be expressed as in (27).

This ends the proof for Proposition 2.

Remark 1: In order to obtain optimal outage performance, $\frac{\partial OP_{i\in\{\mathrm{AF,DF}\}}(\beta,\alpha)}{\partial\beta} = 0$ (fixed α) needs to be solved, while $\frac{\partial OP_{i\in\{\mathrm{AF,DF}\}}(\alpha,\beta)}{\partial\alpha} < 0$ (fixed β) contributes to a decrease in α. However, the closed-form expression of this problem do not exist, hence we solve it in numerical methods.

3.3.3 BER Consideration

In this section, we obtain new expressions for the Bit Error Rate (BER) at the destination. We first consider the outage probability, which was obtained in [24]. Thus, we have

$$BER = E\left[aQ\left(\sqrt{2b\gamma}\right)\right] \tag{30}$$

where $Q(.)$ is the Gaussian Q-Function which is explained by $Q(x) = \frac{1}{\sqrt{2\pi}}\int_x^\infty e^{-\frac{t^2}{2}}dt$ and the modulation formats, i.e. $(a, b) = (1, 2)$ for BPSK, and $(a, b) = (1, 1)$ for QPSK. As a result, before obtaining the BER performance, the distribution function of γ is expected. Then, we begin rewriting the BER expression given in (30) directly in terms of outage probability at the source by using integration, as follows

$$BER_{i\in\{\mathrm{AF,DF}\}} = \frac{a\sqrt{b}}{2\sqrt{\pi}}\int_0^\infty \frac{e^{-bx}}{\sqrt{x}}F_{\gamma_i}(x)dx \tag{31}$$

where $F_{\gamma_i}(x) = \mathrm{OP_i(x)}$ for AF or DF protocol.

4. Numerical Results

In this section, we will examine the throughput performance, the outage probability and BER of the two relaying networks in the presence of channel estimation errors (AF and DF protocol). In particular, let us set the source transmission power, $P_S = 1$ (Joules/sec), power splitting ratio, $\beta = 0.3$, the time fraction, $\alpha = 0.3$, noise variances, $\sigma^2_{\Delta h_2} = \sigma^2_{\Delta h_1} = 0.03$, path loss exponent, $m = 2.7$ and fixed source transmission rate $R_0 = 3$ (bps/Hz), respectively. Unless otherwise stated, the energy harvesting efficiency is set to $\eta = 1$.

In terms of AF and DF relaying networks, the outage probability suffers from different values of $\alpha, \beta \in (0, 0.9)$, Fig. 2 and Fig. 3 illustrate the outage probability under the impact of perfect CSI and imperfect CSI. It can be seen that the outage probability of AF is higher than that of DF. Considering the situation where α and β vary from 0 to 0.9, the outage probability of AF and DF relaying decline substantially when α is at approximately 0.9. Unlike time switching trends in Fig.2, Fig. 3 reveals that the outage probability only decreases as β varies between 0 and 0.7. The reason is that

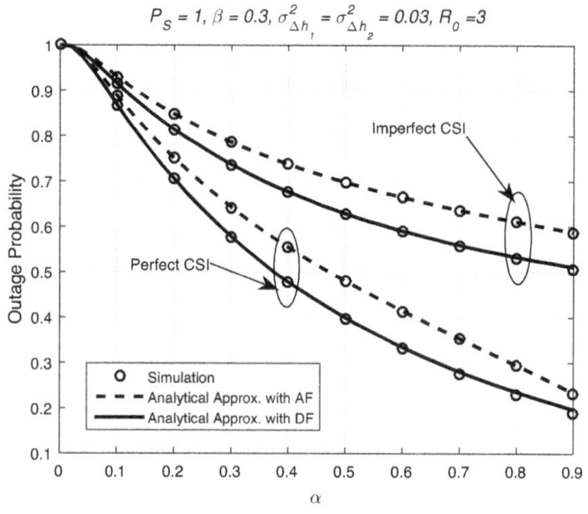

Fig. 2. The outage probability of the perfect and imperfect CSI for AF and DF relaying networks for various values of α.

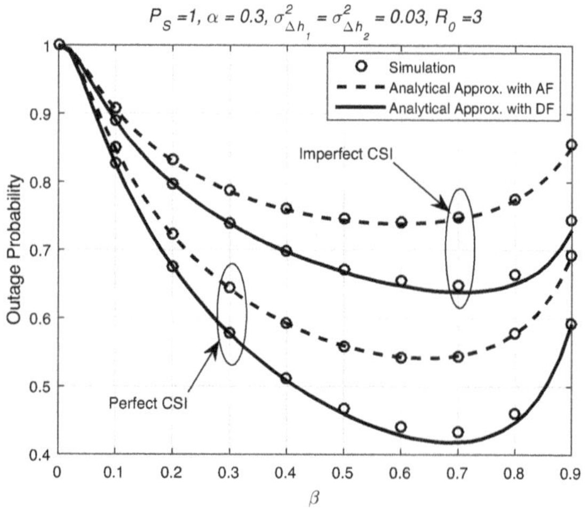

Fig. 3. The outage probability of the perfect and imperfect CSI for AF and DF relaying networks for various values of β.

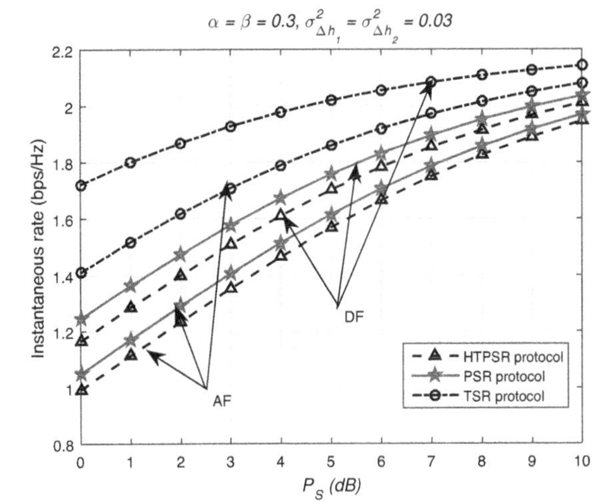

Fig. 4. The instantaneous rate of perfect and imperfect CSI for AF and DF relaying networks for different values of P_S.

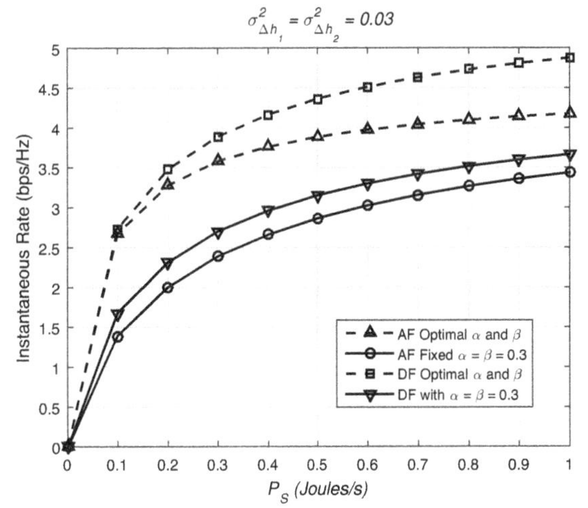

Fig. 6. Impact of optimal time switching and power splitting fraction.

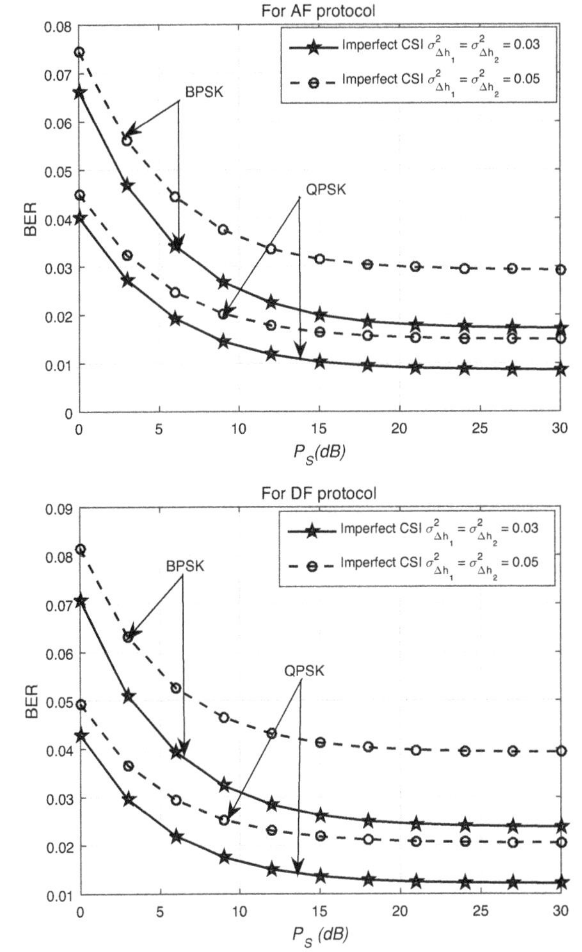

Fig. 5. BER of the AF and DF relaying networks with various values of P_S.

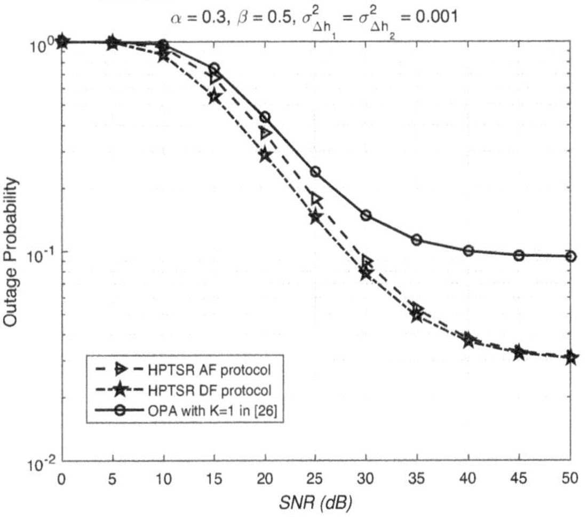

Fig. 7. Comparison between our model with recent work in [26].

more harvested energy for the relay contribute better outage performance. Subsequently, it rises gradually from 0.7 to 0.9 and results in worse outage performance due to less power for information processing in relay-destination link. It can be seen that the performance gap between imperfect CSI and perfect CSI is largest at approximately $\alpha = 0.9$ and $\beta = 0.7$ due to the impact of channel estimation error on the calculation of SNR.

Figure 4 presents the instantaneous rate of imperfect CSI and perfect CSI for AF and DF relaying networks for different values of P_S. In this experiment, we only consider the imperfect CSI and compare the three energy harvesting protocols, namely PSR, TSR [2] with HTPSR. It can be observed that TSR is the best performance in two cases of protocols. In fact, it is worth noting that this performance depends on instantaneous values of the channel, since the transmit power from source, P_S intends to supply the energy harvesting circuit at the relay node in TSR protocol while only small fraction of such power is used for the considered protocol HTPSR. In addition, when the values of P_S increase, the system throughput in the presence of imperfect CSI of the three schemes also rise due to the contribution of P_S to SNR.

The BER of the AF and DF relaying networks was mentioned in (31). As can be seen from Fig. 5, in terms of the BER of AF and DF relaying networks, P_S rises from 0 to 30(dB). We can see that the system with QPSK modulation outperforms BPSK modulation in both AF and DF. In particular, the values of $\sigma^2_{\Delta h_2}$ increase as the values of BER in the imperfect CSI fall. It can be seen that the values of BER in AF network experience the same tendency as DF network.

Impact of optimal time switching and power splitting fraction Fig. 6 introduces the instantaneous rate versus the transmitted power from source. The simulation results prove that the instantaneous rate is the best with optimal α and β. In particular, when P_S increases from 0 to 0.6, there is a rapid increase in the instantaneous rate. Eventually, it gradually rises from 0.6 to 1. It is proved that choosing appropriate optimal values of time switching and power splitting for HTPSR contributes to the optimal instantaneous rate.

Finally, Figure 7 compares the outage probability in our work with the recent similar model under imperfect CSI as presented in [26]. Here, we also conduct extensive simulations considering similar system parameters and error models to evaluate the performance of their proposed framework, such as the related channel estimation error equals to 0.001, the energy conversion efficiency equals to 0.9, and the number of relays $K = 1$. In particular, to minimize the outage probability, an optimal power allocation (OPA) scheme was proposed by the authors in [26]. However, selection of appropriate values of power splitting and time switching coefficients in the HTPSR protocol can enhance outage performance. As can be seen clearly that outage performance of HTPSR is better than the OPA scheme as investigation in [26] in case of considering outage probability versus SNR at the source node.

5. Conclusion

In this paper, we examined both AF and DF relaying networks based on RF energy harvesting systems. Furthermore, the impact of imperfect CSI on the system performance is determined by the harvested power for AF and DF relaying networks. The analytical expressions of achievable throughput, bit error rate (BER) and the impact on imperfect CSI on AF and DF networks were elaborated in the numerical results. Based on the numerical analysis, we provide practical insights into the impact on many various outlines on the energy efficiency of the system by using DF and AF relay nodes. We can see that the throughput of AF relaying networks performs worse compared to the throughput of DF networks. Especially, we obtain the best instantaneous rate for HTPSR with optimal time switching and power splitting fraction.

References

[1] ZHOU, X., ZHANG, R., HO, C. K. Wireless information and power transfer: Architecture design and rate-energy tradeoff. *IEEE Transactions on Communications*, 2013, vol. 61, no. 11, p. 4754–4767. DOI: 10.1109/TCOMM.2013.13.120855

[2] NASIR, A. A., ZHOU, X., DURRANI, S., et al. Relaying protocols for wireless energy harvesting and information processing. *IEEE Transactions on Wireless Communications*, 2013, vol. 12, no. 7, p. 3622–3636. DOI: 10.1109/TWC.2013.062413.122042

[3] THUAN, D. D. Time power switching based relaying protocol in energy harvesting mobile node: Optimal throughput analysis. *Mobile Information Systems*, 2015, p. 1–8. DOI: 10.1155/2015/769286

[4] ZHANG, R., YANG, L. L., HANZO, L. Energy pattern aided simultaneous wireless information and power transfer. *IEEE Journal on Selected Areas in Communications*, 2015, vol. 33, no. 8, p. 1492–1504. DOI: 10.1109/JSAC.2015.2391551

[5] RIIHONEN, T., WERNER, S., WICHMAN, R. Hybrid full-duplex/half-duplex relaying with transmit power adaptation. *IEEE Transactions on Wireless Communications*, 2011, vol. 10, no. 9, p. 3074–3085. DOI: 10.1109/TWC.2011.071411.102266

[6] HUANG, K., LAU, V. K. N. Enabling wireless power transfer in cellular networks: Architecture, modeling and deployment. *IEEE Transactions on Wireless Communications*, 2014, vol. 13, no. 2, p. 902–912. DOI: 10.1109/TWC.2013.122313.130727

[7] JU, H., ZHANG, R. Optimal resource allocation in full-duplex wireless-powered communication network. *IEEE Transactions on Communications*, 2014, vol. 62, no. 10, p. 3528–3540. DOI: 10.1109/TCOMM.2014.2359878

[8] ZHU, Y., WU, X., ZHU, T. Hybrid AF and DF with network coding for wireless two way relay networks. In *Proceedings of the IEEE Wireless Communications and Networking Conference (WCNC)*. 2013, p. 2428–2433. DOI: 10.1109/WCNC.2013.6554941

[9] LIU, C. F., MASO, M., LAKSHMINARAYANA, S., et al. Simultaneous wireless information and power transfer under different CSI acquisition schemes. *IEEE Transactions on Wireless Communications*, 2015, vol. 14, no. 4, p. 1911–1926. DOI: 10.1109/TWC.2014.2376953

[10] AHMED, I., IKHLEF, A., NG, D. W. K., et al. Power allocation for a hybrid energy harvesting relay system with imperfect channel and energy state information. In *Proceedings of the IEEE Wireless Communications and Networking Conference (WCNC)*. 2014, p. 990–995. DOI: 10.1109/WCNC.2014.6952243

[11] CHOI, D., LEE, J. H. Outage probability of two-way full-duplex relaying with imperfect channel state information. *IEEE Communications Letters*, 2014, vol. 18, no. 6, p. 933–936. DOI: 10.1109/LCOMM.2014.2320940

[12] PRASAD, B., ROY, S. D., KUNDU, S. Secondary throughput in underlay cognitive radio network with imperfect CSI and energy harvesting relay. In *Procedings of the IEEE International Conference on Advanced Networks and Telecommunications Systems (ANTS)*. 2015, p. 1–6. DOI: 10.1109/ANTS.2015.7413619

[13] SU, Y., JIANG, L., HE, C. Relay selection for full-duplex cooperative networks with outdated CSI in an interference-limited environment. In *Procedings of the IEEE 83rd Vehicular Technology Conference (VTC Spring)*. 2016, p. 1–5. DOI: 10.1109/VTCSpring.2016.7504443

[14] TOURKI, K., QARAQE, K. A., ALOUINI, M. S. Outage analysis for underlay cognitive networks using incremental regenerative relaying. *IEEE Transactions on Vehicular Technology*, 2012, vol. 62, no. 2, p. 721–734. DOI: 10.1109/TVT.2012.2222947

[15] THUAN, D. D. Energy-aware two-way relaying networks under imperfect hardware: Optimal throughput design and analysis. *Telecommunication System*, 2015, vol. 62, no. 2, p. 449–459. DOI: 10.1007/s11235-015-0085-7

[16] JU, H., ZHANG, R. Throughput maximization in wireless powered communication networks. *IEEE Transactions on Wireless Communications*, 2014, vol. 13, p. 418–428. DOI: 10.1109/TWC.2013.112513.130760

[17] PARK, J., CLERCKX, B. Transmission strategies for joint wireless information and energy transfer in a two-user MIMO interference channel. In *Proceedings of the IEEE International Conference on Communications Workshops (ICC)*. 2013, p. 591–595. DOI: 10.1109/ICCW.2013.6649302

[18] BENKHELIFA, F., ALOUINI, M. S. Simultaneous wireless information and power transfer for MIMO amplify-and-forward relay systems. In *Proceedings of the IEEE Global Communications Conference (GLOBECOM)*. 2015, p. 1–6. DOI: 10.1109/GLOCOM.2015.7417175

[19] LIN, H., ZHAO, R., HE, Y., et al. Secrecy performance of transmit antenna selection with outdated CSI for MIMO relay systems. In *Proceedings of the IEEE International Conference on Communications Workshops (ICC)*. 2016, p. 272–277. DOI: 10.1109/ICCW.2016.7503799

[20] WANG, F., ZHANG, X. Resource allocation for multiuser cooperative overlay cognitive radio networks with RF energy harvesting capability. In *Proceedings of the IEEE Global Communications Conference (GLOBECOM)*. 2016, p. 1–6. DOI: 10.1109/GLOCOM.2016.7842221

[21] LEE, K., KO, J. Adaptive power allocation and splitting with imperfect channel estimation in energy harvesting based self-organizing networks. *Mobile Information Systems*, 2016, p. 1–7. DOI: 10.1155/2016/8243090

[22] NGUYEN, V. D., VAN, S. D., SHIN, O. S. Opportunistic relaying with wireless energy harvesting in a cognitive radio system. In *Proceedings of the IEEE Wireless Communications and Networking Conference (WCNC)*. 2015, p. 87–92. DOI: 10.1109/WCNC.2015.7127450

[23] GRADSHTEYN, I. S., RYZHIK, I. M. *Table of Integrals, Series, and Products*. 4th ed. Academic Press, Inc., 1980. ISBN: 9780123849335

[24] GOLDSMITH, A. *Wireless Communications*. Cambridge (UK): Cambridge Univ. Press, 2005. ISBN: 0521837162

[25] BRAUN, W. J., MURDOCH, D. J. *A First Course in Statistical Programming with R*. Cambridge (UK): Cambridge Univ. Press, 2008. ISBN: 9780521694247

[26] ZHANG, Y., GE, J., MEN, J., et al. Joint relay selection and power allocation in energy harvesting AF relay systems with ICSI. *IET Microwaves, Antennas & Propagation*, 2016, vol. 10, no. 15, p. 1656–1661. DOI: 10.1049/iet-map.2016.0028

About the Authors ...

Dinh-Thuan DO was born in Phu Yen province, Viet Nam. He received his Ph.D. degree from University of Science (VNU-HCM) in 2012. Dr. Thuan was the recipient of the 2015 Golden Globe Award by Ministry of Science and Technology. He is currently Assistant Professor at the Wireless Communications and Signal Processing Lab (WICOM LAB). His research interests include mmWave communications, Massive MIMO, Cooperative communications, Energy harvesting, Full-duplex communications, Cognitive radio.

Hoang-Sy NGUYEN (corresponding author) was born in Binh Duong province, Vietnam. He received the B.S. and MS.c degree from the Department of Computer Science from Ho Chi Minh City University of Information Technology (UIT-HCMC), Vietnam in 2007, 2013, respectively. He is currently pursuing the Ph.D. in School of Electrical Engineering and Computer Science, Technical University of Ostrava, Czech Republic. His research interests include energy-efficient wireless communications, 5G wireless, low-power networks.

Miroslav VOZNAK (born in 1971) is an Associate professor with Department of Telecommunications, VSB-Technical University of Ostrava. He received his Ph.D. degree in telecommunications from the VSB-Technical University of Ostrava in 2002 and he was appointed as an associate professor in 2009. His professional knowledge covers generally Information and Communication technology, in his research, he deals with wireless networks, Voice over IP, security and optimization problems. He is an Associate Editor for: Advances in Electrical and Electronic Engineering, Mobile, Embedded and Distributed Systems. He is an IEEE Senior member and serves as a TPC member of various international conferences.

Thanh-Sang NGUYEN He is currently researcher at Binh Duong University, Vietnam. His research interests include Cooperative communications, Energy harvesting, Cognitive radio, D2D, Computer network and Cloud computing.

14

Wavelet Support Vector Machine Algorithm in Power Analysis Attacks

Shourong HOU, Yujie ZHOU, Hongming LIU, Nianhao ZHU

Dept. of Electronic Engineering, Shanghai Jiao Tong University, 800 Dongchuan Road, Shanghai, China

ruihou@sjtu.edu.cn, mlscagroup@163.com

Abstract. *Template attacks and machine learning are two powerful methods in the field of side channel attacks. In this paper, we aimed to contribute to the novel application of support vector machine (SVM) algorithm in power analysis attacks. Especially, wavelet SVM can approximate arbitrary nonlinear functions due to the multidimensional analysis of wavelet functions and the generalization of SVM. Three independent datasets were selected to compare the performance of template attacks and SVM based on various kernels. The results indicated that wavelet SVM successfully recovered the offset value of the masked AES implementation for each trace, which was obviously 5 to 8 percentage points higher than SVM-RBF. And also, the time required was almost reduced by 40% when using the optimal parameters of wavelet SVM. Moreover, wavelet SVM only required an average of 5.4 traces to break the secret key for the unmasked AES implementation and less than 7 traces for the masked AES implementation.*

Keywords

Power analysis attacks, template attacks, support vector machine, wavelet analysis, kernel function

1. Introduction

Modern cryptographic devices generally implement the cryptographic algorithm and store the corresponding secret key. The cryptographic device must remain the secret key regardless of whether the algorithm itself is public or not. Therefore, it's important that the cryptographic algorithm does not reveal key-related information in the process of execution. Unfortunately, none of the cryptographic devices can eliminate the relevant information about the secret key from various side channels. An attacker may invade the entire security system and break the key through side channel information, which is called side channel attacks (SCAs). Typical SCAs contain power analysis attacks [1], timing attacks [2], acoustic cryptanalysis key extraction attacks [3], electromagnetic attacks [4], and their combinations [5]. In these attacks, power analysis attacks have attracted the attention of industry and academia.

Power analysis attacks were first proposed by Kocher et al. [1], which exposed the fact that the instantaneous power consumption of a cryptographic device depends on the data being processed and operations being performed. Many power analysis attacks methods have sprung up since then, such as Differential Power Analysis (DPA) [1], Template Attacks (TA) [6], [7], Correlation Power Analysis (CPA) [8], Mutual Information Analysis (MIA) [9], and Stochastic Model based Power Analysis (SMPA) [10]. From the viewpoint of engineering, power analysis attacks include two types, namely profiling and non-profiling attacks. For profiling attacks, a basic assumption is that the adversary is free to access target devices and profiling devices. Profiling attacks consist of two phases, called profiling and attacking. During the profiling phase, the adversary analyzes the profiling device by multiple power traces so that the key of the target device can be recovered when the attacking phase is performed. However, non-profiling attacks are single-step attacks that are performed directly on the target device. Over the past decade, various countermeasures against power analysis attacks have been proposed in hardware or software implementations. In general, almost all strategies are divided into two categories, hiding and masking. The masking scheme is very popular due to low cost and high performance. A comprehensive summary of power analysis attacks and countermeasures can refer to the book [11].

Rivest [12] first studied the intersection of machine learning and cryptography. Machine learning is a discipline whose purpose is to build a probability model based on the given data to predict the final result, which includes unsupervised learning, supervised learning and reinforcement learning. Roughly, the purpose of supervised learning is to learn a general function that maps the input space to the desired output space. The application of machine learning algorithms in power analysis attacks has just begun in recent years. Heyszlet et al. [13] used the k-means algorithm to attack the public key cryptosystem. Martinasek, Z. et al. [14] presented power analysis attacks based on multi-layer perceptron (MLP), and the authors [15] improved the MLP approach. The unsupervised clustering algorithm was proposed by Whitnall et al. [16], which was used to build the power consumption leakage model. Zhang et al. [17] researched

DPA attack based on genetic algorithm (GA). In work [18], the generalized CPA based on the k-Nearest Neighbors (k-NN) algorithm was briefly mentioned. Later, Martinasek, Z. et al. [19] improved power analysis attacks based on k-NN.

Hospodar et al. [20], [21] first applied Least Squares SVM (LS-SVM) in power analysis attacks. Although no real attack is performed, it provides a novel perspective on how LS-SVM is used for power analysis attacks. He et al. [22] presented that SVM-based attack recovered the entire secret key of DES performed on an 8-bit Atmel smartcard in a Stanford course project. Profiling attacks based on machine learning algorithms were introduced by Lerman et al. [23], [24]. They first compared TA and learning algorithms, namely Random Forest (RF), SVM, and Self-Organizing Maps (SOM), and then proposed an enhanced brute force algorithm to break the key. Heuser et al. [25] analyzed multiple bits of the key based on the Hamming weight model by using multi-class SVM. The authors divided the intermediate power consumption into several classes and then calculated the probability of belonging to each class. Finally, they adopted the maximum likelihood method to get the correct key. Their results have demonstrated that the performance of SVM is more stable than TA. This probabilistic multi-class SVM approach was later improved by Bartkewitz et al. [26]. More precisely, the authors assumed that the side channel leakage information could split interesting points into a strict order. Lerman et al. [27] presented the attack based on machine learning algorithm against the masked AES implementation. The results declared that SVM required 26 power traces to recover the key and had smaller computational complexity than TA. Hence, it was assumed that sample points of traces may not follow multivariate Gaussian distribution [28].

SVM is the most popular machine learning algorithm in power analysis attacks, while other algorithms have been proved to be feasible [14–19]. The previous work [23–27] focused more on how SVM translates a problem of breaking the key into the classification of machine learning. There is no systematic literature to study the elements that influence the performance of SVM in power analysis attacks. The kernel function (kernel method, kernel trick), hyperparameters (penalty factor, gamma of RBF kernel), feature selection, and other elements have significant effects on the performance of SVM. Wavelet analysis is a powerful tool for signal processing, which is often used to approximate the target function [29]. Although wavelet analysis has been used to process noisy power traces [30–32], the combined effects of wavelet analysis and power analysis attacks through kernel functions have been not yet explored. In order to enhance the sparsity of wavelet approximation and the generalization of SVM, Zhang et al. [33] first proposed a variant SVM algorithm based on wavelet kernel, known as wavelet SVM. It has been widely applied in financial, medical, industrial control, computer vision and other fields.

This paper aims to explore the application of wavelet SVM in power analysis attacks. Our attacks were implemented on three public datasets, including the offset recovery phase and the key recovery phase. We selected the success rate as a measure of the offset recovery phase and the guessing entropy as a metric of recovering the secret key. Furthermore, the results indicate that wavelet SVM is one of the most effective and efficient algorithms in power analysis attacks. This paper attempts to answer the following questions:

- Is wavelet SVM more suitable for power analysis attacks than TA and SVM based on other kernels?
- What is the impact of the optimal parameters of SVM on the classification results?
- What are the effects of the number of power traces (or the number of interesting points) on the performance of SVM based on various kernels?

2. Background

This section provides all the necessary knowledge about the principle and fast implementation of SVM, followed by the brief introduction to TA.

2.1 Binary-Class SVM

Cortes and Vapnik [34] proposed SVM in 1995 to address the binary classification with high generalization. The most basic model of SVM is a linear binary classifier, which aims to determine the separating hyperplane between two classes. Let

$$D_M = \{(\mathbf{X}_i, y_i) \big| \mathbf{X}_i \in R^N, y_i \in \{-1, +1\}, i = 1, 2, ..., M\} \quad (1)$$

represent a training set, where $\mathbf{X}_i$ is a training vector in the feature space, and y_i is the class label of $\mathbf{X}_i$. The separating hyperplane is:

$$f(\mathbf{X}) = \omega^T \phi(\mathbf{X}) + b \quad (2)$$

where $\omega \in R^N$, $b \in R$. By the nonlinear mapping function $\phi(\cdot)$, $\mathbf{X}$ is mapped into a feature space. There are many possible hyperplanes for an SVM classifier. A reasonable choice for the optimal hyperplane is to find the maximum separation (margin) between two classes. Accordingly, the maximum margin classifier can be rewritten as a constrained optimization problem:

$$\begin{gathered} \min_{\omega, b, \xi} \left(\frac{1}{2} \|\omega\|^2 + C \sum_{i=1}^{M} \xi_i \right), \\ \text{s.t. } y_i(\omega^T \phi(\mathbf{X}_i) + b) \geq 1 - \xi_i, \xi_i \geq 0, i = 1, 2, 3, ..., M \end{gathered} \quad (3)$$

where ξ_i is the training error for vector $\mathbf{X}_i$, and C>0 is the regularization parameter which determines a trade-off between training error and margin size, also known as penalty factor.

By introducing the Lagrange multiplier, the optimization problem with constraints in (3) is simplified a dual problem:

$$\begin{gathered} \min_{\alpha} \frac{1}{2} \sum_{i=1}^{M} \sum_{j=1}^{M} \alpha_i \alpha_j y_i y_j K(\mathbf{X}_i, \mathbf{X}_j) - \sum_{i=1}^{M} \alpha_i, \\ \text{s.t. } \sum_{i=1}^{M} \alpha_i y_i = 0, 0 \leq \alpha_i \leq C, i = 1, 2, ..., M \end{gathered} \quad (4)$$

where the kernel function is $K(\mathbf{X}_i, \mathbf{X}_j) = \phi(\mathbf{X}_i)^T \phi(\mathbf{X}_j)$, and α_i are Lagrange multipliers. The optimal solution to this problem is $\alpha^* = (\alpha_1^*, \alpha_2^*, ..., \alpha_M^*)^T$. The ω^* is given as follows:

$$\omega^* = \sum_{i=1}^{M} \alpha_i^* y_i \phi(\mathbf{X}_i) \tag{5}$$

and then a Lagrange multiplier α_j^* that satisfies $0 < \alpha_j^* < C$ is chosen to calculate:

$$b^* = y_j - \sum_{i=1}^{M} \alpha_i{}^* y_i K(\mathbf{X}_i, \mathbf{X}_j) \tag{6}$$

where $K(\mathbf{X}_i, \mathbf{X}_j)$ is the kernel function. Finally, the decision function of hyperplane is [35]:

$$f(\mathbf{X}) = \text{sign}\left(\sum_{i=1}^{M} \alpha_i^* y_i K(\mathbf{X}_i, \mathbf{X}) + b^*\right). \tag{7}$$

2.2 Multi-Class SVM

By adding several extensions, the binary-class SVM can be used to construct multi-class SVM. The mainstream strategies include one-against-one [36], one-against-all [37], directed acyclic graphs (DAG) [38], and error correction output coding [38]. For the sake of training time and accuracy, the machine learning community adopts the one-against-one strategy [39] to train a binary-class SVM classifier for each pair of possible classes. That is, for L classes, $(L-1)L/2$ binary-class SVM classifiers are required to be trained. The prediction results of all binary-class SVM classifiers are combined into the multi-class SVM classifier output, and then the class with the most votes is chosen. For more details, please refer to [36], [39].

2.3 Probabilistic SVM

In order to obtain the maximum likelihood estimate in the key recovery phase, the probability output of the class label c is necessary. Considering the sparsity of SVM, the logistic sigmoid function is usually used to approximate the outputs $y(\mathbf{X}_i)$ of all binary-class SVM classifiers [40]. The posterior conditional probability is given as follows:

$$p\,(c = 1|\mathbf{X}_i) = \frac{1}{1 + \exp(A \cdot y(\mathbf{X}_i) + B)} \tag{8}$$

where vector $\mathbf{X}_i$ belongs to the class $c = 1$. Obviously, $p(c = -1|\mathbf{X}_i) = 1 - p(c = 1|\mathbf{X}_i)$. The parameters A and B are computed by the optimization of cross-entropy error as follows:

$$\underset{A,B}{\arg\min} - \sum_{i=1}^{M} t_i \log(p_i) + (1 - t_i)\log(1 - p_i) \tag{9}$$

where $t_i = (1 + sign[y(\mathbf{X}_i)])/2$ and $p_i = p(c = 1|\mathbf{X}_i)$. The numerical problems of optimization are introduced in [41].

So far, many fast implementations of SVM have been proposed to compute the globally optimal solution. One of the most popular is sequential minimal optimization (SMO), proposed by Platt [42] in 1998. The SMO algorithm decomposes the optimization problem into many smaller-scale problems, which requires only two Lagrange multipliers at one time. This strategy makes it possible to obtain the objective function value of quadratic programming by means of the analytical method, which significantly accelerates the training speed of SVM. The specific implementation of SMO algorithm and numerical problems to be noted, please refer to [42].

2.4 Template Attacks

TA is the most powerful power analysis attack in an information theory sense. The classical TA is based on the multivariate Gaussian distribution $N\left(\mathbf{t}; (\mathbf{m}, \mathbf{C})\right)$ as follows:

$$(2\pi)^{-\frac{N}{2}} |\mathbf{C}|^{-\frac{1}{2}} \exp\left(-\frac{1}{2}(\mathbf{t}-\mathbf{m})^T \mathbf{C}^{-1} (\mathbf{t}-\mathbf{m})\right) \tag{10}$$

where $\mathbf{t}$ represents a N-dimensional vector, $\mathbf{m}$ is the mean vector, $\mathbf{C}$ is the covariance matrix.

In parametric estimation theory, suppose that the number of traces is P_k, the given operation is expressed as O_k, and a power trace is recorded as $\mathbf{t}_{p_k}|O_k$. The estimated template consists of a set of mean vectors $\{\mathbf{m}_k\}$ and a set of covariance matrices $\{\mathbf{C}_k\}$, $k = 1, \ldots, K$. In the maximum likelihood approach, the parameters that maximize the likelihood are selected. Maximizing the likelihood is equal to the log likelihood maximization, which is given by:

$$\log L_k = \log \prod_{p_k=1}^{P_k} p(\mathbf{t}_{p_k}|O_k) = \sum_{p_k=1}^{P_k} \log N\left(\mathbf{t}_{p_k}|\mathbf{m}_k, \mathbf{C}_k\right) \tag{11}$$

where $p(\mathbf{t}_k|O_k)$ is the likelihood probability of power traces $\mathbf{t}_{p_k}$ under the operation O_k performed on the cryptographic device.

In terms of SCAs, where an erroneous environment is assumed, an attacker is more interested in the probability of an instance $\mathbf{X}_i$ belonging to the class c. Hence, instead of predicting the class c, we predict the posterior conditional probability $P_{\text{SVM}}\left(\mathbf{X}_i|c\right)$ of each class c. Since the probability estimate of multi-class SVM is a very specialized discipline, we refer to [41] for the knowledge of how to calculate $P_{\text{SVM}}\left(\mathbf{X}_i|c\right)$. The log likelihood of each possible key k is as follows:

$$\log L_k \equiv \log \prod_{i=1}^{M_k} P_{\text{SVM}}(\mathbf{X}_i|c) \equiv \sum_{i=1}^{M_k} \log P_{\text{SVM}}(\mathbf{X}_i|c) \tag{12}$$

where M_k is the number of power traces belonging to the key k. The key k^* that maximizes the log likelihood in (11) or (12) is chosen, which is given as follows:

$$\underset{k^*}{\arg\max} \log L_{k^*}. \tag{13}$$

3. Wavelet Kernel for Power Analysis

As we all know, the kernel function has been applied in many pattern recognition and machine learning algorithms. By introducing the kernel function, SVM avoids the problem of processing data in high dimensional space and even theoretically infinite dimensional space. The kernel function also maintains the reasonable computational complexity of SVM in the feature space.

3.1 Common Kernel Functions

In general, the kernel function consists of a linear kernel, a polynomial kernel, and an RBF kernel, which must satisfy the Mercer theorem [43]. The Linear kernel function (Linear kernel):

$$K(\mathbf{X}_i, \mathbf{X}_j) = \mathbf{X}_i^T \mathbf{X}_j. \tag{14}$$

The Polynomial kernel function (Poly kernel):

$$K(\mathbf{X}_i, \mathbf{X}_j) = (\mathbf{X}_i^T \mathbf{X}_j + 1)^d. \tag{15}$$

The Gaussian kernel function (RBF kernel):

$$K(\mathbf{X}_i, \mathbf{X}_j) = \exp\left(-\gamma \left\|\mathbf{X}_i - \mathbf{X}_j\right\|^2\right), \gamma > 0 \tag{16}$$

where $d \geq 1$ is the order of polynomials, γ is the hyperparameter of RBF kernel, and the notation $\|\cdot, \cdot\|$ represents the Euclidean distance between two vectors.

The kernel function is equivalent to a similarity function in some feature space. Given two objects, the kernel outputs a similarity score. As long as the kernel functions know how to compare them, the objects can be anything. For the Linear kernel in (14), the similarity is the product of the length of $\mathbf{X}_i$ and the projection length of $\mathbf{X}_j$ in the direction of $\mathbf{X}_i$. The similarity of RBF kernel between two vectors in (16) is reweighted by the hyperparameter γ. A small γ will result in low bias and high variance while a large γ will get higher bias and low variance [35]. Accordingly, when choosing the appropriate kernel function and its hyperparameters to solve practical problems, the expertise in the relevant areas of problems is necessary.

3.2 Wavelet Kernel Functions

From a perspective of signal analysis, one power trace is also a continuous signal in time domain. The traditional signal analysis theory is based on Fourier analysis, which has many deficiencies in the non-stationary signal. Compared with Fourier analysis, wavelet analysis processes signal simultaneously in time domain and frequency domain, which extracts information more effectively from processed signals. Wavelet analysis adopts fast attenuation, known as the wavelet to represent signal waveforms, which can arbitrarily scale and shift the input signal. The wavelet function is:

$$h_{a,b}(x) = \frac{1}{\sqrt{a}} h\left(\frac{x - b}{a}\right) \tag{17}$$

where a is a dilation factor and b is a translation factor. A detailed introduction of wavelet analysis is given in [44], [45]. Zhang et al. [33] proved the method of constructing wavelet kernel. Let $h(x)$ be a wavelet function, respectively $a, b_i, b'_i, x_i, x'_i \in R$, $\mathbf{X}, \mathbf{X}' \in R^N$, and then the wavelet kernel is:

$$K(\mathbf{X}, \mathbf{X}') = \prod_{i=1}^{N} \left(h\left(\frac{x_i - b_i}{a}\right) \cdot h\left(\frac{x'_i - b'_i}{a}\right)\right) \tag{18}$$

where $\mathbf{X} = (x_1, x_2, ..., x_N)$ and $\mathbf{X}' = \left(x'_1, x'_2, x_3..., x'_N\right)$ are N dimensional vectors. and the translation-invariant wavelet kernel is:

$$K(\mathbf{X}, \mathbf{X}') = \prod_{i=1}^{N} h\left(\frac{x_i - x'_i}{a}\right). \tag{19}$$

The Morlet wavelet function is given as follows:

$$h(x) = \cos(1.75x) \cdot \exp(-\frac{x^2}{2}). \tag{20}$$

Thus, the wavelet kernel based on Morlet wavelet function is [33]:

$$K(\mathbf{X}, \mathbf{X}') = \prod_{i=1}^{N} \left(\cos\left(\frac{1.75(x_i - x'_i)}{a}\right) \exp\left(-\frac{\|x_i - x'_i\|^2}{2a^2}\right)\right). \tag{21}$$

Later, many wavelet kernel functions including Gaussian wavelet kernel function [46], [47] were proposed. The Gaussian wavelet function is defined as follows:

$$h(x) = (-1)^{\frac{p}{2}} C_p(x) \exp\left(-\frac{1}{2}x^2\right) \tag{22}$$

where p is a positive even integer, $C_p(x)\exp(-\frac{1}{2}x^2)$ is the pth step's differential coefficient of Gaussian function. The form of Gaussian wavelet function varies with the p value. When p is zero, $C_p(x)$ is 1, which is actually a type of Gaussian function. When the value of p is 2, $C_p(x) = x^2 - 1$ is called the Mexican hat wavelet function. When the value of p is 4, $C_p(x) = x^4 - 6x^2 + 3$ is a four order polynomial that is unstable in the numerical theory. Therefore, we chose Gaussian function and Mexican hat wavelet function as kernel functions. The RBF kernel based on Gaussian function is shown in (16). The wavelet kernel based on Mexican hat wavelet function is:

$$K(\mathbf{X}, \mathbf{X}') = \prod_{i=1}^{N} \left(\left(1 - \frac{(x_i - x'_i)^2}{a^2}\right) \exp\left(-\frac{\|x_i - x'_i\|^2}{2a^2}\right)\right). \tag{23}$$

For the purpose of theoretical completeness and paying tribute to J. Fourier, here the Fourier kernel [48], [49] is given as follows:

$$K(\mathbf{X}, \mathbf{X}') = \prod_{i=1}^{N} \frac{1 - q^2}{2(1 - 2q\cos(x_i - x'_i) + q^2)}. \tag{24}$$

The wavelet kernel approximates the non-stationary signal with high precision, which is impossible for the traditional kernels. The traditional kernel functions such as Gaussian function are related and even redundant. However, the wavelet function is orthonormal, which almost approximates any function in continuous space, thus the generalization of wavelet SVM is improved. Meanwhile, the sparse wavelet kernel accelerates the training speed of SVM. Although the wavelet kernel requires more time to process power traces than other kernels, the overall training time of wavelet SVM is significantly decreased. Consequently, we creatively proposed an assumption that wavelet SVM has better stability and fewer iterations than SVM based on others kernels in power analysis.

4. Experiments

Prof. Lin Chih-Jen of Taiwan University has developed a widely used SVM kit, known as LIBSVM (Library for Support Vector Machines) [50]. LIBSVM is an integrated software for distribution estimation (one-class SVM), C-support vector classification (C-SVC), nu-support vector classification (nu-SVC), epsilon-support vector regression (epsilon-SVR), and nu-support vector regression (nu-SVR). It supports different SVM formulations (multi-class SVM, weighted SVM for unbalanced data), cross-validation, model selection, various kernels, probability estimates, etc. The one-aginst-one strategy is used to predict the probability output P_{SVM}. The highly optimized C-SVC makes it easy to set parameters for the given classification problem.

4.1 How to Set Parameters

It is theoretically possible to test an infinite number of parameters, but in practice, it makes no sense. According to article [51], we selected the regularization parameter C from 0.01 to 10, epsilon (tolerance of termination criterion) from 0.01 to 1 and various kernels including Linear kernel, Poly kernel, RBF kernel, Fourier kernel and wavelet kernels, together it was 3600 of combinations. Besides, TA was selected as a comparison. The Hamming weight model assumes the intermediate power consumption of the entire byte instead of only multiple bits, thus it is selected as the hypothetical power leakage model in this paper.

The first dataset (DS1) aims to break the 16th byte of the first round key of the unmasked AES algorithm. TeSCASE Group has [53] implemented this algorithm on the Sasebo-GII board [52] provided by RCIS [54]. This board has a mechanism to provide users various means to access the reconfiguration function of FPGA. Only 900 power traces were selected to locate interesting points in this dataset. We computed the Pearson correlation between each instant of power traces and the Hamming weight of the S-Box output and then selected the 32 highest correlated points as interesting points. An example of power traces is shown in Fig. 1. The first peak is the plaintext loaded into the register, and the next 10 peaks correspond to 10 rounds of the unmasked AES algorithm.

The second dataset (DS2), which includes 1000 power traces, is prepared for the masked AES implemented in software. Power traces are freely available on the DPA Contest v4 (DPACv4) website [56]. The masking scheme, known as RSM [55], is an additive Boolean masking countermeasure with 16 masked S-Boxes. The mask values are rotated according to the offset value. All power traces were measured during the first round and the beginning of the second round of AES algorithm. The label value corresponds to the offset value (0 to 15). The Pearson correlation between the offset value and each instant of power traces was used to locate interesting points. We selected the two highest correlated points for each mask value, thus the number of interesting points was 32. The offset value is highly correlated with sample points except for the central part of each trace in Fig. 2.

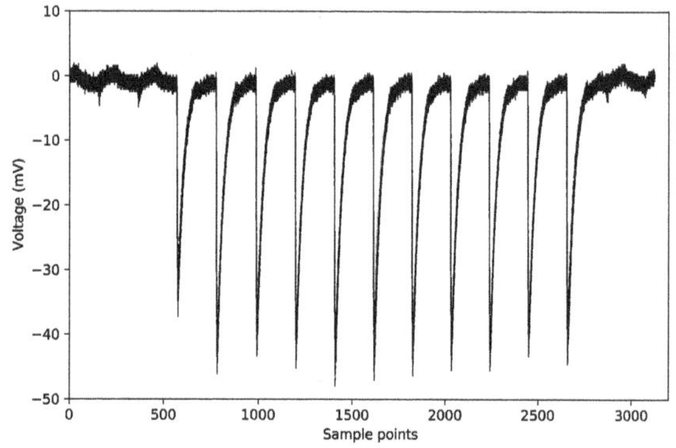

Fig. 1. Power traces of the unmasked AES in DS1.

The third dataset (DS3) concentrates on the second byte of the first round key of the masked AES algorithm. We selected the Hamming weight of the second S-Box (SBox1) output as the label of an SVM classifier. That is, the label value corresponds to the Hamming weight value of the output byte (0 to 8). This dataset includes 3600 power traces, but only 1800 traces are randomly selected in each test. The reason is that the number of traces corresponds to each label may be not the same, namely SVM with imbalanced data. We calculated the Pearson correlation between each instant of 3600 power traces and the Hamming weight of the SBox1 output to locate interesting points. Besides, the 32 highest correlated points were selected as interesting points. Only a particular interesting points have larger power leakage of the S-Box output as illustrated in Fig. 3.

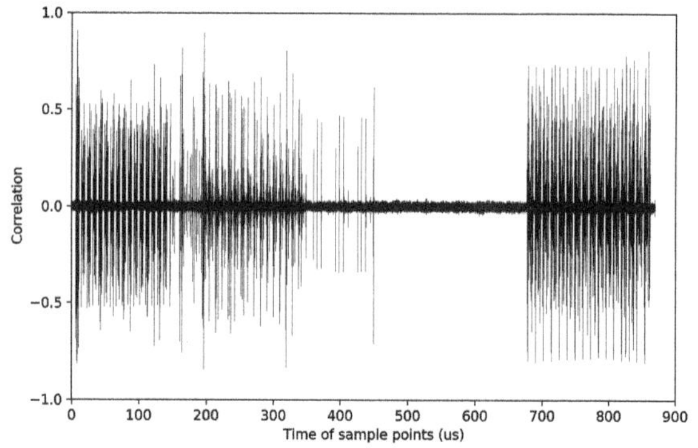

Fig. 2. Correlation between the Hamming weight of the offset value and the power consumption in the 1st round of the masked AES in DS2.

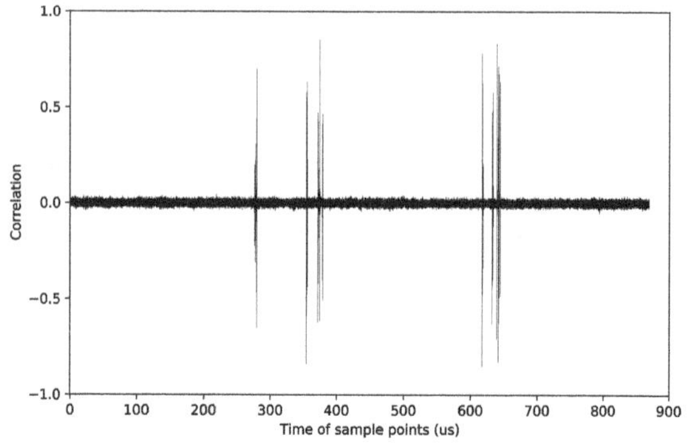

Fig. 3. Correlation between the Hamming weight of the Sbox1 output and the power consumption in the 1st round of the masked AES in DS3.

Here only the Pearson correlation was used for feature selection. In addition, we recommend using other methods, such as the minimum redundancy maximum correlation [57], principal component analysis [58], etc.

4.2 How to Compare Performance

In order to compare TA and several SVM algorithms, we performed different experiments on three datasets. We assumed that the attacker had complete control over the cryptographic device, who measured enough power traces to recover the secret key.

Our experimental methodology was as follows: Given a dataset, a random two-third was reserved as the learning set and the remaining one-thirds was used as the test set. The learning set generated training and validation sets by using 5-fold cross validation. The validation sets of all folds were used to optimize hyperparameters of SVM. The optimal hyperparameter (the one that has the highest average accuracy on the validation folds) was used to train the final SVM algorithm on the training set.

The most popular evaluation method is the accuracy (success rate) of the independent test set. Consequently, we selected the success rate as a measure in the offset recovery phase. In order to make the results more reliable, each experiment was repeated 10 times, and then the success rate of the corresponding test set was recorded. The final result was the average of all success rates.

The success rate is adequate when the number of power traces corresponding to each predicted class is the same. However, for the key recovery phase, which assumes multiple traces, the success rate is not suited as a measure. The problem is that the most likely Hamming weight class has the largest number of power traces when the success rate is selected as a metric. The guessing entropy [59] was selected to evaluate the number of remaining keys. The guessing entropy is defined as follows: let g include the descending probability ranking of all possible keys and i represent the position of the correct key in g. After performing s experiments, one gets a matrix $[g_1, g_2, ..., g_s]$ and a corresponding vector $[i_1, i_2, ..., i_s]$. In other words, the guessing entropy represents the average number of power traces required to recover the correct key. Thus, GE^1 (a guessing entropy of 1) was selected as a measure in the key recovery phase.

5. Results and Analysis

We recovered the offset value of the masked AES by using DS2, especially DS1 and DS3 were used for the key recovery phase. Each attack extracted the offset value before performing the key recovery of DS3. All our experiments were performed on Asus laptop with 2.50GHz Intel Core (TM) i5, 8GB 1067MHz DDR3 (Windows7 x64 Ultimate). The attack lasted about 5 weeks without considering the time to create three datasets.

5.1 Finding the Offset Value

This section explores the performance of different approaches that expose the offset value by using power traces of DS2. We randomly selected 500 traces as a training set and 250 traces as a test set. Moreover, 3~32 interesting points were used in our experiments, which were the most correlated with the offset value. We compared TA and various SVM algorithms such as SVM-Linear, SVM-Poly, SVM-RBF, SVM-Fourier, and wavelet SVM from the success rate and the time required. The impact of the training set size on the success rate was also discussed.

The first experiment explored the effect of various kernels on the success rate of SVM when the training set size was different. Figures 4 and 5 describe the success rate of different numbers of interesting points when using SVM-Linear and SVM-Poly to recover the offset value. First, the performance of SVM-Linear or SVM-Poly is basically not affected by the training set size, which is mainly due to the fact that the feature space of traces is not linearly separable. Furthermore, SVM-Linear and SVM-Poly require a lot of iterations to find the appropriate hyperplane, resulting in very low training efficiency. Second, the number of interesting points per trace significantly determines the success rate. In general, the more interesting points, the higher the success rate. Interestingly, the success rate of 100% size of the training set is obviously lowest for SVM-Poly when the number of interesting points is 16 to 28. One possible explanation is that SVM-Poly appeared overfitting when the training set size is small. We did not give the results of SVM-Linear and SVM-Poly due to the poor performance.

Figures 6, 7, 8, and 9 reveal the corresponding success rates for different numbers of interesting points when SVM-RBF, SVM-Fourier, SVM-Morlet, and SVM-Mexican are used to predict the offset value. As expected, the success rate of SVM increases as the number of interesting points increases. Moreover, the larger training set size, the higher the success rate. This can be explained that the performance of SVM is determined by its parameters, and the training set size is critical to the best parameters of SVM. When the training set size is expanded from 25% to 50%, the success rate of SVM increases significantly, but when the training set size is expanded from 75% to 100%, the success rate of SVM is not obviously improved. Wavelet SVM and SVM-Fourier obtain much higher success rates than SVM-RBF due to the powerful approximation capability.

The purpose of the second experiment was to study the success rate of TA and compare the efficiency of TA and SVM based on various kernels. Figure 10 illustrates the relationship between the success rate of TA and the size of the training set when the number of interesting points is 3 to 32. Generally, the larger number of interesting points, the higher the success rate of TA. However, for the small training set (25%~50% size), the success rate of TA is reduced when the number of interesting points exceeds a certain value. The reason is that when the number of interesting points is

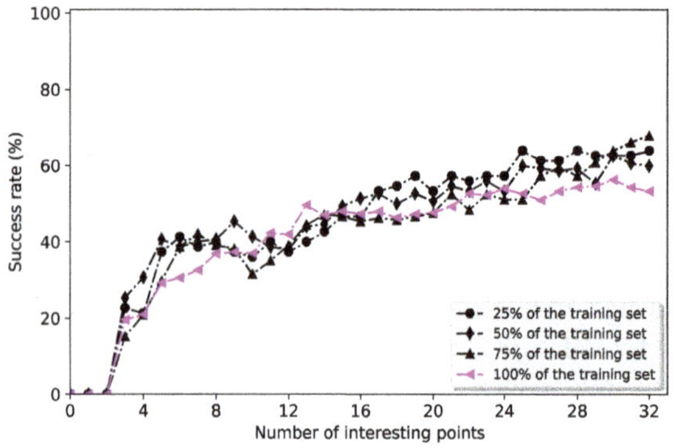

Fig. 4. Success rate of finding the offset value based on SVM-Linear by using power traces of DS2.

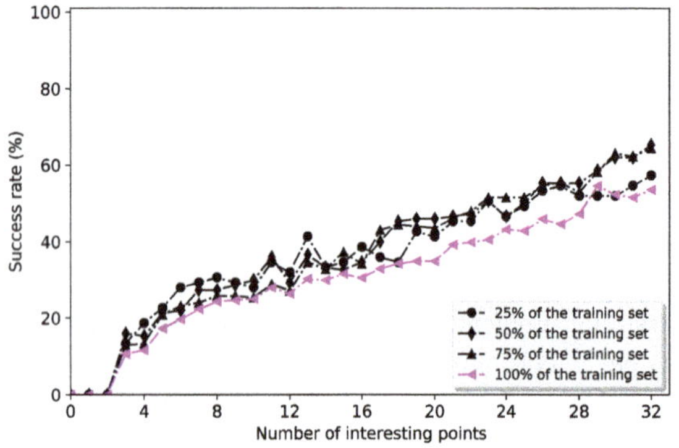

Fig. 5. Success rate of finding the offset value based on SVM-Poly (with degree 2) by using power traces of DS2.

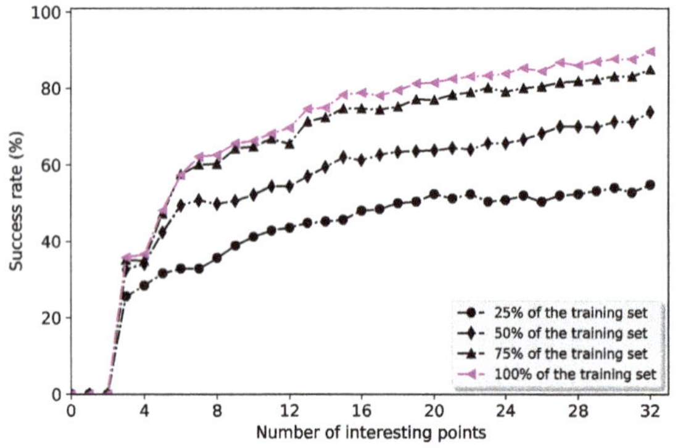

Fig. 6. Success rate of finding the offset value based on SVM-RBF by using power traces of DS2.

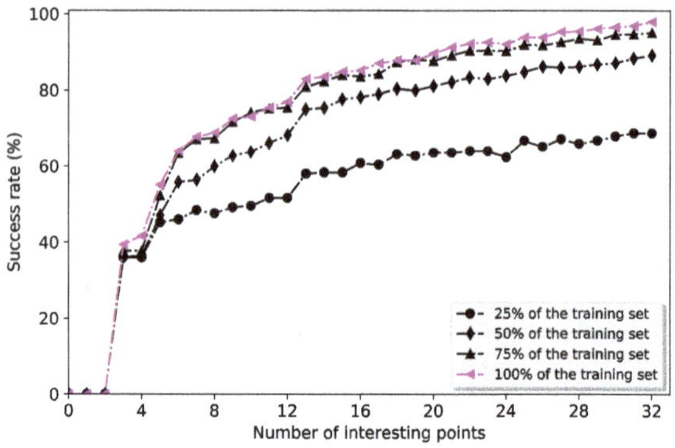

Fig. 7. Success rate of finding the offset value based on SVM-Fourier by using power traces of DS2.

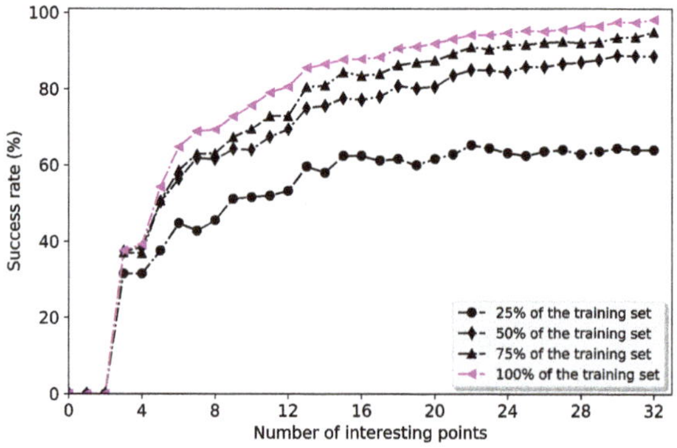

Fig. 8. Success rate of finding the offset value based on SVM-Morlet by using power traces of DS2.

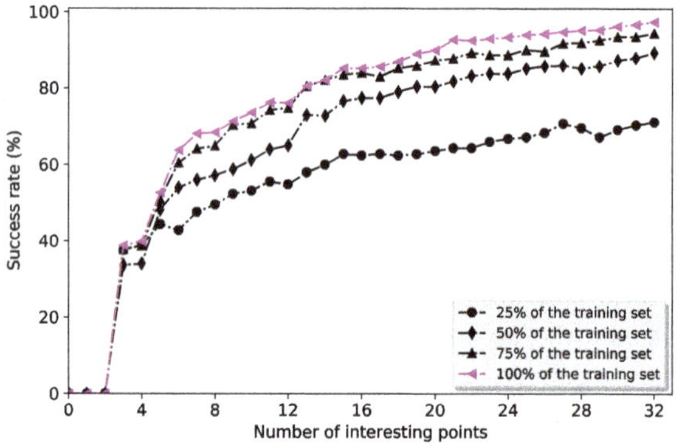

Fig. 9. Success rate of finding the offset value based on SVM-Mexican by using power traces of DS2.

too large, the covariance matrix may be an ill-conditioning matrix [11]. The results illustrate that SVM extracts more information of the offset value than TA. The performance of TA is equivalent to wavelet SVM when the training set size exceeds 75%, but the computational complexity of TA is higher than SVM based on various kernels (see Fig. 11). More precisely, the success rate of TA is very good while its prediction time increases exponentially with the number of interesting points. Although the classical TA does not work well in terms of efficiency, it is still selected for comparison in later experiments.

The third experiment was used to find the optimal dilation factor a in (21) and (23) by comparing the success rate of wavelet SVM. If a is very large, then even if two vectors are quite similar, the kernel function will output a small value. In other words, the support vectors obtained by a training set have little impact on the classification of the test set, which causes the model may be prone to overfitting. When the value of a is small, support vectors have a great effect on the classification. This means that you may not be able to obtain a complex decision boundary. The optimal value of a is 3.2 in our experiments. We used the wavelet function and the value of a to construct a new symbol that represents the type of wavelet kernel. For example, the symbol Mexican1 represents the wavelet function is Mexican hat wavelet function and the value of a is 1. From the perspective of success rate, wavelet SVM is 5~8% higher than SVM-RBF in Fig.12. Note that the success rates of SVM-Morlet1 and SVM-Mexican1 are very low when the number of interesting points exceeds a certain value. This can be interpreted as

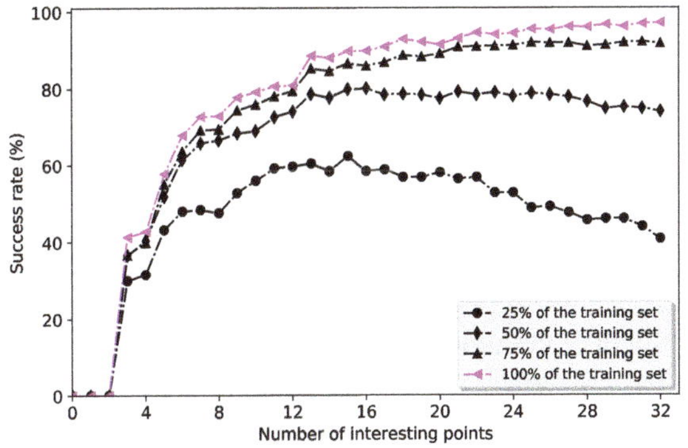

Fig. 10. Success rate of finding the offset value based on TA by using power traces of DS2.

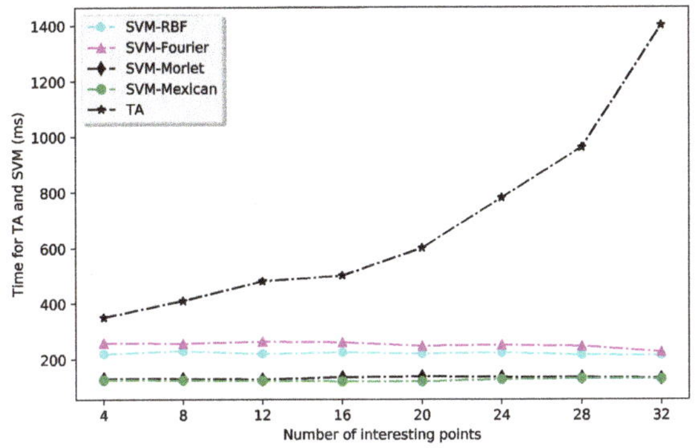

Fig. 11. The training time required of SVM vs TA by using 500 power traces of DS2.

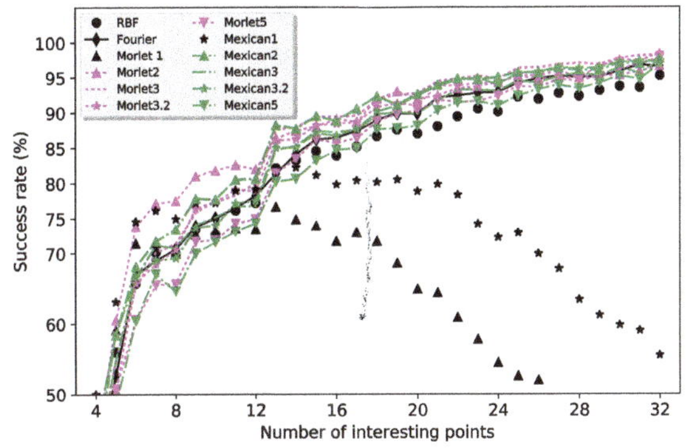

Fig. 12. Success rate of finding the offset value based on different kernels by using 500 power traces of DS2.

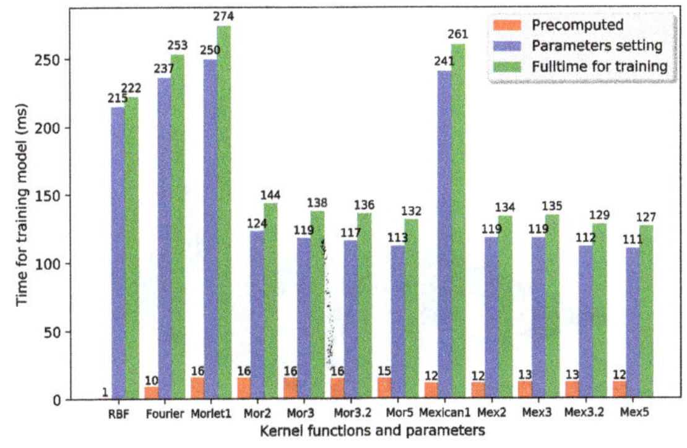

Fig. 13. The efficiency of finding the offset value based on different kernels by using 500 power traces of DS2 [Mor = Morlet, Mex = Mexican].

	C=0.3	C=0.7	C=1	C=7	C=10
RBF	70.3	85.3	90.4	95.6	91.2
Fourier	78	96.8	97.1	98.6	97.1
Morlet1	18.4	35.6	36.7	40.1	40.3
Morlet2	82.6	93.7	95.9	96.7	95.6
Morlet3	86.2	95.8	96.7	97.5	96.5
Morlet3.2	85.8	96.4	97.5	98.4	97.6
Morlet5	78.9	93.5	96.6	97.1	97.6
Mexican1	17.8	58.4	60.3	62.3	60.2
Mexican2	82.7	96.1	97.9	98	97.8
Mexican3	85.6	96.4	97.1	97.8	96.2
Mexican3.2	82.6	97.6	98.8	99.2	97.0
Mexican5	80.1	94.6	95.4	97.3	96.8
TA	83.5	95.3	96.2	97.5	95.1

Tab. 1. The effect of parameters C on success rate of SVM based on different kernels by using power traces of DS2 when the number of interesting points is 32, epsilon=0.32.

an inappropriate value of a. When the value of a is too small, the simple decision boundary deteriorates the generalization of wavelet SVM.

In the fourth experiment, a fixed number of power traces were selected to compare the efficiency of SVM based on various kernels. The efficiency of SVM was evaluated by measuring the time required to use the kernel function to process interesting points (Precomputed), the time required to train parameters of SVM (Parameters setting), and the time required to perform a complete training (Fulltime for training). The results indicate that when the value of a is appropriate, the overall time of wavelet SVM is less than SVM based on other kernels (see Fig. 13). However, wavelet SVM requires more time to process interesting points. Hence, the value of a is crucial for the efficiency of wavelet SVM. When the value of a is near 3.2, the overall training time of wavelet SVM is greatly reduced. Although the Precomputed time is negligible, the overall time of SVM-RBF is still more than wavelet SVM. In the view of time cost, SVM-Fourier and SVM-RBF are similar, but the Precomputed time of SVM-Fourier is equal to wavelet SVM. When the value of a is appropriate, the overall time of wavelet SVM is almost reduced by 30% to 40% compared with SVM-RBF. Consequently, the wavelet kernel accelerates the convergence speed of setting parameters and ultimately reduces the overall training time.

The last experiment was conducted to verify the effect of penalty factor C on the success rate of SVM. The penalty factor controls the cost of misclassification on the training set, which indicates the importance of misclassification to SVM. The large value of C implies the high cost of misclassification (hard margin), which allows SVM to increase the number of iterations to optimize the separating hyperplane. In other words, the generalization of SVM drops due to the large value of C. When the value of C is small, the cost of misclassification is low (soft margin), allowing some misclassifications. The small value of C makes SVM tend to accelerate the speed of training, resulting in a decrease in success rate. The best C is to find a balance between hard margin and soft margin. When the value of C is 0.3, 0.7, 1,

7, and 10, the success rates of SVM based on various kernels are given in Tab. 1. The final results show that the optimal value of C is 7 when the value of epsilon is 0.32.

5.2 Key Recovery Phase

The experiments were carried out to recover the secret key by using power traces of DS1 and DS3. Here, the number of power trace was not limited to only one, thus we made use of various methods to recover the key. In this paper, the Hamming weight leakage model is based on the entire intermediate value of the S-Box output rather than a single bit. Therefore, we adopted the probabilistic multi-class SVM algorithm to distinguish nine different classes. We combined the prediction results of a binary-class SVM classifier from N power traces X_i belonging to the class c by using the posterior probability output $P_{\text{SVM}}(X_i|c)$. We performed the maximum likelihood estimate for each possible key and then selected the key that maximizes the likelihood in (13) by using multiple power traces. Ultimately, a guessing entropy of 1 (GE^1) was selected to measure and compare the performance of different algorithms in the key recovery phase.

In the first experiment, the 16th byte of the last round key of the unmasked AES was extracted by using power traces of DS1. We randomly selected 400 power traces as a training set, 200 power traces as a test set. As can be seen from Figure 14, the success rate of the Hamming weight of the S-Box output increases as the number of interesting points increases. When the value of a is appropriate, the performance of wavelet SVM is still better than SVM-RBF in the key recovery phase. However, with the increase in the number of interesting points, the success rate of SVM-RBF decreases significantly and becomes very unstable. The reason may be that sample points of power traces of DS1 don't obey the multivariate Gaussian distribution. In order to improve the performance of SVM, we adopted cross validation and grid search algorithms to optimize hyperparameters. Overall, the success rate of SVM based on various kernels is maintained at 60~90%. Especially, when the value of a is 2, the success rates of SVM-Morlet and SVM-Mexican reach about 90%. Note that SVM-Fourier obtains a fairly good

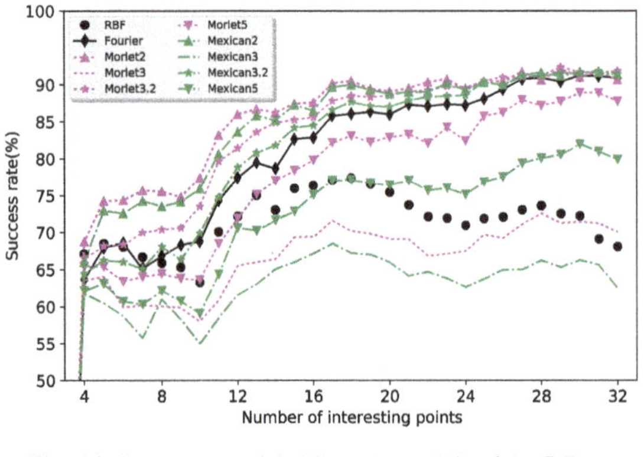

Fig. 14. Success rate of the Hamming weight of the S-Box output based on various kernels by using power traces of DS1.

success rate when the number of interesting points exceeds a certain value, but the time required is more than wavelet SVM. Therefore, wavelet SVM has advantages over TA and SVM based on other kernels when using power traces of DS1 in the key recovery phase.

The second experiment was aimed at the masked AES implementation in terms of the assumption that the offset value of each trace is known. The distribution of the Hamming weight of the S-Box output is not uniformly distributed, which has little effect on the experimental results. After all, the prediction results of SVM are independent of the distribution of the Hamming weight class when the guessing entropy is selected as a measure. As long as the dataset ensures that the number of power traces per Hamming weight class is sufficient, SVM for the unbalanced data is also competent. Here, we performed the maximum likelihood estimate for each possible key by using probabilistic multi-class SVM. A fixed value of the guessing entropy was selected to assess how many power traces of DS3 were required. Figures 15, 16, 17 and 18 describe the maximum likelihood probability of all possible guessing keys (0x00~0xff, and the correct key is 0xec) when using 1 to 20 power traces. SVM-RBF requires the maximum number of power traces to obtain the secret key. Also, SVM-Mexican requires 6 or so power traces to guess the correct key. SVM-Morlet works slightly

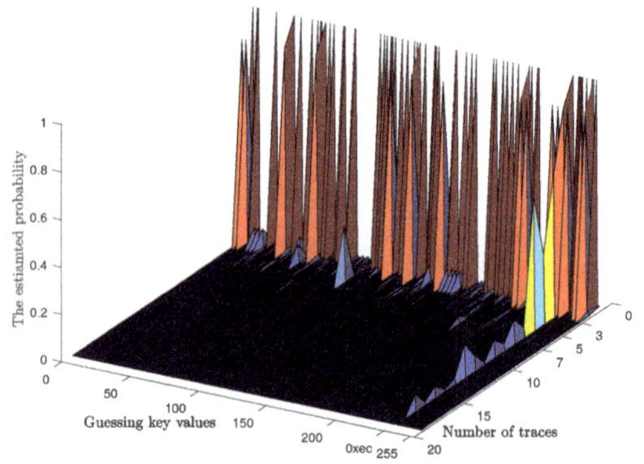

Fig. 15. The maximum likelihood estimation of guessing entropy based on SVM-RBF by using traces of DS3.

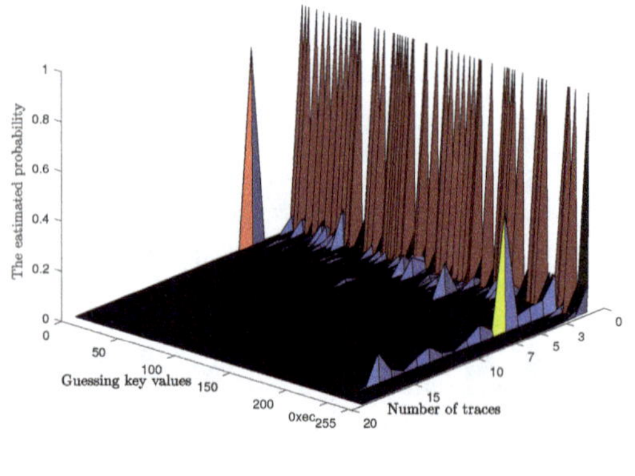

Fig. 16. The maximum likelihood estimation of guessing entropy based on SVM-Fourier by using traces of DS3.

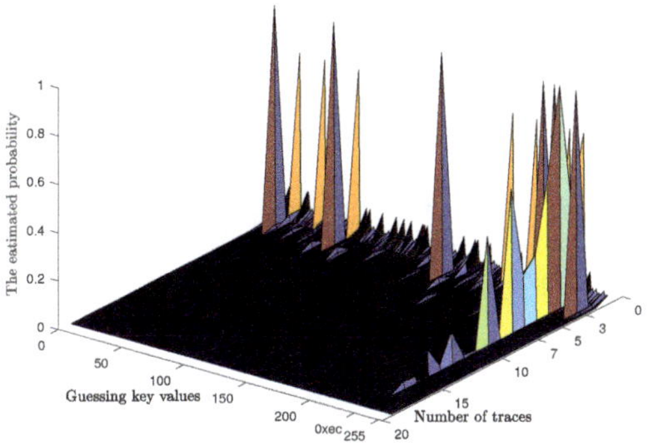

Fig. 17. The maximum likelihood estimation of guessing entropy based on SVM-Morlet by using traces of DS3.

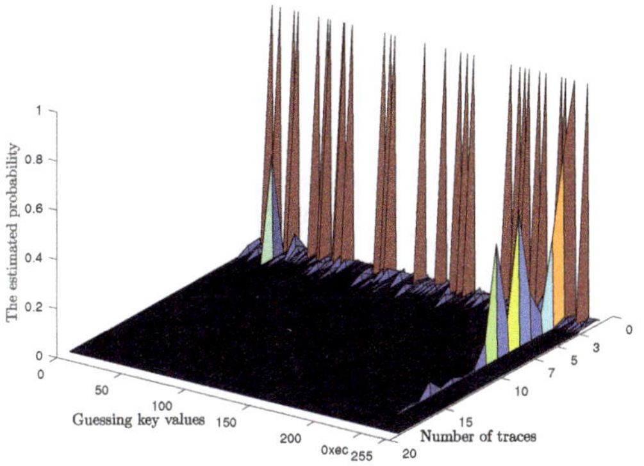

Fig. 18. The maximum likelihood estimation of guessing entropy based on SVM-Mexican by using traces of DS3.

	DS1	DS3
SVM-RBF	9.8	7.2
SVM-Fourier	9.7	6.7
SVM-Morlet	6.7	5.4
SVM-Mexican	6.3	5.3
TA	76	58

Tab. 2. The number of power traces required by SVM and TA when the guessing entropy is set to 1 (GE^1).

better than SVM-Mexican, which needs about 5 power traces. The performance of SVM-Fourier has some instability, which requires over 10 power traces. The results of SVM based on various kernels further confirmed the superiority of wavelet SVM in the key recovery phase.

In the third experiment, TA and SVM were used to recover the secret key by using power traces of DS1 or DS3. For fairness considerations, we repeated hundreds of similar experiments by using TA and SVM based on various kernels and then calculated the average as the final results. The results were recorded in Tab. 2. TA requires about 58 traces to break the key of the unmasked AES implementation by using power traces of DS3. In contrast, wavelet SVM needs a smaller number of power traces for the key recovery of DS3 (5.4 traces in average for SVM-Morlet, 5.3 traces in average for SVM-Mexican). SVM-RBF requires 9.8 traces (using DS1) and 7.2 traces (using DS3) when the guessing entropy is set to 1 (GE^1). The performance of SVM-Fourier is much better than TA, slightly inferior to wavelet SVM. Similar results can be obtained by using power traces of DS1. The results confirm that wavelet SVM is very suitable for the key recovery of the unmasked or masked AES algorithm.

5.3 Comparison with Other Work

In Tab. 3, we summarized the previous work of using non-SVM learning algorithms we have discussed in Sec. 1. We described the results of learning algorithms such as MLP, k-NN, and RF in detail.

The MLP algorithm recovered the key based on one power trace [14]. It achieved 85% empirical success rate and 80% theoretical success rate. They focused on the first byte of the secret key. The authors [15] later proposed averaging of power traces as the preprocessing method, which improved the success rate to 96%. However, they did not give a specific feature selection method and training time. The training process of MLP is very time-consuming in practice.

The k-NN algorithm exhibited great potential in power analysis attacks [19]. The standard CPA and Pearson correlation were used to locate interesting points. They chose 50 sample points for each trace on three datasets as interesting points. The success rate of k-NN (k = 5) was 94.97%. The time required to perform one 10-fold cross validation was less than 1 s. The training time of k-NN is much less than other learning algorithms, but the k nearest neighbor search for all training instances is very time-consuming in the testing phase. Moreover, the k-NN algorithm is very sensitive to neighbor instances. If the neighbor instance happens to be noise, the prediction results will go wrong [35].

The goal was to attack a single bit of 3DES by using the RF algorithm, and its performance went beyond TA [23]. Besides, they used many feature selection methods such as Ranking, PCA, Minimum redundancy maximum relevance (mRMR), and SOM. They selected 20 sample points as interesting points in all experiments. The RF algorithm increased the probability of recovering one byte of the key from 5.80% (TA) to 15.33%. Lerman et. al [27] presented the machine learning attack against the masked AES algorithm by using 1500 power traces of DPACv4. The Pearson correlation coefficient between the offset value and sample points of each trace was used to locate interesting points. They chose 50 points that are most correlated with the offset value as intereting points. Due to the strategy of feature selection, interesting points may be too concentrated to extract more power consumption leakage information. For the RF algorithm, the success rate of recovering the offset value was about 80%. The time to process one trace in the learning phase was less than 1 ms.

In our work, wavelet SVM successfully recovered the offset value of the masked AES algorithm for each power trace, which was obviously 5~8% higher than SVM-RBF and the time required was almost reduced by 40% when

Ref.	Machine learning	Algorithm	No. of traces	Performance
[13]	k-means	ECC	9 traces	recovers the secret scalar from single execution attack
[14]	MLP	AES	2560 traces	the theoretical and empirical success rates are 80% and 85%
[15]	Improved MLP	AES	2560 traces	recovers the key from one trace with accuracy ≥ 96%
[17]	GA	DES	2000 (time), 7000 (freq.)	reduces no. traces by 60% by attacking multiple S-Boxes
[19]	k-NN	AES	2560 traces	recovers the secret offset with accuracy 94.97%
[23]	RF	3DES	400 (a byte of the key)	performs binary bit classification better than TA
[27]	RF	AES	1500 traces	finds the offset value with accuracy 80%

Tab. 3. Machine learning in power analysis attacks.

using the optimal hyperparameters. In order to break the secret key, wavelet SVM only required in average 5.4 traces for the unmasked AES algorithm and less than 7 traces for the masked AES algorithm. Considering the training time and success rate, we firmly believe that the performance of other learning algorithms is slightly inferior to SVM.

6. Conclusion and Future Work

As can be seen from the above description, power analysis attacks are viewed as the classification problems. Power analysis attacks and machine learning create templates (features) to describe power traces of a training set and then calculate the similarity between templates (features) and power traces of a test set. Finally, the results are given with a certain probability. Generally, power analysis attacks assume that sample points of power traces are approximated by a set of finite normal distributions. However, machine learning assumes that sample points are independent and identically distributed, but not restrict to a certain distribution.

TA assumes that sample points of each power trace follow a multivariate Gaussian distribution. Moreover, TA not only describes the power consumption information but also inevitably describes the noise, which makes it need a lot of traces to improve the signal to noise ratio. However, the strategy adopted by SVM is quite different from TA. The key of SVM is to quickly find the separating hyperplane of the offset values or the Hamming weights. TA aims to simulate the real power consumption distribution by considering all sample points of traces. However, SVM focuses on the separation of classes, using only support vectors. Consequently, TA requires more power traces than SVM in the learning (profiling) phase.

Wavelet analysis can approximate any function, which is a powerful tool to process nonlinear and multidimensional signals. SVM is very suitable for solving the classification problems of small-scale dataset. SVM-RBF based on Gaussian function is the most commonly used in engineering, but not necessarily the optimal solution to solve all classification problems. In light of our experiments, wavelet SVM significantly improves the success rate of finding the offset value, which requires less power traces than SVM-RBF in the key recovery phase. Accordingly, wavelet SVM show excellent performance and efficiency in power analysis attacks.

The application of machine learning in power analysis attacks has not been fully explored. An important direction of research is to use the learning algorithms for feature selection. Furthermore, the development of a customized learning algorithm for power analysis attacks will be a challenging area of future work.

References

[1] KOCHER, P. C., JAFFE, J., JUN, B. Differential power analysis. In *Proceedings of the 19th Annual International Cryptology Conference on Advances in Cryptology*. London (UK), 1999, p. 388–397. DOI: 10.1007/3-540-48405-1_25

[2] KOCHER, P. C. Timing attacks on implementations of Diffie-Hellman, RSA, DSS, and other systems. In *Proceedings of the 16th Annual International Cryptology Conference on Advances in Cryptology*. Santa Barbara (USA), 1996, p. 104–113. DOI: 10.1007/3-540-68697-5_9

[3] GENKIN, D., SHAMIR, A., TROMER, E. RSA key extraction via low-bandwidth acoustic cryptanalysis. In *Lecture Notes in Computer Science (including subseries Lecture Notes in Artificial Intelligence and Lecture Notes in Bioinformatics)*. Berlin (Germany), 2014, p. 444–461. DOI: 10.1007/978-3-662-44371-2_25

[4] GANDOLFI, K., MOURTEL, C., OLIVIER, F. Electromagnetic analysis: Concrete results. In *Proceedings of the International Workshop on Cryptographic Hardware and Embedded Systems - CHES 2001*. Paris (France), 2001, p. 251–261. DOI: https://doi.org/10.1007/3-540-44709-1_21

[5] STANDAERT, F., ARCHAMBEAU, C. Using subspace-based template attacks to compare and combine power and electromagnetic information leakages. In *Proceedings of the International Workshop on Cryptographic Hardware and Embedded Systems - CHES 2008*. Washington, D.C (USA), 2008, p. 411–425. DOI: 10.1007/978-3-540-85053-3_26

[6] CHARI, S., RAO, J., ROHATGI, P. Template attacks. In *Proceedings of the International Workshop on Cryptographic Hardware and Embedded Systems - CHES 2002*. Redwood Shores (USA), 2002, p. 13–28. DOI: 10.1007/3-540-36400-5_3

[7] CHOUDARY, O., KUHN, M. G. Efficient template attacks. In *Lecture Notes in Computer Science (including subseries Lecture Notes in Artificial Intelligence and Lecture Notes in Bioinformatics)*. Berlin (Germany), 2014, p. 253–270. DOI: 10.1007/978-3-319-08302-5_17

[8] BRIER, E., CLAVIER, C., OLIVIER, F. Correlation power analysis with a leakage model. In *Proceedings of the International Workshop on Cryptographic Hardware and Embedded Systems - CHES 2004*. Cambridge (USA), 2004, p. 16–29. DOI: 10.1007/978-3-540-28632-5_2

[9] GIRELICHS, B., BATINA, L., TUYLS, P., et al. Mutual information analysis. In *Proceedings of the International Workshop on Cryptographic Hardware and Embedded Systems - CHES 2008*. Washington, D.C. (USA), 2008, p. 426–442. DOI: 10.1007/978-3-540-85053-3_27

[10] SCHINDLER, W., LEMKE, K., PAAR, C. A stochastic model for differential side channel cryptanalysis. In *Proceedings of the International Workshop on Cryptographic Hardware and Embedded Systems - CHES 2005*. Edinburgh (UK), 2005, p. 30–46. DOI: 10.1007/11545262_3

[11] MANGARD, S., OSWALD, E., POPP, T. *Power Analysis Attacks: Revealing the Secrets of Smart Cards*. 1st ed. Secaucus (USA): Springer US, 2007. ISBN: 978-0-387-30857-9

[12] RIVEST, R. L. Cryptography and machine learning. In *Proceedings of the International Conference on the Theory and Applications of Cryptology: Advances in Cryptology - ASIACRYPT'91*. 1991, vol. 739, p. 427–439. DOI: 10.1007/3-540-57332-1_36

[13] HEYSZL, J., IBING, A., MANGARD, S., et al. Clustering algorithms for non-profiled single-execution attacks on exponentiations. In *Proceedings of the 12th International Conference Smart Card Research and Advanced Applications CARDIS 2013*. Berlin (Germany), 2013, p. 79–93. DOI: 10.1007/978-3-319-08302-5_6

[14] MARTINASEK, Z., ZEMAN, V. Innovative method of the power analysis. *Radioengineering*, 2013, vol. 22, no. 2, p. 586–594. ISSN: 1210-2512

[15] MARTINASEK, Z., HAJNY, J., MALINA, L. Optimization of power analysis using neural network. In *Proceedings of the 12th International Conference Smart Card Research and Advanced Applications CARDIS*. Berlin (Germany), 2013, p. 94–107. DOI: 10.1007/978-3-319-08302-5_7

[16] WHITNALL, C., OSWALD, E. Robust profiling for DPA-style attacks. In *Proceedings of the International Workshop Cryptographic Hardware and Embedded Systems - CHES 2015*. Saint-Malo (France), 2015, p. 3–21. DOI: 10.1007/978-3-662-48324-4_1.

[17] ZHANG, Z., WU, L., WANG, A., et al. Improved leakage model based on genetic algorithm. In *IACR Cryptology ePrint Archive*. 2014.

[18] AUMONIER, S. Generalized correlation power analysis. In *Proceedings of the Ecrypt Workshop Tools for Cryptanalysis*. 2007, vol. 518.

[19] MARTINASEK, Z., ZEMAN, V., MALINA, L., et al. k-Nearest neighbors algorithm in profiling power analysis attack. *Radioengineering*, 2016, vol. 25, no. 2, p. 365–382. DOI: 10.13164/re.2016.0365

[20] HOSPODAR, G., GIERLICHS, B., DE MULDER, E., et al. Machine learning in side-channel analysis: A first study. *Journal of Cryptographic Engineering*, 2011, vol. 1, no. 4, p. 293–302. DOI: 10.1007/s13389-011-0023

[21] HOSPODAR, G., DE MULDER, E., GIERLICHS, B., et al. Least squares support vector machines for side-channel analysis. In *Proceedings of the Second International Workshop on Constructive Side-Channel Analysis and Secure Design (COSADE 2011)*. Darmstadt (Germany), 2011, p. 293–302.

[22] HE, H., JAFFE, J, ZOU, L. Side channel cryptanalysis using machine learning: Using an SVM to recover DES keys from a smart card. 2012, Stanford University.

[23] LERMAN, L., BONTEMPI, G., MARKOWITCH, O. Power analysis attack: An approach based on machine learning. *International Journal of Applied Cryptography*, 2014, vol. 3, no. 2, p. 97–115. DOI: https://doi.org/10.1504/IJACT.2014.062722

[24] LERMAN, L., BONTEMPI, G., MARKOWITCH, O. Side channel attack: An approach based on machine learning. In *Proceedings of the Second International Workshop on Constructive Side-Channel Analysis and Secure Design (COSADE 2011)*. Darmstadt (Germany), 2011, p. 29–41. DOI: 10.1504/IJACT.2014.062722

[25] HEUSER, A., ZOHNER, M. Intelligent machine homicide - breaking cryptographic devices using support vector machines. In *Proceedings of the Third International Workshop on Constructive Side-Channel Analysis and Secure Design (COSADE)*. Darmstadt (Germany), 2012, p. 249–264. DOI: 10.1007/978-3-642-29912-4_18

[26] BARTKEWITZ, T., LEMKE-RUST, K. Efficient template attacks based on probabilistic multi-class support vector machines. In *Proceedings of the Smart Card Research and Advanced Applications*. Graz (Austria), 2013, p. 263–276. DOI: 10.1007/978-3-642-37288-9_18

[27] LERMAN, L., MEDEIROS, S. F., BONTEMPI, G., et al. A machine learning approach against a masked AES. In *Proceedings of the 12th International Conference of Smart Card Research and Advanced Applications CARDIS*. Berlin (Germany), 2013, p. 61–75. DOI: 10.1007/978-3-319-08302-5_5

[28] SAEEDI, E., KONG, Y. Side channel information analysis based on machine learning. In *Proceedings of the 8th International Conference on Signal Processing and Communication Systems (ICSPCS)*. 2014, p. 1–7. DOI: 10.1109/ICSPCS 2014.7021075

[29] CHUI, C. K. *Wavelets: A Tutorial in Theory and Applications (Wavelet Analysis and its Applications)*. San Diego, CA (USA): Academic Press, 1992. ISBN: 0323139744, 9780323139748

[30] SOUISSI, Y., ELAABID, M. A., DEBANDE, N., et al. Novel applications of wavelet transforms based side-channel analysis. In *Proceedings of the Non-Invasive Attack Testing Workshop*. 2011.

[31] PARK, A., RYOO, J., HAN, D. G. CPA performance comparison based on wavelet transform. In *Proceedings of the IEEE International Carnahan Conference on Security Technology*. 2012, p. 201–206. DOI: 10.1109/CCST.2012.6393559

[32] DEBANDE, N., SOUISSI, Y., ABDELAZIZ, M., et al. Wavelet transform based pre-processing for side channel analysis. In *Proceedings of the 45th Annual IEEE/ACM International Symposium on Microarchitecture Workshops*. Vancouver (Canada), 2012, p. 32–38. DOI: 10.1109/ MICROW.2012.15

[33] ZHANG, L., ZHOU, W., JIAO, L. Wavelet support vector machine. *IEEE Transactions on Systems, Man, & Cybernetics, Part B (Cybernetics)*, 2004, p. 34–39. DOI: 10.1109/TSMCB.2003.811113

[34] CORTES, C., VAPNIK, V. Support-vector networks. *Machine Learning*, 1995, p. 273–297. DOI: 10.1007/BF00994018

[35] HANG, L. *Statistical Learning Method*. 1st ed. Beijing (China): Tsinghua University Press, 2012. ISBN: 978-7-302-27595-4

[36] FRANC, V., HLAVAC, V. Multi-class support vector machine. In *Proceedings of the 16th International Conference on Pattern Recognition*. Quebec (Canada), 2002, vol. 2, p. 236–239. DOI: 10.1109/ICPR.2002.1048282

[37] DIETTERICH, T. G., BAKIRI, G. Solving multiclass learning problems via error correcting output codes. *Journal of Artificial Intelligence Research*, 1994, vol. 2, no. 1, p. 263–286. DOI: 10.1.1.72.7289

[38] HASTIE, T., TIBSHIRANI, R. Classification by pairwise coupling. In *Proceedings of the Conference on Neural Information Processing Systems*. 1998, vol. 26, p. 451–471. DOI: 10.1214/aos/1028144844

[39] HSU, C. W., LIN, C. J. A comparison of methods for multi-class support vector machines. *IEEE Transactions on Neural Networks*, 2002, p. 415–425. DOI: 10.1109/72.991427

[40] PLATT, J. C. Probabilistic outputs for support vector machines and comparisons to regularized likelihood methods. *Advances in Large Margin Classifiers*, p. 61–74. DOI: 10.1.1.41.1639

[41] LIN, H. T., LIN, C. J., WENG, R. C. A note on platt's probabilistic outputs for support vector machines. *Machine Learning*, 2007, p. 267–276. DOI: 10.1007/s10994-007-5018-6

[42] PLATT, J. C. Fast training of support vector machines using sequential minimal optimization. *Advances in Kernel Methods*, Cambridge (USA): MIT Press, 1999, p. 185–208. ISBN: 0-262-19416-3

[43] MERCER, J. Functions of positive and negative type, and their connection with the theory of integral equations. *Philosophical Transactions of the Royal Society A: Mathematical, Physical and Engineering Sciences*. DOI: 10.1098/rsta.1909.0016.

[44] WICKERHAUSER, M. V. Adapted wavelet analysis from theory to software. *SIAM Review*, 1996. DOI: 10.1137/1038018

[45] TORRENCE, C., COMPO, G. P. A practical guide to wavelet analysis. *Bulletin of the American Meteorological Society*, 1998, p. 61–78.

[46] PARIKH, U. B., DAS, B., MAHESHWARI, R. P. Combined wavelet-SVM technique for fault zone detection in a series compensated transmission line. *IEEE Transactions on Power Delivery*, p. 1789–1794. DOI: 10.1109/TPWRD.2008.919395

[47] DU, P., TAN, K., XING, X. Wavelet SVM in reproducing kernel Hilbert space for hyperspectral remote sensing image classification. *Optics Communications*. 2010, p. 4978–4984. DOI: 10.1016/j.optcom. 2010.08.009

[48] RUPING, S. SVM kernels for time series analysis. *Universitätsbibliothek Dortmund*, 2001. DOI: 10.1.1.23.9841

[49] EDUARD, G. B., LI, F., SMINCHISESCU, C. Fourier kernel learning. In *Lecture Notes in Computer Science (Including Subseries Lecture Notes in Artificial Intelligence and Lecture Notes in Bioinformatics)*. 2012, p. 459-–473. DOI:10.1007/978-3-642-33709-3_33

[50] CHANG, C. C., LIN, C. J. LIBSVM: A library for support vector machines. *ACM Transactions on Intelligent Systems and Technology (TIST)*, 2011, vol. 2, no. 27. DOI: 10.1145/1961189.1961199

[51] HSU, C. W., CHANG, C.C., LIN, C. J. A practical guide to support vector classification. *BJU International*, 2003.

[52] RESEARCH INSTITUTE FOR SECURE SYSTEMS. *Side-Channel Attack Standard Evaluation Board Sasebo-gii Specification*. [Online] Cited 2017-04-08. Available at: http://www.rcis.aist.go.jp/files/special/SASEBO/SASEBO-GII-en/SASEBO-GII_Spec_Ver1.01_English.pdf

[53] NUEESS LAB NORTHEASTERN UNIVERSITY. *TeSCASE - Testbed for Side Channel Analysis and Security Evaluation*. [Online] Cited 2017-04-08. Available at: http://tescase.coe.neu.edu/?current_page=homepage

[54] RESEARCH INSTITUTE FOR SECURE SYSTEMS. *Evaluation Environment for Side-Channel Attacks*. [Online] Cited 2017-04-08. Available at: https://www.risec.aist.go.jp/project/sasebo/

[55] NASSAR, M., SOUISSI, Y., GUILLEY, S., et al. RSM: A small and fast countermeasure for AES, secure against 1st and 2ndorder zero-offset SCAs. In *Proceedings of the Conference Exhibition on Design, Automation Test in Europe (DATE)*. Dresden (Germany), 2012, p. 1173–1178. DOI: 10.1109/DATE.2012.6176671

[56] GUILLEYHO, S. *DPA contest v4*. [Online] Cited 2017-04-08. Available at: http://www.dpacontest.org/v4/rsm_doc.php

[57] PENG, H., LONG, F., DING, C. Feature selection based on mutual information criteria of max-dependency, max-relevance, and min-redundancy. *IEEE Transactions on Pattern Analysis and Machine Intelligence*, 2005, p. 1226–1238. DOI: 10.1109/TPAMI.2005.159

[58] GIERLICHS, B., LEMKE-RUST, K., PAAR, C. Templates vs. stochastic methods. In *Proceedings of the International Workshop on Cryptographic Hardware and Embedded Systems - CHES 2006*. Yokohama (Japan), 2006, p. 15–29. DOI: 10.1007/11894063_2

[59] STANDAERT, F. X., MALKIN, T. G., YUNG, M. A unified framework for the analysis of side-channel key recovery attacks. In *Proceedings of the 28th Annual International Conference on Advances in Cryptology: The Theory and Applications of Cryptographic Techniques*. Berlin (Germany), 2009, p. 443–461. DOI: 10.1007/978-3-642-01001-9_26

About the Authors . . .

Shourong HOU was born in Shandong, China. He received his B.Eng. from Xidian University in 2015. His research interests include machine learning and side channel attack. He is now a Ph.D. candidate at Shanghai Jiao Tong University.

Yujie ZHOU was born in Zhejiang, China. She received his Ph.D. from University of Science and Technology of China in 1997. Her research interests include digital integrated circuit design, embedded system, trusted computing and digital copyright protection.

Hongming LIU was born in Jiangxi, China. He received his Ph.D. from Shanghai Jiao Tong University in 2014. His research interests include cryptographic chip design, blockchain and machine learning.

Nianhao ZHU was born in Jiangshu, China. She received his Ph.D. from Shanghai Jiao Tong University in 2014. His research interests include cryptographic chip design and side channel attack.

15

Evaluation of Computing Symmetrical Zolotarev Polynomials of the First Kind

Jan KUBAK [1], *Pavel SOVKA* [1], *Miroslav VLCEK* [2]

[1] Dept. of Circuit Theory, Faculty of Electrical Engineering, Czech Technical University in Prague
[2] Dept. of Applied Mathematics, Faculty of Transportation Sciences, Czech Technical University in Prague

jankubak@gmail.com, sovka@fel.cvut.cz, vlcek@fd.cvut.cz

Abstract. *This report summarize and compares various methods for computing the symmetrical Zolotarev Polynomial of the first kind and its spectrum with each other. Suitable criteria are suggested for the comparison. The best numerical stability shows the method employing Chebyshev polynomial recurrence. In case of the polynomial spectrum computation the best method is the one using the difference backward recursion introduced by M. Vlček. Both methods are able to generate the polynomial of high degree up to, at least, 2000, using 32-bit IEEE floating point arithmetics.*

Keywords

Chebyshev polynomial, symmetrical Zolotarev polynomial of the first kind, spectrum of Zolotarev polynomial, power expansion, trigonometric functions, forward and backward recursion, binomial coefficients

1. Introduction

Zolotarev polynomials (ZP) are important generalization of Chebyshev polynomials. The practical utilization of ZP was not possible till recently. The reason was the absence of method able to generate ZP of high enough degree, as was emphasized in [1]. The ZP play, among others, the key role in analytical design of the optimal FIR filters [2], [3]. Particularly, the symmetrical ZP of the first kind (ZP1S) are utilized in the design of almost equiripple half-band FIR filters [4]. The utilization of ZP1S for nonstational signal analysis also seems to be promisable [5]. Recently ZP were employed for design of circular antenna array [6]. In spectral analysis the degree of the ZP1S is related to the number of samples of analyzed signal segment, and it is also directly related to the order of a filter. Therefore, utilizing ZP1S in both applications requires to generate polynomials with relatively high degree, typically, degree of hundreds. There are two typical tasks in computing ZP: One is the computation of actual ZP, while the ZP spectral coefficients computation. No method had been able to generate the spectrum of polynomial with satisfying high degree [2], until the recursive algorithm was developed [7]. Nevertheless, based on findings in [7], and others, there are several methods of the computation of Chebyshev polynomials or ZP or their coefficients (spectra) e.g. [8–12]. However, the usability boundaries of these methods and a detailed numerical comparison with each other have not yet been systematically explored. It is also unclear which degree of ZP1S or its spectrum can be achieved.

The goal of this article is to explore and compare selected methods for computation of ZP1S and its spectrum. In task of actual ZP1S computation we decided to use two often used methods: Chebyshev recursion [13] and power expansion with binomial coefficients [13]. For computation of ZP1S spectral coefficients we employed backward recursion [7]. We developed an empirical technique determining if the results is not a valid polynomial. Using this technique we distinguish between selected ZP1S computational methods in range of ZP1S parameters. We also explore dynamic range during the computation and its stability. The paper is organized as follows: an introduction is in Sec. 1; the Chebyshev polynomials are briefly described in Sec. 2; the symmetrical Zolotarev polynomial of the first kind is outlined in the Sec. 3; Section 4 summarizes various methods of the polynomial computation, the methods are evaluated in Sec. 5; a brief discussion and conclusion is in Sec. 6 and 7, respectively.

2. Chebyshev Polynomials

Before describing the Zolotarev polynomials we briefly summarize the Chebyshev polynomials. The Chebyshev polynomial $T_n(x)$ of the first kind is defined by the relation [13]

$$T_n(x) = \cos(n\theta) \quad \text{when} \quad x = \cos(\theta). \qquad (1)$$

Variable x on interval $\subset [-1, 1]$ corresponds to variable θ on interval $\subset [0, \pi]$. Nonlinear transformation of variable x onto variable θ converts $T_n(x)$ into function $\cos n\theta$ as illustrated in Fig. 1a and Fig.1b, respectively.

The Chebyshev polynomial $T_n(x)$ of degree n can be generated iteratively by following forward recursion [13]

$$T_n(x) = 2x\,T_{n-1}(x) - T_{n-2}(x), \quad n = 2, 3, \ldots, \qquad (2a)$$

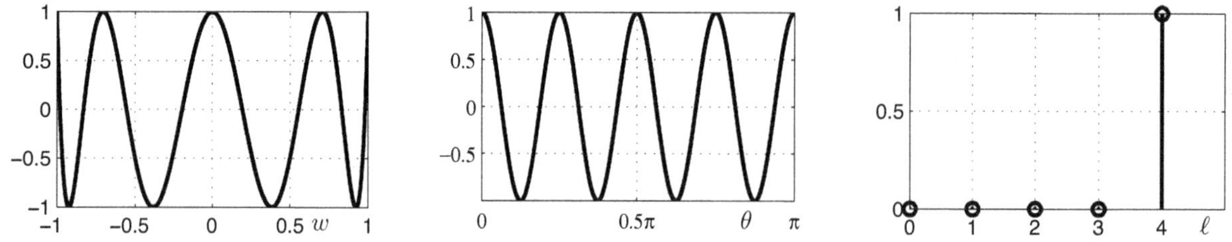

(a) $T_n(x)$ on range $[-1, 1]$; $n = 4$.

(b) Corresponding function $\cos n\theta$ on range $[0, \pi]$; $n = 4$.

(c) The spectrum of $\cos n\theta$; $n = 4$.

Fig. 1. An illustration of mapping of $T_n(x)$ onto $\cos n\theta$ by transformation $x = \cos\theta$ together with its spectrum.

with initial condition

$$T_0(x) = 1, T_1(x) = x. \tag{2b}$$

Substituting for $x = \cos\theta$ in (2) results in well-known trigonometric identity

$$\cos n\theta = 2\cos\theta\,\cos(n-1)\theta - \cos(n-2)\theta, \quad n = 2, 3, \ldots. \tag{3}$$

Nevertheless, this formula can not be used for computing ZP1S.

The solution of the recursion (2) results in a formula which describes the Chebyshev polynomial of the first kind of degree n in terms of power of x as [13]

$$T_n(x) = \sum_{k=0}^{n/2} c_k^{(n)} x^{n-2k}, \quad \frac{n}{2} \subset \mathbb{N} \tag{4a}$$

where coefficients

$$c_k^{(n)} = (-1)^k\, 2^{n-2k-1}\left[2\binom{n-k}{k} - \binom{n-k-1}{k}\right] \quad (2k < n) \tag{4b}$$

and

$$c_k^{(2k)} = (-1)^k \quad (k \geq 0). \tag{4c}$$

Thus, the Chebyshev polynomial $T_n(x)$ of order n can be also written as power expansion with coefficients given by binomial expansion.

2.1 Interpretation of Chebyshev Polynomial and its Spectrum

For the purpose of this study it is valuable to introduce the following interpretation of Chebyshev polynomial. If we regard the continuous-time variable x (or $theta$) as "time" then Chebyshev polynomial $T_n(x)$ (or $\cos(n\theta)$) can be regarded as a signal waveform. Chebyshev polynomial $\cos(n\theta)$ can be converted to a spectral domain using Fourier series concept. Because we deal with an even symmetrical polynomial the Fourier transform degrades into cosine transform. The resulting spectrum of Chebyshev polynomial $\cos(n\theta)$ of degree n is given

$$a_\ell = \frac{1}{\pi}\int_0^\pi \cos(n\theta)\mathrm{e}^{-\mathrm{j}\ell\theta 2\pi/T_0}\mathrm{d}\theta \tag{5}$$

where $T_0 = \pi$ for this case. Using the principle of the orthogonality we obtain only one nonzero spectral coefficient on frequencies $\frac{\ell 2\pi}{T_0}$. The Chebyshev polynomial and its corresponding spectrum is shown in Fig. 1b and Fig. 1c, respectively. This interpretation will be generalized later for Zolotarev polynomials.

It should be noted, it is possible to introduce the opposite interpretation, i.e. spectral coefficient a_ℓ, $\ell = 0, 1, \ldots, n$ represent the impulse response of a filter in time domain and Chebyshev polynomial can be considered as the frequency response of the respective filter. This approach is used in filter design using ZP [3], [4], [7], [14].

3. Symmetrical Zolotarev Polynomial of the First Kind

The ZP1S can be expressed using the Chebyshev polynomial of the first kind as

$$Z_p(w, k') = (-1)^{\frac{p}{2}} T_p\left(\sqrt{\frac{w^2 - k'^2}{1 - k'^2}}\right), \quad \frac{p}{2} \subset \mathbb{N} \tag{6}$$

where p is the degree of Chebyshev polynomial as well as of the ZP1S, see [7]. The ZP1S is defined on standard interval for polynomial approximation $w \subset [-1, 1]$. The k' is the modulus of elliptical functions. The ZP1S main features are equipple behavioral on two disjoint intervals $w \subset [-1, k'] \cup [k', 1]$ and elevated central lobe on the interval $w \subset [-k', k']$. Thus, the k' can be interpreted as the half-width of the central lobe bounded by intersections with the absolute values[1] of one, see Fig. 2. The parameter k' is given on the interval $[0, 1)$. The polynomial degree must be even as the polynomial $Z_p(w, k')$ is even symmetrical. Note that the polynomial has p zeros.

[1]The central lobe of the ZP1S of degree p is positive or negative when $\frac{p}{2}$ is even or odd, respectively.

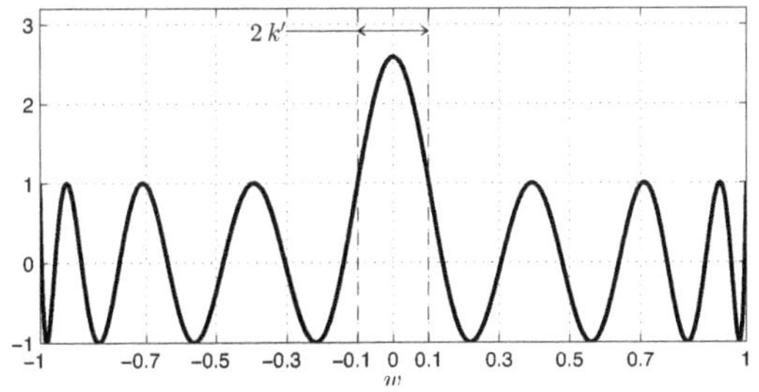

Fig. 2. The symmetrical Zolotarev polynomial of the first kind $Z_p(w, k')$ for $p = 8$ and $k' = 0.1$.

The ZP are members of elliptical functions family of which mathematical discipline is rather difficult. To understand the ZP1S nature with ease we can approach to the Zolotarev polynomial as to trigonometric function. Employing this approach the ZP1S is expressed by the Chebyshev polynomial of the first kind: by using the substitution $w = \cos\theta$ for ZP1S, denoted as $Z_p(w, k')|w = \cos\theta$, we obtain function, which can be interpreted as a cosine with the central lobe elevated, see Fig. 3a. Similar approach is used also in [7] for the approximation of FIR filters by a ZP. For $w = \cos\theta$ and $k' = 0$ the $Z_p(w, k')$ degenerates to the trigonometric cosine of the degree p. This can be easily justified: by substituting $k' = 0$ in (6) we obtain $Z_p(w, k') = (-1)^{\frac{p}{2}} T_p(w)$, and by using the substitution $w = \cos\theta$ we obtain $\cos p\theta$.

The global extreme of the ZP1S is influenced by both k' and p: its absolute value gets larger when k' or p get larger. This phenomena become obvious by looking at Fig. 3. Compare Fig. 3a with Fig. 3b, where the degree is the same and k' is 0.10 or 0.11, respectively. Furthermore, compare Fig. 3a with Fig. 3c, where the k' is the same and the degree is 16 and 20, respectively. Recent analysis shows the ZP1S are orthogonal in both disjoint intervals $w \subset [-1, k'] \cup [k', 1]$ for fixed k' and degrees $p = 2, 4, 6, \ldots$. These findings are yet to be published.

The equation (6) can be rewritten using

$$T_{2n}(x) = T_n(2x^2 - 1)$$

as

$$Z_p(w, k') = (-1)^{\frac{p}{2}} T_{\frac{p}{2}}\left(\frac{2w^2 - 1 - k'^2}{1 - k'^2}\right), \quad \frac{p}{2} \subset \mathbb{N}. \tag{7}$$

Use of this equation for future generation of ZP1S is beneficial compared with use of (6) for three reasons. Firstly, it lowers the upper index of the sum in equation (4a) to the half. Secondly, the number of iterations of recurrence (2) is also reduced to the half. Thirdly, the argument of $T_{\frac{p}{2}}(x)$ is free of the square root witch is rather problematic computational operation. For all three reasons, the use of equation (7) enables to enumerate the ZP1S with less errors compared with equation (6).

It is also worthy to note that the ZP1S can be alternatively expressed as the linear combination of the Chebyshev polynomials $T_n(x)$ as [7], [15].

$$Z_p(w, k') = (-1)^{\frac{p}{2}} \sum_{\ell=0}^{\frac{p}{2}} a(2\ell) T_{2\ell}(w), \quad \frac{p}{2} \subset \mathbb{N}. \tag{8}$$

By using the substitution $w = \cos\theta$ in (8) we obtain

$$Z_p(w, k')|_{w=\cos\phi} = \sum_{\ell=0}^{\frac{p}{2}} a(2\ell)\, \cos(2\ell\theta), \quad \frac{p}{2} \subset \mathbb{N}. \tag{9}$$

The $a(2\ell)$ coefficients[2] are, in fact, the coefficients of Fourier series[3] [18]. Therefore, the $a(2\ell)$ coefficients can be regarded as the spectral coefficients of $Z_p(w, k')$. Equation (9) expresses the spectrum of ZP which involves generally $\frac{p}{2}$ spectral lines. The expansion (8) can be generally used for the approximation of an arbitrary continuous function over a finite interval $[-1, 1]$. One possible solution how to generate the coefficients $a(\ell)$ can be found in e.g. [8]. Another very effective and simple approach, in which we are interested, can derived from the linear differential equation for ZP [7] resulting in backward recursion with time-varying coefficients.

3.1 Discretization

This section introduces a discretization of continuous-time variables x and θ which is needed for computing ZP1S on a computer. All above defined formulas shall be used in a discrete-time manner requiring proper sampling of continues-time variables x or θ. The discretization can be done by letting $x = iT$, where $i = 0, 1, 2, \ldots$ and T is the sample step for variable x. Similarly, the variable θ is discretized by letting $\theta = Ti$, where $i = 0, 1, 2, \ldots$. From this point forward the discussion will be supposing that all schemes of ZP1S are discrete-time functions. Let us remind that when a proper sample step is used then the behaviour of a discrete-time method is close to a behavior of original continuous-time one. We use sample step approximately $T = 1/10^4$ for interval $[-1, 1]$ or $[0, \pi]$.

4. An Approach to FIR Design Using Zolotarev Polynomial

Almost equiripple half-band filter originates from ZP1S differential equation [4]; However, in this article we show simplified possible approach to FIR filter design employing ZP1S. ZP1S can form frequency response of a pass band FIR filter with passband around central frequency. A FIR filter transfer function can be written in general form [16] as

$$H(z) = \sum_{v=0}^{N-1} h(v) z^{-v} \tag{10}$$

[2]Since the ZP1S is even symmetrical function only even coefficients a are nonzero.
[3]In this case the Fourier cosine series.

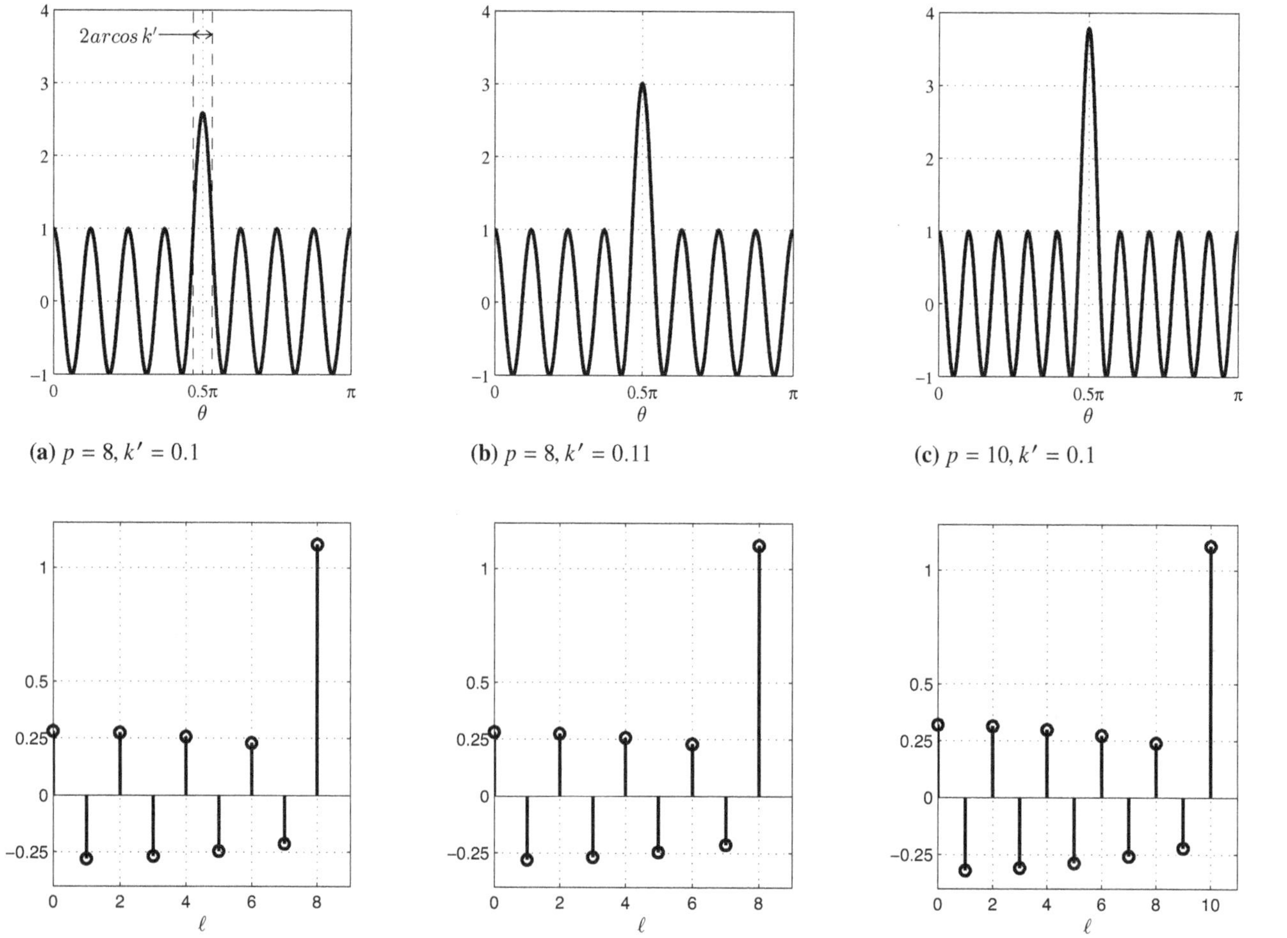

(a) $p = 8, k' = 0.1$ (b) $p = 8, k' = 0.11$ (c) $p = 10, k' = 0.1$

(d) $p = 8, k' = 0.1$ (e) $p = 8, k' = 0.11$ (f) $p = 10, k' = 0.1$

Fig. 3. Symmetrical Zolotarev polynomial of the first kind $Z_p(w, k')|w = \cos\theta$ (a–c); and together with its spectral coefficients (d–g).

where $h(v)$ is the impulse response, $N = 2M + 1$ is the filter length and M is the filter order. The filter transfer function employing ZP1S polynomial is given [7] as

$$H(z) = z^{-\frac{M}{2}} \frac{1}{A} \left[\sum_{\ell=0}^{M-1} a(\ell) T_\ell(w) \right] = \\ = z^{-\frac{M}{2}} \frac{1}{A} \left[Z_M(w, k') \right] \tag{11}$$

where A is the scaling coefficient. By comparing (10) with (11) the impulse response can assembled from a coefficients as

$$\begin{aligned} h(2v) &= \frac{1}{2A} a(M - v), \\ h(2M + 2v) &= \frac{1}{2A} a(v), \\ h(2v + 1) &= 0, \\ v &= 0, \ldots, M. \end{aligned} \tag{12}$$

Such an impulse response is even symmetrical and forms a linear phase type I. FIR filter. The coefficients $a(v)$ are direct output of the backward recursion algorithm [7].

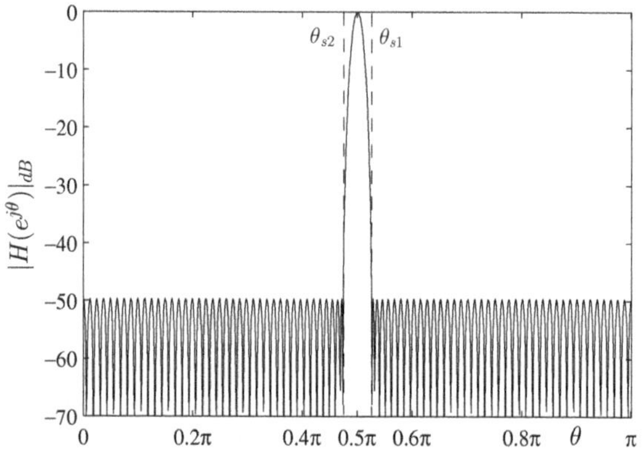

Fig. 4. FIR filter frequency response based on the symmetrical Zolotarev polynomial of the first kind $Z_M(w, k')$ for $M = 80$ and $k' = 0.08$ yielding left and right stopband edge $\theta_{S1} = 0.4745\pi$ and $\theta_{S2} = 0.5255\pi$, respectively, and stopband ripple $\delta = -49.7$ dB.

The variable w in (11) is related to complex variable z by transformation $w = \frac{1}{2}(z + z^{-1})$. The transformation degrades to $w = \cos(\theta)$ in case of frequency response $H(z = e^{j\theta})$:

$w = \frac{1}{2}(z + z^{-1}) = \frac{1}{2}(e^{j\theta} + e^{-j\theta}) = \cos(\theta)$. Hence,

$$H(e^{j\theta}) = z^{-\frac{M}{2}} \frac{1}{A} \left[Z_M(\cos(\theta), k') \right]. \quad (13)$$

The k' is a function of θ_S which is symmetrical stopband edge. Having in mind the transformation of w in case of frequency response $w = \cos(\theta)$ the k' is simply given as

$$k' = \cos(\theta_{S1}) \quad (14)$$

where θ_{S1} is left stopband edge. The stopband ripple is gives as

$$\delta = 20 \log_{10} \frac{1}{A}. \quad (15)$$

The scaling coefficient A is virtually maximum value of ZP1S. The maximal value is localized in the center of the polynomial and can be derived from (8) by substituting $w = 0$ as

$$A = Z_M(w = 0, k')_{\max} = \sum_{\ell=0}^{M-1} a(\ell) T_\ell(0). \quad (16)$$

The derivation of degree equation $p = f(\delta, \theta_S)$ is out of scope of this paper. The example of symmetrical pass band FIR filter using ZP1S is in Fig. 4. Also a high pass filter can be formed by discarding zero samples of filter impulse response. A low pass filter can be easily converted from a high pass one. Various filter types based on Zolotarev polynomials are described in [3], [4], [7], [14] and others.

5. Possible Schemes of ZP1S Computation

This section defines possible schemes of ZP1S computation. Any ZP1S of order p can be computed using two basic approaches.

Firstly, direct computation in "time" domain: by using power expansion with binomial coefficients (4a) or by implementing the forward recursion (2). Both methods in "time" generate a Chebyshev polynomial of degree p along with the transformation of parameter x onto θ. The parameter x is transformed as

$$x = \sqrt{\frac{cos^2\theta - k'^2}{1 - k'^2}} \quad (17)$$

in terms of (6) and as

$$x = \frac{2\cos^2\theta - 1 - k'^2}{1 - k'^2} \quad (18)$$

in terms of (7).

Secondly, by computing polynomial in "spectral" domain: evaluating spectral coefficients a_ℓ using difference backward recursion by Vlcek [7] followed by the inverse (Fourier) cosine transform[4] yielding the "waveform" $Z_p(w, k')|_{w=\cos\phi}$ (9).

There are other possibilities of computing ZP1S, as mentioned before. For example, using the factorized form given by ZP1S zeros [7], or rather complicated methods using Remez algorithm or other approaches based on least deviation and polynomials with weights in "time" domain [10] and [9]. But we are interested in very effective recursive method [7] promising the computation of ZP spectra of high degree and in its counterpart in the "time" domain (2). The aim is to compare the numerical behaviour of these selected methods.

[4]For discrete-time functions by discrete time cosine transform.

5.1 ZP1S Computation in "Time" Domain

This subsection briefly discusses and compares the numerical behaviour of linear time variant (LTV) recursion (2) and the power expansion with binomial coefficients (4a), both used for computing a Chebyshev polynomial of degree p.

5.1.1 Computation Using Expansion of $\mathbf{T_n(x)}$

This method uses binomial coefficients given by equation (4a) to generate Chebyshev polynomials $T_n(x)$. Due to use of binomial coefficients the computation demands rather big dynamic range [17] and then poor computational precision can be expected. In consequence only the low degrees of ZP1S can be generated using this technique.

5.1.2 Computation by Using LTV Recursion

This method of computing Chebyshev polynomials is based on using recursion (2) with the substitution (18). The discretized version of equation (2) can be written as

$$T_n[i] = 2x[i]T_{n-1}[i] - T_{n-2}[i], \quad n = 2, 3, \ldots, \\ i = 0, 1, \ldots, N-1, \quad (19)$$

with the initial condition

$$T_0[i] = 1, T_1[i] = x[i] \quad (20)$$

where N is the number of points at which ZP1S is sampled. Parameter N is given by used sample step T and the definition scope T_0 of variable x or θ: $N = T_0/T$. In our case we use $N = 10^4$.

Recursive schemes similar to (2) no matter if they are time-invariant or time-varying are known to be very effective and stable when used for limited number of samples even when fixed-point arithmetics is used [18–20]. But to our knowledge, no systematic study of numerical behaviour of this type of LTV recursion has not yet been performed for the special case of computing of ZP1S of very high degree. If recursions (2) or (3) are used then the zeros of resulting Chebyshev polynomials are almost equidistantly spread in the intervals $[-1, 1]$ or $[0, \pi]$. In the latter case of $T_n(x) = \cos(n\theta)$ interval between zeros is precisely equidistant. However, if the Chebyshev recursion (2) is applied to compute ZP1S, especially for higher values of k' or higher ZP1S degree, its zeros are pushed towards the edges of both the definition scopes $[-1, 1]$ and $[0, \pi]$. Hence, higher errors in computing ZP1S can be expected. The recursive schemes (2) or (3) are known to be very effective and stable when used for limited number of samples. The scheme is stable even when fixed-point arithmetics is used. Detailed numerical study of the recursive equation of this type can be found in e.g. [19].

Conclusion. We briefly summarize comparison of methods of $T_n(x)$ computing using either the Chebyshev recursion (2) or using power expansion with binomial coefficients (4a). The latter method is expected to has worse numerical behavior. The evaluation of the binomial coefficients by equation (4b) causes severe problems due to high dynamic range. We can see in (4a) the values of coefficients $c_k^{(n)}$ can be up to twice greater than the values of binomial coefficients $\binom{n-k}{k}$, which results in additional increasing of dynamical range. Nevertheless, there are some modified schemes of computing binomial coefficients (4a) which offers better numerical performance, e.g. [21]; however, we did not implemented these schemes. Due to described reasons we favor the use of Chebyshev recursion (2).

5.2 Generation of ZP1S Using Spectral Coefficients

The algorithm to compute ZP1S spectral coefficients is derived by Vlcek [7]. The algorithm for computing ZP1S is just one of a family, of which algorithms are able to generate various kinds of ZP. The algorithm for ZP1S computation is a backward recursion with time-varying coefficients, and it can be described by

$$d_1[\ell]\,a(2\ell-6) = d_2[\ell]\,a(2\ell-4) + d_3[\ell]\,a(2\ell-2) + d_4[\ell]\,a(2\ell) \qquad (21)$$

$$\ell = m+2, m+1, m, \ldots, 3$$

where $d_1 = m^2-(\ell-3)^2$, $d_2 = 3(m^2-(\ell-2)^2)+(2\ell-4)(2\ell-5)k'$, and $d_3 = 3(m^2-(\ell-1)^2)+(2\ell-2)(2\ell-1)k'$ are time-varying coefficients. The $m = \frac{p}{2}$ is half of degree p. The computational algorithm based on the formula is in Tab. 1. The result is set of spectral coefficients a_ℓ, $\ell = 1,\ldots,p$. The spectral coefficients are transformed into ZP1S using the inverse cosine series[5].

given	$m = p/2,\ k'$
init-alisation	$a\,(2m) = (1-k'^2)^{-m},\ a\,(2m+2) = 0,$ $a\,(2m+4) = 0$
body (for	$\ell = m+2$ *to* $3)$
	$\left[m^2-(\ell-3)^2\right]a(2\ell-6) =$
	$-\left[3(m^2-(\ell-2)^2)+(2\ell-4)(2\ell-5)\,k'^2\right]a(2\ell-4)$
	$-\left[3(m^2-(\ell-1)^2)+(2\ell-2)(2\ell-1)\,k'^2\right]a(2\ell-2)$
	$-\left[m^2-\ell^2\right]a(2\ell)$
end)	

Tab. 1. Recursive evaluation of the coefficients $a(2\ell)$ for symmetrical Zolotarev polynomial of the first kind $Z_p(w,\kappa) = (-1)^m T_m\left(\sqrt{\frac{w^2-k'^2}{1-k'^2}}\right) = (-1)^m \sum_{\ell=0}^{m} a(2\ell) T_{2\ell}(w)$.

[5]The discrete inverse cosine transform (iDCT)

6. Evaluation of ZP1S Generation Methods Numerical Behavior

In this section we explore numerical behavior of the ZP1S computational methods described in previous section. Firstly we show that both LTV recursion (2) and (21) are stable. Secondly, we explore dynamic range of the methods. And at thirdly, we apply empirical criterion onto the methods results.

In case of recursion stability determination we employ following criteria. The first stability criterion states: if cumulated error throughout recurrence iterations is bounded by a linear function than the recursive system is considered stable. A recurrence system has to have a limited number of iterations. This approach is based on the one described in [19]. As a second stability criterion we chose more analytical approach. We consider each iteration of LTV system as a LTI system. We show that for every iteration of LTV system every particular LTI system is stable.

The empirical criterion is based on fundamental theorem of algebra: the number of zeros of every polynomial is uniquely given by its degree. We facilitate this polynomial property by counting number of zeros in waveform generated by tested method. If a number of zeros differs from given polynomial degree than the generated waveform is definitely not the desired ZP1S polynomial. This criterion is not able to validate a method, but it is capable to point at one which is definitely unusable in a subset of ZP1S parameter space.

6.1 Stability of Chebyshev LTV Forward Recursion

In this section we show illustrative results confirming the stability of Chebyshev LTV forward recursion (2). We use previously drafted stability criterion: we adopt and slightly modify the approach suggested in [19]. We evaluate accumulative sums of quantization errors of quantized recurrence. The computation using 32-bit floating-point IEEE number format is considered as "precise". We compare the computation of quantized recurrence in 32-bit fix point arithmetics with "precise" one. For the recursion (2) error is defined as

$$\mathrm{er_n}[i] = T_n[i] - \hat{T}_n[i] \qquad i = 0,1,2,\ldots,N-1 \qquad (22)$$

where $\hat{T}_n[i]$ represents the quantized computation version, while $T_n[i]$ represents the "precise" one. The variable n is an index of recurrence iteration. The p stands for the given polynomial order and N for the number of samples taken from respective definition scope. The cost function given by the sum of errors is estimated according to

$$J[n] = \frac{1}{n}\sum_{i=0}^{N-1} |\mathrm{er_n}[i]|,\ n = 1,2,\ldots,p. \qquad (23)$$

The stability criterion is as follows: if the function $J[n]$ is bounded by a liner function then the recursion in question is stable for limited number of iterations within the finite interval $[1, p]$, where $p < \infty$.

The accumulative error $J[n]$ for Chebyshev forward recurrence for $k' = 0.1$ and degree $p = 100$ is shown in Fig. 5. The function $J[n]$ is bounded, that is, any linear function with properly chosen slope gives greater values than $J[n]$; therefore, the recursion can be considered stable. We verified this criterion to be valid up to degree 2000 with various values of k'.

Another interesting property of computed ZP1S waveform by the Chebyshev recursion method reveals the shape of cumulated error throughout θ, "time" dimension. This error which is than cumulated is given by

$$\mathrm{er2}_n[j] = \sum_{j=0}^{j} T_n[j] - \hat{T}_n[j] \quad xj = 0, 1, 2, ..., N-1, \qquad n = 1, 2, \ldots, p. \tag{24}$$

The cumulative sum of error between "precise" computing $T_n[i]$ and quantized computing $\hat{T}_n[i]$ is shown in Fig. 6 for different polynomial degree within the interval [3, 100].

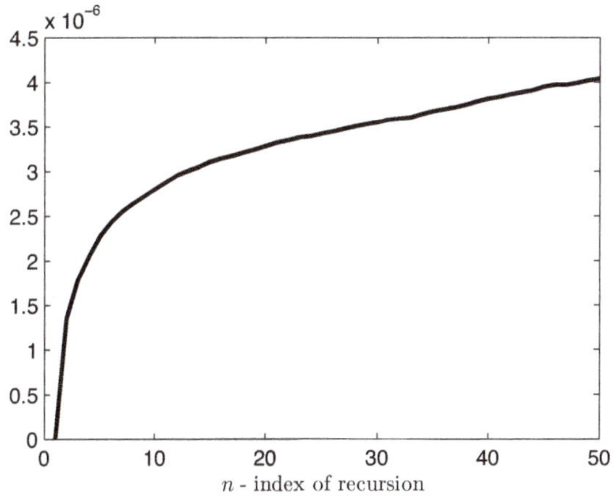

Fig. 5. The cumulative error (23) comparing the quantized ZP1S computation using Chebyshev forward recursion (2) with the "precise" computation version; for $k' = 0.1$ and degree $p = 50$.

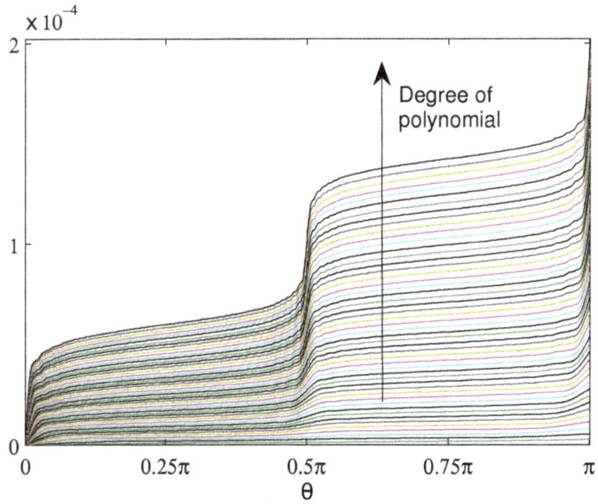

Fig. 6. The cumulative error between "precise" and quantized ZP1S waveform cumulated in "time" domain for $k' = 0.1$ and degrees of $p = 3, 4, \ldots, 100$.

One can see that the error rapidly grows at the centre of definition scope, where the main lobe of ZP1S is placed. The error at the centre of definition scope gets bigger as polynomial degrees or values of k' increases. For higher polynomial degree are errors the bigger as can be seen in Fig. 6 when we examine curves from bottom to up. This error increasing is the consequence of the central lobe maximum value and of ZP1S zeros pushing towards the edges of the definition scope. For higher polynomial degree and constant elliptical modulus k' the main lobe of ZP1S is higher and wider pushing zeros more to the ends of definition scope as can be seen in Fig. 3a and 3c.

6.2 Stability of LTV Backward Recursion

The backward LTV recursion is given by (21) and Tab. 1. To analyze its stability we employ the first of described stability criteria. For recursion (21) we use the error of spectral coefficients given by

$$\mathrm{err}_n[\ell] = a[\ell] - \hat{a}[\ell], \quad \ell = 3, 4, ..., p+2 \tag{25}$$

where the $\hat{a}[\ell]$ represents the quantized version with 32-bit fix-point arithmetics, and the $a[\ell]$ represents version using 32-bit IEEE floating point arithmetics, which is again considered as "precise". The p stands for the given polynomial degree and the n for iteration index of the recursion. The cost function is now given by the cumulative sum of errors (25) as

$$\tilde{J}[n] = \sum_{\ell=3}^{n+2} |\mathrm{err}[\ell]|, \quad n = 3, 4, \ldots, p+2, \tag{26}$$

The stability of backward LTV recursion (21) can be verified similarly to the stability of forward Chebyshev recursion (2). Again, if $\tilde{J}[n]$ is a bounded (or linear) function of the index iteration n then the algorithm stability is ensured for limited number of iterations within the finite interval $[3, p+2]$, where $p < \infty$. The shape of accumulated error $\tilde{J}[n]$ (26) is shown in Fig. 7. This function is clearly bounded and has a similar shape as function $J[l]$ in previous Fig. 5. The stability was experimentally proved for different polynomial degrees within the interval [10, 2000]. Furthermore, one can see that the recursion (21) gives less maximum of accumulated error than the Chebyshev forward recursion (2): compare values 6×10^{-7} with 4×10^{-6}. Also the slope of function $J[l]$ at the beginning and at the end is greater than the slope of function $\tilde{J}[l]$. Thus, it can be concluded, the backward recursion (21) is more robust with respect to stability and gives more precise results than the forward Chebyshev recursion (2). It is interesting to note that the stability of recursion improves with increasing number of iteration index n. This is illustrated by decreasing the slope (or "saturation") of $\tilde{J}[n]$ with increasing iteration index n. It can be concluded that recursion (1) is stable. Note that the cumulative error of the backward recursion increases as the polynomial degree increases in linear, not exponential, rate.

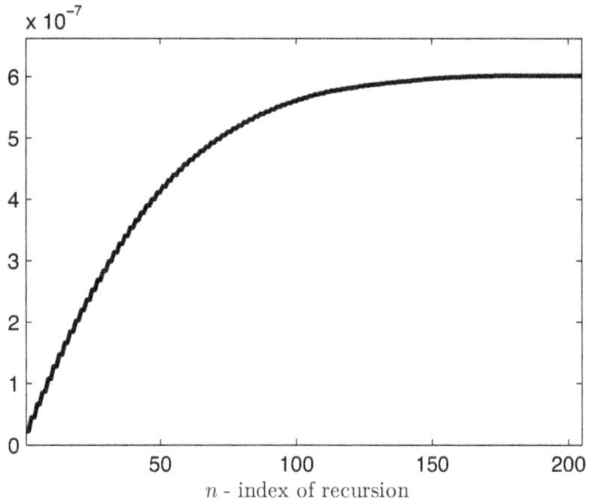

Fig. 7. The cumulative error (26) comparing the quantized ZP1S computation using backward recursion (21) with the "precise" computation version; for $k' = 0.1$ and degree $p = 200$.

The second stability criterion examination follows. The analysis of this type of recursion can be performed using state-space equation and transition matrix (e.q. [20]). The dependency of coefficients $d[\ell]$ on both $m = \frac{p}{2}$ and ℓ parameters is nonlinear. The dynamical ranges corresponding to time-varying coefficients $d[\ell]$ are significantly different. For both these reasons we approach to LTV system is to set of LTI systems fixed for each iteration rather than to use the analysis of transition matrix behaviour.

Analyzed backward LTV recursion can be interpreted as IIR filter of order 6 with time-varying coefficients. When we trace parameters of the recursion in each iteration then the stability of this recursion during the whole iteration process can be verified using pole positions given by the denominator of the transfer function. The idea behind this approach is very simple. We substitute the LTV system by linear time-invariant one. That means for each fixed iteration instant the system is supposed to be LTI with transfer function

$$H_\ell(z) = \frac{z^6}{1 - \frac{d_2[\ell]}{d_1[\ell]}z^2 - \frac{d_3[\ell]}{d_1[\ell]}z^4 - \frac{d_4[\ell]}{d_1[\ell]}z^6}, \quad \ell = 3, \ldots, n+2 \tag{27}$$

where $d_x[\ell]$ are time varying coefficients defined in (21). Resulting denominator of the transfer function represents a polynomial for each iteration index ℓ. Roots of this polynomial, poles of transfer function $H(z)$, are inspected and pole positions and the maxima of pole modulus are checked. This evaluation process is repeated throughout whole iteration process. One chosen result is illustrated for ZP1S parameters of $k' = 0.1$ and $p = 100$ in Fig. 8, which illustrates the trace of pole positions during the whole iteration process. All pole modulus during the iteration process are less then one: hence, ensuring the stability. The values of poles modulus decreases with iteration index.

The best test results (the lowest values of poles modulus: the greatest distances of poles from the unit circle) are obtained for low values of k': $k' \leq 0.2$. Nevertheless, for all permissible values of $k' \subset [0, 1]$ the solution is stable. The stability was experimentally proved for different polynomial degrees within the interval [10, 2000].

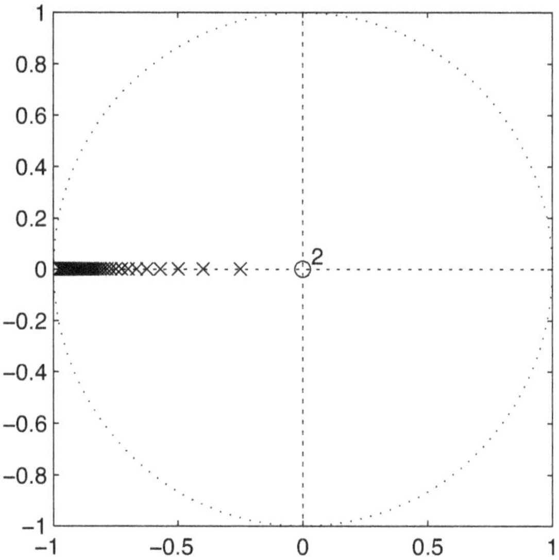

Fig. 8. Pole positions of LTI transfer function 27 for all iterations of LTV backward recursion for given ZP1S parameters: $k' = 0.1$ and degree $p = 200$.

6.3 Dynamic Range

Dynamic range needed for all selected method for the computing ZP1S was checked for all polynomial degrees within the interval [10, 2000]. The result is as expected. Dynamic range needed for binomial coefficients computing (4a) is much greater than for the LTV forward Chebyshev recursion (2) and also greater than backward LTV recursion (21).

For example, for degree $N = 100$ and $k' = 0.1$, the range of binomial coefficients $c_k^{(n)}$ in eq. (4b) values is approximately 10^{37}, while for LTV forward recursion is about 10^6 and for LTV backward recursion and its coefficients $a(2\ell)$ approximately 10^5. When we compare the dynamic range of the floating-point 32-bit IEEE number format with the range needed for the computing binomial coefficients we can see that the 32-bit floating-point number format is not sufficient for safe computing these coefficients. On the other hand dynamic ranges for both LTV recursions are comparable and much less than the dynamic range of the 32-bit floating-point number format. This is the reason, why both recursions work properly while binomial coefficients does not.

6.4 Empirical Test Criterion

The empirical criterion tests if a generated waveform representing ZP1S satisfies the number of zeros. The number of zeros is uniquely given by the polynomial order p. Therefore, we can find the number of zeros of computed polynomial ZP1S and compare it with the polynomial order p. If the number of found zeros differs from the given polynomial degree then the tested waveform ZP1S can not be a proper Zolotarev polynomial. The simple and robust enough way of enumerating the number of zeros is the zero crossings count (ZCC)

$$\mathrm{ZCC}(y) = \sum_{i=0}^{N-1} \left| \frac{\mathrm{diff}\,\{\mathrm{sgn}(y[i])\}}{2} \right| \tag{28}$$

where $y[i]$ is tested discrete generated polynomial. Using (28) we can define the zero count error (ZCE) as

$$\mathrm{ZCE}(y, p) = -p + \sum_{i=0}^{N-1} \left| \frac{\mathrm{diff}\,\{\mathrm{sgn}(y[i])\}}{2} \right|. \tag{29}$$

Thus, if the ZCE computed by (29) is bigger then zero than the generated waveform do not meet ZP1S properties. If the number of zeros of computed polynomial ZP1S is equal to the given polynomial order then this ZP1S might be the proper polynomial. Tests for all selected methods of ZP1S computation are using 32-bit IEEE floating number format. All waveforms are represented by $N = 10^4$ samples.

The results of ZCE evaluated for polynomial generated using power expansion with binomial coefficients (4a) is in Fig. 9. We can see that the method (4a) definitely fails if the polynomial degree is greater than 80. Tested method

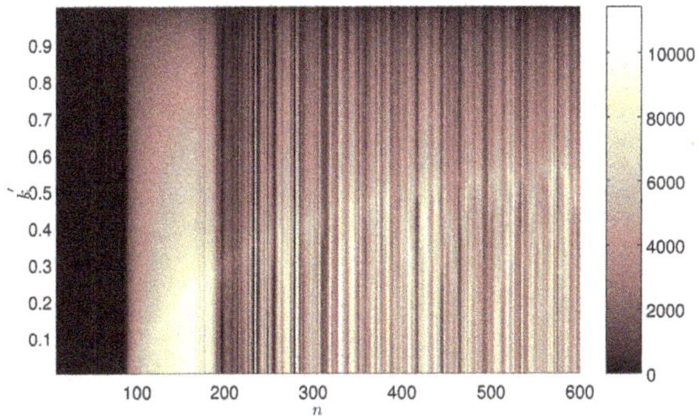

Fig. 9. The ZCE evaluated for a waveform generated using power expansion with binomial coefficients (4a); non black areas denote the ZP1S parameter space where the generated waveform definitely does not meet ZP1S properties.

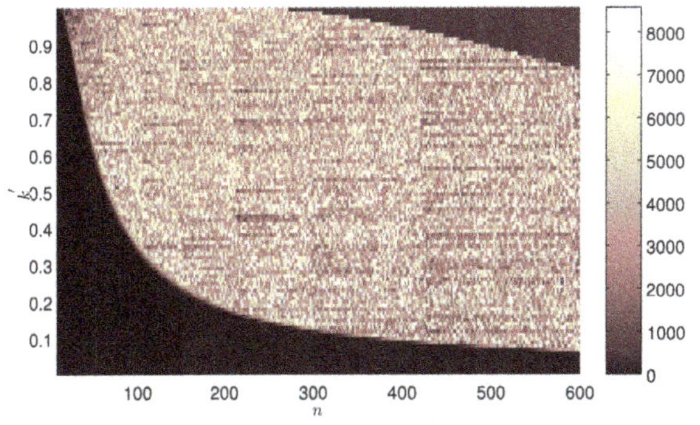

Fig. 10. The ZCE evaluated for a waveform generated using the LTV backward recursion followed by the iDCT; non black areas denote the ZP1S parameter space where the generated waveform definitely does not meet ZP1S properties.

computes waveform based on (7). If the waveform is computed based on (6) the method fails if degree reaches ≈ 35. In case a waveform generated by LTV Chebyshev forward recursion (2) the results show correct number of zeros throughout all values of tested ZP1S parameter space: $k' \subset (0, 1)$ and $p = 3, \cdots, 2000$.

The results of the ZCE evaluated for a waveform computed using the LTV backward recursion (21) followed by the iDCT are in Fig. 10. We can see that it should be possible to generate ZP1S of high degrees; however the interval of k' is being severally limited with increasing degree.

7. Discussion

The study of selected ZP1S computation methods reveals that the one employing the Chebyshev LTV forward recursion (2) is the most suitable method for computing ZP1S: the empirical criterion shows the number of zeros matching the polynomial degree throughout whole parameter space $k' \subset (0, 1)$ and $p = 3, \cdots, 2000$; furthermore, the study confirms the stability of the LTV forward recursion in the same ZP1S parameter space.

In case of method using the LTV backward recursion followed by the iDCT the empirical criterion shows much worse performance, compared with forward recursion. However, the analysis of stability shows the LTV backward recursion more stable than the forward one. This seemingly contradicts, but it can be explained as follows. Both the LTV backward recursion and the iDCT are not free of the error propagation caused by quantization used in the numerical computation of ZP1S on a digital system. The problem is, we are unable to separate errors of the LTV backward recursion from errors caused by iDCT. Thus, one can expect that this method has worse results than direct computing ZP1S by forward LTV recursion (2). Since the LTV backward recursion is stable, even more stable than forward, it is obvious that the iDCT introduces greater errors during the computation compared with the LTV backward recursion. The iDCT errors are the result of the accumulation of cosines weighed by the spectral coefficients $a(2\ell)$ in (9). More detailed analysis of iDCT errors is out of the scope of this paper; however, it could prove interesting.

In case of employing of the ZP for design of FIR filters we have shown that the methods based on backward recursion are stable and capable of generate a filter coefficients of high orders up to thousands. However, corresponding filter frequency response, which is bounded with a filter coefficients by inverse Fourier series, might be problematic in some cases. Therefore, it seems to be advisable to use the LTV forward Chebyshev type of recursion to generate a filter frequency response.

We intend to focus future research on errors of iDCT, and generally on reducing of errors of latter methods for computing ZP1S.

8. Conclusion

Numerical study of computing symmetrical Zolotarev polynomials of the first kind shows that both recursions, the forward recursion for direct computing ZP1S and the backward recursion for computing spectrum of ZP1S, are stable. Method employing forward LTV Chebyshev recursion is the one most suitable one for computing the polynomial. For computing spectrum of the polynomial the Vlcek backward recursion has the best performance with respect to errors. This study also reveals that the method of computing ZP1S by backward recursion followed by iDCT has much worse numerical behavioral than the one using forward recursion.

Acknowledgments

This work is supported by Student Grant Contest (SGS) grant Advanced Algorithms of Digital Signal Processing and their Applications, grant number SGS17/183/OHK3/3T/13.

References

[1] LEVY, R. Generalized rational function approximation in finite intervals using Zolotarev functions. *IEEE Transactions on Microwave Theory and Techniques*, 1970, vol. 18, no. 12, p. 1052–1064. DOI: 10.1109/tmtt.1970.1127411

[2] CHEN, X., PARKS, T. Analytic design of optimal FIR narrow-band filters using Zolotarev polynomials. *IEEE Transactions on Circuits and Systems*, 1986, vol. 33, p. 1065–1071. DOI: 10.1109/tcs.1986.1085868

[3] ZAHRADNIK, P., VLCEK, M. Perfect decomposition narrow-band FIR filter banks. *IEEE Transactions on Circuits and Systems II: Express Briefs*, 2012, vol. 59, no. 11, p. 805–809. DOI: 10.1109/tcsii.2012.2218453

[4] VLCEK, M., ZAHRADNIK, P. Almost equiripple low-pass FIR filters. *Circuits, Systems, and Signal Processing*, 2013, vol. 32, no. 2, p. 743–757. DOI: 10.1007/s00034-012-9484-0

[5] VLCEK, M., JANIK, J., TURON, V., et al. A way to a new multispectral transform. In *Proceedings of the 11th WSEAS International Conference on Signal Processing, Computational Geometry and Artificial Vision and 11th WSEAS International Conference on Systems Theory and Scientific Computation (GAVTASC'11)*. Stevens Point, Wisconsin (USA), 2011, p. 177–182. ISBN: 978-1-61804-027-5

[6] MOHAN, K. N., KANNADASSAN, D., ZINKA, S. R. Design and implementation of Dolph, Chebyshev and Zolotarev circular antenna array. *Indian Journal of Science and Technology*, 2016, vol. 9, no. 36.

[7] VLCEK, M., UNBEHAUEN, R. Zolotarev polynomials and optimal FIR filters. *IEEE Transactions on Signal Processing*, 1999, vol. 47, no. 3, p. 717–730. DOI: 10.1109/78.747778

[8] COOPER, G. J. The evaluation of the coefficients in the Chebyshev expansion. *The Computer Journal*, 1967, vol. 10, no. 1, p. 94–100. DOI: 10.2307/2003301

[9] SKLYAROV, V. P., A numerical experiment related to zolotarev polynomials for weighted sup-norm. *Computational Mathematics and Mathematical Physics*, 2011, vol. 51, no. 10, p. 1679. DOI: 10.1134/s0965542511100149

[10] BOGYTAROV, B., Effective approach to least deviation problems. *Sbornik: Mathematics*, 2002, vol. 193, no. 12, p. 1749–1769. DOI: 10.1070/sm2002v193n12abeh000698

[11] MILEV, L. Numerical computation of the Markov factors for the systems of polynomials with the Hermite and Laguerre weights. In *Proceedings of the 6th Conference on Numerical Methods and Applications*. 2006, p. 386–393. DOI: 10.1007/978-3-540-70942-846

[12] GRASEGGER, G., VO, N. T. An algebraic-geometric method for computing Zolotarev polynomials. In *Proceedings of the International Symposium on Symbolic and Algebraic Computation*. Kaiserslautern (Germany), 2016, p. 173–180. DOI: 10.1145/3087604.3087613

[13] MASON, J. C. *Chebyshev Polynomials*. Boca Raton, FL (USA): Chapman & Hall/CRC, 2003. ISBN: 978-0849303555

[14] ZAHRADNIK, P., VLCEK, M., Fast analytical design algorithms for FIR notch filters. *IEEE Transactions on Circuits and Systems I: Regular Papers*, 2004, vol. 51, no. 3, p. 608–623. DOI: 10.1109/iscas.1994.408963

[15] SPETIK, R. *The Discrete Zolotarev Transform* (Ph.D. thesis). Prague: Department of Circuit Theory, Faculty of Electrical Engineering, CTU Prague, 2009.

[16] OPPENHEIM, A. V., SCHAFER, R. W., BUCK, J. R. *Discrete-Time Signal Processing*. 2nd ed. Prentice Hall, 1999. ISBN: 0137549202

[17] LAMPRET, V., Estimating the sequence of real binomial coefficients. *Journal of Inequalities in Pure & Applied Mathematics*, 2006, vol. 7, no. 5.

[18] OPPENHEIM, A. V., WILLSKY, A. S., HAMID, S. *Signals and Systems*. 2nd ed. Prentice Hall, 1996. ISBN: 0138147574

[19] PANJER, H., WANG, S. On the stability of recursive formulas. *ASTIN Bulletin*, 1993, vol. 23, p. 227–258. DOI: 10.2143/ast.23.2.2005093

[20] DACUNHA, J. J. Stability for time varying linear dynamic systems on time scales. *Journal of Computational and Applied Mathematics*, 2005, no. 2, vol. 176, p. 381–410. DOI: 10.1016/j.cam.2004.07.026

[21] LOADER, C. *Fast and Accurate Computation of Binomial Probabilities*. [Online] Cited 2017-07-13. Available at: http://octave.1599824.n4.nabble.com/attachment/3829107/0/loader2000Fast.pdf

About the Authors . . .

Jan KUBÁK received the M.S. degree in electrical engineering from the Faculty of Electrical Engineering of the Czech Technical University (FEE CTU), Prague, in 2013. He is currently studying doctoral degree at the Department of Circuit Theory, FEE CTU. He is engaged in research on selective spectral transforms. He is FPGA digital circuit designer.

Pavel SOVKA received the M.S. and Ph.D. degrees in electrical engineering from the Faculty of Electrical Engineering of the Czech Technical University (FEE CTU), Prague, in 1981 and 1986, respectively. From 1985 to 1991 he worked in the Institute of Radioengineering and Electronics of the Czech Academy of Sciences, Prague. In 1991 he joined the Department of Circuit Theory, FEE CTU. He worked on the application of adaptive systems to noise and echo cancellation, speech analysis, changepoint detection, and signal separation. Presently he is engaged in research on biomedical signal processing and selective spectral transforms.

Miroslav VLČEK was born in Prague, The Czech Republic, in 1951. He received the graduate degree in theoretical physics from Charles University, Prague, in 1974, and the Ph.D. degree in communication engineering and the D.Sc. degree from the Czech Technical University (CTU), Prague, in 1979 and 1994, respectively.From 1974 to 1993, he was with the Department of Circuit Theory, Faculty of Electric Engineering, CTU, where he is currently a Professor of theory of electrical engineering. Since 1995, he has been the Head of the Department of Applied Mathematics, Faculty of Transportation Sciences, CTU. He was the Alexander-von-Humboldt Fellow at the University of Erlangen, Nürberg, Germany, in 1988 and 1997. He currently teaches courses in system theory and digital filter design. His scientific interests include filter design and digital signal processing, and theory of approximation and higher transcendental functions.

16

Design of Second Order Recursive Digital Integrators with Matching Phase and Magnitude Response

Kriti GARG , Dharmendra K. UPADHYAY

Dept. of Electronics and Communication Engineering, Netaji Subhas Institute of Technology, Sector-3, Dwarka, 110078 New Delhi, India

kritigarg13@gmail.com, upadhyay_d@rediffmail.com

Abstract. *Location of poles and zeroes greatly affect phase response and magnitude response of a system. Recently, pole-zero optimization emerged as an effective approach to approximately match magnitude response of a system with that of an ideal one. In this brief, a methodology for the design of linear phase integrators and ones with constant phase of −90 degree is proposed.*

The aim of this method is to simultaneously attain multiple objectives of magnitude and phase optimization. In this method, magnitude response error is minimized under the constraint that the maximum passband phase-response error is below a prescribed level. Examples are included to illustrate the proposed design technique.

Keywords

Digital integrator, genetic algorithm, linear phase, pole-zero optimization, recursive, simulated annealing

1. Introduction

An integrator is a system whose output signal is the time integral of its input signal. It can be mathematically modelled as,

$$H_i(\omega) = 1/j\omega$$

where $j = \sqrt{-1}$ and ω is the angular frequency in radians per second. Integrators find immense applications in the areas of signal processing, bio-medical engineering, radar engineering, sonar engineering, control system etc. Multitude of techniques have been developed for the design of recursive and non-recursive digital integrators. It is interesting to note that for linear phase integrators, Maximum Percentage Relative Error (MPRE) in magnitude response has rolled down from 5 % to 0.48 % in the decade 2005 – 2015.

Several designs of digital integrator have been proposed in the literature [1–7]. Among these, the simplest approximations to the desired frequency response are Rectangular, Trapezoidal and Simpson digital integrators. It can be easily noticed that, as $\omega \to 0$, $H(e^{j\omega}) \to 1/j\omega$ for these approximations. These digital integrators can be sufficient for the integration of oversampled signals (signals sampled much above the Nyquist rate), but a deeper investigation suggests that there is still a possibility of better designs for signals that are critically sampled at the Nyquist rate.

Digital integrators have been designed using quadrature rules, such as the Newton-Cotes and Gauss- Legendre rules by Ngo and Tseng [3–6]. These methods, however, are complex to design and implement due to the usage of fractional sampling rates, hence lacking computational efficiency. Lagrange interpolators have been suggested to elevate some of these problems. Most recently Tseng et al. have proposed to implement the fractional delays in the Hartley transform domain [7].

One of the design methods which has become very popular among researchers is Iterative Optimization Method [8–20]. It is widely used to improve the performance of a system by reducing its runtime, bandwidth, memory requirement, or some other property. Optimization methods such as Linear Programming, Simulated Annealing, Genetic Algorithm, and Pole-Zero optimization have been used earlier to design Infinite Impulse Response (IIR) digital integrators and differentiators. Papamarkos-Chamzas [8] have used Linear Programming optimization method to design digital integrators. Al-Alaoui [9] has also proposed a family of digital integrators by using interpolation and Simulated Annealing optimization method. Upadhyay-Singh (US) have proposed recursive wideband digital integrator for 0.48 % MPREs in magnitude responses over almost the full Nyquist band except near to $\omega = \pi$ [10]. Genetic Algorithm has been exploited in [16] to obtain a class of second order linear phase integrators. In [17], Jain-Gupta-Jain have used Minimax and Pole, Zero, and Constant Optimization Methods to obtain second, third and fourth order IIR digital integrators. Later, in [18], Gupta et al. (GET) have proposed recursive wideband digital integrators using Modified Particle Swarm Optimization (MPSO). In [19], Jalloul and Al-Alaoui have employed Particle Swarm Optimization to propose designs of integrators and differentiators. Upadhyay, in [20], has used Pole-Zero Optimization to achieve integrators with relative error in magnitude response not exceeding 0.37 %. In [21], efficient design of FIR digital differentiator using the L_1-optimality criterion is proposed.

Often application in controls, wave shaping, oscillators and communication require a constant 90° phase for differentiators and −90° phase for integrators. In [22], Al-Alaoui cascaded differentiator and integrator operator with fractional advance and delay respectively to obtain constant 90° phase for differentiators and constant −90° phase for integrators. Al-Alaoui also showed that doubling the sampling rate improves the magnitude response. Combining the two actions improves both the magnitude and phase responses. It should be noted that the approach applies to other differentiators and integrators with linear phases, or approximately linear phases, such as the second-order Al-Alaoui integrators and differentiators. However, the approach is limited only to linear phase differentiator and integrator. Also, fractional delays and advances are complex to implement. In this brief, our goal is to improve the performance and computational efficiency of digital integrators as a standalone system.

Iterative constrained optimization has already been proven a superior method of obtaining close-to-perfect magnitude responses. However, cost function defined in these methods only considers magnitude error with no account of phase error. For real-time applications, further reduction in magnitude error is not as important as reduction in phase error is.

Every recursive digital filter can be completely specified by its poles and zeros with suitable scaling. Poles and zeros give useful insights into a filter's frequency response, and can be used as the basis for digital filter design. This paper attempts at developing a design method which governs constraints on the location of poles and zeroes in order to achieve multi-objectives of phase and magnitude responses, both together.

The rest of the paper is organized as follows. Section 2 gives a brief account of problem under consideration. Solution Methodology is developed in Sec. 3. Design Steps are presented in Sec. 4. Design Examples are given in Sec. 5 of the paper. In Sec. 6, Performance Results are elucidated. Conclusions are drawn in Sec. 7.

2. Problem Formulation

The paper addresses two design objectives. Firstly, the design of stable recursive linear phase digital integrator and secondly, the design of stable recursive digital integrator with a constant phase of −90°. These designs, however, involve non-converging objectives of magnitude and phase optimization and there exist a trade-off among the different objectives. The trade-off parameters considered in the designs are: passband magnitude error and passband phase deviation, passband cut-off frequency in phase response (ω_p): which is defined as range of frequency for which phase deviation $\leq \delta$.

Design objectives for linear phase integrators can hence be summarized as:

1. Low wideband magnitude error
2. Low passband phase deviation (constant group delay)

Design objectives for integrators with constant phase of −90° are given below:

1. Low wideband magnitude error
2. High Passband cut-off frequency in phase response
3. Low passband phase deviation

Let, $H_l(z)$ denotes the transfer function of linear phase integrator and $H_c(z)$ denotes the transfer function of integrator with constant phase of −90°.

A generalized digital recursive transfer function of second order is of the form, given in (1).

$$I(z) = I_o \frac{(z - z_1)(z - z_2)}{(z - p_1)(z - p_2)} \tag{1}$$

where z_1, z_2 are zeroes and p_1, p_2 are poles of the considered transfer function. Here, I_o is a multiplier constant or scaling factor. The frequency response $I(\mathrm{e}^{\mathrm{j}\omega})$ of the considered system can be written as in (2),

$$(\mathrm{e}^{\mathrm{j}\omega}) = |I(\mathrm{e}^{\mathrm{j}\omega})|\mathrm{e}^{\mathrm{j}\theta(\omega)} = I_o \frac{\prod_{i=1}^{2} |I_i(\mathrm{e}^{\mathrm{j}\omega})|\mathrm{e}^{\mathrm{j}\phi_i(\omega)}}{\prod_{i=3}^{4} |I_i(\mathrm{e}^{\mathrm{j}\omega})|\mathrm{e}^{\mathrm{j}\phi_i(\omega)}} \tag{2}$$

where $|I(\mathrm{e}^{\mathrm{j}\omega})|$ and $\theta(\omega)$ are the magnitude and phase response of the digital integrator. $|I_1(\mathrm{e}^{\mathrm{j}\omega})|$, $|I_2(\mathrm{e}^{\mathrm{j}\omega})|$, $|I_3(\mathrm{e}^{\mathrm{j}\omega})|$, $|I_4(\mathrm{e}^{\mathrm{j}\omega})|$ and $\phi_1(\omega)$, $\phi_2(\omega)$, $\phi_3(\omega)$, $\phi_4(\omega)$ are the magnitude responses and phase characteristics of z_1, z_2, p_1, p_2 respectively. Now, phase response of the overall transfer function is given as, $\theta(\omega) = \phi_1(\omega) + \phi_2(\omega) - \phi_3(\omega) - \phi_4(\omega)$.

Error function to be minimized is given in (3).

$$E = \int_0^{\pi} \left(\left|I(\mathrm{e}^{\mathrm{j}\omega})\right| - \frac{1}{\omega} \right)^2 \mathrm{d}\omega. \tag{3}$$

The main reason for expressing the magnitude constraints in terms of magnitude-square instead of magnitude is to avoid the appearance of a denominator for the gradients of the magnitude constraints, which the use of the simple magnitude would have made unavoidable.

A necessary and sufficient condition for the stability of a causal Linear Time Invariant (LTI) digital IIR filter is that all poles of its irreducible transfer function lie strictly inside the unit circle.

3. Solution Methodology

Consider a polynomial equation $x = (z - a)$, where $a \in \mathbf{R}$. Substituting $z = \mathrm{e}^{\mathrm{j}\omega}$, we get phase of x, given in (4).

$$\phi_x(\omega) = \tan^{-1} \frac{\sin \omega}{\cos \omega - a}. \tag{4}$$

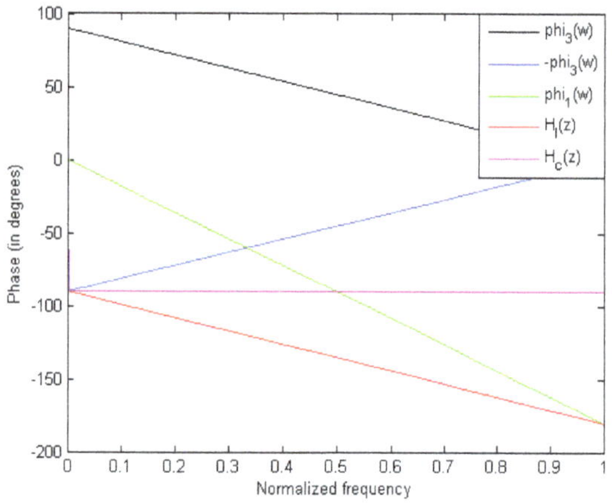

Fig. 1. Illustration of phase effects of zeroes and poles to obtain linear phase characteristics for a system.

location of root	$\phi_x(\omega)\|_{\omega=0}$	$\phi_x(\omega)\|_{\omega=\pi}$
$a > 1$	π	0
$a < -1$	0	$-\pi$
$a = 1$	$\pi/2$	0
$a = -1$	0	$-\pi/2$

Tab. 1. Values of $\phi_x(\omega)$ for different root locations.

Groupdelay of x is defined as:

$$\tau_x(\omega) = -\frac{\mathrm{d}\phi_x(\omega)}{\mathrm{d}\omega} = -\frac{1 - a\cos\omega}{(1 - a\cos\omega) + a(a - \cos\omega)}. \quad (5a)$$

From (5a), it is noted that:

$$\lim_{a\to\infty} \tau_x(\omega) = 0. \quad (5b)$$

Therefore, indicating that greater the value of a, better the phase linearity of the system.

We next require to compute values of $\phi_x(\omega)$ at extreme frequency points in the Nyquist Band (i.e. at $\omega = 0$ and at $\omega = \pi$) for different locations of root. Table 1 summarizes values of $\phi_x(\omega)$ for different root locations.

3.1 Design of Linear Phase Integrators

Figure 1 illustrates the phase effects of zeroes and poles to obtain linear phase characteristics for a discrete time system. It is observed from this figure that in order to have a linear phase response for a system with transfer function $H_l(z)$, phase function of one of the poles p_1, i.e, $\phi_3(\omega)$ should be equal to 90° at $\omega = 0$ and decrease monotonically henceforth with constant slope for the complete range of Nyquist frequency. Another pole p_2 should be located strictly inside the unit circle to satisfy stability constraints for the system. One of the zeroes z_2 should be so chosen that it neutralizes the phase effects of p_2. The phase response of zero z_1, i.e, $\phi_1(\omega)$ is required to be monotonically decreasing, linear function of frequency with slope much greater than that of $\phi_3(\omega)$, so that, when phase characteristic of p_1 is subtracted from that of z_1, i.e, $\phi_1(\omega) - \phi_3(\omega)$, linear phase characteristic is obtained for the overall transfer function $H_l(z)$.

Based on this, required phase response of p_1 can be mathematically modelled by the following conditions:

1. $\phi_3(\omega)|_{(\omega=0)} = \pi/2$,
2. $\phi_3(\omega)|_{(\omega=\pi)} = 0$,
3. $\left|\frac{\mathrm{d}}{\mathrm{d}\omega}\phi_3(\omega)\right| = \frac{1}{2}$.

From Tab. 1, it is observed that for pole location $z = 1$, first two of the above mentioned conditions are fulfilled.

Differentiating $\phi_3(\omega)$, we get

$$\frac{\mathrm{d}\phi_3(\omega)}{\mathrm{d}\omega} = \frac{1 - a\cos\omega}{1 + a^2 - 2a\cos\omega}.$$

Put $a = 1$

$$\left|\frac{\mathrm{d}\phi_3(\omega)}{\mathrm{d}\omega}\right| = \frac{1}{2}$$

which satisfies the third condition.

Hence, it is concluded that pole p_1 is located at $z = 1$. Isolated poles on the unit circle may be called marginally stable. The impulse response corresponding to a single pole on the unit circle never decays, but neither does it grow.

Next, constraints on phase function of z_1 are described below:

1. $\phi_1(\omega)|_{(\omega=0)} = 0$,
2. $\tau_1(\omega)$ = constant (for phase linearity).

For the first condition to hold true, zero z_1 can take any value in the interval $(-\infty, -1)$, as concluded from Tab. 1. Assume z_1 lies in the interval $(-\alpha, -1)$. Solution for α can be obtained by solving the equation:

$$\tau_x(\omega)|_{x=\alpha,\omega=0} = \kappa \qquad \kappa \in (0, 1), \quad (6)$$

$\kappa = 1$ results into a perfect linear phase response while $\kappa = 0$ leads to a constant phase response of $-90°$.

Different solutions of α can be obtained for different defined values of κ.

An LTI filter is stable if and only if its poles are strictly inside the unit circle ($|z| = 1$) in the complex z plane. In particular, a pole p outside the unit circle ($|p| > 1$) gives rise to an impulse-response component proportional to p^n which grows exponentially over time n. Therefore, pole p_2 must lie in the interval (-1,1). To achieve linear phase characteristics for the system, p_2 and z_2 should lie close enough to each other. Let Δp be the distance between the two. For faithful neutralization of the two phase effects, $|\Delta p|$ should not exceed 0.1

In conclusion, to obtain linear phase characteristics for a second order system, p_1 should be located at $z = 1$. p_2 should lie in the interval (-1,1). $z_2 = p_2 + |\Delta p|$, where $|\Delta p|$ does not exceed 0.1. z_1 should lie in the interval $(-\alpha, -1)$ where α is linearly related to user-defined parameter κ.

3.2 Design of Integrators with Constant Phase of $-90°$

Now that we have estimated location of poles and zeroes of $H_l(z)$ to obtain linear phase characteristics for the system, we would now like to investigate location of poles and zeroes to achieve a constant phase response of $-90°$ for a system with transfer function $H_c(z)$. Let z_1', z_2', p_1' and p_2' denote the zeroes and poles of $H_c(z)$.

From Fig. 1, it is observed that a positive angular shift of 26.56° in phase response of $H_l(z)$ causes its linear phase characteristics to transform into a constant phase characteristics of $-90°$. It can henceforth be argued that cumulative sum of angular phase shifts contributed by poles and zeroes of $H_c(z)$ should be equal to 26.56°.

Let,

1. Δ_1 be the positive angular phase shift caused by z_1', measured w.r.t $\phi_1(\omega)$ (see Fig. 1)
2. Δ_2 be the negative phase shift caused by z_2', measured w.r.t frequency axis.
3. Δ_3 be the negative phase shift caused by p_1', measured w.r.t $-\phi_3(\omega)$.
4. Δ_4 be the positive phase shift caused by p_2', measured w.r.t frequency axis.

It is to be noted that phase function of pole p_1' suffers the same constraints as p_1 and it continues to be located at $z = 1$. Hence, $\Delta_3 = 0$.

Now, summation of phase shifts caused by z_1', z_2' and p_2' should be equal to 26.56°, given by (7),

$$|\Delta_1| - |\Delta_2| + |\Delta_4| = 26.56° + \Delta \tag{7}$$

where Δ is tolerable angular phase shift. At first, consider $|\Delta_2|$ and Δ small enough to be neglected. So, (7) becomes,

$$|\Delta_1| + |\Delta_4| = 26.56°. \tag{8a}$$

In (8a), $z_1' \geq -1$. So, maximum value of Δ_1 is achieved when z_1' is located at $z = -1$.

$$\max(|\Delta_1|) = 18.44° \tag{8b}$$

This implies that

$$\min(|\Delta_4|) = 8.12°. \tag{8c}$$

Relationship between root location a and corresponding maximum angular phase deviation (Δ) caused by it is derived as follows:

Maximum Angular Phase Deviation (Δ) can be defined as slope of a tangent to the phase response curve at $\omega = 0$. By above definition,

$$\left.\frac{d\phi_x(\omega)}{d\omega}\right|_{\omega=0} = \Delta. \tag{9}$$

Similarly, relationship between root location a and corresponding maximum deviation caused (δ) by it is derived as follows:

Maximum Phase Deviation (δ) can be defined as the value of phase obtained at a stationary point on the phase curve at which the tangent changes from a positive value on the left of this point to a negative value on the right.

Solving for ω in the equation $\frac{d\phi_x(\omega)}{d\omega} = 0$ we get

$$\frac{1 - a\cos\omega}{1 + a^2 - 2a\cos\omega} = 0$$
$$\Rightarrow a = \frac{1}{\cos\omega}$$
$$\Rightarrow \omega = \cos^{-1}(1/a).$$

By definition

$$\phi_x(\omega)|_{\omega=\cos^{-1}(1/a)} = \delta. \tag{10a}$$

This gives us

$$\tan^{-1}\frac{\sin(\cos^{-1}(1/a))}{(1/a) - a} = \delta. \tag{10b}$$

Solution of (9) and (10b) gives relationship between Δ and δ.

For $\Delta = 8.12°$, solving (9) a comes out to be -0.3. This implies that pole p_2' is located at $z = -0.3$

As location of zero z_1' goes beyond $z = -1$ (towards $-\infty$), $|\Delta_1|$ decreases requiring $|\Delta_4|$ to increase and pole to shift beyond $z = -0.4$ (towards -1). Equation (8) suggests that $|\Delta_4|$ can not exceed 26.56°. This marks maximum bound on $|\Delta_4|$. This is achieved when pole p_2' is located at $z = -0.6$. Hence, pole p_2' is restricted to be located in the interval $(-0.6, -0.3)$.

Practically, neither Δ nor $|\Delta_2|$ can be 0. From (7), we conclude the following:

$$\max(|\Delta_1| - |\Delta_2|) = 18° + \Delta, \tag{11}$$
$$\min(|\Delta_1| - |\Delta_2|) = \Delta. \tag{12}$$

From above analysis, we investigate that p_1' is located at $z = 1$. p_2' lies in the interval $(-0.6, -0.3)$. z_1' lies in the interval $(-\beta, -1)$ and z_2' lies in the interval $(-\gamma, 0)$.

For different values of δ (and hence Δ) chosen, different solution sets for (β, γ) can be derived.

4. Design Approach

The typical approach in designing of a digital IIR integrator is to minimize the maximum amplitude-response error and maximum phase-response error. The optimization can be carried out by minimizing the magnitude error under the constraint that the passband phase error δ is within prescribed level.

Iterative optimization methods are often the only choice for non linear equations. It is a mathematical procedure that generates a sequence of improving approximate solutions until convergence is reached. Genetic Algorithm (GA) is already used in the literature for designing digital differentiators and integrators [13]. In this work, GA is used in a similar fashion as in [13]. However, the optimization works on optimizing the gain factor and the locations of poles and zeroes of the digital filter as opposed to optimizing the coefficients of the numerator and the denominator.

Steps followed in the design of $H_l(z)$ and $H_c(z)$ are given below:

Step 1: For the design of $H_l(z)$, define a value of κ. Then, solution of (6) gives the value of α.

For the design of $H_c(z)$, choose a required value for δ. Simultaneous solution of (9), (10) and (11) yields solution set for (β, γ).

Step 2: Run the iterative optimization with poles and zeroes set to vary within the respective prescribed intervals for both the designs, as discussed in previous section. Cost function is given in (3). The optimization runs until convergence is reached.

The constraints on the location of poles and zeroes ensure that the phase-response error is under a prescribed level. The frequency range is defined as $0 \le \omega \le \pi$ radians/second.

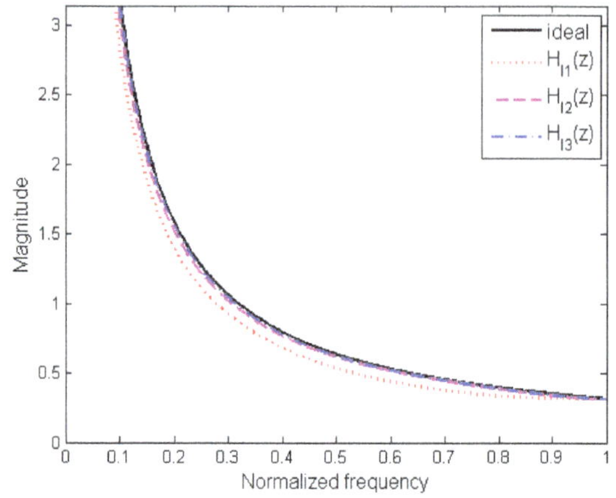

Fig. 2. Magnitude responses of proposed integrators $H_{l1-l3}(z)$.

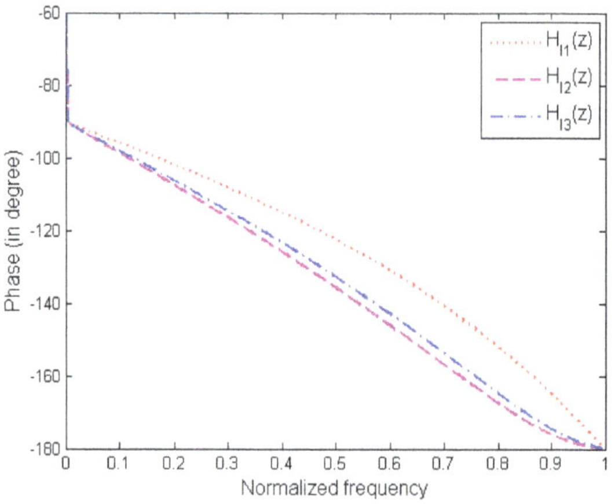

Fig. 3. Phase responses of proposed integrators $H_{l1-l3}(z)$.

5. Design Examples

In real world applications, most of the optimization problems involve more than one objective to be optimized. The objectives in most of engineering problems are often conflicting. In the case, one extreme solution would not satisfy both objective functions and the optimal solution of one objective will not necessarily be the best solution for other objective(s). Therefore, different solutions will produce trade-off between different objectives and a set of solutions is required to represent the optimal solutions of all objectives. Several design examples are included which demonstrate the effectiveness of the design technique.

5.1 Linear Phase Integrators

In this section, three designs of linear phase digital integrators have been considered for different values of κ. Figures 2 and 3 illustrate the magnitude and phase responses of proposed integrators $H_{l1-l3}(z)$.

Example 1: $\kappa = 0.8$

Solving (6) for $\kappa = 0.8$, α comes out to be 5. Proposed integrator is given in (13).

$$H_{l1}(z) = 0.1518\frac{z^2 + 5.4866z + 2.5897}{z^2 - 0.4521z - 0.5479}. \quad (13)$$

The design has poor magnitude response and lacks phase linearity.

Example 2: $\kappa = 0.95$

Solving (6) for $\kappa = 0.95$, α comes out to be 20. Proposed integrator is given in (14).

$$H_{l2}(z) = 0.05578\frac{z^2 + 15.5885z + 8.4692}{z^2 - 0.5361z - 0.4639}. \quad (14)$$

The design out performs other linear phase design examples proposed in this section in terms of phase linearity over the full Nyquist band and has reasonably well magnitude response for complete Nyquist frequency range.

Example 3: $\kappa = 0.9$

Solving (6) for $\kappa = 0.9$, α comes out to be 10. Proposed integrator is given in (15).

$$H_{l3}(z) = 0.08504\frac{z^2 + 10.5789z + 5.8587}{z^2 - 0.4929z - 0.5071}. \quad (15)$$

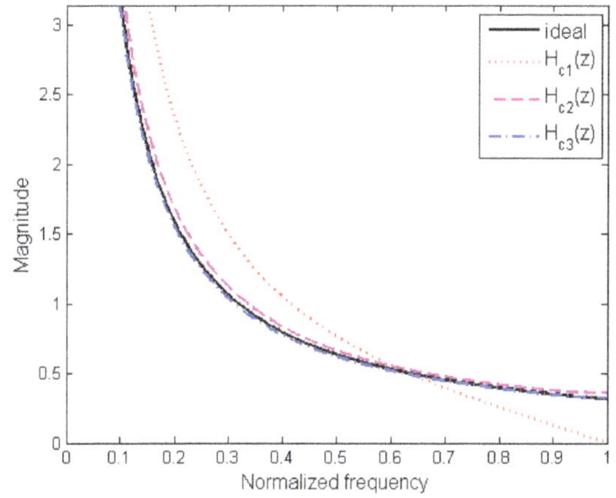

Fig. 4. Magnitude responses of proposed integrators $H_{c1-c3}(z)$.

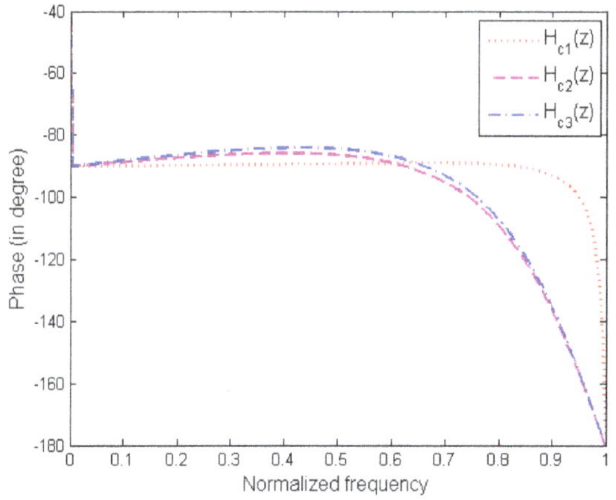

Fig. 5. Phase responses of proposed integrators $H_{c1-c3}(z)$.

The design outperforms other design examples in terms of magnitude response and also has considerably good phase response over wideband.

5.2 Integrators with Constant Phase of −90°

This section demonstrates three designs of integrators with constant phase of −90° for different values of tolerable maximum phase deviation δ. Figures 4 and 5 show the magnitude and phase responses of designed integrator $H_{c1-c3}(z)$.

Example 1: $\delta = 2°$

For $\delta = 2°$, β and γ are found to be -1.05 and -0.6 respectively. Proposed integrator is given in (16).

$$H_{c1}(z) = 0.7679\frac{z^2 + 1.5715z + 0.5560}{z^2 - 0.4204z - 0.5796}. \quad (16)$$

Phase deviation for this design does not exceed 2° in the frequency range $0 \le \omega \le 0.88\pi$. So, cut-off frequency in phase response (ω_p) turns out to be 0.88π. The design, however, can not be considered magnitude efficient with Maximum Magnitude Relative Error (MMRE) of 0.5025 over wideband.

Example 2: $\delta = 5°$

For $\delta = 5°$, β and γ come out to be -1.9 and -0.2 respectively. Proposed integrator is given in (17).

$$H_{c2}(z) = 0.5089\frac{z^2 + 1.9341z + 0.2312}{z^2 - 0.4902z - 0.5098}. \quad (17)$$

The design has considerably accurate magnitude response within $\delta = 5°$ in phase response in the frequency range $0 \le \omega \le 0.69\pi$. Here, ω_p comes out to be 0.69π, making it attractive for real-time applications.

Example 3: $\delta = 10°$

For $\delta = 10°$, solution of β and γ are found to be −2.3 and −1.8. Proposed integrator is given in (18).

$$H_{c3}(z) = 0.4857\frac{z^2 + 1.8514z + 0.1751}{z^2 - 0.4976z - 0.5024}. \quad (18)$$

The design excels in magnitude response and has $\delta = 10°$ in phase response in the frequency range $0 \le \omega \le 0.76\pi$. ω_p for this design is 0.76π

Though, it is disappointing that none of the design example achieves a constant phase response of −90° over full Nyquist band, it is worth noticing that passband cut-off frequency ω_p or bandwidth of the system largely depends on location of zero z_1'. The dependence is depicted below:

$$\phi_1(\omega)|_{\omega=\omega_p} = -90°.$$

For a given location of z_1', approximate passband for the system could be predicted. Conversely, location of z_1' could be determined for a prescribed value of ω_p.

6. Performance Measure

SA is another popular optimization technique that is widely used in the literature for obtaining designs for integrators and differentiators [9]. In this work, SA is implemented to simulate integrators proposed in this section. $H_{l1}(z)$ and $H_{c2-c3}(z)$ are used as initial guess in the optimization algorithm.

Designs of digital integrators proposed using SA are of the general form, as given in (19).

$$H(z) = k_o\frac{z^2 + a_o z + a_1}{z^2 + b_o z + b_1}. \quad (19)$$

Proposed linear phase integrators are tabulated in Tab. 2 and ones with constant phase of −90° are tabulated in Tab. 3.

To verify the accuracy of proposed linear phase integrators, recently published Jain-Gupta-Jain [16], $H_{2MO+PZC}(z)$ [17] , Gupta et al. [18] and Upadhyay recursive digital integrators are considered. Their transfer functions are given in (20–23), respectively.

Integrators	k_o	a_o	a_1	b_o	b_1
$H_{I4}(z)$	0.09522	9.66190	5.19090	−0.49750	−0.50250
$H_{I5}(z)$	0.08515	10.57370	5.81810	−0.50220	−0.49780
$H_{I6}(z)$	0.08506	10.54160	5.48710	−0.54160	−0.45840
$H_{I7}(z)$	0.08860	10.07480	5.39990	−0.48330	−0.51670

Tab. 2. Coefficients of proposed linear phase integrators obtained using SA.

Integrators	k_o	a_o	a_1	b_o	b_1
$H_{c4}(z)$	0.5068	1.8037	0.1678	−0.4924	−0.5076
$H_{c5}(z)$	0.4875	1.8312	−0.5032	0.2274	−0.4968
$H_{c6}(z)$	0.5246	1.9445	0.2629	−0.4488	−0.5512

Tab. 3. Coefficients of proposed integrators with constant phase of -90 degree obtained using SA.

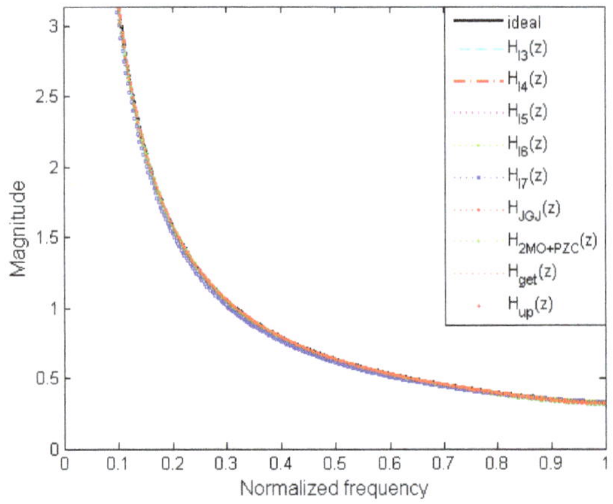

Fig. 6. Magnitude responses of proposed and the existing linear phase integrators with the ideal integrator.

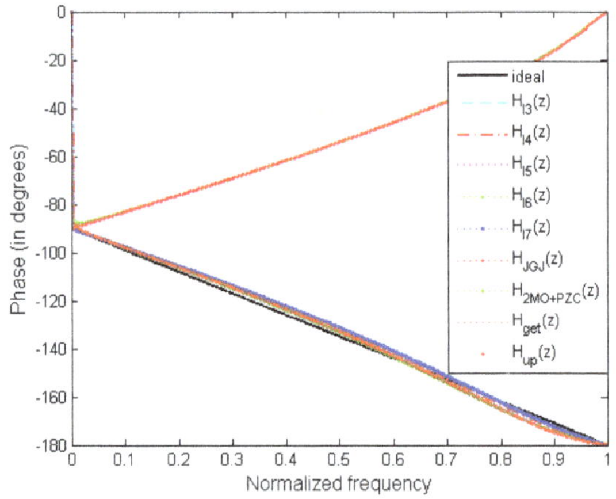

Fig. 7. Phase responses of proposed and the existing linear phase integrators with the ideal integrator.

$$H_{\mathrm{JGJ}}(z) = 0.0868\frac{z^2+0.9148z+0.5122}{z^2-0.4881z-0.5107}, \quad (20)$$

$$H_{\mathrm{2MO+PZC}}(z) = 0.8655\frac{z^2+0.6933z+0.0623}{z^2-0.4834z-0.5151}, \quad (21)$$

$$H_{\mathrm{get}}(z) = \frac{0.0901z^2+0.9216z+0.5429}{z^2-0.4445z-0.5543}, \quad (22)$$

$$H_{\mathrm{up}}(z) = 0.8642\frac{z^2+0.6848z+0.0577}{z^2-0.4930z-0.5070}. \quad (23)$$

And as a comparison benchmark for proposed recursive integrators with constant phase of −90°, Ngo [3] and Tseng [5] recursive digital integrators are considered. Their transfer functions are given in (24) and (25), respectively.

$$H_{\mathrm{Ngo}}(z) = \frac{9+19z^{-1}-5z^{-2}+z^{-3}}{24(1-z^{-1})}, \quad (24)$$

Integrators	Max δ	max MRE
$H_{I3}(z)$	3.6089	0.0273
$H_{I4}(z)$	3.4775	0.0433
$H_{I5}(z)$	3.9078	0.0341
$H_{I6}(z)$	3.7995	0.0398
$H_{I7}(z)$	3.8579	0.0580
$H_{\mathrm{JGJ}}(z)$	38.4682	0.2153
$H_{\mathrm{2MO+PZC}}(z)$	45.0071	0.2909
$H_{\mathrm{get}}(z)$	37.6768	0.2082
$H_{\mathrm{up}}(z)$	10.4348	0.0149

Tab. 4. Max Phase Deviation (in degrees) and Max MRE of linear phase integrators.

$$H_{\mathrm{Tseng}}(z) = z\frac{-3693+67260z^{-1}+88650z^{-2}-14388z^{-3}+2139z^{-4}}{139968(1-z^{-1})}. \quad (25)$$

Figures 6 and 7 show the magnitude and phase responses of proposed linear phase integrators $H_{I3\text{-}I7}(z)$, $H_{\mathrm{JGJ}}(z)$, $H_{\mathrm{2MO+PZC}}(z)$, $H_{\mathrm{get}}(z)$, $H_{\mathrm{up}}(z)$ digital integrator with the

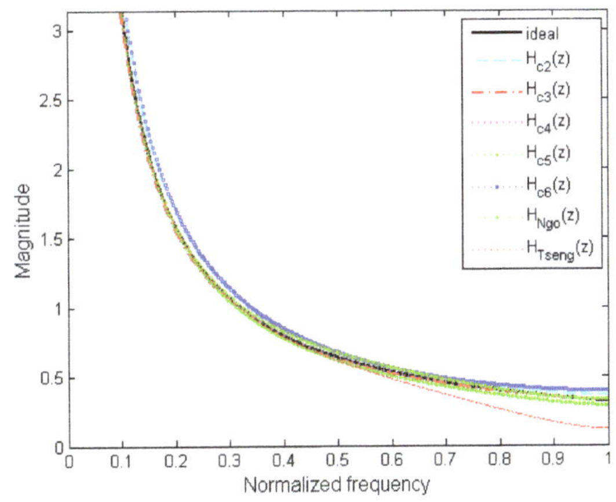

Fig. 8. Magnitude responses of proposed and the existing −90° phase integrators with the ideal integrator.

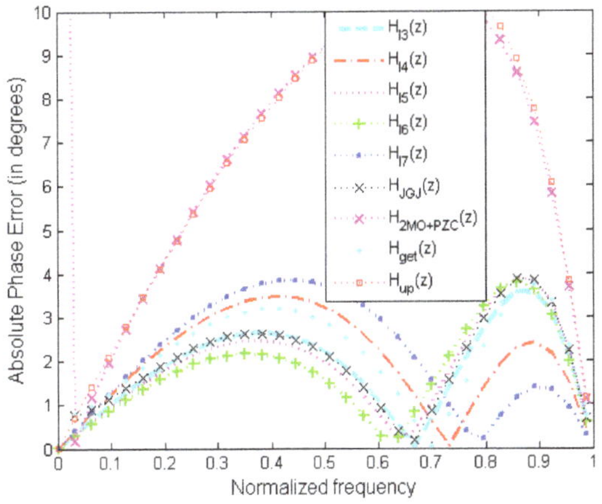

Fig. 11. Relative Phase Errors of proposed and the existing linear phase integrators.

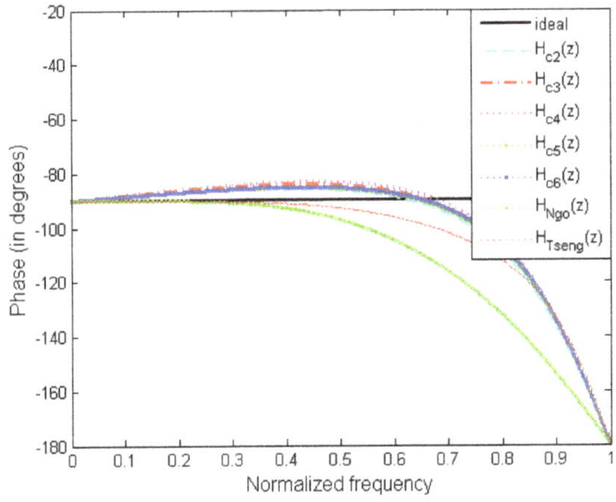

Fig. 9. Phase responses of proposed and the existing −90° phase integrators with the ideal integrator.

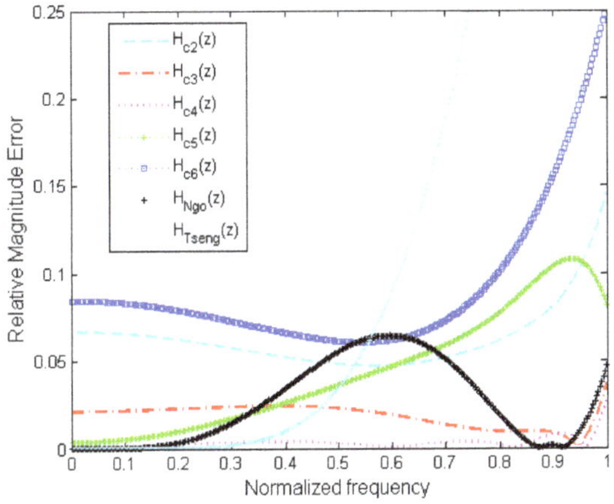

Fig. 12. Relative Magnitude Errors of proposed and the existing −90° phase integrators.

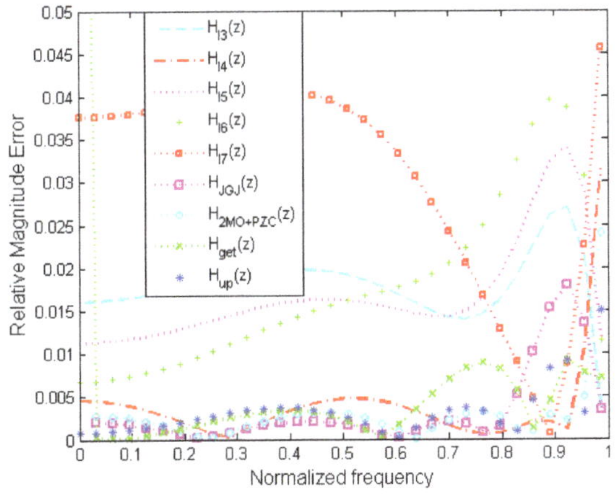

Fig. 10. Relative Magnitude Errors of proposed and the existing linear phase integrators.

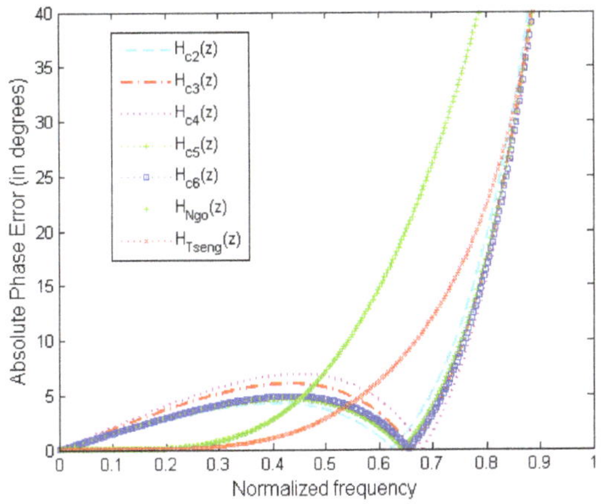

Fig. 13. Relative Phase Errors of proposed and the existing −90° phase integrators.

ideal one for the complete Nyquist frequency range, respectively. Further, Figures 8 and 9 show the magnitude and phase responses of proposed integrators with constant phase of −90° $H_{c2\text{–}c6}(z)$, Ngo [3] and Tseng [5] digital integrator with the ideal one for the complete Nyquist frequency range, respectively.

For proposed and existing linear phase integrators, δ and MMRE, over full Nyquist band, are given in Tab. 4 and that for proposed and existing integrators with constant phase of −90°, for $0 \leq \omega \leq 0.71\pi$ and $0 \leq \omega \leq 0.76\pi$ are given in Tab. 5.

Integrators	Max $\delta(\omega_c = 0.71\pi)$	Max $\delta(\omega_c = 0.76\pi)$	Max MRE
$H_{c2}(z)$	5.9806	12.1286	0.1460
$H_{c3}(z)$	6.0427	10.0212	0.0367
$H_{c4}(z)$	6.8929	8.3278	0.0279
$H_{c5}(z)$	4.6879	10.5810	0.1082
$H_{c6}(z)$	4.9046	9.6289	0.2514
$H_{\text{Ngo}}(z)$	27.0425	34.8119	0.0643
$H_{\text{Tseng}}(z)$	12.5687	17.1973	0.6180

Tab. 5. Max Phase Deviation and Max MRE of Integrators with constant phase of -90 degree.

Figures 10 and 11 show plots of the variation of magnitude error function and phase error function with frequency for the proposed linear phase integrators $H_{l3-l7}(z)$, $H_{\text{JGJ}}(z)$, $H_{\text{2MO+PZC}}(z)$, $H_{\text{get}}(z)$, $H_{\text{up}}(z)$ digital integrators

Figures 12 and 13 show plots of the variation of magnitude error function and phase error function with frequency for integrators with constant phase of $-90°$ designed using our method $H_{c2-c6}(z)$, Ngo and Tseng integrator.

From Figures 10 and 11, it is observed that all the proposed linear phase integrator designs $H_{l3-l7}(z)$ have an MMRE not more than 0.06 and maximum phase deviation less than 4° over wideband. All these designs clearly outperform any of the existing integrators in phase response over complete Nyquist range. $H_{\text{JGJ}}(z)$, $H_{\text{2MO+PZC}}(z)$, $H_{\text{get}}(z)$ and $H_{\text{up}}(z)$ have a maximum phase deviation of 38.4682°, 45.0071°, 37.6768° and 10.4348° for $0 \le \omega \le \pi$ and an MMRE of 0.2153, 0.2909, 0.2082 and 0.0149, respectively.

$H_{l3}(z)$ has an MMRE of 0.0273 over wideband, better than the existing or proposed integrator designs. It exhibits a maximum phase deviation of just 3.6089° over complete Nyquist range. $H_{l4}(z)$ exhibits an MMRE of 0.0433 and a maximum phase deviation of 3.4775°. $H_{l5}(z)$ and $H_{l6}(z)$ have an MMRE < 0.02 for $0 \le \omega \le 0.8\pi$ and an MMRE of 0.0341 and 0.0398 over wideband, respectively. They exhibit a maximum phase deviation of 3.9078° and 3.7995° over complete Nyquist band, respectively. With an MMRE of 0.0580 over wideband, $H_{l7}(z)$ has relatively low magnitude error for $0.7\pi \le \omega \le 0.95\pi$. It exhibits maximum phase deviation of 3.8579° over wideband and outperforms all proposed and existing integrator designs for $0.7\pi \le \omega \le \pi$ in phase response.

With reference to Figures 12 and 13, $H_{\text{Ngo}}(z)$ and $H_{\text{Tseng}}(z)$ perfectly approximate ideal phase response of $-90°$ for ω upto 0.3π after which their phase deviation drastically increases for the rest of the Nyquist band. $H_{\text{Tseng}}(z)$ exhibits excellent magnitude response upto $\omega = 0.3\pi$ but then fails to maintain this for the rest of the Nyquist frequency range. $H_{\text{Ngo}}(z)$, however, has an MMRE of 0.0643 over wideband. All proposed designs $H_{c2-c6}(z)$ have MMRE better than that of $H_{\text{Tseng}}(z)$ over wideband.

$H_{c4}(z)$ outperforms existing and proposed integrators in magnitude response with an MMRE of just 0.0279 over wideband. It is followed by $H_{c3}(z)$ which has a reasonably well magnitude response for full Nyquist band with an MMRE of 0.0367. These designs, however, do not stand best in phase response but manage to maintain $\delta < 10°$ for $0 \le \omega \le 0.77\pi$ and $0 \le \omega \le 0.76\pi$ respectively. These designs could be termed as magnitude-dominant designs.

$H_{c5}(z)$ and $H_{c2}(z)$ out stand existing and other proposed integrator designs in phase response for $0 \le \omega \le 0.71\pi$ with $\delta < 5°$ up to $\omega = 0.71\pi$ and δ approximately equal to 5° upto $\omega = 0.70\pi$, respectively. $H_{c5}(z)$ outperforms $H_{\text{Ngo}}(z)$ for $0 \le \omega \le 0.69\pi$ in magnitude response and exhibits an MMRE of 0.1082 over wideband. $H_{c2}(z)$ lacks in magnitude response with an MMRE of 0.1460 for complete Nyquist band. These designs could be called phase optimal designs.

Amongst proposed integrator designs, $H_{c6}(z)$ exhibits worst magnitude response over wideband and has an MMRE less than 0.1 up to $\omega = 0.8\pi$ and an MMRE of 0.2514 over full Nyquist band. But it is interesting to note that it has $\delta < 5°$ for $0 \le \omega \le 0.7\pi$ and $\delta < 10°$ for $0 \le \omega \le 0.76\pi$. This can be considered as best trade off between δ and ω_c among all other proposed designs which either have $\delta < 5°$ upto $\omega = 0.7\pi$ and $\delta < 10°$ for $\omega < 0.76\pi$ or have $\delta < 10°$ for $\omega = 0.76\pi$ and $\delta < 5°$ for $\omega < 0.7\pi$. It exhibits a unique bandwidth optimality which none of the other design does.

Extensive simulations have been carried out to verify the system's response of proposed integrators but the actual characteristics of the system, in hardware implementation, is governed by the bit-resolution of the processor. The systems coefficients are required to be converted in binary for processing. The coefficients which are multiple of 5 could be represented in binary by finite number of bits while others require infinite number of bits for their binary representation, hence undergoing digital word length effect, owing to which, systems response might worsen. Now a days, 64-bit processors are commonly used for real-time applications. In such a scenario, impact of digital word length effect of coefficients on system's response will be minimal. The poles of all the proposed integrators have been constraint to be located at $z = 1$, which ensures the response of the proposed integrators match the ideal one near the origin ($\omega = 0$).

7. Conclusion

Methodology for the design of second order linear phase integrators and ones with constant phase of $-90°$ is presented.

Proposed integrators $H_{I3-I7}(z)$ show an excellent phase linearity with $\delta < 4°$ over full Nyquist band while simultaneously maintaining low magnitude error. Constant phase of $-90°$ is achieved for proposed integrators $H_{c2-c6}(z)$ for $0 \leq \omega \leq 0.73\pi$ and they exhibit considerably low magnitude error over wideband. Trade-off among magnitude error, phase deviation and passband cut-off frequency is analysed and results show attractive real-time signal processing applications of proposed integrators as per the requirement of accuracy and bandwidth.

The performances of the proposed integrators are compared with few other state of the art integrators and proven of considerable high accuracy. As a future scope of the research, impact of digital word length of the coefficients on the system's response of proposed integrators could be studied further.

References

[1] KUMAR, B., CHAOUDHURY, D. R., KUMAR, A. On the design of linear phase, FIR integrators for midband frequencies. *IEEE Transactions on Signal Processing*, Oct. 1996, vol. 44, no. 10, p. 2391–2395. DOI: 10.1109/78.539024

[2] AL-ALAOUI, M. Novel digital integrator and differentiator. *Electronics Letters*, Feb. 1993, vol. 29, no. 4, p. 376–378. DOI: 10.1049/el:19930253

[3] NGO, N. Q. A new approach for the design of wideband digital integrator and differentiator. *IEEE Transactions on Circuits and Systems II: Express Briefs*, Sep. 2006, vol. 53, no. 9, p. 936–940. DOI: 10.1109/TCSII.2006.881806

[4] TSENG, C. C. Closed-form design of digital IIR integrators using numerical integration rules and fractional sample delays. *IEEE Transactions on Circuits and Systems I: Regular Papers*, Mar. 2007, vol. 54, no. 3, p. 643–655. DOI: 10.1109/TCSI.2006.887641

[5] TSENG, C. C., LEE, S. L. Digital IIR integrator design using Richardson extrapolation and fractional delay. *IEEE Transactions on Circuits and Systems I: Regular Papers*, Sep. 2008, vol. 55, no. 8, p. 2300-2309. DOI: 10.1109/TCSI.2008.920099

[6] TSENG, C. C. Digital integrator design using Simpson rule and fractional delay filter. textitIEE Proceedings - Vision, Image, and Signal Processing, Feb. 2006, vol. 153, no. 1, p. 79–86. DOI: 10.1049/ip-vis:20045208

[7] TSENG, C. C., LEE, S. L. Design of digital IIR integrator using discrete Hartley transform interpolation method. In *Proceedings of the 2009 IEEE International Symposium on Circuits and Systems*. Taipei, May 2009, p. 2181–2184. DOI: 10.1109/ISCAS.2009.5118229

[8] PAPAMARKOS, N., CHAMZAS, C. A new approach for the design of digital integrators. *IEEE Transactions on Circuits System Part I: Fundamental Theory and Applications*, 1996, vol. 43, no. 9, p. 785-791, DOI: 10.1109/81.536749

[9] AL-ALAOUI, M. A. Class of digital integrators and differentiators. *IET Signal Processing*, 2011, vol. 5, no. 2, p. 251–260. DOI: 10.1049/iet-spr.2010.0107

[10] UPADHYAY, D. K., SINGH, R. K. Recursive wideband digital differentiator and integrator. *Electronics Letters*, 2011, vol. 47, no. 11, p. 647–648. DOI: 10.1049/el.2011.0420

[11] JIANG, A., KWAN, H. K. Minimax design of IIR digital filters using SDP relaxation technique. *IEEE Transactions on Circuits and Systems I: Regular Papers*, 2010, vol. 57, no. 2, p. 378–390. DOI: 10.1109/TCSI.2009.2023770

[12] LAI, X. P., LIN, Z. P. Minimax design of IIR digital filters using a sequential constrained least-squares method. *IEEE Transactions on Signal Processing*, 2010, vol. 58, no. 7, p. 3901–3906. DOI: 10.1109/TSP.2010.2046899

[13] ABABNEH, J. I., BATAINEH, M. H. Linear phase FIR filter design using particle swarm optimization and genetic algorithms. *Digital Signal Processing*, Jul. 2008, vol. 18, no. 4, p. 657–668. DOI: 10.1016/j.dsp.2007.05.011

[14] UPADHYAY, D. K. Recursive wideband digital differentiators. *Electronics Letters*, Dec. 2010, vol. 46, no. 25, p. 1661–1662. DOI: 10.1049/el.2010.2113

[15] UPADHYAY, D. K. Class of recursive wideband digital differentiators and integrators. *Radioengineering*, Sep. 2012, vol. 21, no. 3, p. 904–910. ISSN: 1805-9600

[16] JAIN, M., GUPTA, M., JAIN, N. Linear phase second order recursive digital integrators and differentiators. *Radioengineering*, Jun. 2012, vol. 21, no. 2, p. 712–717. ISSN: 1805-9600

[17] JAIN, M., GUPTA, M., JAIN, N. Analysis and design of digital IIR integrators and differentiators using minimax and pole, zero, and constant optimization methods. *Hindawi Publishing Corporation ISRN Electronics*, vol. 2013, p. 1-14. DOI: 10.1155/2013/493973

[18] GUPTA, M., RELAN, B., YADAV, R., et al. Wideband digital integrators and differentiators designed using particle swarm optimisation. *IET Signal Processiing*, 2014, vol. 8, no. 6, p. 668–679. DOI: 10.1049/iet-spr.2013.0011

[19] JALLOUL, M. K., AL-ALAOUI, M. A. Design of recursive digital integrators and differentiators using particle swarm optimization. *International Journal of Circuit Theory and Applications*, Jul. 2015, vol. 44, no. 5, p. 948–967. DOI: 10.1002/cta.2115

[20] UPADHYAY, D. K. Recursive wideband linear phase digital differentiators and integrators. In *Proceedings of the International Conference on Computing Communication Control and Automation (ICCUBEA)*. 2015, p. 927–931. DOI: 10.1109/ICCUBEA.2015.184

[21] AGGARWAL, A., RAWAT, T. K., KUMAR, M., et al. Efficient design of digital FIR differentiator using L_1-method. *Radioengineering*, Jun. 2016, vol. 25, no. 2, p. 383–389. DOI: 10.13164/re.2016.0383

[22] AL-ALAOUI, M. A. Using fractional delay to control the magnitude and phases of integrators and differentiators. *IET Signal Processing*, Jun. 2007, vol. 1, no. 2, p. 107–119. DOI: 10.1049/iet-spr:20060246

About the Authors . . .

Kriti GARG received her Bachelor of Engineering in Electronics and Communication Engineering from Panjab University, Chandigarh in 2014 and Master of Technology in Signal Processing from Netaji Subhas Institute of Technology, University of Delhi, New Delhi in 2016. She is currently working as Design Engineer at NXP Semiconductors, Noida. Her research interests include Digital Signal Processing, Signal System, Circuit System and Digital Design.

Dharmendra K. UPADHYAY holds Bachelor of Engineering in Electronics and Communication Engineering from Kumaon University, Nainital and Master of Technology in Communication and Information Systems from Aligarh Muslim

University, Aligarh, U.P. He obtained Ph.D. from Uttarakhand Technical University, Dehradun, Uttarakhand. He is currently working as Professor in Division of Electronics and Communication Engineering, Netaji Subhas Institute of Technology, University of Delhi, New Delhi. His research interests include Digital Signal Processing, Microwave Filter and Antenna Designs. He has published several papers in various International Journals including IET, Electronics Letters. He is a life member of ISTE. He is also the reviewer of several reputed International Journals namely IEEE Transactions on Circuits and Systems-I; Circuits and Systems-II; Signal Processing; Circuits Systems and Signal Processing-Springer, ETRI Journal etc.

17

Wi-Fi Influence on LTE Downlink Data and Control Channel Performance in Shared Frequency Bands

Jiri MILOS, Ladislav POLAK, Stanislav HANUS, Tomas KRATOCHVIL

Dept. of Radio Engineering, Brno University of Technology, Technická 3082/12, 616 00 Brno, Czech Republic

{milos, polakl, hanus, kratot}@feec.vutbr.cz

Abstract. *Nowadays, providers of wireless services try to find appropriate ways to increase user data throughput mainly for future 5G cellular networks. Utilizing the unlicensed spectrum (ISM bands) for such purpose is a promising solution: unlicensed frequency bands can be used as a complementary data pipeline for UMTS LTE (Universal Mobile Telecommunication System - Long Term Evolution) and its advanced version LTE-Advanced, especially in pico- or femtocells. However, coexisting LTE and WLAN services in shared ISM bands at the same time can suffer unwanted performance degradation. This paper focuses predominantly on co-channel coexistence issues (worst case) between LTE and WLAN (IEEE 802.11n) services in the ISM band. From the viewpoint of novelty, the main outcomes of this article are follows. Firstly, an appropriate signal processing approach for coexisting signals with different features in the baseband is proposed. It is applied in advanced link-layer simulators and its correctness is verified by various simulations. Secondly, the influence of IEEE 802.11n on LTE data and control channel performance is explored. Performance evaluation is based on error rate curves, depending on Signal-to-Interference ratio (SIR). Presented results allow for better understanding the influence of IEEE 802.11n on the LTE downlink physical control channels (PCCH) and are valuable for mobile infrastructure vendors and operators to optimize system parameters.*

Keywords

LTE, WLAN, LTE physical channels, coexistence, interference, ISM band, 5G

1. Introduction

The increasing demand high mobile data rates and the growing number of user equipments (e.g. Internet-of-Things services) brings forth the question of how to improve the performance or extend the functionality of existing 3G/4G cellular networks, mainly in small cells. The licensed spectrum is the first choice for mobile operators thanks to its predictable behavior ensuring Quality of Service (QoS), mobility and system control. Naturally, the amount of available licensed spectrum is limited. Some former television bands have been sold to mobile operators in vendue. These frequency bands are almost fully occupied, especially in Europe [1–3]. Thus, there is a need for further expansion or a different solution.

Currently, the companies Qualcomm and Huawei are driving innovation to transfer Long Term Evolution (LTE) technology and its advanced version (LTE-A) to Industry-Science-Medical (ISM) unlicensed bands [4–6]. This concept is also called LTE-Unlicensed (LTE-U). Such innovation has been planned as a complementary or supporting data pipeline in small cells where demands on user data are higher. Both above mentioned companies have utilized the 2.4 GHz and 5 GHz ISM bands for LTE/LTE-A and take advantage of respective signal propagation possibilities. The considered frequency bands are used especially for Wireless and Personal Local Area Network (WLAN and WPAN). Some mobile operators are building picocells and utilizing Wireless-Fidelity (Wi-Fi) networks in the 2.4 and 5 GHz radio frequency (RF) bands as a supporting data pipeline in city centers or places with high density of user equipments [7], [8]. These Wi-Fi networks are controlled centrally by mobile operators. Locally, they have the potential to provide higher data throughput than 3G/4G small cells.

Several recent papers deal with the study of coexistence of LTE and WLAN/WPAN networks in ISM bands [9]. Abinader et al. in [10] presented average user throughput results for coexisting LTE and Wi-Fi in indoor environment in the 5.8 GHz unlicensed band and introduced a generalized collaborative coexistence algorithm. The presented results show that the average LTE user throughput value is similar to the case of non-coexistence scenarios and contrasts with degraded Wi-Fi average user throughput (for high Access Point (AP) density). Cavalcante in [11] provided a coexistence analysis of LTE and Wi-Fi in the 900 MHz RF band under TGah indoor environment for an LTE system with system bandwidth of 20 MHz. Other studies [12–15] explored possible LTE average user throughput and performance decreasing under different LTE and Wi-Fi coexistence scenarios, mainly in 5 GHz unlicensed bands. Appropriate simulation tools and methods for study of LTE and WLAN coexistence play a key role. A simulation-based study of the effect of LTE interference on an IEEE 802.11a system has been briefly discussed

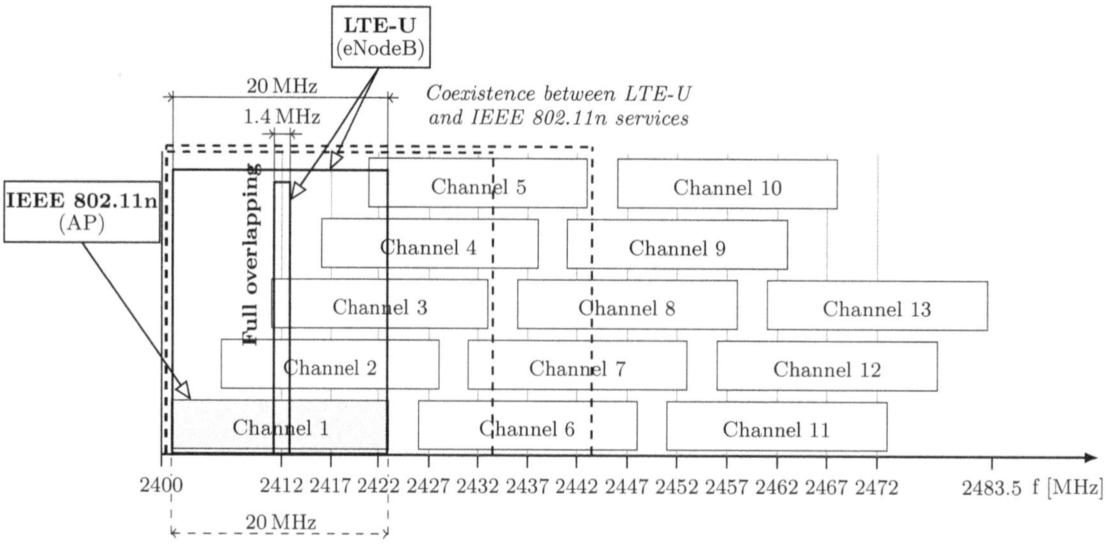

Fig. 1. Graphical representation of LTE-Unlicensed (1.4 and 20 MHz system bandwidth marked by blue and red color, respectively) and Wi-Fi (IEEE 802.11n, system bandwidth 20 MHz) services in frequency domain, full overlapping in ISM 2.4 GHz band.

in [16]. Comprehensive evaluation of the Licensed-Assisted Access (LAA) LTE and IEEE 802.11 coexistence scenarios based on system-level simulations were presented in [17]. In [18] the discrete-event Established Network Simulator (ns-3) has been extended to study LTE and Wi-Fi coexistence in 5 GHz band.

From this state-of-the-art review is evident that performance degradation of LTE and Wi-Fi communication systems caused by their coexistence on link-level has not been investigated in details. Hence, in this work we present an approach to analyse LTE including LTE downlink physical control channel (PCCH) performance at the link level influenced by Wi-Fi (based on IEEE 802.11n) in the same ISM frequency bands. To the best of our knowledge, no similar exploration in this form has been presented yet. The proposed methodology is appropriate especially for worst-case coexistence scenarios.

The main contributions of the article are follows:

1) An appropriate approach to emulate and evaluate coexistence between LTE (downlink) and Wi-Fi (IEEE 802.11n) at the link-level in ISM bands is proposed.
2) Based on LTE and Wi-Fi co-channel coexistence evaluation, general conclusions for non-critical and critical coexistence scenarios are formulated.

The remaining part of this paper is composed as follows. The considered co-channel coexistence scenario between LTE and Wi-Fi is outlined in Section 2. Furthermore, Section 2 presents the proposed signal processing of coexisting LTE and Wi-Fi signals and its implementation to simulators. Section 3 states and discusses the simulated performance results of LTE (downlink) under the presented coexistence scenario. Finally, the main outputs of the work are clearly summarized in Conclusion.

2. Coexistence Scenario

Integration with the licensed spectrum, minimum changes in LTE air-interface and ensuring coexistence in unlicensed bands [19] are general design principles and prerequisites for LTE-U. Aggregation of licensed and unlicensed carriers brings better network performance, longer range and increasing capacity. Furthermore, unification of the LTE network with common authentication, security and management is an advantage. There are two main approaches for LTE-U. First one is supplemental downlink (SDL) which increases throughput only in downlink (main option for LTE Frequency Division Duplex (FDD)) whereas second one is carrier aggregation (CA) which increases throughput in both downlink and uplink (an option for LTE Time Division Duplex (TDD)). According to Huawei and its concept for unlicensed secondary carrier design for FDD [19], the following option is adopted: the primary cell (Pcell) FDD in downlink (user data+control information) is transmitted in the licensed band and the secondary cell (Scell) FDD in downlink (user data+control information) is transmitted in the unlicensed band. In this work, we only consider SDL due to the major use of LTE-FDD in the European region.

Currently, the 2.4 GHz and 5 GHz ISM bands are utilized by WLAN (Wi-Fi) and WPAN (e.g. Bluetooth, ZigBee) technologies. Wi-Fi, built on IEEE 802.11n and IEEE 802.11ac standards, is the dominant system in the 2.4 GHz and 5 GHz ISM bands, respectively [20]. Hence, modeling, measurement and suppression of interferences from mutual coexistence between WLAN/WPAN and LTE in ISM bands is becoming a key issue in future 5G networks. In Fig. 1, we can see the graphical representation of LTE-U and Wi-Fi services overlapping in the 2.4 GHz ISM band. All possible WLAN channels with system bandwidth $BW_{sys}^{\text{Wi-Fi}} = 20\,\text{MHz}$ are depicted with marked central frequencies.

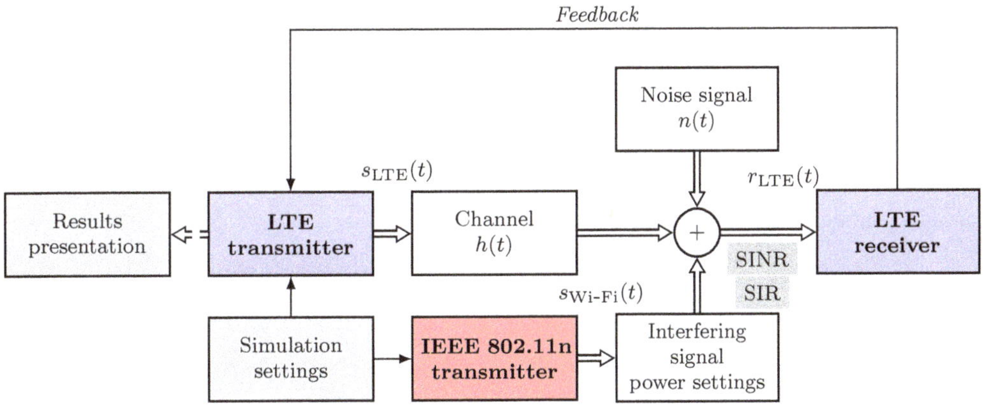

Fig. 2. Block diagram of the proposed LTE vs. IEEE 802.11n coexistence link-level simulator.

Band [GHz]	**Frequency range** [MHz]	**Bandwidth** [MHz]
2.4	2400–2483.5	83.5
5.1	5150–5350	200
5.3	5350–5470	120
5.4	5470–5825	355
5.8	5725–5875	150
5.9	5850–5925	75

Tab. 1. Summary of worldwide available 2.4 GHz and 5 GHz ISM frequency bands.

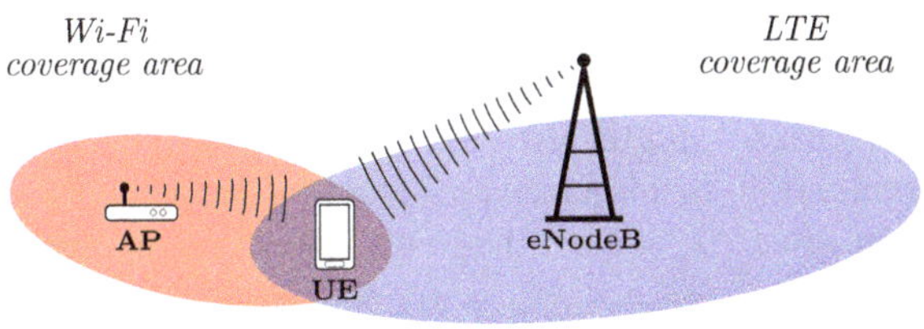

Fig. 3. Graphical representation of the LTE and Wi-Fi coexistence scenario in the same geographical area.

We consider that LTE services are provided on the first channel central frequency (f_c = 2412 MHz). LTE with $BW_{\mathrm{sys}}^{\mathrm{LTE}} = 1.4$ MHz is fully overlapped by Wi-Fi channels 1, 2 and 3 whereas, LTE with $BW_{\mathrm{sys}}^{\mathrm{LTE}} = 20$ MHz is fully overlapping with Wi-Fi channel 1 and partially overlapping with channels 2, 3, 4 and 5 (see Fig. 1). According to the described scenario, we consider only co-channel inter-system coexistence (CIC) which represents the worst-case scenario. In such scenario the RF spectrum of one communication system is completely overlapping with the RF spectrum of other communication system (central frequencies of both systems are the same). Consequently, mutual interference of the systems is the greatest. It is equivalent to overlapping of the baseband signals [25]. A list of available unlicensed frequency bands is summarized in Tab. 1.

2.1 Coexistence Link Level Simulator

A block diagram of the proposed LTE-U/WLAN link level coexistence analysis model in downlink direction is depicted in Fig. 2. The LTE downlink link level simulator, developed at TU Vienna, was adopted as the basic simulation tool [26], [27]. We extended this MATLAB-based simulator by adding physical downlink control channel models. Thus, the impact of interferences from coexsitence between LTE and Wi-Fi is explored for all LTE downlink physical control channels, namely: Physical Control Format Indicator Channel (PCFICH), Physical Downlink Control Channel (PDCCH) and Physical Hybrid ARQ Control Channel (PHICH) [28].

LTE control channels are used for signaling and transferring system information from the base station to individual user equipment (e.g. HARQ processing, user equipment power level settings, type of precoding) [29]. For emulation of the interfering Wi-Fi signal, we have proposed a universal WLAN link-level simulator supporting IEEE 802.11n technology. More details about this simulator can be found in [30].

Transceiver blocks in LTE downlink are marked with blue color and the interfering IEEE 802.11n system is marked with red color. The simulator was adjusted for simulations of inter-system coexistence. Motivated by processing time constraints, the simulator is implemented in baseband only, working with a complex envelope of the signal $s(t) = s_{\mathrm{I}}(t) + \mathrm{j}s_{\mathrm{Q}}(t)$, where s_{I} and s_{Q} are in-phase and quadrature components, respectively [31]. In this case, baseband signals in the frequency domain have a frequency spectra concentrated near zero frequency, which fits for the presented co-channel coexistence scenario. A simple graphical representation of the investigated LTE-U and Wi-Fi collision scenario in the same geographical area is shown in Fig. 3.

Coexistence Channel Model

The presented LTE vs. IEEE 802.11n coexistence analysis model is based on general coexistence model described in time domain by the following equation:

$$r_0(t) = \underbrace{s_0(t) * h_0(t)}_{\text{useful signal}} + \underbrace{n_0(t)}_{\text{noise}} + \underbrace{\sum_{k=1}^{N_{\mathrm{I}}} (s_k(t) * h_k(t))}_{\text{interfering transmitters}} \quad (1)$$

where $r_0(t)$ is the received useful signal, $s_0(t)$ is the transmitted useful signal, $h_0(t)$ is the impulse response of the useful channel (LTE branch), $*$ represents discrete convolution and $n_0(t)$ is an additive white Gaussian noise added to the investigated receiver input. Interfering branch $s_k(t)$ is modeled as a sum of the signals from N_{I} interfering transmitters. Each interfering signal is led through a fading channel described by channel impulse response.

According to the presented coexistence scenario (see Fig. 2), the number of interferers $N_{\mathrm{I}} = 1$ (IEEE 802.11n) and $h_1(t) = \delta(t)$ (delta function, no fading channel in the interfering branch). Thus we can rewrite (1) to

$$r_{\mathrm{LTE}}(t) = \underbrace{s_{\mathrm{LTE}}(t) * h(t)}_{\text{LTE signal}} + \underbrace{n(t)}_{\text{noise}} + \underbrace{s_{\text{Wi-Fi}}(t)}_{\text{Wi-Fi signal}} \qquad (2)$$

where $r_{\mathrm{LTE}}(t)$ is the received LTE signal, $s_{\mathrm{LTE}}(t)$ is the transmitted LTE signal and $s_{\text{Wi-Fi}}(t)$ is the interfering Wi-Fi signal. The channel block adjusts the LTE output signal $s_{\mathrm{LTE}}(t)$ in accordance with the used channel model and its impulse response $h(t)$. In this paper, no fading channel model is considered, thus $h(t) = \delta(t)$. Hence, (2) could be simplified to

$$r_{\mathrm{LTE}}(t) = s_{\mathrm{LTE}}(t) + n(t) + s_{\text{Wi-Fi}}(t). \qquad (3)$$

The output signal ($s_{\mathrm{LTE}}(t)$) from the LTE transmitter block (*victim*), is an OFDMA-based baseband signal in time domain with average power per symbol $\sigma^2_{s_{\mathrm{LTE}}(t)} = 1$. The output signal from IEEE 802.11n (Wi-Fi, *interferer*) transmitter is an OFDM-based baseband signal in time domain $s_{\text{Wi-Fi}}(t)$ with average power per symbol $\sigma^2_{s_{\text{Wi-Fi}}(t)} = 1$. After that, the $s_{\mathrm{LTE}}(t)$ signal passes through the channel model block, it enters the signal adder (see Fig. 2). Here, a random noise vector $n(t)$ and the interfering signal $s_{\text{Wi-Fi}}(t)$ are added to the useful signal $s_{\mathrm{LTE}}(t)$ according to the defined Signal-to-Interference plus Noise Ratio (SINR) and Signal-to-Interference Ratio (SIR), respectively. In the presented scenario, SINR for each sample is defined as

$$\mathrm{SINR} = \frac{P_{\mathrm{tx}}^{(\mathrm{LTE})}}{\sigma_n^2 + P_{\mathrm{tx}}^{(\text{Wi-Fi})}} \qquad (4)$$

where σ_n^2 is the average power of the noise signal $n(t)$ which is modeled as a vector of normally-distributed random values with zero mean $\mu_{n(t)} = 0$, $P_{\mathrm{tx}}^{(\mathrm{LTE})}$ is the power of LTE signal at the transmitter output and $P_{\mathrm{tx}}^{(\text{Wi-Fi})}$ is the power of Wi-Fi signal in transmitter output. In the coexistence scenario considered, $\sigma_n^2 = 0$; hence, the SINR from (4) simply leads to SIR

$$\mathrm{SIR} = \frac{P_{\mathrm{tx}}^{(\mathrm{LTE})}}{P_{\mathrm{tx}}^{(\text{Wi-Fi})}}. \qquad (5)$$

Both SINR and SIR are defined as the post-FFT ratio at the receiving antenna (after FFT operation in the LTE receiver) [26]. Each sample of the received signal r_{LTE} influenced by interfering singal is calculated as

$$r_{\mathrm{LTE}} = s_{\mathrm{LTE}} + \frac{1}{\sqrt{2}} \frac{N_{\mathrm{FFT}}}{N_{\mathrm{tot}}} s_{\text{Wi-Fi}} 10^{\left(\frac{-\mathrm{SIR}_{(\mathrm{dB})}}{20}\right)} \qquad (6)$$

where N_{FFT} is the FFT size used in LTE transmitter/receiver and N_{tot} is the total number of subcarriers in a single LTE OFDMA symbol. In case of non-zero noise signal average power σ_n^2, the received signal r_{LTE} definition is obtained by adding noise sample to (6), i.e.,

$$r_{\mathrm{LTE}} = s_{\mathrm{LTE}} + \frac{1}{\sqrt{2}} \frac{N_{\mathrm{FFT}}}{N_{\mathrm{tot}}} \left[s_{\text{Wi-Fi}} 10^{\left(\frac{-\mathrm{SIR}_{\mathrm{dB}}}{20}\right)} + n\, 10^{\left(\frac{-\mathrm{SNR}_{\mathrm{ch\,(dB)}}}{20}\right)} \right] \qquad (7)$$

where $\mathrm{SNR}_{\mathrm{ch}}$ is the SNR in considered AWGN channel model ($\mathrm{SNR}_{\mathrm{ch}} = P_{\mathrm{tx}}^{(\mathrm{LTE})}/\sigma_n^2$) with constant value in whole simulation. In case of added AWGN (7), SINR defined in (4) is an independent value in the coexistence model and numerically SINR $\approx$ SIR. As mentioned above, two baseband signals (LTE and Wi-Fi) are added due to the property of linearity considered for this scenario. Interpolation is not used since it has negligible influence to performance results [35].

The interfered LTE signal $r_{\mathrm{LTE}}(t)$ with additive noise enters the LTE receiver, where inverse OFDMA operations are performed. According to the used SISO transmission mode, the multi-antenna decoding process is not necessary. Due to using the AWGN channel model, an estimation of used transmission channel is not provided here, thus the LTE receiver has perfect knowledge of the channel. Furthermore, Soft-Sphere Decision (SSD) is used as the demodulation algorithm [26]. We also assume a static transmission scenario (no movement of LTE receiver).

In the presented LTE downlink simulator, the physical control channel (PCFICH, PHICH and PDCCH) signal processing at the receiving side is performed separately from the physical downlink shared channel (PDSCH) which carries user data [29]. Channel decoding and evaluation of the received information is performed. The received user data, control information, data acknowledgement information and Channel Quality Indicator (CQI) are reported to the LTE transmitter block. Of course, the influenced signal $r(t)$ could also lead to the IEEE 802.11n receiver where inverse OFDM operations are performed and the obtained results are evaluated [30].

Physical Downlink Shared Channel

Quality of reception of user data in LTE downlink transmitted via PDSCH depends mainly on quality of the respective transmission channel, used modulation and channel coding, the use of Hybrid-ARQ (HARQ) retransmissions and link adaptation algorithms and user data scheduling. The modulation and channel coding scheme used in PDSCH is determined from CQI parameter. The list of these parameters is presented in Tab. 2. The CQI index [26] defines the corresponding modulation type and code rate, including channel coding efficiency. Due to only having a single user in the simulation scenario, there is no demand for user data scheduling (user data occupies all available resource elements (REs)). In the presented simulations, CQI is always defined as fixed.

CQI index	Modulation	Code rate × 1024	Efficiency
1	QPSK	78	0.1523
2		120	0.2344
3		193	0.3770
4		308	0.6016
5		449	0.8770
6		602	1.1758
7	16-QAM	378	1.4766
8		490	1.9141
9		616	2.4063
10	64-QAM	466	2.7305
11		567	3.3223
12		666	3.9023
13		772	4.5234
14		873	5.1152
15		948	5.5547

Tab. 2. The list of LTE Channel Quality Indicator (CQI) parameters [26].

Physical Control Format Indicator Channel

The Control Format Indicator (CFI) parameter is transmitted via PCFICH. CFI contains two-bit value which defines mapping of PDCCH in LTE downlink subframe [28]. PCFICH uses block channel coding with code rate 1/16. PCFICH is always placed in the first OFDM symbol in downlink subframe. In the frequency domain, PCFICH is spread into four parts (additional frequency diversity) and its position is also given by cell identification number $N_{\text{cell}}^{\text{ID}}$.

Physical Downlink Control Channel

PDCCH is the most important control channel in LTE downlink since it carries system information aboout scheduling grants, MIMO settings, modulation and channel coding (CQI) etc. PDCCH symbols are located at the beginning of each subframe while the number of OFDM symbols for PDCCH is defined by CFI value [28]. PDCCH uses convolutional coding with code rate 1/3. At the receiver, the PDCCH is processed via combination of Blind and Viterbi decoding process.

Physical Hybrid-ARQ Indicator Channel

PHICH transmits the Hybrid-ARQ Indicator (HI) which contains acknowledge or non-acknowldge message of previous user data sent in uplink [29]. PHICH uses simple repetition channel coding with code rate 1/3. Repeated symbols are spread using up to 8 orthogonal sequences. Number of PHICH transmitted in single subframe depends mainly on system bandwidth and PHICH scaling factor N_g which defines number of PHICH groups in single subframe. PHICH is situated either in the first or the first and second OFDM symbol. Each PHICH group is also spread into three parts. Their position is given by $N_{\text{cell}}^{\text{ID}}$ as well [29].

3. Coexistence Analysis

LTE and Wi-Fi (IEEE 802.11n) general system parameters considered in the presented simulation scenario are sumarized in Tab. 3. For LTE we consider two extreme system bandwidths $BW_{\text{sys}}^{\text{LTE}} = 1.4\,\text{MHz}$ (narrowest) and 20 MHz (widest) and subcarrier spacing $\Delta_{\text{f}}^{\text{LTE}} = 15\,\text{kHz}$. The Wi-Fi system model is considered for system $BW_{\text{sys}}^{\text{Wi-Fi}} = 20\,\text{MHz}$ only and $\Delta_{\text{f}}^{\text{Wi-Fi}} = 312.5\,\text{kHz}$. Extended Wi-Fi system bandwidth of 40 MHz is not modeled.

System parameter			LTE		Wi-Fi
System bandwidth	BW_{sys}	[MHz]	1.4	20	20
Number of occupied subcarriers	N_{sc}	[-]	72	1200	56
IFFT/FFT size	N_{FFT}	[-]	128	2048	64
Subcarrier space	Δ_{f}	[kHz]	15	15	312.5

Tab. 3. General physical layer parameters of the LTE and Wi-Fi link-level models.

Simulation parameter		Value
Signal-to-Interference ratio (SIR) range		⟨−30, 30⟩ dB
Number of transmitted subframes		2000
Channel Quality Indicator range		⟨1, 15⟩
PDSCH:	Number of allocated users	1
	Scheduling algorithm	static and fixed
PHICH:	Group scaling factor	$N_g = \{1/6, 1/2, 1, 2\}$
	HI (users) per group	1
PCFICH:	CFI message range	⟨1, 4⟩ or ⟨0, 3⟩
PDCCH:	Transmitted DCI format	F0

Tab. 4. Simulation parameters used in the presented link-level coexistence scenario.

The number of occupied subcarriers equals to 56 according to the definition of the Greenfield Wi-Fi mode [20]. We consider single-user SISO (SU-SISO) transmission mode for both systems. For LTE it means a single base station (eNodeB) and a single user equipment (UE).

Before the coexistence simulation starts, it is necessary to define simulation parameters. Parameters common to the LTE/Wi-Fi trnansmitter and LTE receiver are listed in Tab. 4. The independent SIR variable should be set together with the number of transmitted LTE subframes ($N_{\text{subf}} = 2000$). The PDSCH performance is simulated for CQI in range from 1 to 15 (see Tab. 2) as a fixed value. PHICH error rate performance is simulated for all available scaling factor N_g values with single user per PHICH group. The performance results (BER, BLER, throughput) in all physical channels are computed by averaging over all their allocated resource elements.

3.1 Simulation Results

In this section, results of the LTE vs. Wi-Fi coexistence analysis are presented and discussed. The LTE PDSCH link performance represented as a dependence of Block Error Rate (BLER) on SIR is shown in Fig. 4. Results were obtained for various CQI values and system bandwidths of 1.4 and 20 MHz. As we can see from presented BLER results, LTE receiver with CQI value set in the range from 13 to 15 is unable to reach the required 10 % BLER and thus, these CQI are unusable for user data transmission under presented LTE/WLAN coexistence scenario. The BLER reference level $10^{-1} = 10\,\%$ BLER was determined according to the required target quality for LTE shared physical channel user data reception [21].

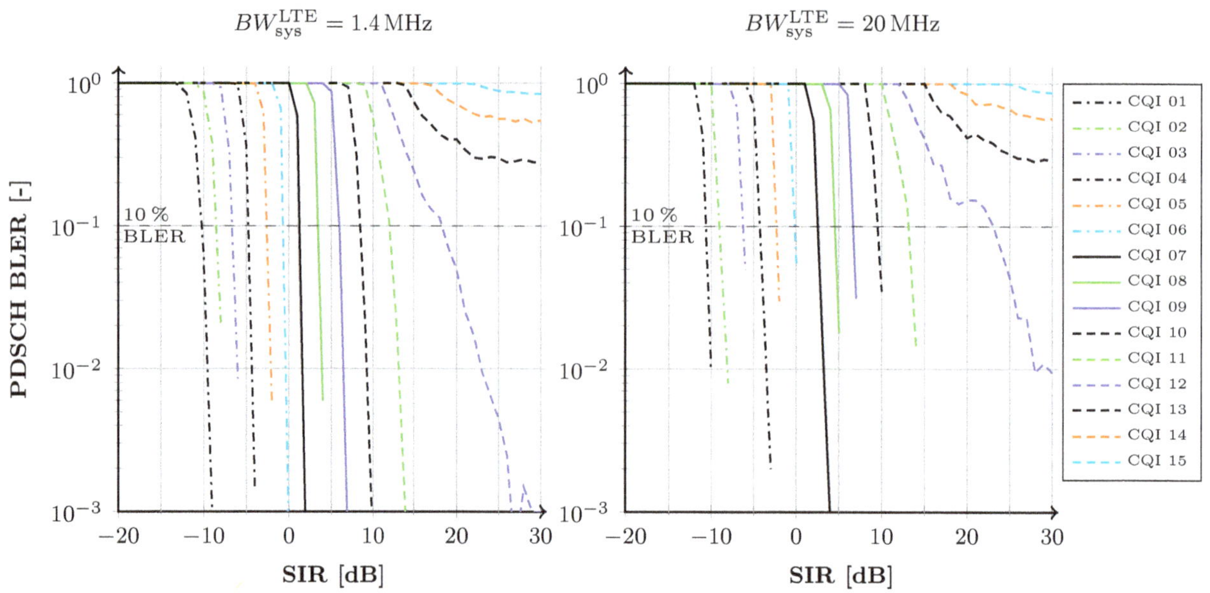

Fig. 4. LTE PDSCH Block error rate link performance for various CQI's under the Wi-Fi co-channel coexistence scenario ($BW_{\text{sys}}^{\text{LTE}} = 1.4\,\text{MHz}$ and 20 MHz, $BW_{\text{sys}}^{\text{Wi-Fi}} = 20\,\text{MHz}$).

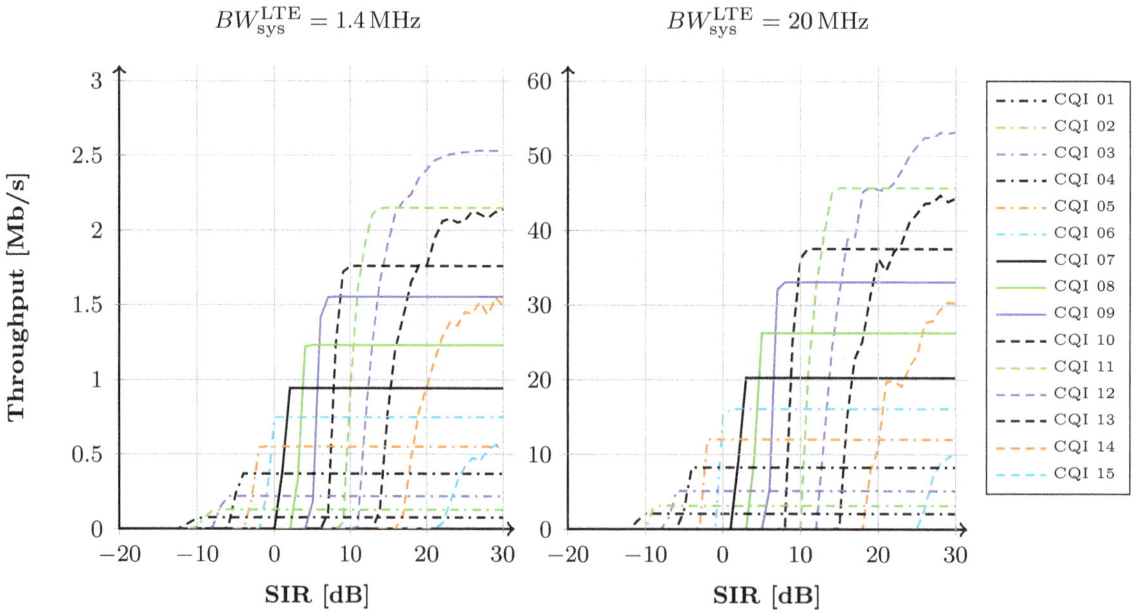

Fig. 5. LTE PDSCH user data throughput for various CQI's under the Wi-Fi co-channel coexistence scenario ($BW_{\text{sys}}^{\text{LTE}} = 1.4\,\text{MHz}$ on the left, $BW_{\text{sys}}^{\text{LTE}} = 20\,\text{MHz}$ on the right), $N_{\text{subf}} = 2000$, $BW_{\text{sys}}^{\text{Wi-Fi}} = 20\,\text{MHz}$.

LTE PDSCH throughput curves versus SIR for $BW_{\text{sys}}^{\text{LTE}} = 1.4\,\text{MHz}$ and 20 MHz are shown in Fig. 5. From the results it is seen that LTE throughput is measurable for SIR higher than −12 dB, independently of the considered LTE system bandwidth. Using of CQI = 13, 14 and 15 could bring unpredictable behavior. Using of CQI = 12 is possible but the throughput is slightly decreased compared to non-coexistence scenario results [27].

The LTE PDSCH to SIR mapping for 10 % BLER is shown in Fig. 6. These mapping curves are very important for modeling and simulation methodology. This mapping is termed effective SINR/SIR mapping (ESM). It serves as the basis for physical layer abstraction for link-to-system level mapping, including MIMO technology [22], [23]. The concept is also used in Vienna LTE System-level simulator for LTE intra-system interference evaluation [24]. The CQI to SIR or SINR mapping is approximately linear in non-coexistence scenarios due to equidistant points of intersection [24]. Here, the equidistance is not obtained for CQI using 64QAM (10, 11 and 12). As we can see from the figure, CQI = 13, 14 and 15 are not mapped to SIR (these curves do not reach 10 % BLER).

Figure 7 presents the BER performance results of the co-channel inter-system interference, depending on SIR, in the PCFICH and PDCCH physical control channel. PCFICH missdetection could bring packet loss and even short-term loss of connection. Hence, exploring of its performance is important. One percent PCFICH bit error rate (see Fig. 7 on the left) is reached for SIR = −14.1 dB ($BW_{\text{sys}}^{\text{LTE}}$ = 1.4 MHz) and −12.8 dB ($BW_{\text{sys}}^{\text{LTE}}$ = 20 MHz). PDCCH information is transmitted with one percent bit error at SIR = 0 dB (see Fig. 7 on the right). Evaluated results reveal the fact that the LTE system bandwidth used has negligible effect on the LTE PDCCH physical downlink control channel performance at its co-channel coexistence with IEEE 802.11n.

Finally, LTE PHICH control channel performances affected by co-channel Wi-Fi interferences was investigated and the obtained results described as BER vs. SIR dependence are shown in Fig. 8. The purpose PHICH channel in the downlink is to carry HARQ acknowledgements for uplink data transfers. PHICH performance requirements are defined by 3GPP [33]. PHICH BER (acknowledge to non-acknowledge error and vice versa) should be within the range from 10^{-3} to 10^{-4} (see gray filled area in Fig. 8). Firstly, PHICH performances were obtained for various scaling factors (N_g) at $BW_{\text{sys}}^{\text{LTE}}$ = 1.4 MHz. In such case (see blue curves in Fig. 8), parameter (N_g) has no influence on PHICH performance.

Mapping of LTE downlink physical control channels depends mainly on cell identification number ($N_{\text{cell}}^{\text{ID}}$), system bandwidth $BW_{\text{sys}}^{\text{LTE}}$, PHICH scaling factor N_g and amount of system information to transmit (PDCCH). Example of the LTE downlink physical control channels mapping (symbols 1 and 2) for the highest system bandwidth $BW_{\text{sys}}^{\text{LTE}}$ = 20 MHz, $N_{\text{cell}}^{\text{ID}}$ = 0 is depicted in Fig. 9. Here, the LTE spectrum is fully overlapping with Wi-Fi spectrum (IEEE 802.11n, $BW_{\text{sys}}^{\text{Wi-Fi}}$ = 20 MHz). The part of bandwidth used for user data and control information transmitting (BW_{used}) is usually less than the system bandwidth in OFDM-based systems. LTE with system bandwidth 20 MHz uses 1200 subcarriers in frequency domain and 1 DC subcarrier. Due to Δ_f^{LTE} = 15 kHz distance between subcarriers, the used bandwidth $BW_{\text{used}}^{\text{LTE}} = (1200 + 1) \times 15\,\text{kHz} = 18.015\,\text{MHz}$.

The Wi-Fi (IEEE 802.11n) with the same system bandwidth uses only 56 subcarriers and 1 DC subcarrier. The distance between subcarriers in Wi-Fi system is much higher ($\Delta_f^{\text{Wi-Fi}}$ = 312.5 kHz), thus $BW_{\text{used}}^{\text{Wi-Fi}} = (56+1)\times 312.5\,\text{kHz}$ = 17.8125 MHz. There is 98.88 % overlap in LTE spectrum. As we can see from Fig. 9, amount of PHICH influenced by Wi-Fi interference also depends on scaling factor N_g, where a specific part of PHICH is not fully overlapped by Wi-Fi spectrum (see magnified image).

The same dependences were investigated for $BW_{\text{sys}}^{\text{LTE}}$ = 20 MHz. At the widest LTE system bandwidth, there is a significant difference of PHICH performance for various N_g values. When $N_g = 1/6$ is considered then to fulfil BER range from 10^{-3} to 10^{-4} the coding gain for $BW_{\text{sys}}^{\text{LTE}}$ = 20 MHz is approximately 11 dB less in comparison with $BW_{\text{sys}}^{\text{LTE}}$ = 1.4 MHz. Such behaviour is given by different PHICH mapping and different amounts of interfered resource elements in the resource grid (see gray filled areas in Fig. 9).

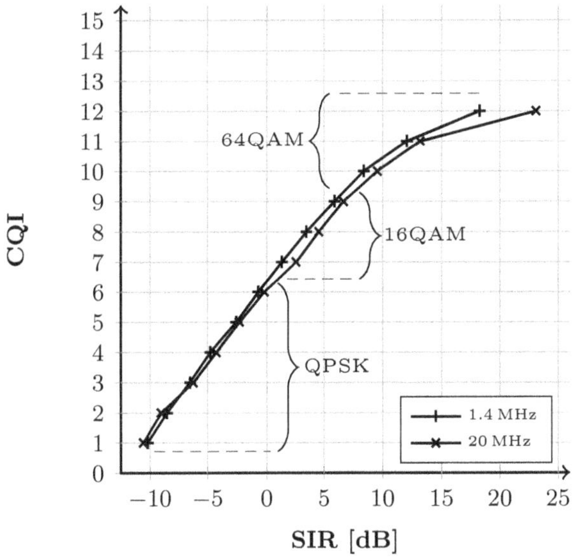

Fig. 6. LTE PDSCH CQI to SIR ESM mapping under the Wi-Fi co-channel coexistence scenario.

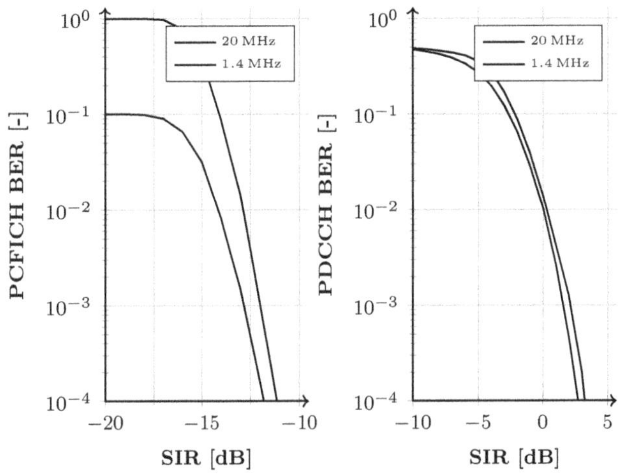

Fig. 7. LTE PCFICH and PDCCH control channel performance under the Wi-Fi co-channel coexistence scenario (PCFICH on the left, PDCCH on the right).

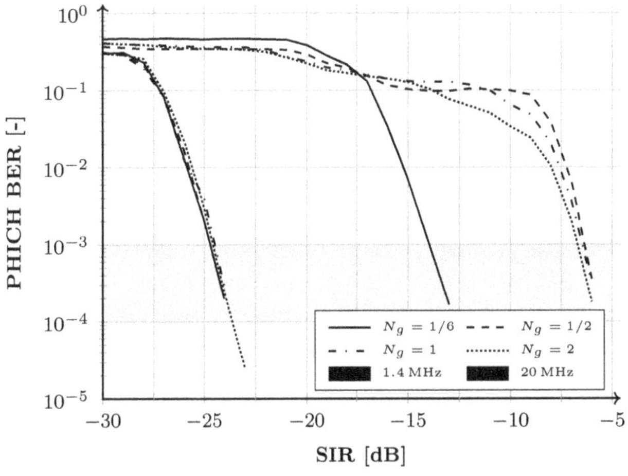

Fig. 8. LTE PHICH control channel performance for various scaling factors N_g under the Wi-Fi co-channel coexistence scenario ($BW_{\text{sys}}^{\text{LTE}}$ = 1.4 MHz blue, $BW_{\text{sys}}^{\text{LTE}}$ = 20 MHz red).

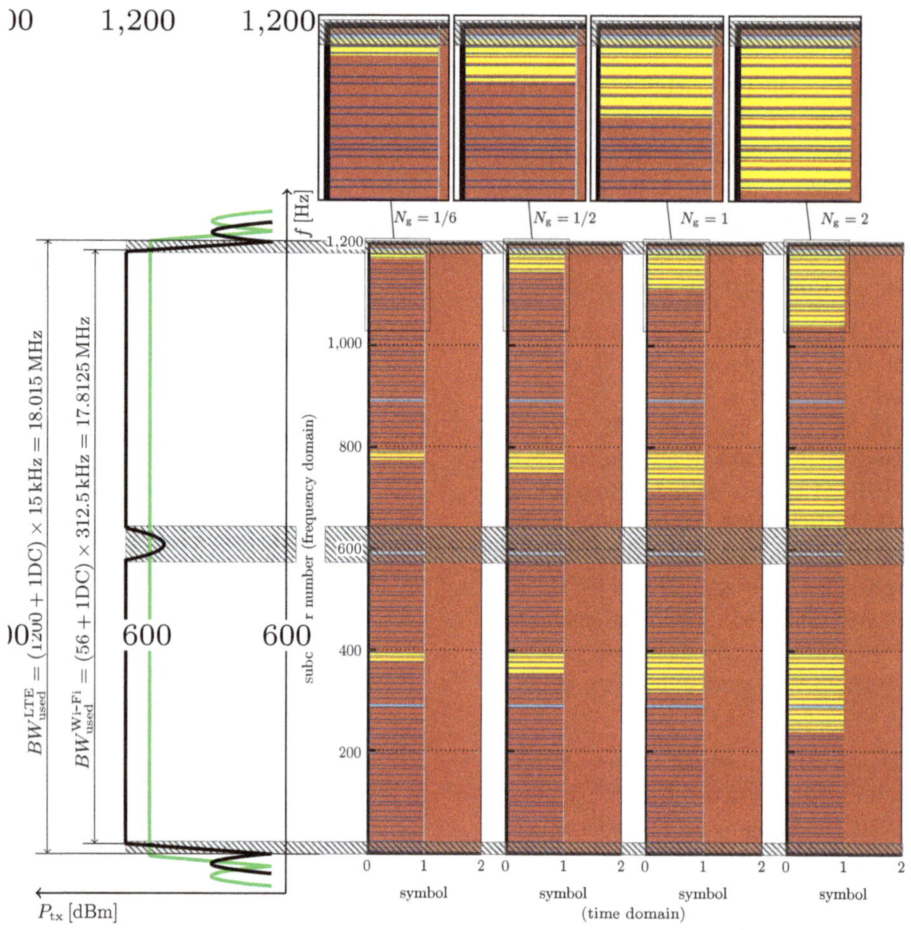

Fig. 9. LTE physical control channel mapping in resource grid $BW_{sys}^{LTE} = 20\,MHz$, $N_{cell}^{ID} = 0$ (example case) and scaling factor N_g (red = PDCCH, blue = PCFICH, yellow = PHICH) and LTE spectrum (green) fully overlapped by a Wi-Fi spectrum (black, IEEE 802.11n, $BW_{sys}^{Wi\text{-}Fi} = 20\,MHz$) in frequency domain (on the left). Magnified part of the physical control channels mapping is on the top.

Overall, BER PHICH curves reach the reference value 10^{-3} at SIR = −24.6 dB for $BW_{sys}^{LTE} = 1.4\,MHz$, SIR = −14 dB for $BW_{sys}^{LTE} = 20\,MHz$, $N_g = 1/6$ and SIR = −6.6 dB for $BW_{sys}^{LTE} = 20\,MHz$, $N_g \neq 1/6$. From these values (lower SIR) is evident that LTE PHICH downlink channel has good resistance features against interferences from coexistence with Wi-Fi.

4. Conclusion

This paper introduces a novel approach in analysing LTE link-level coexistence issues in shared frequency bands. It presents the results of LTE data and control channel performance under co-channel inter-system coexistence with WLAN (Wi-Fi) services. For the analysis, the LTE-U vs. Wi-Fi coexistence link level simulator was created. The presented simulator could provide a basic tool for research in the field of controlled heterogenous networks and cooperative algorithms. The analysis of the obtained results from the presented co-channel coexistence scenario leads to the following general conclusions:

1. LTE transmission is robust and resistant against interference from Wi-Fi services (IEEE 802.11n) in shared frequency bands.
2. User data transmitted via LTE PDSCH are well protected for the link with lower CQI index (from 1 to 11). PDSCH link with CQI = 12 operates properly for SIR greater than 15 dB. Using the PDSCH link with CQI higher than 12 is not suitable for LTE vs. Wi-Fi (IEEE 802.11n) co-channel coexistence scenarios.
3. LTE PDSCH CQI to SIR ESM mapping is linear in the range of CQI from 1 to 9 and non-linear in the range of CQI from 10 to 12 (CQI = 13, 14 and 15 are unusable for transmission in the LTE vs. WLAN coexistence scenario).
4. Scalable bandwidth in LTE has inconsiderable impact on the PDCCH physical control channel performance (convolutional channel coding with code rate 1/3). The performance of PCFICH is more influenced by scalable bandwidth, where $\Delta SIR \approx 1.8\,dB$ at PCFICH BER equaling to 10^{-2} (block channel coding with code rate 1/16).
5. For the presented scenario, the scaling factor N_g highly affects PHICH control channel performance mainly for LTE system bandwidth 20 MHz. The PHICH BER results for $N_g = 1/6$ show a gain 7 dB (at PHICH BER 10^{-3}) comparing to other N_g values.

Coexistence methodology and results presented in this paper enable to better understand the LTE performance and reliability at the same frequency bands and location area with WLAN. This paper defines conditions and preliminary recommendations for LTE and Wi-Fi operation of uncontrolled LTE-U networks. The results could be valuable for developers, vendors and mobile operators to optimize system parameters.

Acknowledgments

This work is supported by CATRENE under the project named CORTIF CA116, the MEYS of the Czech Republic no. LF14033 and by the BUT project no. FEKT-S-14-2177 (PEKOS). The described research was performed in laboratories supported by the SIX project; no. CZ.1.05/2.1.00/03.0072, the operational program Research and Development for Innovation. The research described in this paper was financed by the Czech Ministry of Education in the frame of the National Sustainability Program under grant LO1401. For the research, the infrastructure of the SIX Center was used.

References

[1] FUENTES, M., GARCIA-PARDO, C., GARRO, E., et al. Coexistence of digital terrestrial television and next generation cellular networks in the 700 MHz band. *IEEE Wireless Communications*, 2014, vol. 21, no. 6, p. 63–69. DOI: 10.1109/MWC.2014.7000973

[2] POLAK, L., KALLER, O., KLOZAR, L., et al. Coexistence between DVB-T/T2 and LTE standards in common frequency bands. *Wireless Personal Communications*, 2016, vol. 88, no. 3, p. 669–684. DOI: 10.1007/s11277-016-3191-2

[3] POLAK, L., KLOZAR, L., KALLER, O., et al. Study of coexistence between indoor LTE femtocell and outdoor-to-indoor DVB-T2-Lite reception in a shared frequency band. *EURASIP Journal on Wireless Communications and Networking*, 2015, no. 114, p. 1–14. DOI: 10.1186/s13638-015-0338-x

[4] HUAWEI TECHNOLOGIES CO., LTD., CHINA. *The Unlicensed Spectrum Usage for Future IMT Technologies (white paper)*. 18 pages. [Online] Cited 2015-05-15. Available at: http://www.huawei.com/

[5] QUALCOMM TECHNOLOGIES, INC., USA. *Extending the benefits of LTE Advanced to unlicensed spectrum (white paper)*. 19 pages. [Online] Cited 2015-05-15. Available at: http://www.qualcomm.com/

[6] QUALCOMM TECHNOLOGIES, INC., USA. *LTE in Unlicensed Spectrum: Harmonious Coexistence with Wi-Fi (white paper)*. 19 pages. [Online] Cited 2015-05-15. Available at: http://www.qualcomm.com/

[7] FUJITSU NETWORK COMMUNICATIONS INC., TEXAS, USA. *High-Capacity Indoor Wireless Solutions: Picocell or Femtocell (white paper)*. 10 pages. [Online] Cited 2015-05-15. Available at: http://www.us.fujitsu.com/telecom/

[8] 3rd GENERATION PARTNERSHIP PROJECT, TECHNICAL SPECIFICATION GROUP SERVICES AND SYSTEM ASPECTS, FRANCE *Architecture enhancements for non-3GPP accesses (technical specification)*. 131 pages. [Online] Cited 2015-05-16. Available at: http://www.3gpp.org/DynaReport/23402.htm

[9] JOENGHO, J., HUANING, N., LI, Q. C., et al. LTE in the unlicensed spectrum: Evaluating coexistence mechanisms. In *Proceedings of the Globecom Workshops*. 2014, p. 740–745. DOI: 10.1109/GLOCOMW.2014.7063521

[10] ABINADER, F. M., ALMEIDA, E. P. L., CHAVES, F. S., et al. Enabling the coexistence of LTE and Wi-Fi in unlicensed bands. *IEEE Communications Magazine*, 2014, vol. 52, no. 11, p. 54–61. DOI: 10.1109/MCOM.2014.6957143

[11] CAVALCANTE, A. M., ALMEIDA, E., VIEIRA, R. D., et al. Performance evaluation of LTE and Wi-Fi coexistence in unlicensed bands. In *Proceedings of the 77th IEEE Vehicular Technology Conference (VTC Spring)*. Jun. 2013, p. 1–6. DOI: 10.1109/VTCSpring.2013.6692702

[12] RONGRONG, S., XINGLIN, W. Analysis and test for co-site of LTE and WiFi system. In *Proceedings of the IEEE International RF and Microwave Conference (RFM)*. Dec. 2011, p. 315–319. DOI: 10.1109/RFM.2011.6168757

[13] BHORKAR, A., IBARS, C., PINGPING, Z. On the throughput analysis of LTE and WiFi in unlicensed band. In *Proceedings of the 48th Asilomar Conference on Signals, Systems and Computers*. Nov. 2014, p. 1309–1313. DOI: 10.1109/ACSSC.2014.7094671

[14] AL-DULAIMI, A., AL-RUBAYE, S., QIANG, N., et al. 5G communications race: Pursuit of more capacity triggers LTE in unlicensed band. *IEEE Vehicular Technology Magazine*, Mar. 2015, vol. 10, no. 1, p. 43–51. DOI: 10.1109/MVT.2014.2380631

[15] JOENGHO, J., LI, Q. C., HUANING, N., et al. LTE in the unlicensed spectrum: A novel coexistence analysis with WLAN systems. In *Proceedings of the IEEE Global Communications Conference (GLOBECOM)*. Dec. 2014, p. 3489–3464. DOI: 10.1109/GLOBCOM.2014.7037343

[16] MA, Y., KUESTER, D. G., CODER, J., et al. A simulation study of the LTE interference on WiFi signal detection. In *Proceedings of the URSI National Radio Science Meeting (URSI NRSM)*. Boulder (USA), 2016, p. 1–2. DOI: 10.1109/USNC-URSI-NRSM.2016.7436233

[17] MUKHERJEE, A., CHENG, J. F., FALAHATI, S., et al. System architecture and coexistence evaluation of licensed-assisted access LTE with IEEE 802.11. In *Proceedings of the IEEE International Conference on Communications Workshop (ICCW)*. London (United Kingdom), 2015, p. 2350–2355. DOI: 10.1109/ICCW.2015.7247532

[18] GIUPPONI, L., HENDERSON, T., BOJOVIC, B., et al. *Simulating LTE and Wi-Fi Coexistence in Unlicensed Spectrum with ns-3*. 12 pages. [Online] Cited 2016-07-25. Available at: www.arxiv.org/abs/1604.06826

[19] HUAWEI TECHNOLOGIES CO., LTD., CHINA. *U-LTE: Unlicensed Spectrum Utilization of LTE (white paper)*. 20 pages. [Online] Cited 2015-05-15. Available at: http://www.huawei.com/

[20] IEEE COMPUTER SOCIETY STD., USA. *802.11n, Part 11: Wireless LAN Medium Access Control (MAC) and Physical Layer Specifications*. 536 pages. [Online] Cited 2015-05-30. Available at: http://standards.ieee.org/getieee802/download/802.11n-2009.pdf

[21] CABAN, S., RUPP, M., MEHLFÜHRER, C., et al. *Evaluation of HSDPA and LTE: From Testbed Measurements to System Level Performance*. 1st ed. Wiley, 2011. ISBN: 9780470711927

[22] BRUENINGHAUS, K., ASTELY, D., SALZER, T., et al. Link performance models for system level simulations of broadband radio access systems. In *Proceedings of the 16th IEEE International Symposium on Personal, Indoor and Mobile Radio Communications*. Sep. 2005, p. 2306–2311. DOI: 10.1109/PIMRC.2005.1651855

[23] IEEE TGm. *IEEE 802.16m Evaluation Methodology Document (EMD): Evaluation Methodology for P802.16m-Advanced Air Interface*. 162 pages. [Online] Cited 2016-09-26. Available at: http://ieee802.org/16/tgm/docs/80216m-08_004r2.pdf

[24] IKUNO, J. C., WRULICH, M., RUPP, M. System level simulation of LTE networks. In *Proceedings of the 71st IEEE Vehicular Technology Conference (VTC 2010-Spring)*. May 2010, p. 1–5. DOI: 10.1109/VETECS.2010.5494007

[25] GLEISSNER, F., HANUS, S. Co-channel and adjacent channel interference measurement of UMTS and GSM/EDGE systems in 900 MHz radio band. *Radioengineering*, 2008, vol. 17, no. 3, p. 74–80. ISSN: 1805-9600

[26] MEHLFÜHRER, C., WRULICH, M., IKUNO, J. C., et al. Simulating the Long Term Evolution physical layer. In *Proceedings of the 17th European Signal Processing Conference (EUSIPCO 2009)*. Glasgow, 2009, p. 1471-1478. ISBN: 978-161-7388-76-7

[27] VIENNA UNIVERSITY OF TECHNOLOGY, AUSTRIA. *LTE Downlink Link Level Simulator*. [Online] Cited 2015-05-29. Available at: http://www.nt.tuwien.ac.at/research/mobile-communications/vienna-lte-a-simulators/

[28] MILOS, J., HANUS, S. Performance analysis of PCFICH and PDCCH LTE control channels. *Radioengineering*, 2014, vol. 23, no. 1, p. 445–451. ISSN: 1805-9600

[29] MILOS, J., HANUS, S. Analysis of LTE physical Hybrid-ARQ control channel. *Advances in Electrical and Computer Engineering*, 2014, vol. 14, no. 2, p. 97–100. DOI: 10.4316/AECE.2014.02016

[30] MILOS, J., POLAK, L., SLANINA, M., et al. Link-level simulator for WLAN networks. In *Proceedings of the 1st International Workshop on Link- and System Level Simulations (IWSLS^2)*. Vienna (Austria), 2016, p. 1–4.

[31] TRANTER, W. H. *Principles of Communication Systems Simulation with Wireless Applications*. Upper Saddle River (USA): Prentice Hall, 2004. ISBN: 0-13-494790-8

[32] MILOS, J., POLAK, L., SLANINA, M., et al. Measurement setup for evaluation the coexistence between LTE downlink and WLAN networks. In *Proceedings of the 10th International Symposium on Communication Systems, Networks and Digital Signal Processing (CSNDSP)*. Prague (Czech Republic), 2016, p. 1–4.

[33] 3rd GENERATION PARTNERSHIP PROJECT. *Evolved Universal Terrestrial Radio Access (E-UTRA) and Evolved Universal Terrestrial Radio Access Network (E-UTRAN); Overall Description; Stage 2. Ver. 8.12.0 TS 36.300*. 147 pages. [Online] Cited 2015-06-01. Available at: www.3gpp.org/ftp/Specs/archive/36_series/36.300/36300-8c0.zip

[34] LTE-U Forum. *LTE-U Forum: Coexistence Study for LTE-U SDL (technical paper V1.0)*. 52 pages. [Online] Cited 2016-07-26. Available at: www.lteuforum.org/documents.html

[35] KUDER, Z., MILOS, J., HANUS, S. Radio coexistence of major and upcoming wireless standards in the ISM bands. In *Proceedings of the 26th IEEE International Conference Radioelektronika*. Košice (Slovakia), 2016, p. 1–4. DOI: 10.1109/RADIOELEK.2016.7477426

[36] GOLDSMITH, A. *Wireless Communications*. 1st ed. London (UK): Cambridge University Press, 2005. ISBN: 0521837162

About the Authors . . .

Jiří MILOŠ was born in Uherské Hradiště, the Czech Republic in 1986. He received his M.Sc. and Ph.D. degrees at the Faculty of Electrical Engineering and Communications from the Brno University of Technology. His research interests are modeling, simulation and measurement of wireless and cellular communication technologies.

Ladislav POLÁK was born in Štúrovo, Slovakia in 1984. He received his M.Sc. degree in 2009 and Ph.D. degree in 2013, both in Electronics and Communication from the Brno University of Technology (BUT), the Czech Republic. Currently he is an assistant professor at the Department of Radio Electronic (DREL), BUT. His research interests are Digital Video Broadcasting (DVB) standards, wireless communication systems, simulation and measurement of the coexistence between wireless communication systems, signal processing, video image quality evaluation and design of subjective video quality methodologies. He has been an IEEE member since 2010.

Stanislav HANUS was born in Brno, the Czech Republic, in 1950. He received his M.Sc. and Ph.D. degrees from the Brno University of Technology. He is Professor at the Department of Radio Electronics, Faculty of Electrical Engineering and Communication in Brno. His research is concentrated on Mobile Communications and Television Technology.

Tomáš KRATOCHVÍL was born in Brno, Czech Republic, in 1976. He received the M.Sc. degree in 1999, Ph.D. degree in 2006 and Assoc. Prof. position in 2009, all in Electronics and Communications from the Brno University of Technology. He is currently a full professor at the Department of Radio Electronics, Brno University of Technology. His research interests include digital television and audio broadcasting, its standardization and video and multimedia transmission including video image quality evaluation. He has been an IEEE member since 2001 and senior member since 2016.

18

Waveform Agile Sensing Approach for Tracking Benchmark in the Presence of ECM using IMMPDAF

Gnane Swarnadh SATAPATHI, Srihari PATHIPATI

Dept. of Electronics and Communication Engineering, National Institute of Technology Karnataka, Mangalore, 575 025 Karnataka, India

ec13f02@nitk.edu.in, srihari@nitk.ac.in

Abstract. *This paper presents an efficient approach based on waveform agile sensing, to enhance the performance of benchmark target tracking in the presence of strong interference. The waveform agile sensing library consists of different waveforms such as linear frequency modulation (LFM), Gaussian frequency modulation (GFM) and stepped frequency modulation (SFM) waveforms. Improved performance is accomplished through a waveform agile sensing technique. In this method, the selection of waveform to be transmitted at each scan is determined, by jointly computing ambiguity function of waveform and Cramer-Rao Lower Bound (CRLB) matrix of measurement errors. Electronic counter measures (ECM) comprises of stand-off jammer (SOJ) and self-screening jammer (SSJ). Interacting multiple model probability data association filter (IMMPDAF) is employed for tracking benchmark trajectories. Experimental results demonstrate that, waveform agile sensing approach require only* 39.98 *percent lower mean average power compared to earlier studies. Further, it is observed that the position and velocity root mean square error values are decreasing as the number of waveforms are increasing from 5 to 50.*

Keywords

Clutter, electronic countermeasures, root mean square error, target tracking

1. Introduction

The chief objective of target tracking is to increase the probability of detection (i.e. to detect and track true targets) and reduce false alarms. The presence of strong interference sources (ECM, clutter, false alarm (FA) and multipath) significantly degrade the performance of radar target tracking. SSJ and SOJ are regarded as effective ECM techniques. In SOJ, the enemy aircrafts are present outside the surveillance region of the radar. It radiates high power jamming signals into the radar main lobe or side lobe, to misguide the enemy target, which is entering into the surveillance region. In SSJ, the jammer itself is present on the enemy target and sends erroneous signals to the radar. The main objective of both these ECM techniques is to corrupt radar measurements, and hence deceive the radar to track false targets. In addition to ECM, clutter and multipath in the environment collectively increases the complexity of radar to detect and track the true targets. Therefore, there is a strong need to adapt radar parameters to get improved observations, which will increase the performance of tracking in the presence of strong interference.

Adaptive waveform selection is considered as an important electronic counter counter measure (ECCM) for tracking targets in the presence of ECM [1]. Varying waveforms from scan to scan can give significant information about true targets and helps to locate them accurately by improving radar measurements. The main objective here, is to enhance the tracking performance by building waveform library with multiple waveforms and select the waveform to be transmitted based on prediction of next state of the target.

Benchmark problem with six standard target trajectories has been proposed in [2]. Interacting Multiple Model (IMM) adaptive estimator was presented as an efficient target tracking algorithm for maneuvering targets [3]. Significant further research was conducted on tracking benchmark trajectories in the presence of ECM [4–7]. In all these, SOJ and range gate pull off (RGPO) were viewed as major ECM techniques. An important contribution for benchmark target trajectories has been reported [1], [8] and applied IMMPDAF as a track filter for all six benchmark trajectories in the presence of ECM. A number of potential research problems that have to be carried out in future were recommended in [8]. Significant among them are adaptive waveform selection using various radar waveforms, tracking closely spaced targets, considering on-board jammer, including background clutter and incorporating multipath effects.

Alternatively, interacting multiple model/ multiple hypothesis tracker (IMM/ MHT) solution to the radar benchmark problem was proposed in [9], which requires less radar resources than [8]. But IMM/MHT was computationally complex when it was applied to practical situations.

Benchmark problem with IMMPDAF was further extended by incorporating background clutter with fixed LFM waveform [10]. A new set of algorithms for radar management were introduced in [11], [12], which presented post detection integration techniques for benchmark problem to jointly reduce radar energy and to improve accuracy of target measurements in the presence of ECM. In all the above proposed algorithms, only single LFM waveform was used for benchmark tracking.

Waveform agile sensing algorithms were developed to select a particular waveform from a bank of waveforms, which aims to maximize the probability of detection and minimize the mean square error. A novel approach was recommended using adaptive waveform selection for linear target tracking using Kalman filter in clutter free environment and was extended to include clutter [13], [14]. Different optimization techniques for waveform design were explored in [15–17]. Improved version of tracking non-linear model by using dynamic waveform selection was suggested in [18]. Generalized frequency modulated waveforms for non-linear scenario were presented in [19], [20]. Both these approaches yielded better results for various scenarios. Adaptive waveform selection was recommended for multi static radars using IMMPDAF model in [21].

This paper presents waveform agile sensing based approach to track benchmark trajectories in the presence of strong interference. Current work focuses on improving tracking performance by constructing a bank of radar waveforms (LFM, GFM and SFM) up to 50, including SSJ, SOJ, false alarm and incorporating multipath effects with background clutter.

The paper is organized as follows. Section 2 presents problem formulation. Waveform agile sensing for neutralizing ECM and clutter is developed in Sec. 3. Section 4 discusses IMMPDAF algorithm. Simulation results and discussion are included in Sec. 5. Conclusion and future work has been incorporated in Sec. 6.

2. Problem Formulation

In the proposed waveform agile sensing approach, a phased array radar (10 GHz operating frequency and a rectangular array with 900 elements) with minimum variance distortion less response (MVDR) adaptive beamformer has been used to obtain the measurements. The transmit waveform is selected based on CRLB of scan-to-scan measurement errors from a bank of frequency coded radar waveforms (LFM, GFM and SFM) up to 50. A point target is assumed and closely spaced targets are ignored. Cell Averaging-Constant False Alarm Rate (CA-CFAR) adaptive thresholding technique is employed. Various electronic counter measures are included such as SOJ and SSJ. In addition, background clutter, multipath effects and false alarms are also incorporated. IMMPDAF is applied to track all the six benchmark trajectories in this complex scenario. The work flow of entire simulation process is briefly illustrated in Fig. 1.

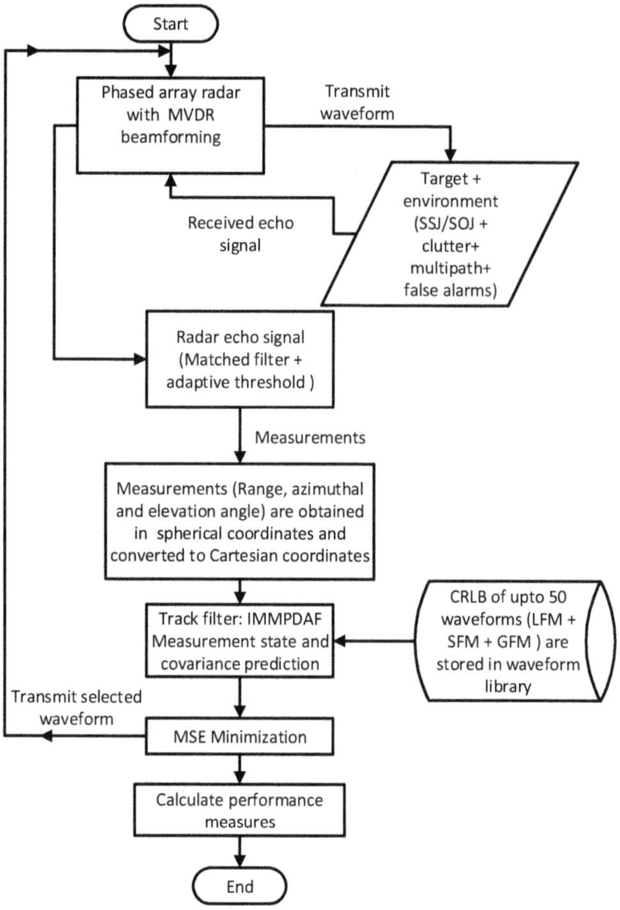

Fig. 1. Flowchart of entire simulation process.

2.1 Measurement Model

The measurements received from the radar are in spherical coordinates (range, azimuthal angle and elevation angle). Phased array radar scans the entire region in both azimuth and elevation direction simultaneously. The measurements from spherical coordinates are converted in to Cartesian coordinates with reference to radar position. The measurements from the radar is given by

$$Z_i = [Z_{ix}, Z_{iy}, Z_{iz}] \tag{1}$$

where Z_{ix}, Z_{iy} and Z_{iz} are the positions in x, y & z directions respectively of ith element of the measurement vector with respect to origin of the radar. The environment consists combination of clutter, ECM, false alarm and multipath. Hence, the resultant measurements obtained from the radar is

$$Z = Z_{\text{target}} + Z_{\text{clutter}} + Z_{\text{jammer}} + Z_{\text{false alarm}} + Z_{\text{multipath}} \tag{2}$$

where Z_{target}, Z_{clutter}, Z_{jammer}, $Z_{\text{false alarm}}$, and $Z_{\text{multipath}}$ are measurements due to target, clutter, jamming, false alarm and multipath at the radar receiver respectively.

2.2 Jammers

The main task of jammer is to hamper the radar functionality [22]. Jammers send high power radio frequency signals to fool the radar throughout its operating bandwidth.

These jammers can be on-board or with an escort to the enemy target. Generally there are two types of jammers namely self-screening jammer (SSJ) and stand-off jammer (SOJ). The noise generated by jammer is measured in terms of effective radiated power (ERP) and is formulated as

$$\mathrm{ERP} = \frac{G_J P_J}{L_J} \tag{3}$$

where G_J is gain in jammer antenna, P_J is jammer transmitted power and L_J is total loss in jammer. The following subsections describe briefly about SSJ and SOJ.

2.2.1 Self-Screening Jammer

Self screening jammers (SSJs) are well known as self-protection jammers and are positioned on-board the target. They take advantage in the vicinity of radar surveillance area and send noise echoes to the radar main beam so as to crack the lock of radar. Signal to Jamming ratio (S/J) for self screening jammer case [22] is given as

$$\frac{S}{J} = \frac{P_t \tau G \sigma B_J}{\mathrm{ERP} \cdot 4\pi R^2 L} \tag{4}$$

where P_t is peak transmit power, G is antenna gain, τ is radar pulse width, σ is radar cross section, B_J is Jammer bandwidth, R is range, B_r is receiver bandwidth, L is receiver loss and ERP represents effective radiated power. Generally, it is considered that jammer power is greater than signal power transmitted by the radar i.e. $S/J < 1$. Yet, when the target approaches radar, at a certain range signal power of radar will be equal to jamming power and this range is called as cross-over range. Beyond this cross-over range the jammer power is ineffective and is given by

$$(R_{co})_{SSJ} = \left[\frac{P_t G \sigma B_J}{4\pi B_r L \cdot \mathrm{ERP}}\right]^{1/2}. \tag{5}$$

2.2.2 Stand-Off Jammer

Stand off jammers (SOJs) transmits noise signals from lethal range of the radar so as not to detect the enemy targets that are entering into the radar surveillance area. Signal to Jamming ratio for stand-off jammer is given by

$$\frac{S}{J} = \frac{P_t \tau G^2 R_J^2 \sigma B_J}{4\pi \cdot \mathrm{ERP} \cdot R^4 L} \tag{6}$$

and cross-over range where jamming power is equal to signal power is

$$(R_{co})_{SOJ} = \left[\frac{P_t G^2 R_J^2 \sigma B_J G_{pc}}{4\pi \cdot \mathrm{ERP} \cdot G' B_r L}\right]^{1/4} \tag{7}$$

where R_J is range of jammer from the radar and G_{pc} represents time bandwidth product. The target has to stay beyond this $(R_{co})_{SOJ}$ range, and try to send spurious signals to the radar for improper detection of the targets.

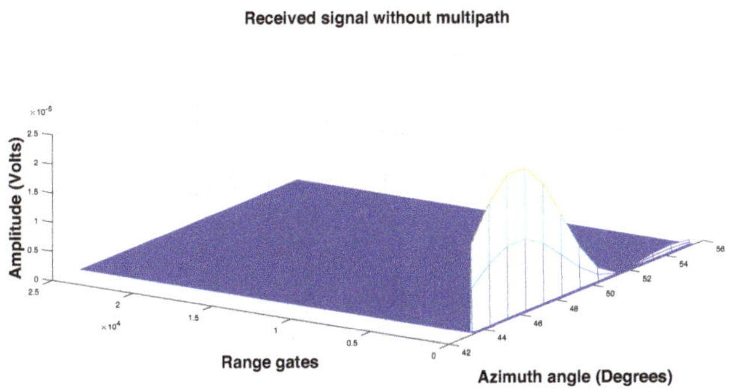

(**a**) Received signal without multipath.

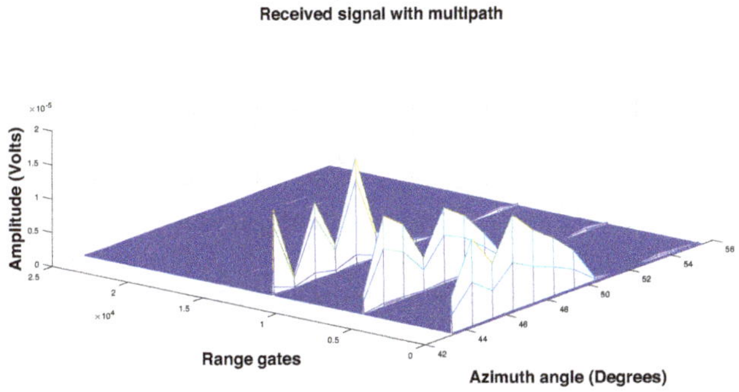

(**b**) Received signal with multipath.

Fig. 2. Received signal with and without multipath.

2.3 Clutter Model

Let $Z_{k,1}, Z_{k,2} \ldots Z_{k,n}$ be the measurements obtained from radar for a particular instant of time k. These measurements consists of false alarms with true measurements of the target due to the presence of clutter. If ρ is considered as clutter density and V_k as validation gate volume then false alarms are assumed to be Poisson distributed with mean ρV_k. The observations from radar, which will fall in the validation region are only considered for tracking. Poisson probability for attaining n false alarms is

$$\mu(n) = \frac{e^{-\rho V_k}(\rho V_k)^n}{n!}. \tag{8}$$

It is assumed that clutter is distributed uniformly in volume V_k.

2.4 Multipath

The echoes from the target return to the receiver other than direct path is known as multipath. Multipath generates false targets to appear and misleads the radar receiver. These multipath creates false targets, which are very difficult to distinguish from actual targets. Figure 2a depicts a received signal without multipath and Fig. 2b shows two false targets are appearing due to multipath. If the echo is reflected from the rough surface, then error occurs in both azimuth and elevation angles due to diffuse scattering. Further, if the echo is reflected from building or non-flat land then error occurs significantly in azimuth angle. The major problem in tracking targets due to multipath effect is that, the false targets and actual targets seems to be coherent. The envelop sum of signals that are received at the receiver is considered to be Rayleigh distributed. The signals arriving at the receiver may have destructive or constructive interference. Let R_n and ϕ_n be electric field and relative phase of N multipath signals. Then, the total electric field at the receiver is given by

$$\tilde{R} = \sum_{n=1}^{N} R_n e^{j\phi_n}. \tag{9}$$

It is assumed that R_n and ϕ_n are independent and distributed uniformly. The probability density function of Rayleigh distribution is given by

$$f_{\mathrm{R}}(r) = \frac{r}{\sigma^2} e^{-\frac{r^2}{2\sigma^2}}. \tag{10}$$

Equation (10) is for slow fading and is valid $\forall r \geq 0$.

2.5 Performance Measures

In the target tracking literature, various performance measures have been proposed [23]. The following performance metrics (Tab. 1 in Appendix) are considered for evaluating the track performance of benchmark targets in the presence of ECM, clutter, multipath and false alarms.

3. Neutralizing Techniques for ECM, Clutter and Multipath Effects

In this section several mechanisms to neutralize ECM, clutter and multipath effects have been described. Waveform agile sensing, adaptive beamforming and adaptive thresholding techniques are successfully applied to neutralize these undesirable effects. The following subsections briefly describe these methods.

3.1 Waveform Agile Sensing

The main objective of adaptive waveform selection is to minimize tracking mean square error. Waveform selection process is based on signal to noise ratio, signal to clutter ratio and type of tracking algorithm applied. The parameter (Ω_j) of the waveform is selected so as to minimize the tracking MSE and is given by

$$J(\Omega_j) = E_{X_j,Z_j|Z_{1:j}}\left[(X_j - \hat{X}_j)^T (X_j - \hat{X}_j)\right] \tag{11}$$

where $E(.)$ is the expectation operation over predicted and original state of measurements. $\hat{X}_j$ is the state estimate of X_j for j observations. Equation (11) is regarded as the cost function and aims to select the waveform parameter which yields minimum MSE at a particular scan j.

Single dynamic model may not exactly represent the motion of the target, as targets may maneuver. For this reason three PDAF filters are fused to IMM algorithm. Each filter is subjected for a particular target motion and will be running in parallel. The updated state estimate and covariance is the weighted combination of each individual PDAF state estimate and covariance respectively. The comprehensive derivation of IMM filter is done in [24].

The mean square error of specific track can be reduced by selecting waveform adaptively and this is accomplished by reducing trace of the updated error covariance matrix. The updated error covariance matrix is given by

$$\begin{aligned} P_{j+1|j+1}(\Omega_{j+1}) =& P_{j+1|j} - \left[1 - \beta_{j+1}^{0}\right] W_{j+1}(\Omega_{j+1}) \\ & S_{j+1}(\Omega_{j+1}) W_{j+1}^{T}(\Omega_{j+1}) + \tilde{P}_{j+1}(\Omega_{j+1}). \end{aligned} \tag{12}$$

From (12), it is evident that observation error covariance is a function of waveform parameter Ω_{j+1}. Thus, the updated error covariance is also a function of Ω_{j+1}. Since each PDAF in IMMPDAF is independently depended on Ω_{j+1}, then the updated error covariance $P_{j+1|j+1}^{\mathrm{IMM}}$ is a function of Ω_{j+1} .The waveform library contains multiple waveforms with different combinations of parameters.

$$\Omega_{j+1} = \min\left(\mathrm{Trace}(P_{j+1|j+1}^{i*})\right) \tag{13}$$

where $i^* = \arg\max_i \mu_{j+1|j+1}^{i}$.

The cost function for selecting waveform is briefly described in [13]. This is computed by evaluating the trace of updated covariance matrix for Kalman filter. The same procedure was extended to IMM filter in [21], [25] and (13) is chosen as one of the cost function. The waveform which gives highest mode probability of ith model of the filter and minimum covariance value is chosen for next scan. This technique has been well applied with various waveforms (LFM, GFM and SFM) to neutralize the undesirable effects.

3.2 Adaptive Beamforming

The beamforming aims to focus the main beam of radar in the intended direction and fix a null in the undesired direction. As the targets in practical scenario will not be stationery and the signal may arrive in any direction. Hence, conventional beamforming finds little use in these situations. The array weights needs to adapt continuously with change in environment and this process is known as adaptive beamforming. In the proposed approach, minimum variance distortion less response (MVDR) has been used to adapt weights of the beamformer. MVDR beamformer requires only direction of arrival (does not require any other information) to determine the beamformer weight vector

$$W = \frac{R^{-1}S(\theta)}{S(\theta)^H R S(\theta)} \tag{14}$$

where $S(\theta)$ is steering vector corresponding to a specific direction and R is spatial covariance matrix. The key functionality of MVDR beamformer is to minimize the total output noise power by setting the gain as unity in a particular direction and this is done by adapting the weight vector of beamformer.

Figure 3a shows the received signal with clutter, multipath and jamming before adaptive beamforming. It is very difficult to distinguish the target as it is embedded in clutter, ECM, multipath and false alarms. Figure 3b shows

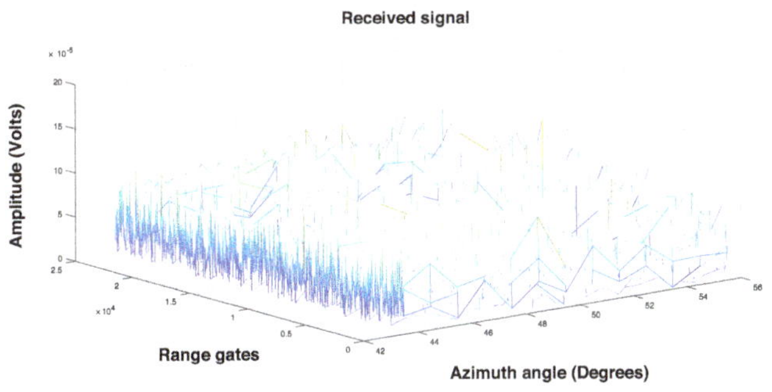

(a) Received signal before adaptive beamforming.

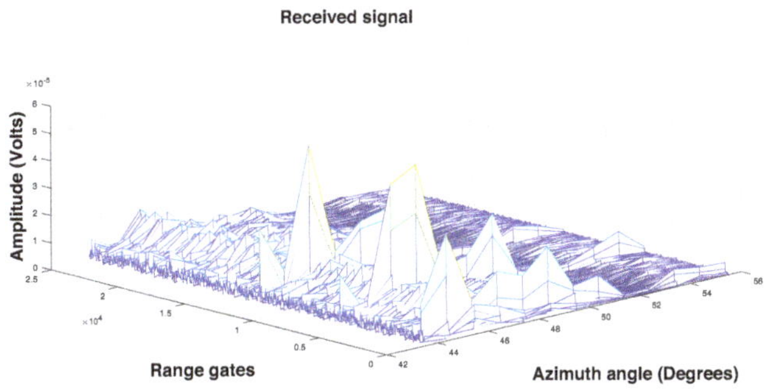

(b) Received signal after adaptive beamforming.

Fig. 3. Received signal before and after adaptive beamforming.

the received signal with adaptive beam forming. It can be visualized clearly that the noise is suppressed and targets can be identified along with multipath. MVDR beamformer is employed to suppress the ECM, clutter and false alarms in undesired direction.

3.3 Adaptive Thresholding

This subsection briefly describes about adaptive thresholding technique employed as a counter measure to ECM, clutter and multipath effects. Generally the received radar echo signal is compared to a fixed threshold value and hence declared the presence of a target based on, weather the matched filter output exceeds the threshold or not. In order to apply a fixed threshold in Neyman-Pearson detector, it seeks complete statistical information of the received echo signal. But, in practical situations, the statistical information of the received echo signal (corrupted with ECM, clutter and multipath effects) changes rapidly due to the change in environmental conditions. Therefore, there is a need for adaptive thresholding which varies according to change in environment and hence counter these unwanted effects.

The main function of the radar detector is to maximize the probability of detection and minimize the probability of false alarms. CA-CFAR is chosen for adaptive thresholding. In CA-CFAR, the cell under detection is known as cell under test (CUT). The neighboring cells of CUT are used to estimate the noise power. Some of the neighboring cells (lagging and leading) of CUT are referred as guard cells. These cells are ignored while estimating noise to avoid signal energy leakage from the CUT into training cells. The noise is estimated from the training cells which can be represented as

$$U_1 = \frac{1}{N_1}\sum_{i=1}^{N_1} x_i \tag{15a}$$

$$U_2 = \frac{1}{N_2}\sum_{j=1}^{N_2} x_j \tag{15b}$$

where N_1 and N_2 represents the number of training cells, x_i and x_j denote the samples in each training cell respectively. Both U_1 and U_2 are combined to estimate the total background noise P_n. The noise estimate is multiplied with relevant scaling factor α, so as to maintain probability of false alarm as constant. The detection threshold T is given as

$$T = \alpha P_n. \tag{16}$$

Threshold (T) is compared with CUT in order to decide the presence of a target. Figure 4 shows detection of targets using adaptive thresholding. It clearly depicts that the target is detected along with multipath and few false alarms. CA-CFAR is applied to suppress ECM, clutter, false alarms and multipath effects.

The echo waveform from the target is corrupted by ECM, clutter, false alarms and multipath effects. Combination of the above effective techniques such as; Waveform agile sensing, adaptive beamformer and adaptive thresholding have been successfully applied to minimize the effects of these interferences.

4. Interacting Multiple Model Probability Data Association Filter

This section provides the brief outline of the IMM-PDAF track filter. The integration of interacting multiple model (IMM) with probability data association (PDA) makes an effective tracker for tracking maneuvering targets in the presence of ECM and false alarms [8].

4.1 Probability Data Association Filter (PDAF)

The PDA filter calculates the associated probability for each measurement based on validating a particular target. An assumption is made such that m observations are validated for a particular scan. The equations required to update state and covariance for PDAF are given below:

Prediction of the state and measurement is given by

$$\hat{x}_{j+1|j} = F\hat{x}_{j|j}, \tag{17}$$
$$\hat{z}_{j+1|j} = H\hat{x}_{j+1|j}. \tag{18}$$

Covariance of the predicted state

$$P_{j+1|j} = FP_{j|j}F' + Q. \tag{19}$$

Covariance with respect to measurement

$$S_{j+1}\left(\Omega_{j+1}\right) = HP_{j+1|j}H' + C_{j+1}\left(\Omega_{j+1}\right) \tag{20}$$

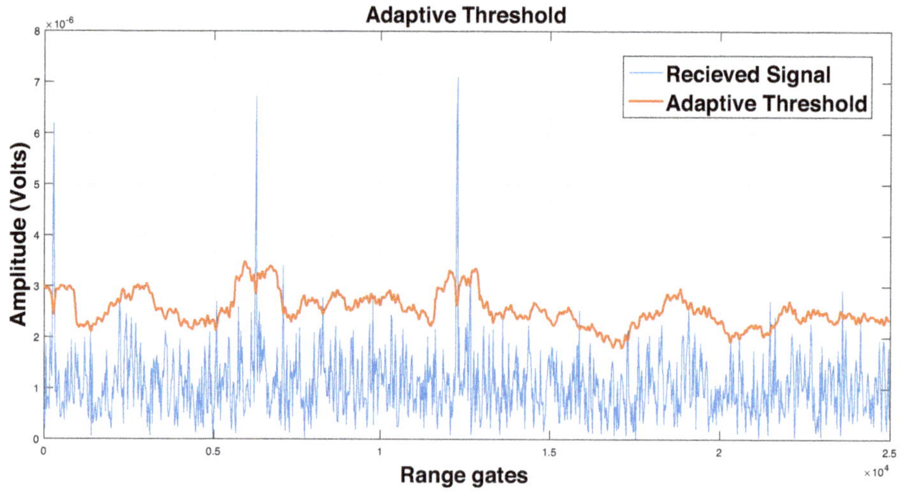

Fig. 4. Illustrating adaptive thresholding mechanism.

where C_{j+1} is the observation error covariance matrix corresponding to Ω_{j+1} waveform. The validation region (gate)-the ellipsoid can be written as

$$\mathcal{V}_{j+1} = \left\{ z : \left[z - \hat{z}_{j+1|j} \right]^T S_{j+1}\left(\Omega_{j+1}\right)^{-1} \left[z - \hat{z}_{j+1|j} \right] \leq \gamma \right\} \tag{21}$$

where γ is the gate threshold determined by the chosen gate probability P_G. Innovation corresponding to the i-th validated measurement

$$v^i_{j+1} = z^i_{j+1} - \hat{z}_{j+1|j} \quad i = 1, \ldots, m_{j+1}. \tag{22}$$

Volume of the validation region is given as

$$V_{j+1}\left(\Omega_{j+1}\right) = c_{n_z} \left| \gamma S_{j+1}\left(\Omega_{j+1}\right) \right|^{1/2} \tag{23}$$

where c_{n_z} represents unit hypersphere volume with dimension n_z (*i.e.* $[c_1, c_2, c_3] = [2, \pi, 4\pi/3]$). Probability of the i-th validated measurement is

$$\beta^i_{j+1}(\Omega_{j+1}) = \begin{cases} \frac{e^i(\Omega_{j+1})}{b + \sum_{l=1}^{m(j+1)} e_l(\Omega_{j+1})}, & i = 1, \ldots, m(j+1), \\ \frac{b}{b + \sum_{l=1}^{m(j+1)} e_l(\Omega_{j+1})}, & i = 0, \end{cases} \tag{24}$$

$\beta_0(k+1)$ is association probability which represents that none of the measurement is correct

$$\mathrm{e}^i(\Omega_{j+1}) \triangleq \mathrm{e}^{-\frac{1}{2} {v^i_{j+1}}^T S_{j+1}(\Omega_{j+1})^{-1} v^i_{j+1}}, \tag{25}$$

$$b \triangleq \left(\frac{2\pi}{\gamma} \right)^{\frac{n_z}{2}} m(j+1) c_{n_z}^{-1} \frac{1 - P_D P_G}{P_D}. \tag{26}$$

State update

$$\hat{x}_{j+1|j+1}(\Omega_{j+1}) = \hat{x}_{j+1|j} + W_{j+1}(\Omega_{j+1}) v_{j+1}(\Omega_{j+1}) \tag{27}$$

where $W_{j+1}(\Omega_{j+1})$ is filter gain and $v_{j+1}(\Omega_{j+1})$ is known as combined innovation which is calculated as

$$v_{j+1} \triangleq \sum_{i=1}^{m(j+1)} \beta^i_{j+1} v^i_{j+1}, \tag{28a}$$

$$W_{j+1}(\Omega_{j+1}) \triangleq P_{j+1|j} H' S_{j+1}(\Omega_{j+1}). \tag{28b}$$

Covariance associated to update state is given as

$$\begin{aligned} P_{j+1|j+1}(\Omega_{j+1}) = & P_{j+1|j} - \left[1 - \beta^0_{j+1} \right] W_{j+1}(\Omega_{j+1}) \\ & S_{j+1}(\Omega_{j+1}) W^T_{j+1}(\Omega_{j+1}) + \tilde{P}_{j+1}(\Omega_{j+1}), \end{aligned} \tag{29a}$$

$$\begin{aligned} \tilde{P}_{j+1}(\Omega_{j+1}) \triangleq & \left[\sum_{i=1}^{m(j+1)} \beta^i_{j+1}(\Omega_{j+1}) v^i_{j+1} {v^i_{j+1}}^T - v_{j+1} v^T_{j+1} \right] \\ & \times W^T_{k+1}(\Omega_{j+1}). \end{aligned} \tag{29b}$$

Equations (27) and (29) represent the final update equation of state and covariance of PDAF respectively.

4.2 Interacting Multiple Model Estimator (IMM)

This subsection presents an outline of IMM method. Single targets can effectively be tracked by probability data association filter (PDAF) following a linear trajectory. But in practical scenarios, the targets may have higher acceleration turns, yielding abrupt maneuvers. Thus, there is a need to apply IMM technique, where PDAF filters are functioning in parallel and each PDAF filter having corresponding system matrix. The updated state estimate and covariance is the weighted sum of individual PDAF filters. Comprehensive explanation and derivation on IMM is given in [24]. In IMM, the state estimate is calculated by using different filters that run in parallel with the weighted combination of previous updated state estimates or with initial condition.

Updated state estimate $\hat{x}^l_{j/j}$, updated error covariance $\hat{P}^l_{j/j}$ and model probability value $\mu^l_{j/j}$ of individual filter information is present for updating next iteration $j+1$ with observation value z_{j+1}. A concise derivation of IMM algorithm is given below:

- Computing mixed input to tracking filter

Predicted model probability is calculated by

$$\mu^r_{j+1|j} = \sum_{l=1}^{n} p_{lr}\mu^l_{j|j} \tag{30}$$

where the model probability conditioned on$j$ is

$$\mu^{l|r}_{j|j} = (1/\mu^r_{j+1|j})p_{lr}\mu^l_{j|j}. \tag{31}$$

The mixed state estimate and covariance which is given as input to the PDAF filter is calculated by

$$\hat{x}^{0r}_{j|j} = \sum_{l}^{n} \mu^{l|r}_{j|j}\hat{x}^l_{j|j}, \tag{32a}$$

$$P^{0r}_{j|j} = \sum_{l=1}^{n} \mu^{l|r}_{j|j}\{P^l_{j|j} + [\hat{x}^l_{j|j} - \hat{x}^{0r}_{j|j}][\hat{x}^l_{j|j} - \hat{x}^{0r}_{j|j}]^T\}. \tag{32b}$$

- Updating mixed state estimate and covariance

The state estimate and covariance of each rth filter is updated from the input (32a) and (32b) to obtain updated state estimate $(\hat{x}^r_{j+1|j+1})$ and covariance $(\hat{P}^r_{j+1|j+1})$.

- Calculating model likelihood function

$$\Lambda^m_{j+1} = \frac{1}{\sqrt{|2\pi S^r_{j+1}|}} e^{-\frac{1}{2}[\tilde{z}_{j+1}]^T [S^r_{j+1}]^{-1}[\tilde{z}_{j+1}]}. \tag{33}$$

- Updating model probability of each filter

$$\mu^r_{j+1|j+1} = \frac{1}{b}\mu^r_{j+1|j}\Lambda^r_{j+1} \tag{34}$$

where b is normalization factor

$$b = \sum_{l=1}^{r} \mu^l_{j+1|j}\Lambda^l_{j+1}. \tag{35}$$

- Combining state estimate

$$\hat{x}^{\text{IMM}}_{j+1|j+1} = \sum_{r=1}^{n} \mu^r_{j+1|j+1}\hat{x}^r_{j+1|j+1}, \tag{36a}$$

$$\begin{aligned} P^{\text{IMM}}_{j+1|j+1} = &\sum_{r=1}^{n} \mu^r_{j+1|j+1} \\ &\times \{P^r_{j+1|j+1} + \left[\hat{x}^r_{j+1|j+1} - \hat{x}^{\text{IMM}}_{j+1|j+1}\right] \\ &\left[\hat{x}^r_{j+1|j+1} - \hat{x}^{\text{IMM}}_{j+1|j+1}\right]^T\}. \end{aligned} \tag{36b}$$

Equations (36a) and (36b) represent the final update equations for state and covariance of IMM estimator. At any instant of time, any one of the state dynamic model in the parallel filters will be equal to the target trajectory and it gets automatically switched to it by evaluating model probability.

5. Simulation Results

This section deals with the simulation results and discussion. The following subsections describe six benchmark trajectories used in simulations and experimental results.

5.1 Benchmark Trajectories

Six benchmark trajectories from [8] have been used for testing waveform agile sensing in combined jamming (SOJ and SSJ), clutter and multipath scenario by applying IMM-PDAF. Each trajectory turn rates, trajectory simulated time, constant velocity and target type is explained below:

5.1.1 Benchmark Trajectory-1

Initially the target is at position [75, 30, 1.26] km from radar and finally it reaches [73.54, 4.7, 1.26] km with $2g$ and $3g$ turns. The trajectory is simulated for 165 s. The constant speed limit is maintained at 290 m/s. The trajectory shows large aircraft.

5.1.2 Benchmark Trajectory-2

Benchmark trajectory-2 is taken as second target trajectory for comparison of tracking performance with adaptive waveform selection. Initially the target is at a position [47, −45, 4.57] km from the radar and finally reaches to [34, −36.54, 3.760] km with $2.5g$ and $4g$ turns. Trajectory is simulated for 150 s. The trajectory shows small maneuverable commercial jet. The true and estimated trajectory of benchmark-2 is shown in Fig. 5.

5.1.3 Benchmark Trajectory-3

The target makes $4g$ acceleration turns with 45° and 90° at first 30 s and 60 s respectively. The maximum and minimum speed it travels is 457 m/s and 274 m/s respectively. The trajectory shows medium bomber. The simulation time of the trajectory is 145 s.

5.1.4 Benchmark Trajectory-4

Target makes turn 45° with $4g$ and $6g$ acceleration with a minimum speed of 251 m/s. The simulation time of trajectory is 184 s with maneuvering density of 9.92 %. The trajectory shows medium bomber. The true and estimated trajectory of benchmark-4 is shown in Fig. 6.

5.1.5 Benchmark Trajectory-5

The target takes complex manuvers with $5g$, $6g$ and $7g$ acceleration turns. The trajectory shows fighter aircraft. The simulation time of trajectory is 182 s with maneuvering density of 17.5 %.

5.1.6 Benchmark Trajectory-6

The target makes two $6g$ and two $7g$ turns and maintains a constant speed of 426 m/s. The trajectory shows fighter aircraft. The simulation time of trajectory is 188 s with maneuvering density of 18 %. The trajectory shows medium bomber.

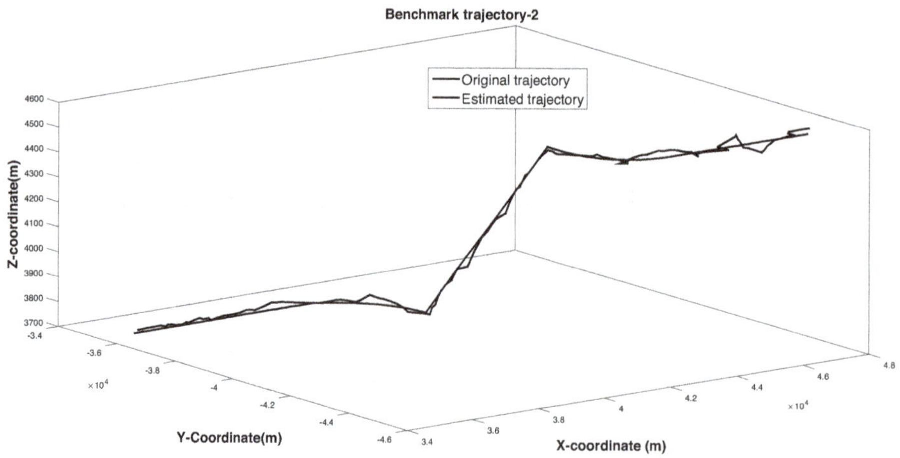

Fig. 5. Tracking of Benchmark target trajectory-2 in the presence of FA, SOJ, clutter and multipath.

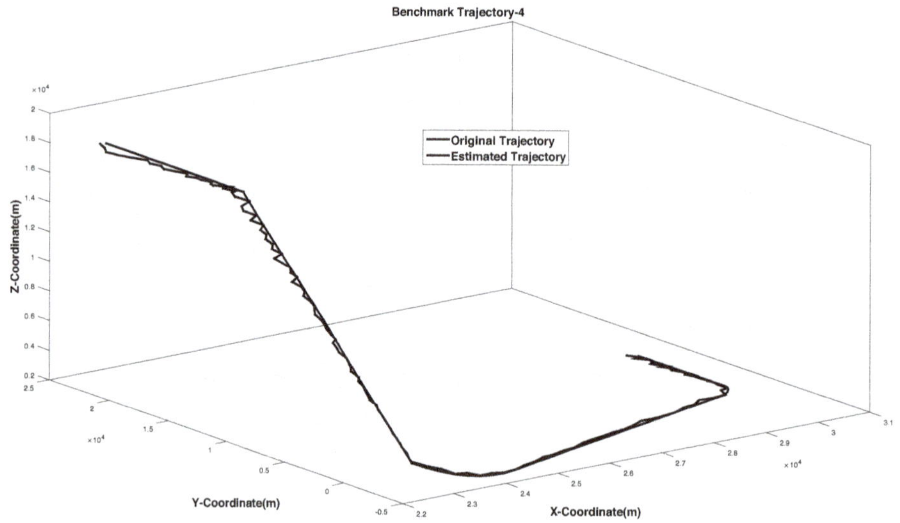

Fig. 6. Tracking of Benchmark target trajectory-4 in the presence of FA, SOJ, clutter and multipath.

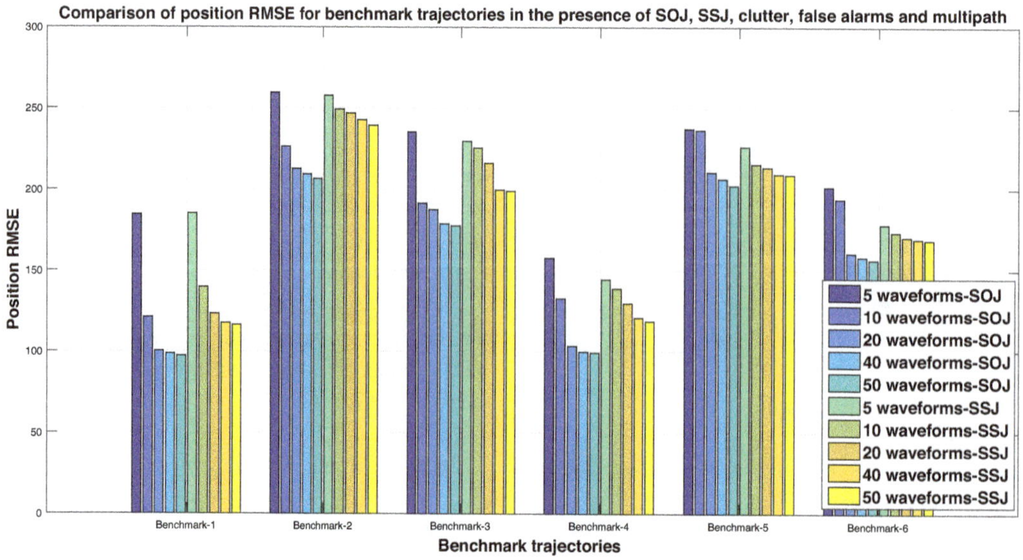

Fig. 7. Comparison of position RMSE for benchmark trajectories in the presence of SOJ, SSJ, clutter, false alarms and multipath.

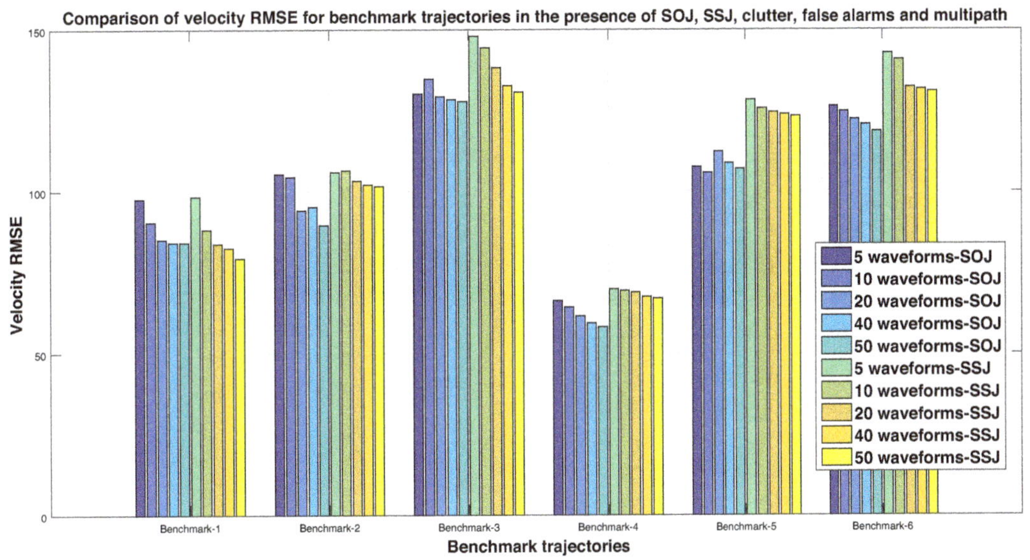

Fig. 8. Comparison of velocity RMSE for benchmark trajectories in the presence of SOJ, SSJ, clutter, false alarms and multipath.

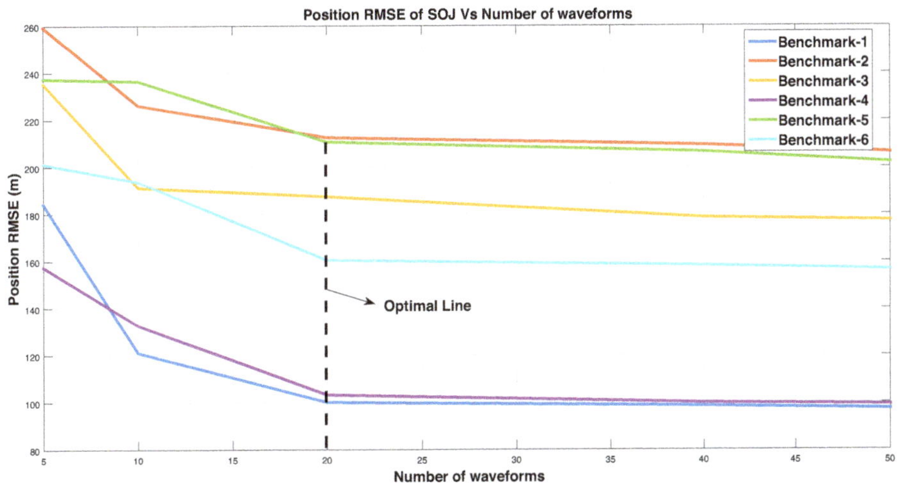

Fig. 9. Performance of position RMSE Vs number of waveforms for Benchmark Trajectories in the presence of SOJ, clutter, false alarm and multipath.

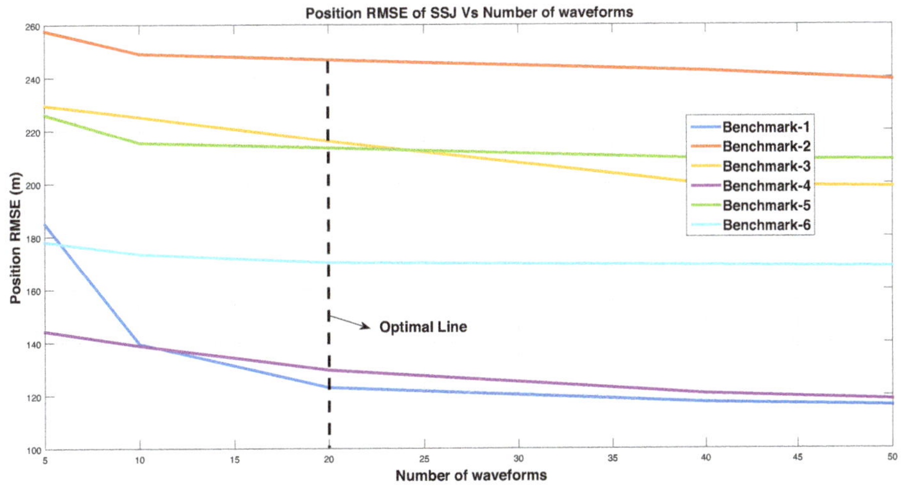

Fig. 10. Performance of position RMSE Vs number of waveforms for Benchmark Trajectories in the presence of SSJ, clutter, false alarm and multipath.

5.2 Simulation Results and Discussion

Results for the above six benchmark trajectories with complex scenarios have been tabulated in Tab. 2 and Tab. 3. Table 2 shows results in the presence of SOJ, clutter, multipath and FA are compared with earlier research study [10], which ignored multipath effects. On the other hand, results in the presence of SSJ, clutter, multipath and FA are tabulated in Tab. 3. Jammer on board the target (SSJ) have not been considered by any other earlier studies conducted, therefore these values have not been compared with any other results.

The tabulated results in Tab. 2 demonstrate that average power in current research study is significantly reduced for all benchmark trajectories except for benchmark-4. The simulation results reveals that proposed waveform agile sensing technique requires only 39.98 % lower mean average power compared to past studies [10]. Besides that, the cost function C_1, which is associated with radar energy is significantly lowered in the current method. However, the cost function C_2, which is associated with radar time is little higher. The track loss performance is zero for benchmark-1, 2 trajectories and is 1.6 %, 0.8 % , 1.1 % and 1.94 % for benchmark-3, 4, 5 and 6 trajectories respectively. From Tab. 2, it is also observed that position and velocity root mean square errors are higher. This clearly indicates that the multipath effect increased the measurement error along with the background clutter.

Furthermore, as the number of waveforms are increased from 5 to 50, both position RMSE and velocity RMSE values are decreasing (same is illustrated in Figures 7 to 8) [26]. It can also be observed that position RMSE with 50 waveforms in the current study is less than previous work for benchmark-1 and for other benchmark trajectories its mean difference is 52.124 m for 50 waveforms. From Tab. 2, velocity RMSE in current study is higher than previous work with mean difference of 40.56 m/s. The earlier studies [10] ignored multipath in the environment. In the current research study multipath is incorporated in addition to the clutter and jamming, thus presenting a strong interference in the environment. Hence, the performance of velocity RMSE is slightly degraded when compared to previous studies.

The performance evaluation of IMMPDAF in the presence of SSJ, clutter, multipath and false alarm is tabulated in Tab. 3. It indicates the position and velocity RMSE values are decreasing as the number of waveforms increased for all benchmark trajectories. The average power for benchmark -1, 2 and 6 are lower than benchmark-3, 4 and 5 trajectories. Track loss for benchmark 1, 2 and 5 is zero, whereas average track loss for benchmark 3, 4 and 6 is 1.86 % . The C_1 and C_2 cost functions are lower for benchmark 1, 2 and 6 compared to benchmark 3, 4 and 5 trajectories.

Plots are obtained by varying the number of waveforms in adaptive waveform selection with respect to position and velocity RMSE values for SOJ and SSJ scenarios. From Fig. 7 to 8, it is evident that the position and velocity RMSE of SSJ is significantly higher than SOJ for all benchmark trajectories. Further, as the number of waveforms in waveform library is increasing from 5 to 50, the position and velocity RMSE values are decreasing.

Performance of position RMSE with number of waveforms considered for SOJ and SSJ cases are shown in Fig. 9 and 10 respectively. Optimal number of waveforms for both the cases are iluustrated with a dotted line as shown in Figures 9 and 10. In case of SOJ, the optimal RMSE observed for benchmark 1, 2, 4, 5 and 6 is at 20 waveforms. The optimal number of waveforms to be considered is 10 for benchmark-3 in case of SOJ scenario. On the other hand, in case of SSJ, the optimal number of waveforms is widely varied for benchmark trajectories. For benchmark trajectory 2,5 and 6 the optimal number of waveforms to be considered is 10 in case of SSJ. Where as for benchmark trajectory 1 and 4 the optimality is achieved at 20 waveforms. For benchmark trajectory-3, the optimality is attained for 40 waveforms. Therefore, on an average the optimal performance is achieved at 20 waveforms for all such benchmarks in both the cases.

In this research work, closely spaced targets have not been considered and assumed only point targets. Besides, only frequency coded waveforms are used for waveform agile sensing. Further, ECM techniques such as; range gate pull off (RGPO) and velocity gate pull off (VGPO) are not examined. Future work can be carried out with these problem formulations.

The achieved results exhibits enhanced performance using waveform agile sensing approach in the presence of ECM, clutter, false alarm, and multipath effects. Hence, this method may be applied to track highly manuvering targets in the presence of strong interference.

6. Conclusion

Improved performance for tracking benchmark has been demonstrated using waveform agile sensing technique in the presence of strong interference. A number of frequency coded waveforms (LFM, SFM and GFM) up to 50 have been stacked in the waveform bank to enhance the track performance. IMMPDAF is applied to track benchmark with ECM (SOJ/SSJ), clutter and multipath effects. The simulation results show that proposed waveform agile sensing method requires only 39.98 % lower mean average power compared to previous studies [10]. It is also evident that, as the number of waveforms are increasing from 5 to 50, both position and velocity RMSE values are decreasing. Furthermore, position and velocity RMSE values are higher for SSJ than SOJ. This work can be further extended to incorporate phase coded waveforms to track benchmark. To conclude, potential research problems to be carried in future enlisted by Blair [8], have been addressed here and achieved significant results.

Acknowledgments

This research work is supported by Young Scientist Research Grant (No.SB/FTP/ETA-110/2014) of Department of Science and Technology, Govt. of India.

References

[1] KIRUBARAJAN,T., BAR-SHALOM, Y., BLAIR, W., et al. IMM-PDAF for radar management and tracking benchmark with ECM. *IEEE Transactions on Aerospace and Electronic Systems*, 1998, vol. 34, no. 4, p. 1115–1134. DOI: 10.1109/7.722696

[2] BLAIR, W., WATSON, G., HOFFMAN, S. Benchmark problem for beam pointing control of phased array radar against maneuvering targets. In *Proceedings of the IEEE American Control Conference*. 1994, vol. 2, p. 2071–2075. DOI: 10.1109/ACC.1994.752441

[3] DAEIPOUR, E., BAR-SHALOM, Y., LI, X. Adaptive beam pointing control of a phased array radar using an imm estimator. In *Proceedings of the IEEE American Control Conference*. 1994, vol. 2, p. 2093–2097. DOI: 10.1109/ACC.1994.752445

[4] BLAIR, W., WATSON, G., GENTRY, G., et al. Benchmark problem for beam pointing control of phased array radar against maneuvering targets in the presence of ECM and false alarms. In *Proceedings of the IEEE American Control Conference*. 1995, vol. 4, p. 2601–2605. DOI: 10.1109/ACC.1995.532318

[5] KIRUBARAJAN, T., BAR-SHALOM, Y., DAEIPOUR, E. Adaptive beam pointing control of a phased array radar in the presence of ECM and false alarms using IMMPDAF. In *Proceedings of the IEEE American Control Conference*. 1995, vol. 4, p. 2616–2620. DOI: 10.1109/ACC.1995.532321

[6] SLOCUMB, B., WEST, P., SHIREY, T., et al. Tracking a maneuvering target in the presence of false returns and ECM using a variable state dimension Kalman filter. In *Proceedings of the IEEE American Control Conference*. 1995, vol. 4, p. 2611–2615. DOI: 10.1109/ACC.1995.532320

[7] RAGO, C., MAHRA, R. Target tracking in the presence of ECM: A filter design tool. In *Proceedings of the IEEE Twenty-Ninth Southeastern Symposium on System Theory*. 1997, p. 514–518. DOI: 10.1109/SSST.1997.581720

[8] BLAIR, W., WATSON, G., KIRUBARAJAN, T., et al. Benchmark for radar allocation and tracking in ECM. *IEEE Transactions on Aerospace and Electronic Systems*, 1998, vol. 34, no. 4, p. 1097–1114. DOI: 10.1109/7.722694

[9] BLACKMAN, S., DEMPSTER, R., BUSCH, M., et al. IMM/MHT solution to radar benchmark tracking problem. *IEEE Transactions on Aerospace and Electronic Systems*, 1999, vol. 35, no. 2, p. 730–738. DOI: 10.1109/7.766953

[10] ANGELOVA, D., SEMERDJIEV, E., MIHAYLOVA, L., et al. An IMMPDAF solution to benchmark problem for tracking in clutter and standoff jammer. In *Proceedings of the International Conference on EuroFusion*. 1999, p. 123–128.

[11] BEHAR, V., ANGELOVA, D., VASSILEVA, B., et al. A set of algorithms for radar management and target tracking in the presence of SOJ. *Comptes Rendus de l'Academie Bulgare des Sciences*, 2001, vol. 54, no. 7, p. 17.

[12] BEHAR, V., KABAKCHIEV, C. Radar waveform allocation by using post detection integration. *Comptes Rendus de l'Academie Bulgare des Sciences*, 2002, vol. 55, no. 1, p. 63.

[13] KERSHAW, D. J., EVANS, R. J. Optimal waveform selection for tracking systems. *IEEE Transactions on Information Theory*, 1994, vol. 40, no. 5, p. 1536–1550. DOI: 10.1109/18.333866

[14] KERSHAW, D. J., EVANS, R. Waveform selective probabilistic data association. *IEEE Transactions on Aerospace and Electronic Systems*, 1997, vol. 33, no. 4, p. 1180–1188. DOI: 10.1109/7.625110

[15] RAGO, C., WILLETT, P., BAR-SHALOM, Y. Detection-tracking performance with combined waveforms. *IEEE Transactions on Aerospace and Electronic Systems*, 1998, vol. 34, no. 2, p. 612–624. DOI: 10.1109/7.670395

[16] NIU, R., WILLETT, P., BAR-SHALOM, Y. Tracking considerations in selection of radar waveform for range and range-rate measurements. *IEEE Transactions on Aerospace and Electronic Systems*, 2002, vol. 38, no. 2, p. 467–487. DOI: 10.1109/TAES.2002.1008980

[17] HONG, S. M., EVANS, R. J., SHIN, H. S. Optimization of waveform and detection threshold for range and range-rate tracking in clutter. *IEEE Transactions on Aerospace and Electronic Systems*, 2005, vol. 41, no. 1, p. 17–33. DOI: 10.1109/TAES.2005.1413743

[18] SIRA, S. P., PAPANDREOU-SUPPAPPOLA, A., MORRELL, D. Characterization of waveform performance in clutter for dynamically configured sensor systems. In *Proceedings of the Waveform Diversity and Design Conference*. 2006.

[19] SIRA S. P., MORRELL, D. Dynamic configuration of time-varying waveforms for agile sensing and tracking in clutter. *IEEE Transactions on Signal Processing*, 2007, vol. 55, no. 7, p. 3207–3217. DOI: 10.1109/TSP.2007.894418

[20] SIRA S. P., LI, Y., PAPANDREOU-SUPPAPPOLA, A., et al. Waveform-agile sensing for tracking. *IEEE Signal Processing Magazine*, 2009, vol. 26, no. 1, p. 53–64. DOI: 10.1109/MSP.2008.930418

[21] NGUYEN, N., DOGANCAY, K., DAVIS, L. Adaptive waveform selection for multistatic target tracking. *IEEE Transactions on Aerospace and Electronic Systems*, 2015, vol. 51, no. 1, p. 688–701. DOI: 10.1109/TAES.2014.130723

[22] MAHAFZA, B. R. *Radar Signal Analysis and Processing Using MATLAB*. 1st ed. CRC Press, 2008. ISBN: 9781420066432

[23] GORJI, A. A., THARMARASA, R., KIRUBARAJAN, T. Performance measures for multiple target tracking problems. In *Proceedings of the 14th IEEE International Conference on Information Fusion (FUSION)*. 2011, p. 1–8. ISBN: 978-1-4577-0267-9

[24] BAR-SHALOM, Y., LI, X. R., KIRUBARAJAN, T. *Estimation with Applications to Tracking and Navigation: Theory Algorithms and Software*. John Wiley & Sons, 2004. ISBN: 978-0-471-41655-5

[25] SAVAGE, C. O., MORAN, B. Waveform selection for maneuvering targets within an IMM framework. *IEEE Transactions on Aerospace and Electronic Systems*, 2007, vol. 43, no. 3, p. 1205–1214. DOI: 10.1109/TAES.2007.4383612

[26] MIAO, L., ZHANG, J. J., CHAKRABARTI, C. et al. Algorithm and parallel implementation of particle filtering and its use in waveform-agile sensing. *Journal of Signal Processing Systems*, 2011, vol. 65, no. 2, p. 211–227. DOI: 10.1007/s11265-011-0601-2

[27] BLAIR, W., WATSON, G., et al. *Benchmark for Radar Resource Allocation and Tracking Targets in the Presence of ECM*. Sep. 1996, 67 pages. [Online] Cited 2016-07-28. Available at: http://www.dtic.mil/dtic/tr/fulltext/u2/a286909.pdf

About the Authors . . .

Gnane Swarnadh SATAPATHI (corresponding author) was born in Visakhapatnam, India. He is currently pursuing Ph.D. in National Institute of Technology Karnataka, Surathkal in the area of radar target tracking.

Srihari PATHIPATI was born in Nellore, India. He obtained masters degree in communication engineering and signal processing from University of Plymouth, UK and recieved his Ph.D. from Andhra University, India in 2012. He is currently working as Assistant Professor in National Institute of Technology Karnataka, Surathkal. His research interests include radar signal processing and target tracking. He is a senior member of IEEE.

Appendix:

Sl.No	Performance metrics	Description	Reference
1	Root Mean Square Error (RMSE)	It measures the difference between actual value and estimated value. $$\text{RMSE} = \sqrt{\frac{1}{K}\sum_{i=1}^{k}(x_i - \hat{x_i})^2} \quad \text{(A.37)}$$ where x_i=Actual value; $\hat{x_i}$=Predicted value; K=No of observations.	[23]
2	Track Loss	The track is declared to be lost if the error in the estimated value of the target is greater than 1.5 range gates in range. It measures the percentage of tracks that are lost during simulation.	[8], [23]
3	Cost functions (C_1 & C_2)	C_1: It corresponds to period of operation when radar energy is critical. $$C_1 = \bar{E}_{\text{ave}} + 10^3\bar{T}_{\text{ave}} \quad \text{(A.38)}$$ C_2: It corresponds to period of operation when radar time is critical. $$C_2 = \bar{E}_{\text{ave}} + 10^5\bar{T}_{\text{ave}} \quad \text{(A.39)}$$ Where, $\bar{E}_{\text{ave}}$= Average energy per second; $\bar{T}_{\text{ave}}$ = Average radar time per second.	[8]
4	Average Power	Rate of energy flow averaged over one full period. $$P_{\text{avg}} = \frac{\text{Pulse Width } (\tau)}{\text{PRT } (T)} \cdot \text{Peak Power} \quad \text{(A.40)}$$	[8], [27]

Tab. 1. Performance measures.

SlNo	Track Length (s)	Max. Acc. (m/s^2)	Man. Density (%)	Number of Waveforms	Ave.Power (W)		Pos.RMSE (m)		Vel.RMSE (m/s)		Cost C_1		Cost C_2		Track loss (%)	
				Benchmark Trajectory -1												
1				5	3.14		184.26		97.64		3.61		50.15		0	
2				10	2.85		121.12		90.58		3.31		48.53		0	
3	165	29.4	24.24	20	2.60	7.28[1]	100.17	115[1]	85.15	50.27[1]	3.04	7.63[1]	47.31	41.65[1]	0	0[1]
4				40	2.55		98.62		84.31		2.78		46.89		0	
5				50	2.52		97.31		84.31		2.78		46.83		0	
				Benchmark Trajectory -2												
6				5	3.09		259.48		105.36		3.56		50.52		0	
7				10	3.42		226.29		104.51		3.91		52.03		0	
8	150	39.2	34.66	20	2.96	6.16[1]	212.60	100.3[1]	94.24	52.8[1]	3.43	6.51[1]	49.77	40.84[1]	0	0[1]
9				40	2.88		209.31		95.28		3.37		48.64		0	
10				50	2.79		206.42		89.65		3.31		48.59		0	
				Benchmark Trajectory -3												
11				5	7.34		235.37		130.26		7.93		66.13		2	
12				10	7.16		191.24		134.79		7.82		66.02		1.5	
13	145	39.2	20.83	20	7.03	10.36[1]	187.41	148.7[1]	129.32	79.15[1]	7.78	10.71[1]	65.67	45.18[1]	1.5	0[1]
14				40	6.93		178.63		128.47		7.32		65.21		1.5	
15				50	6.72		177.38		127.91		7.01		65.07		1.5	
				Benchmark Trajectory -4												
16				5	7.29		157.62		66.31		7.89		65.72		0.8	
17				10	7.19		132.73		64.28		7.75		65.61		0	
18	184	58.8	9.92	20	7.13	3.07[1]	103.28	45.81[1]	61.49	36.55[1]	7.71	3.42[1]	65.47	37.37[1]	0	0[1]
19				40	7.10		99.86		59.32		7.68		65.39		0	
20				50	7.06		99.27		58.14		7.63		65.30		0	
				Benchmark Trajectory -5												
21				5	7.41		237.38		107.61		7.92		66.01		1.8	
22				10	7.30		236.49		105.74		7.84		65.92		1.3	
23	182	68.6	17.5	20	7.26	15.91[1]	210.71	171.5[1]	112.31	74.49[1]	7.79	16.27[1]	65.49	51.94[1]	0.8	0[1]
24				40	7.19		206.46		108.72		7.74		65.37		0.8	
25				50	7.12		202.13		106.97		7.68		65.22		0.8	
				Benchmark Trajectory -6												
26				5	4.47		201.28		126.37		4.78		57		2.3	
27				10	4.36		193.79		124.81		4.63		56.24		2.1	
28	188	68.6	18	20	4.29	7.62[1]	160.64	114.8[1]	122.29	72.44[1]	4.57	7.99[1]	56.08	44.48[1]	2.1	1[1]
29				40	4.16		158.31		120.76		4.46		55.91		1.6	
30				50	4.01		156.52		118.62		4.39		55.71		1.6	

[1]-Authors [10] ignored multipath effects.

Tab. 2. Comparison results for Benchmark Targets in the presence of SOJ+ Clutter + multipath + FA.

SlNo	Track Length (s)	Max. Acc. (m/s^2)	Man. Density (%)	Number of Waveforms	Ave.Power (W)	Pos.RMSE (m)	Vel.RMSE (m/s)	Cost C_1	Cost C_2	Track loss (%)
				Benchmark Trajectory -1						
1				5	3.03	185	98.47	3.49	49.47	0
2				10	2.71	139.66	88.21	3.16	47.65	0
3	165	29.4	24.24	20	2.63	123.28	83.94	3.08	47.45	0
4				40	2.58	117.62	82.61	2.93	46.28	0
5				50	2.51	116.31	79.42	2.82	46.21	0
				Benchmark Trajectory -2						
6				5	3.13	257.74	105.98	3.61	50.62	0
7				10	3.06	249.07	106.53	3.53	50.28	0
8	150	39.2	34.66	20	3	246.85	103.35	3.47	49.98	0
9				40	2.98	242.61	102.15	3.36	47.32	0
10				50	2.73	239.37	101.61	3.32	45.23	0
				Benchmark Trajectory -3						
11				5	7.43	229.67	147.94	8.76	68.01	2.6
12				10	7.39	225.36	144.45	8.41	67.82	2.4
13	145	39.2	20.83	20	7.35	216.13	138.27	8.32	67.39	2.0
14				40	7.16	199.62	132.69	7.83	66.97	1.7
15				50	7.12	198.79	130.71	7.78	66.83	1.7
				Benchmark Trajectory -4						
16				5	7.32	144.38	69.86	7.92	65.87	1.2
17				10	7.28	138.92	69.37	7.86	65.70	0
18	184	58.8	9.92	20	7.21	129.76	68.81	7.74	65.59	0
19				40	7.07	120.89	67.43	7.43	65.06	0
20				50	7.04	118.67	66.94	7.39	64.93	0
				Benchmark Trajectory -5						
21				5	7.32	226.17	128.39	7.91	66.03	0
22				10	7.26	215.48	125.73	7.88	65.95	0
23	182	68.6	17.5	20	7.19	213.62	124.54	7.76	65.82	0
24				40	7.09	209.36	123.89	7.69	65.62	0
25				50	7.03	208.97	123.28	7.42	65.54	
				Benchmark Trajectory -6						
26				5	4.79	178.19	142.69	5.06	55.86	3.1
27				10	4.63	173.50	140.72	5.01	55.29	2.2
28	188	68.6	18	20	4.54	170.42	132.26	4.82	55.07	2.2
29				40	4.46	169.38	131.63	4.78	54.91	2.2
30				50	4.32	168.74	130.91	4.71	54.86	2.2

Tab. 3. Comparison results for Benchmark Targets in the presence of FA + SSJ+ Clutter + Multipath.

Min-Max MSE-based Interference Alignment for Transceiver Designs in Cognitive Radio Networks

Ha Hoang KHA[1], *Hung Quang TA*[2]

[1] Faculty of Electrical and Electronics Engineering, Ho Chi Minh City University of Technology, VNU-HCM, Vietnam
[2] Faculty of Information Technology, Hanoi University, Vietnam

hhkha@hcmut.edu.vn, hta@hanu.edu.vn

Abstract. *This paper is concerned with an optimal design of the precoders and receive filters for cognitive radio (CR) networks in which multiple secondary users (SUs) share the same frequency band with multiple primary users (PUs). To cope with interference and to achieve fairness among users, we develop an interference alignment (IA) scheme by minimizing the maximum mean squared error (Min-Max MSE) of the received signals. Since the Min-Max MSE design problems are nonconvex in the design matrix variables of the precoders and receive filters, we develop an alternating optimization algorithm with provable convergence to iteratively find the optimal solutions. In each iteration, the precoder design problems can be recast as second order cone program (SOCP) while the optimal receive filters can be derived in closed-form solutions. Finally, numerical results are provided to demonstrate the superiority of the proposed method as compared to previous work in terms of the information rate and bit error rate.*

Keywords

Multiuser MIMO, cognitive radio, interference alignment, Min-Max MSE, transceiver design

1. Introduction

With the growing demand of high data rate wireless services, radio frequency spectrum resources have recently become scare and precious. Thus, spectrum utilization efficiency has been of great concern in wireless communication network designs. Cognitive radio (CR) has been known as a wireless communication technique which renders radio spectrum exploitation more effective [1–4]. In CR networks, secondary users (SUs) can opportunistically access the spectrum licensed to primary users (PUs). However, the SUs are allowed to transmit their signals only if their transmission does not cause adverse interference to the PUs [5].

Of interest in this paper is a CR network in which *multiple* PUs and *multiple* SUs transmit at the same time in the same frequency band. In such a CR network, inter-user interference can be severe and significantly degrade the system performance. As the users are equipped with multiple antennas, additional spatial dimensions can be exploited to deal with interference. To deal with interference and improve the system performance in CR networks, the optimal transmission strategy designs for the SUs in the physical layer are crucial issues. Various performance metrics which are widely used for the transceiver designs in CR networks are the sum-rate and sum of mean square error (MSE) [5–8]. The authors in [6] designed the SU transmission strategies to maximize the SU weighted sum-rate while maintaining reliable communication for a single PU. Reference [7] designed the precoders and receive filters to maximize the sum-rate for cognitive MIMO ad-hoc networks with a single PU by using the weighted minimum MSE method. The authors in [9] developed a space alignment algorithm for a CR network with a single PU and single SU. Alternatively, reference [2] introduced the cooperative game theory for spectrum sharing in multiuser multiple-input multiple output (MIMO) cognitive radio networks.

Recently, interference alignment (IA) has known as an efficient approach to deal with interference in multiuser MIMO wireless networks [10–13]. The underlying idea of IA is to construct the transmitted signals such that the interference signals at each receiver are aligned into a reduced dimension subspace which is orthogonal to the one spanned by the desired signals [10–12]. IA has been applied into K-user interference channels [10], [12], wireless X networks [14], and heterogeneous networks (HetNets) [15]. Recently, it has been adapted and extended to CR networks, see [15–18] and references therein. In [18], the SU transmission is aligned into unused spatial directions of the PU channel. The authors in [15] applied an IA scheme for HetNets in which the macro-cell and small-cells operate in an underlay CR mode and the secondary terminals exploit unused space dimensions of the primary terminal for interference cancellation. Additionally, the authors in [19] developed IA techniques with different levels of network coordination for spectral coexistence of HetNets in terrestrial and satellite paradigms. Reference [20] employed IA for HetNets in which IA was developed for a quantized version of the alignment direction and joint IA and space-frequency block codes schemes were proposed. However, the models in almost all previous works are restricted to *single* primary user and multiple secondary users.

In this paper, we consider the optimal transceiver design for MIMO CR networks with *multiple* PUs and SUs. In order to efficiently mitigate interference and obtain fairness among users, we seek the transmission strategies to minimize the maximum user MSE (Min-Max MSE) among the PUs and SUs. To the best of our knowledge, the problem of Min-Max MSE based IA for interference mitigation in multiuser MIMO CR networks to achieve MSE fairness among users has not been considered yet. Since Min-Max MSE based design problems are highly nonconvex with respect to the matrix variables, the globally optimal solutions are difficult to obtain. This paper will propose the structures of the transceivers and iterative algorithms to obtain the suboptimal solutions. The main contributions of the paper are listed as follows:

- We formulate the transceiver designs of the PUs and SUs for the CR network as the Min-Max MSE IA problems to achieve fairness among the users in the same network. To guarantee the higher priority to the PUs, the PU transmission is designed without awareness of the presence of the SUs while the SUs must confine their transmitted signals into unused communication directions of the PU receivers to restrict interference to the PUs.
- We propose a structure of each SU transmit precoder as a cascade of two precoder matrices. The first matrix is selected to align the SU signals into a subspace orthogonal with a desired signal subspace at each PU receiver while the second matrix is designed to minimize the maximum user MSE with the individual transmit power constraint. To overcome difficulties associated with the nonconvexity of the design problems, we adopt the alternating optimization to decompose the design problems into tractable convex optimization subproblems. Specifically, we reformulate the precoder designs into the second-order cone programming (SOCP) problems which can be efficiently solved by interior-point methods. On the other hand, for the receive filter designs, we derive closed-form expressions for the optimal receive filters.
- We show the convergence of the proposed algorithm by theoretical analysis and numerical results. We also provide the computational complexity of the proposed algorithms.
- We provide numerical results to examine the performance of the proposed method in terms of user rates and BER in comparison with the IA scheme in [17] which uses the interference leakage minimization. The numerical results demonstrate the superior performance of the proposed method.

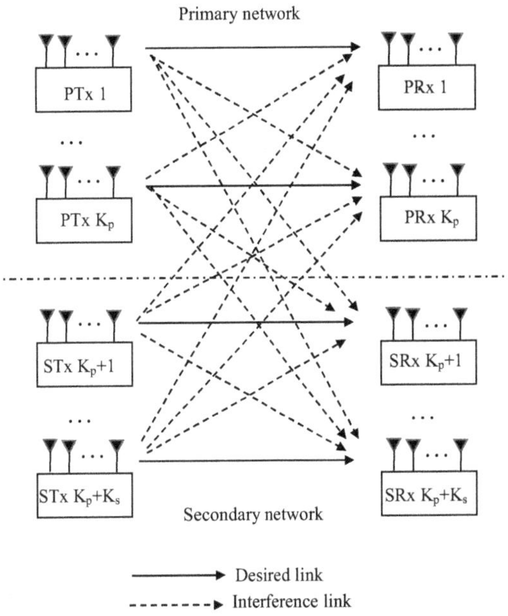

Fig. 1. A cognitive radio network with multiple PUs and SUs.

The rest of the paper is organized as follows. Section 2 describes the CR model considered in the paper. In Sec. 3, the transceiver design problems are formulated and the proposed method for the optimal transceiver designs is developed. Numerical results are provided in Sec. 4. Finally, Section 5 provides concluding remarks.

The following notations are used in the paper. Boldface upper (lower) case letters represent matrices (vectors). $\mathbf{X}^H$, $\mathbf{X}^\dagger$, rank($\mathbf{X}$), tr($\mathbf{X}$), and $\mathcal{N}(\mathbf{X})$ denote the Hermitian transposition, conjugate, rank, trace and null space of matrix $\mathbf{X}$. $[\mathbf{X}]_k^\ell$ is a matrix taking $\ell - k + 1$ columns from column k of matrix $\mathbf{X}$. vec($\mathbf{X}$) is the column vector obtained by stacking the columns of the matrix $\mathbf{X}$. $||\cdot||_F$ and $||.||_2$ are the Frobenius norm of a matrix and the norm-2 of a vector, respectively. $\otimes$ stands for the Kronecker product. blkdiag$(\mathbf{X}_1, \mathbf{X}_2, ..., \mathbf{X}_n)$ is a block diagonal matrix in which the main diagonal blocks are $\mathbf{X}_1, \mathbf{X}_2, ..., \mathbf{X}_n$. $\mathbf{I}_d$ is a $d \times d$ identity matrix while $\mathbf{0}_{n\times m}$ is the $n \times m$ null matrix. A complex Gaussian random vector $\mathbf{x}$ with mean $\bar{\mathbf{x}}$ and covariance $\mathbf{R}_\mathbf{z}$ is represented by $\mathbf{x} \sim \mathcal{CN}(\bar{\mathbf{x}}, \mathbf{R}_\mathbf{x})$.

2. System Model

Consider a wireless CR network in which K_s pairs of the secondary transmitters (STxs) and secondary receivers (SRxs) share spectrum resources with K_p pairs of the primary transmitters (PTxs) and primary receivers (PRxs) in an underlay method, as illustrated in Fig. 1. For ease of presentation, we define the set of PUs as $\mathcal{K}_p = \{1, 2, ...K_p\}$ while that of SUs is denoted as $\mathcal{K}_s = \{K_p + 1, K_p + 2, ..., K_p + K_s\}$. The total number of users in the entire network is $K = K_p + K_s$ and the set of all the users in the network is $\mathcal{K} = \mathcal{K}_p \cup \mathcal{K}_s$. Assume that transmitter k is equipped with N_k antennas while receiver k is equipped with M_k antennas. Transmitter k sends the signal $\mathbf{x}_k \in \mathbb{C}^{d_k \times 1}$, where d_k is the number of data streams, to its associated receiver k. To be able to recover d_k data streams at user k, it is required that $d_k \leq \min\{N_k, M_k\}$. We assume the transmitted symbols of $\mathbf{x}_k$ are independently Gaussian encoded symbols with $\mathbb{E}(\mathbf{x}_k\mathbf{x}_k^H) = \mathbf{I}$, and $\mathbb{E}(\mathbf{x}_k\mathbf{x}_\ell^H) = \mathbf{0}$ for $k \neq \ell$. To transmit d_k data streams over N_k antennas, transmitter k performs a linear processing operation on the signal $\mathbf{x}_k$ by a linear precoder $\mathbf{F}_k \in \mathbb{C}^{N_k \times d_k}$. The transmitted power at user k is constrained by

$$||\mathbf{F}_k||_F^2 \leq P_{k,\max}, \quad k \in \mathcal{K} \tag{1}$$

where $P_{k,\max}$ is the maximum transmit power at user k . The received signal at user k [1]

$$\mathbf{y}_k = \mathbf{H}_{k,k}\mathbf{F}_k\mathbf{x}_k + \sum_{\ell \in \mathcal{K}\setminus\{k\}} \mathbf{H}_{k,\ell}\mathbf{F}_\ell\mathbf{x}_\ell + \mathbf{z}_k, \quad k \in \mathcal{K} \tag{2}$$

where $\mathbf{z}_k \in \mathbb{C}^{M_k \times 1}$ is an additive while Gaussian noise vector with $\mathbf{z}_k \sim \mathcal{CN}(\mathbf{0}, \sigma_k^2\mathbf{I})$. $\mathbf{H}_{k,\ell} \in \mathbb{C}^{M_k \times N_\ell}$ is the channel matrix between transmitter ℓ and receiver k. The channels are assumed to be time-invariant over the transmission time period under consideration [2]. To retrieve d_k data streams, receiver k performs a linear signal processing operation on the received signal by a receive filter matrix $\mathbf{W}_k \in \mathbb{C}^{M_k \times d_k}$. The resultant signal at receiver k is given by

$$\begin{aligned}\hat{\mathbf{x}}_k &= \mathbf{W}_k^H\mathbf{y}_k \\ &= \mathbf{W}_k^H\mathbf{H}_{k,k}\mathbf{F}_k\mathbf{x}_k + \sum_{\ell \in \mathcal{K}\setminus\{k\}} \mathbf{W}_k^H\mathbf{H}_{k,\ell}\mathbf{F}_\ell\mathbf{x}_\ell + \mathbf{W}_k^H\mathbf{z}_k.\end{aligned} \tag{3}$$

It can be observed from (3) that the received signal suffers inter-user interference not only from the secondary but also from the primary network. Different from the general interference channels in which all the users in the network cooperate to design their transmission strategies in order to optimize the system performance, the cooperation between PUs and SUs is in general limited in CR networks. In such, the design of the precoders $\mathbf{F}_k$ and receive filter matrices $\mathbf{W}_k$ to efficiently handle interference is a fundamental challenge.

To design the transmission strategies of the users, we adopt the CSI assumptions similar to [2], [18], [21–23]. The perfect CSI of the primary links and secondary links is assumed to be available at the SUs. In practice, the secondary receiver can estimate its associated CSI and then feedback to its transmitter. In addition, the SU can obtain the cross-channels to the primary receivers by sensing the signal transmitted from the primary receiver in time-division duplexing (TDD). Since the channels are assumed be fixed in the processing period, the cross-channel matrices can be exchanged in the secondary networks [2]. Furthermore, we assume the CSI is perfectly known at the users. For the cases in which the CSI is imperfectly known, the results in this paper are served as a benchmark [18], [21], [23].

To perfectly eliminate all interference in (3), a full IA scheme can be applied. Then, the precoders and receive matrices are designed to meet the following conditions [10], [11]:

$$\mathbf{W}_k^H\mathbf{H}_{k,\ell}\mathbf{F}_\ell = \mathbf{0}, \quad k, \ell \in \mathcal{K}, k \neq \ell, \tag{4}$$

$$\text{rank}(\mathbf{W}_k^H\mathbf{H}_{k,k}\mathbf{F}_k) = d_k, \quad k \in \mathcal{K}. \tag{5}$$

The transceivers satisfying the above conditions can be designed by an iterative algorithm using the interference leakage minimization in [10]. As discussed in [16], full IA can result in zero inter-user interference but the sum-rate performance of the PUs can be significantly degraded since the PUs sacrifice their own transmission rate to assist the SUs to achieve the interference-free transmission. In this paper, we will propose the Min-Max MSE based IA in which interference from the SUs to the PUs is completely eliminated while the SUs adapt their transmission strategies to improve the performance.

3. Problem Formulation and Proposed Method

In this section, we consider optimizing the precoding and receiver matrices to improve the worst user MSE performance of the system subject to individual transmitted power constraints. Since the PUs have higher priority to access the spectrum than the SUs, the PUs can selfishly design their transmission strategies without awareness of the presence of the SUs. On the other hand, the SU transmission strategies must guarantee no harmful interference to the PUs.

3.1 Primary Transmission Strategies

It is worth mentioning that the PUs typically offer limited cooperation with the SUs, the PUs determine their transmission strategies based on the knowledge of channels between the PUs [5]. From (3), the post-processed signal at PU k, $k \in \mathcal{K}_p$, can be rewritten as

$$\begin{aligned}\hat{\mathbf{x}}_k = &\underbrace{\mathbf{W}_k^H\mathbf{H}_{k,k}\mathbf{F}_k\mathbf{x}_k}_{\text{desired signal}} + \underbrace{\sum_{\ell \in \mathcal{K}_p\setminus\{k\}} \mathbf{W}_k^H\mathbf{H}_{k,\ell}\mathbf{F}_\ell\mathbf{x}_\ell}_{\text{interference from PUs}} \\ &+ \underbrace{\sum_{\ell \in \mathcal{K}_s} \mathbf{W}_k^H\mathbf{H}_{k,\ell}\mathbf{F}_\ell\mathbf{x}_\ell}_{\text{interference from SUs}} + \underbrace{\mathbf{W}_k^H\mathbf{z}_k}_{\text{noise}}, \quad k \in \mathcal{K}_p.\end{aligned} \tag{6}$$

The SU transmission must not cause any interference to the PUs. Thus, the zero-interference conditions should be imposed.

$$\mathbf{W}_k^H\mathbf{H}_{k,\ell}\mathbf{F}_\ell = \mathbf{0}, \ell \in \mathcal{K}_s, k \in \mathcal{K}_p. \tag{7}$$

Accordingly, the MSE of PU k, $k \in \mathcal{K}_p$, from (6) is defined by

$$\begin{aligned}\xi_k(\mathbf{F}_k, \mathbf{W}_k) &= \mathbb{E}[||\hat{\mathbf{x}}_k - \mathbf{x}_k||_2^2] \\ &= ||\mathbf{I} - \mathbf{W}_k^H\mathbf{H}_{k,k}\mathbf{F}_k||_F^2 + \sigma_k^2||\mathbf{W}_k||_F^2 \\ &+ \sum_{\ell \in \mathcal{K}_p\setminus\{k\}} ||\mathbf{W}_k^H\mathbf{H}_{k,\ell}\mathbf{F}_\ell||_F^2, \quad k \in \mathcal{K}_p.\end{aligned} \tag{8}$$

It can be seen that minimizing the MSE in (8) results in the interference leakage minimization in IA schemes [24].

[1]Note that user k is referred to as a PU when $k \in \mathcal{K}_p$ and user k with $k \in \mathcal{K}_s$ is a SU while user k is either a PU or a SU for $k \in \mathcal{K}$.

Thus, the MSE minimization is also known as MSE-based IA [24], [25]. To provide fairness among users, we design the precoders and receive filters at PUs to minimize the worst user MSE among PUs. The design problem can be mathematically formulated as

$$\min_{\{\mathbf{W}_k,\mathbf{F}_k\}_{k\in\mathcal{K}_p}} \max_{k\in\mathcal{K}_p} \xi_k(\mathbf{F}_k,\mathbf{W}_k), \tag{9a}$$

$$\text{s.t.} \quad ||\mathbf{F}_k||_F^2 \le P_{k,\max}, \quad k\in\mathcal{K}_p. \tag{9b}$$

It can be shown that the optimization problem (9) is not jointly convex in the precoder and receive matrices. In addition, the transceiver matrices $\{\mathbf{F}_k\}$ and $\{\mathbf{W}_k\}$ are coupled with each other in the objective function. Thus, it is difficult to optimize the transceivers simultaneously. Fortunately, the optimization problem (9) is convex for individual variables while the others are fixed. Thus, we adopt the alternating optimization algorithm to iteratively solve (9) [24], [25]. First, given fixed receive matrices, we design the precoders which can be reformulated as

$$\min_{\{\mathbf{F}_k\}_{k\in\mathcal{K}_p}} \max_{k\in\mathcal{K}_p} \xi_k(\mathbf{F}_k,\mathbf{W}_k), \tag{10a}$$

$$\text{s.t.} \quad ||\mathbf{F}_k||_F^2 \le P_{k,\max}, \quad k\in\mathcal{K}_p. \tag{10b}$$

By introducing an auxiliary variable η_p, the optimization problem (10) can be equivalently rewritten as

$$\min_{\{\mathbf{F}_k\}_{k\in\mathcal{K}_p},\eta_p} \eta_p, \tag{11a}$$

$$\sqrt{\xi_k(\mathbf{F}_k,\mathbf{W}_k)} \le \eta_p, \quad k\in\mathcal{K}_p, \tag{11b}$$

$$\text{s.t.} \quad ||\mathbf{F}_k||_F^2 \le P_{k,\max}, \quad k\in\mathcal{K}_p. \tag{11c}$$

Next, we shall reformulate the optimization problem (11) into a tractable one. Note that $\xi_k(\mathbf{F}_k,\mathbf{W}_k)$ can be equivalently rewritten as

$$\xi_k(\mathbf{F}_k,\mathbf{W}_k) = ||\mathbf{W}_k^H\mathbf{\Phi}_k\mathbf{H}_p\mathbf{F}_p - \mathbf{\Delta}_k||_F^2 + \sigma_k^2||\mathbf{W}_k||_F^2, \quad k\in\mathcal{K}_p \tag{12}$$

where we have defined

$$\mathbf{H}_p = \begin{bmatrix} \mathbf{H}_{1,1} & \mathbf{H}_{1,2} & \cdots & \mathbf{H}_{1,K_p} \\ \mathbf{H}_{2,1} & \mathbf{H}_{2,2} & \cdots & \mathbf{H}_{2,K_p} \\ \vdots & \vdots & \vdots & \vdots \\ \mathbf{H}_{K_p,1} & \mathbf{H}_{K_p,2} & \cdots & \mathbf{H}_{K_p,K_p} \end{bmatrix},$$

$$\mathbf{\Phi}_k = \left[\ \mathbf{0}_{M_k\times\sum_{\ell=1}^{k-1}M_\ell} \quad \mathbf{I}_{M_k} \quad \mathbf{0}_{M_k\times\sum_{\ell=k+1}^{K_p}M_\ell}\ \right],$$

$$\mathbf{\Delta}_k = \left[\ \mathbf{0}_{d_k\times\sum_{\ell=1}^{k-1}d_\ell} \quad \mathbf{I}_{d_k} \quad \mathbf{0}_{d_k\times\sum_{\ell=k+1}^{K_p}d_\ell}\ \right],$$

and

$$\mathbf{F}_p = \text{blkdiag}(\mathbf{F}_1,\mathbf{F}_2,\cdots,\mathbf{F}_{K_p}).$$

Applying formulas $||\mathbf{X}||_F^2 = (\text{vec}(\mathbf{X}))^H\text{vec}(\mathbf{X}) = ||\text{vec}(\mathbf{X})||_2^2$ and $\text{vec}(\mathbf{ABC}) = (\mathbf{I}\otimes\mathbf{AB})\text{vec}(\mathbf{C})$ into (12), one has

$$\begin{aligned}\xi_k(\mathbf{F}_k,\mathbf{W}_k) = &\ ||[\mathbf{I}_{d_p}\otimes(\mathbf{W}_k^H\mathbf{\Phi}_k\mathbf{H}_p)]\text{vec}(F_p) - \text{vec}(\mathbf{\Delta}_k)||_2^2, \\ &+\sigma_k^2||\mathbf{W}_k||_F^2 \\ = &\left\| \begin{array}{c} \delta_k \\ \mathbf{\Gamma}_k\text{vec}(F_p) - \text{vec}(\mathbf{\Delta}_k) \end{array} \right\|_2^2 \end{aligned} \tag{13}$$

with $d_p = \sum_{k\in\mathcal{K}_p} d_k$, $\delta_k = \sigma_k||\mathbf{W}_k||_F$, $\mathbf{\Gamma}_k = \mathbf{I}_{d_p}\otimes(\mathbf{W}_k^H\mathbf{\Phi}_k\mathbf{H}_p)$, for $k\in\mathcal{K}_p$. The precoder design in (11) becomes

$$\min_{\{\mathbf{F}_k\}_{k\in\mathcal{K}_p},\eta_p} \eta_p, \tag{14a}$$

$$\text{s.t.} \quad \left\| \begin{array}{c} \delta_k \\ \mathbf{\Gamma}_k\text{vec}(F_p) - \text{vec}(\mathbf{\Delta}_k) \end{array} \right\|_2 \le \eta_p, \ k\in\mathcal{K}_p, \tag{14b}$$

$$||\text{vec}(\mathbf{F}_k)||_2 \le \sqrt{P_{k,\max}}\ k\in\mathcal{K}_p. \tag{14c}$$

It can be observed from (14) that the objective is linear while the constraints are second-order cones. Thus, the optimization problem (14) is known as SOCP which can be efficiently solved by available solvers such as SeDuMi, CVX [27], [28].

Next, by fixing the precoders, we design the receive filter matrices. Note that the receive filter matrix at one PU only affects its own MSE. Thus, the receive matrix design for the PRx k, $k\in\mathcal{K}_p$, from (9) is recast as

$$\min_{\mathbf{W}_k} \xi_k(\mathbf{F}_k,\mathbf{W}_k). \tag{15}$$

Taking the partial derivative of $\xi_k(\mathbf{F}_k,\mathbf{W}_k)$ with respect to $\mathbf{W}_k^\dagger$ and setting it to zero, one can obtain the optimal receive filter as

$$\mathbf{W}_k = \left[\sum_{\ell\in\mathcal{K}_p} \mathbf{H}_{k,\ell}\mathbf{F}_\ell\mathbf{F}_\ell^H\mathbf{H}_{k,\ell}^H + \sigma_k^2\mathbf{I}\right]^{-1} \mathbf{H}_{k,k}\mathbf{F}_k, \ k\in\mathcal{K}_p. \tag{16}$$

Denote the objective function in (9) as $\psi_p(\mathbf{F}_k,\mathbf{W}_k) = \max_{k\in\mathcal{K}_p}\xi_k(\mathbf{F}_k,\mathbf{W}_k)$. ϵ denotes an acceptable error tolerance. The iterative algorithm for the design of the PU precoding and receive matrices is summarized in Algorithm 1.

Algorithm 1: Iterative algorithm for the transceiver designs of the primary network.

Data: K_p, M_k, N_k, d_k, $P_{k,\max}$ for $k\in\mathcal{K}_p$, CSI of the PU links, ϵ;

Initialization: $\{\mathbf{W}_k^{(0)}\}_{k\in\mathcal{K}_p}$, $\{\mathbf{F}_k^{(0)}\}_{k\in\mathcal{K}_p}$, $n = 0$;

Calculate MSE ξ_k from (8) and obtain $\psi_p(\mathbf{F}_k^{(0)},\mathbf{W}_k^{(0)})$;

repeat

  Update $n = n+1$;

  Solve (14) to obtain $\{\mathbf{F}_k^{(n)}\}_{k\in\mathcal{K}_p}$;

  Calculate $\{\mathbf{W}_k^{(n)}\}_{k\in\mathcal{K}_p}$ from (16) ;

  Calculate MSE ξ_k from (8) and obtain $\psi_p(\mathbf{F}_k^{(n)},\mathbf{W}_k^{(n)})$;

until $|\psi_p(\mathbf{F}_k^{(n)},\mathbf{W}_k^{(n)}) - \psi_p(\mathbf{F}_k^{(n-1)},\mathbf{W}_k^{(n-1)})| \le \epsilon$;

Result: Optimal solutions $\{\mathbf{W}_k^*\}_{k\in\mathcal{K}_p}$, $\{\mathbf{F}_k^*\}_{k\in\mathcal{K}_p}$.

Note that to design the precoders and receive filters at the PUs, the PUs are oblivious of the presence of the SUs and do not need to know the CSI related to the SUs. The PUs require only the CSI of the PU links. Let $\mathbf{F}_k^{(n)}$ and $\mathbf{W}_k^{(n)}$ be the solutions to (14) and (15) at iteration n, respectively. Since the optimization problem (14) minimizes the maximum MSE while the receive filters are fixed, we have $\psi_p(\mathbf{F}_k^{(n)}, \mathbf{W}_k^{(n)}) \geq \psi_p(\mathbf{F}_k^{(n+1)}, \mathbf{W}_k^{(n)})$. Similarly, the maximum MSE is minimized with the receiver filter in (16), i.e., $\psi_p(\mathbf{F}_k^{(n+1)}, \mathbf{W}_k^{(n)}) \geq \psi_p(\mathbf{F}_k^{(n+1)}, \mathbf{W}_k^{(n+1)})$. Thus, one has

$$\psi_p(\mathbf{F}_k^{(n)}, \mathbf{W}_k^{(n)}) \geq \psi_p(\mathbf{F}_k^{(n+1)}, \mathbf{W}_k^{(n)}) \geq \psi_p(\mathbf{F}_k^{(n+1)}, \mathbf{W}_k^{(n+1)}). \tag{17}$$

This means that the worst user MSE is monotonically reduced at each iteration. In addition, the MSE is lower bounded by zero. Thus, Algorithm 1 is convergent at least to a local optimal solution.

With regard to computational complexity, the major computational complexity of Algorithm 1 comes from solving the SOCP (14) to find $\{\mathbf{F}_k^{(n)}\}_{k\in\mathcal{K}_p}$ and calculating $\{\mathbf{W}_k^{(n)}\}_{k\in\mathcal{K}_p}$ from (16). The SOCP is solved with the computational complexity of $O((\sum_{k\in\mathcal{K}_p} N_k d_k)^2(2K_p)^{2.5} + (2K_p)^{3.5})$ [28] while the computational complexity of finding $\{\mathbf{W}_k^{(n)}\}_{k\in\mathcal{K}_p}$ is $O(\sum_{k\in\mathcal{K}_p} M_k^3)$. Thus, the overall computational complexity of Algorithm 1 is approximated by $O((\sum_{k\in\mathcal{K}_p} N_k d_k)^2(2K_p)^{2.5}+(2K_p)^{3.5}+\sum_{k\in\mathcal{K}_p} M_k^3)$ for each iteration.

3.2 Secondary Transmission Strategies

Given the PU transmission strategies, the SUs adapt their transmission to optimize their transmission performance while not creating harmful interference to the PUs. From (3), the estimated signal at SU k, $k \in \mathcal{K}_s$, can be rewritten as

$$\begin{aligned}\hat{\mathbf{x}}_k = &\underbrace{\mathbf{W}_k^H \mathbf{H}_{k,k}\mathbf{F}_k\mathbf{x}_k}_{\text{desired signal}} + \underbrace{\sum_{\ell\in\mathcal{K}_s\setminus\{k\}} \mathbf{W}_k^H \mathbf{H}_{k,\ell}\mathbf{F}_\ell\mathbf{x}_\ell}_{\text{interference from SUs}} \\ &+ \underbrace{\sum_{\ell\in\mathcal{K}_p} \mathbf{W}_k^H \mathbf{H}_{k,\ell}\mathbf{F}_\ell\mathbf{x}_\ell}_{\text{interference from PUs}} + \underbrace{\mathbf{W}_k^H \mathbf{z}_k}_{\text{noise}}, \quad k \in \mathcal{K}_s.\end{aligned} \tag{18}$$

Different from the PUs, the SUs must cope with interference from both SUs and PUs. To enhance the system performance, the SUs exploit the CSI of the channel links from the SUs to PUs and between the SUs to design the optimal transceivers. From (18), the MSE for SU k is given by

$$\begin{aligned}\xi_k(\mathbf{F}_k, \mathbf{W}_k) &= \mathbb{E}[||\hat{\mathbf{x}}_k - \mathbf{x}_k||_2^2], \\ &= ||\mathbf{I} - \mathbf{W}_k^H \mathbf{H}_{k,k}\mathbf{F}_k||_F^2 + \sum_{\ell\in\mathcal{K}_s\setminus\{k\}} ||\mathbf{W}_k^H \mathbf{H}_{k,\ell}\mathbf{F}_\ell||_F^2 \\ &+ \sum_{\ell\in\mathcal{K}_p} ||\mathbf{W}_k^H \mathbf{H}_{k,\ell}\mathbf{F}_\ell||_F^2 + \sigma_k^2||\mathbf{W}_k||_F^2, \quad k \in \mathcal{K}_s.\end{aligned} \tag{19}$$

With the conditions of zero-interference from the SUs to the PUs given in (7), the fairness Min-Max MSE transceiver design for the SUs is expressed as

$$\min_{\{\mathbf{W}_k,\mathbf{F}_k\}_{k\in\mathcal{K}_s}} \quad \max_{k\in\mathcal{K}_s} \quad \xi_k(\mathbf{F}_k, \mathbf{W}_k), \tag{20a}$$

$$\text{s.t.} \quad ||\mathbf{F}_k||_F^2 \leq P_{k,\max}, \quad k \in \mathcal{K}_s, \tag{20b}$$

$$\mathbf{W}_k^H \mathbf{H}_{k,\ell}\mathbf{F}_\ell = \mathbf{0}, \; k \in \mathcal{K}_p, \ell \in \mathcal{K}_s. \tag{20c}$$

Note that condition (20c) guarantees that interference from the SUs does not spill into the desired signal subspace at the PU receivers. To satisfy condition (20c), the SU precoder, $\mathbf{F}_\ell, \ell \in \mathcal{K}_s$ must lie in the null space of $\mathbf{W}_k^H \mathbf{H}_{k,\ell}$ for all $k \in \mathcal{K}_p$. It implies that

$$\mathbf{F}_\ell \in \mathcal{N}(\mathbf{W}_1^H \mathbf{H}_{1,\ell}) \cap \mathcal{N}(\mathbf{W}_2^H \mathbf{H}_{2,\ell}) \cap \ldots \cap \mathcal{N}(\mathbf{W}_{K_p}^H \mathbf{H}_{K_p,\ell}). \tag{21}$$

By defining matrix $\mathcal{H}_\ell \in \mathbb{C}^{(\sum_{k\in\mathcal{K}_p} d_k)\times M_\ell}$ as

$$\mathcal{H}_\ell = [(\mathbf{W}_1^H \mathbf{H}_{1,\ell})^H, (\mathbf{W}_2^H \mathbf{H}_{2,\ell})^H, \ldots, (\mathbf{W}_{K_p}^H \mathbf{H}_{K+p,\ell})^H]^H, \tag{22}$$

equation (21) is rewritten as

$$\mathbf{F}_\ell \in \mathcal{N}(\mathcal{H}_\ell). \tag{23}$$

Then, the necessary condition for existence of nullity of matrix $\mathcal{H}_\ell$ is $M_\ell \geq \sum_{k\in\mathcal{K}_p} d_k$. Accordingly, a null space of $\mathcal{H}_\ell$ can be found by [26]

$$\mathbf{\Psi}_\ell = \left[\mathbf{I}_{M_\ell} - \mathcal{H}_\ell^H(\mathcal{H}_\ell\mathcal{H}_\ell^H)^{-1}\mathcal{H}_\ell\right]_1^{M_\ell - \sum_{k\in\mathcal{K}_p} d_k}. \tag{24}$$

Then, to fulfill condition (23), we propose to design each precoder as a product of two matrices given by

$$\mathbf{F}_\ell = \mathbf{\Psi}_\ell \tilde{\mathbf{F}}_\ell \tag{25}$$

where $\tilde{\mathbf{F}}_\ell \in \mathbb{C}^{(M_\ell - \sum_{k\in\mathcal{K}_p} d_k)\times d_\ell}$ is designed to improve the performance of the SUs. It should be noted that with the structure in (25), the signals transmitted from the SUs are not pilled into the desired signal subspace at the PUs. Them, the optimization problem (20) can be rewritten as

$$\min_{\{\tilde{\mathbf{F}}_k,\mathbf{W}_k\}_{k\in\mathcal{K}_s}} \quad \max_{k\in\mathcal{K}_s} \quad \xi_k(\tilde{\mathbf{F}}_k, \mathbf{W}_k), \tag{26a}$$

$$\text{s.t.} \quad ||\mathbf{\Psi}_k\tilde{\mathbf{F}}_k||_F^2 \leq P_{k,\max}, \; k \in \mathcal{K}_s. \tag{26b}$$

Similar to the PU transceiver designs, the optimization problem (26) is nonconvex with respect to the design variables. However, it is convex with respect to the precoder matrix variables when the receive filter matrices are fixed, and vice versa. Thus, we employ the alternating optimization. By fixing the receive filters, the precoder design can be represented as

$$\min_{\{\tilde{\mathbf{F}}_k\}_{k\in\mathcal{K}_s}} \quad \max_{k\in\mathcal{K}_s} \quad \xi_k(\tilde{\mathbf{F}}_k, \mathbf{W}_k), \tag{27a}$$

$$\text{s.t.} \quad ||\mathbf{\Psi}_k\tilde{\mathbf{F}}_k||_F^2 \leq P_{k,\max}, \; k \in \mathcal{K}_s. \tag{27b}$$

To transform problem (27) into a tractable convex optimization, we define

$$\mathbf{H}_s = \begin{bmatrix} \mathbf{H}_{K_p+1,K_p+1} & \mathbf{H}_{K_p+1,K_p+2} & \cdots & \mathbf{H}_{K_p+1,K} \\ \mathbf{H}_{K_p+2,K_p+1} & \mathbf{H}_{K_p+2,K_p+2} & \cdots & \mathbf{H}_{K_p+2,K} \\ \vdots & \vdots & \vdots & \vdots \\ \mathbf{H}_{K,K_p+1} & \mathbf{H}_{K,K_p+2} & \cdots & \mathbf{H}_{K,K} \end{bmatrix};$$

$$\mathbf{\Psi}_s = \text{blkdiag}(\mathbf{\Psi}_{K_p+1}, \mathbf{\Psi}_{K_p+1}, \cdots, \mathbf{\Psi}_{K_p+1}),$$

$$\boldsymbol{\Phi}_k = \left[\begin{array}{ccc} \mathbf{0}_{M_k \times \sum_{\ell=K_p+1}^{k-1} M_\ell} & \mathbf{I}_{M_k} & \mathbf{0}_{M_k \times \sum_{\ell=k+1}^{K} M_\ell} \end{array} \right],$$

$$\boldsymbol{\Delta}_k = \left[\begin{array}{ccc} \mathbf{0}_{d_k \times \sum_{\ell=K_p+1}^{k-1} d_\ell} & \mathbf{I}_{d_k} & \mathbf{0}_{d_k \times \sum_{\ell=k+1}^{K} d_\ell} \end{array} \right],$$

and

$$\mathbf{F}_s = \text{blkdiag}(\tilde{\mathbf{F}}_{K_p+1}, \tilde{\mathbf{F}}_{K_p+2}, \cdots, \tilde{\mathbf{F}}_K).$$

Then, the MSE in (19) of the SUs can be rewritten as

$$\xi_k(\tilde{\mathbf{F}}_k, \mathbf{W}_k) = ||\mathbf{W}_k^H \boldsymbol{\Phi}_k \mathbf{H}_s \boldsymbol{\Psi}_s \mathbf{F}_s - \boldsymbol{\Delta}_k||_F^2 + \sum_{\ell \in \mathcal{K}_p} ||\mathbf{W}_k^H \mathbf{H}_{k,\ell} \mathbf{F}_\ell||_F^2 + \sigma_k^2 ||\mathbf{W}_k||_F^2, \ k \in \mathcal{K}_s, \quad (28)$$

or, equivalently,

$$\xi_k(\tilde{\mathbf{F}}_k, \mathbf{W}_k) = \left\| \begin{array}{c} \delta_k \\ \boldsymbol{\Gamma}_k \text{vec}(F_s) - \text{vec}(\boldsymbol{\Delta}_k) \end{array} \right\|_2^2, \quad (29)$$

where $\delta_k = (\sum_{\ell \in \mathcal{K}_p} ||\mathbf{W}_k^H \mathbf{H}_{k,\ell} \mathbf{F}_\ell||_F^2 + \sigma_k^2 ||\mathbf{W}_k||_F^2)^{1/2}$, and $\boldsymbol{\Gamma}_k = \mathbf{I}_{d_s} \otimes (\mathbf{W}_k^H \boldsymbol{\Phi}_k \mathbf{H}_s \boldsymbol{\Psi}_s)$ with $d_s = \sum_{k \in \mathcal{K}_s} d_k$. Then, the precoder design for the SUs is written as the following SOCP

$$\min_{\{\tilde{\mathbf{F}}_k\}_{k \in \mathcal{K}_s}, \eta_s} \quad \eta_s, \quad (30a)$$

$$\text{s.t.} \quad \left\| \begin{array}{c} \delta_k \\ \boldsymbol{\Gamma}_k \text{vec}(F_s) - \text{vec}(\boldsymbol{\Delta}_k) \end{array} \right\|_2 \le \eta_s, \quad k \in \mathcal{K}_s, \quad (30b)$$

$$||(\mathbf{I}_{d_k} \otimes \boldsymbol{\Psi}_k) \text{vec}(\tilde{\mathbf{F}}_k)||_2 \le \sqrt{P_{k,\max}} \quad k \in \mathcal{K}_s, \quad (30c)$$

where η_s is an additional auxiliary variable. After achieving the optimal precoders, the next step is to design the receive filters. It can be noted that each receive filter of the user affects to its own MSE rather than the other users. With the obtained precoders from (30), we can express the design problem of the receive filter for the SRx k as follows

$$\min_{\mathbf{W}_k} \quad \xi_k(\tilde{\mathbf{F}}_k, \mathbf{W}_k), \quad k \in \mathcal{K}_s. \quad (31)$$

The optimal receive matrices can be obtained by solving $\partial \xi_k / \partial \mathbf{W}_k^\dagger = 0$ which results in

$$\mathbf{W}_k = \left[\sum_{\ell \in \mathcal{K}} \mathbf{H}_{k,\ell} \boldsymbol{\Psi}_\ell \tilde{\mathbf{F}}_\ell \tilde{\mathbf{F}}_\ell^H \boldsymbol{\Psi}_\ell^H \mathbf{H}_{k,\ell} + \sigma_k^2 \mathbf{I} \right]^{-1} \mathbf{H}_{k,k} \boldsymbol{\Psi}_k \tilde{\mathbf{F}}_k, \quad k \in \mathcal{K}_s. \quad (32)$$

The detailed alternating optimization for the SU transceiver designs is given in Algorithm 2 in which we have defined $\psi_s(\mathbf{F}_k, \mathbf{W}_k) = \max_{k \in \mathcal{K}_s} \xi_k(\mathbf{F}_k, \mathbf{W}_k)$. Note that from (19) the SUs need to estimate the total interference from the PUs to design their transceivers in the secondary network [6]. In addition, it is observed from (20c) that the SUs require the knowledge of the receive filters at the PUs. The SUs can obtain the knowledge of the PU receive filters by sensing the information exchanged from the PRx to the PTxs or by exploiting the central fusion center [22] or the spectrum manager [29].

Algorithm 2: Iterative algorithm for the transceiver designs of the secondary network.

Data: $K_p, K_s, M_k, N_k, d_k, P_{k,\max}$ for $k \in \mathcal{K}$, $\{\mathbf{F}_k, \mathbf{W}_k\}_{k \in \mathcal{K}_p}$, global CSI, ϵ;

Initialization: $\{\mathbf{W}_k^{(0)}\}_{k \in \mathcal{K}_s}$, $\{\mathbf{F}_k^{(0)}\}_{k \in \mathcal{K}_s}$, $n = 0$;

Calculate MSE ξ_k from (19) and obtain $\psi_s(\mathbf{F}_k^{(0)}, \mathbf{W}_k^{(0)})$;

repeat

- Update $n = n + 1$;
- Calculate matrix $\boldsymbol{\Psi}_k^{(n)}$ for $k \in \mathcal{K}_s$ from (24);
- Solve (30) to obtain $\{\tilde{\mathbf{F}}_k^{(n)}\}_{k \in \mathcal{K}_s}$, and then achieve $\mathbf{F}_k^{(n)} = \boldsymbol{\Psi}_k^{(n)} \tilde{\mathbf{F}}_k^{(n)}$;
- Calculate $\{\mathbf{W}_k^{(n)}\}_{k \in \mathcal{K}_s}$ from (32);
- Calculate MSE ξ_k from (19) and obtain $\psi_s(\mathbf{F}_k^{(n)}, \mathbf{W}_k^{(n)})$;

until $|\psi_s(\mathbf{F}_k^{(n)}, \mathbf{W}_k^{(n)}) - \psi_s(\mathbf{F}_k^{(n-1)}, \mathbf{W}_k^{(n-1)})| \le \epsilon$;

Result: Optimal solutions $\{\mathbf{W}_k^*\}_{k \in \mathcal{K}_s}$, $\{\mathbf{F}_k^*\}_{k \in \mathcal{K}_s}$.

Similar to Algorithm 1, it is easy to show that Algorithm 2 is also guaranteed to be convergent at least to a local optimal solution since we have

$$\psi_s(\mathbf{F}_k^{(n)}, \mathbf{W}_k^{(n)}) \ge \psi_s(\mathbf{F}_k^{(n+1)}, \mathbf{W}_k^{(n)}) \ge \psi_s(\mathbf{F}_k^{(n+1)}, \mathbf{W}_k^{(n+1)}). \quad (33)$$

The major computational complexity of Algorithm 2 to find $\{\mathbf{F}_k^{(n)}\}_{k \in \mathcal{K}_s}$ by calculating $\boldsymbol{\Psi}_k^{(n)}$ for $k \in \mathcal{K}_s$ from (24) and solving the SOCP (30) is $O((\sum_{\ell \in \mathcal{K}_s}(M_\ell - \sum_{i \in \mathcal{K}_p} d_i) d_\ell)^2 (2K_s)^{2.5} + (2K_s)^{3.5} + (\sum_{k \in \mathcal{K}_p} d_k)^3)$ [28]. The computational complexity of finding $\{\mathbf{W}_k^{(n)}\}_{k \in \mathcal{K}_s}$ is $O(\sum_{k \in \mathcal{K}_s} M_k^3)$. Thus, the overall computational complexity for each iteration in Algorithm 2 is about $O((\sum_{\ell \in \mathcal{K}_s}(M_\ell - \sum_{i \in \mathcal{K}_p} d_i) d_\ell)^2 (2K_s)^{2.5} + (2K_s)^{3.5} + (\sum_{k \in \mathcal{K}_p} d_k)^3 + \sum_{k \in \mathcal{K}_s} M_k^3)$.

4. Simulation Results

In this section, we provide simulation results to validate the performance of the proposed Min-Max MSE based IA. In all simulations, the MIMO Rayleigh channel coefficients are randomly generated with zero mean and unit variance. We consider that $\sigma_k^2 = \sigma_n^2 = 1$, $k \in \mathcal{K}$, $P_{k,\max} = P_{p,\max}$ for $k \in \mathcal{K}_p$ and $P_{k,\max} = P_{s,\max}$ for $k \in \mathcal{K}_s$. We define the signal to noise ratio (SNR) of the PUs as $\gamma_p = \frac{P_{p,\max}}{\sigma_n^2}$ and the SNR of the SUs as $\gamma_s = \frac{P_{s,\max}}{\sigma_n^2}$. We consider the CR network with $K_p = 2$ PUs and $K_s = 2$ SUs. We set $N_k = 6$, $M_k = 2$, $d_k = 2$ for $k \in \mathcal{K}$, and $\epsilon = 10^{-6}$. We carry out Monte-Carlo simulations over 200 channel realizations.

First, we study the convergence characteristics of the proposed Min-Max MSE method. We set $\gamma_p = \gamma_s = 10$ dB. We choose initial precoder and receive filter matrices to be

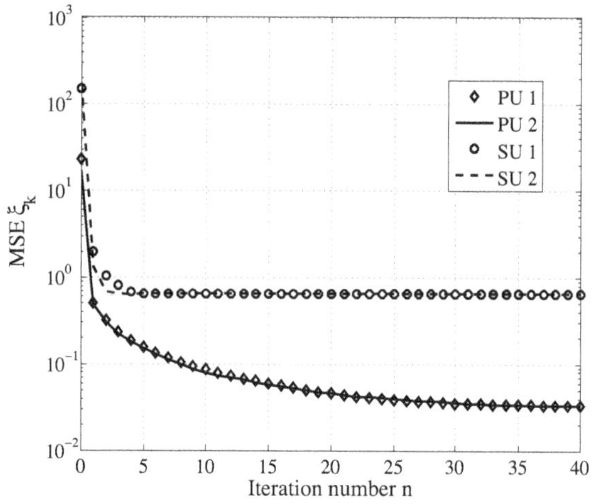

Fig. 2. Convergence of the proposed alternating optimization for the Min-Max MSE transceiver design problems.

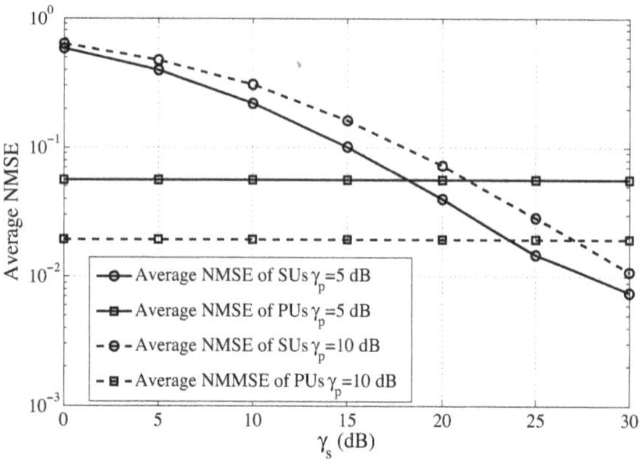

Fig. 3. Average NMSEs of the PUs and SUs versus the SU SNR.

right and left singular matrices of the corresponding desired signal links. The evolution of the user MSEs obtained from each iteration is plotted in Fig. 2 for a channel realization. As observed from the figure that the algorithm is convergent in less than 40 iterations. It should be noted that at the optimum, the MSEs of the SUs are the same. Similarly, the MSEs of all PUs are identical at the optimum. This confirms the MSE fairness among users in the same network. In addition, with the same SNR, the MSE performance of the PUs is better than that of the SUs. That is because the PUs have higher priority to access the spectrum while the SUs must sacrifice their performance to maintain no harmful interference to the PUs.

Next, we investigate the MSE performance of the proposed method in terms of the average MSE normalized by the number of the data streams (NMSE) [30] and we study the impact of the SU transmission on the performance of the PUs. Figure 3 displays the average worst-user NMSE versus the SU SNR for the primary and secondary networks with $\gamma_p = \{5, 10\}$ dB. It can be seen from Fig. 3 that when the transmitted power of the PUs increases, i.e., γ_p increases from 5 dB to 10 dB, the MSE performance of the PUs is improved. On the other hand, an increase in the PU transmitted power results in the MSE performance degradation of the SUs. That is because that an increase in transmitted power of

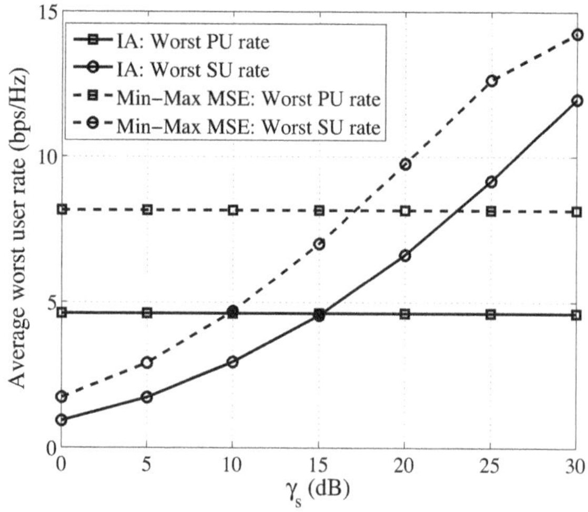

Fig. 4. Average worst user rates of the PUs and SUs versus the SU SNR for the proposed Min-Max MSE in comparison with those for IA in [17].

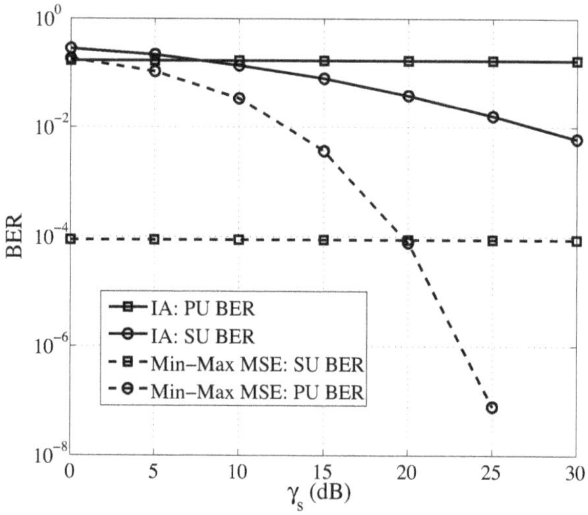

Fig. 5. BER of the PUs and SUs versus the SU SNR for the proposed Min-Max MSE in comparison with those for IA in [17].

the PUs causes severer interference to the SUs. In contrast, the MSEs of the PUs are unchanged while the transmitted power of the SUs increases. This confirms that the SUs do not cause harmful interference to the PUs.

Now, we compares the user information rate and bit eror rate (BER) performance of the proposed Min-Max MSE algorithm with the IA scheme in [17] in which the interference leakage minimization is adopted to find the transceivers. We plot the average worst-user information rate of the SUs and PUs versus the SU SNR in Fig. 4 for $\gamma_p = 5$ dB. It can be observed from Fig. 4 that the proposed method outperforms the IA scheme in [17] in terms of the average worst user rate. The worst user rate of the primary network is unchanged when the SU SNR γ_s increases. This result reveals that the PU performance is not affected by the transmission of the SUs. To investigate the BER performance, the transmitted signals are modulated by quadrature phase shift keying (QPSK) signal constellations. Figure 5 plots the average BER of all the users for the primary and secondary networks. It is clear from Fig. 5 that the proposed method offers the superior BER performance when compared to the IA scheme in [17].

5. Conclusion

This paper has presented the optimal designs of the transmission strategies for the multiuser MIMO CR network. Our approach is to adopt the Min-Max MSE performance metric as the objective function for the optimization problems. We have derived the SOCP problems for the precoder designs and the closed form solutions for the receive filter designs. We have introduced the structures of the precoders to guarantee that the SU transmission do not cause adverse interference to the PUs. The iterative algorithms for the transceiver designs are showed to be convergent in few tens of iteration and their computational complexity is low. The numerical results demonstrate that the proposed method provide MSE fairness among the users. They also show that the proposed method outperforms the previous IA approach in terms of the worst user rate and bit error rate.

Acknowledgments

This research is funded by Vietnam National Foundation for Science and Technology Development (NAFOSTED) under grant number 102.04-2013.46.

References

[1] WANG, B., LIU, K. R. Advances in cognitive radio networks: A survey. *IEEE Journal of Selected Topics in Signal Processing*, 2011, vol. 5, no. 1, p. 5–23. DOI: 10.1109/JSTSP.2010.2093210

[2] LIU, Y., DONG, L. Spectrum sharing in MIMO cognitive radio networks based on cooperative game theory. *IEEE Transaction on Wireless Communications*, 2014, vol. 13, no. 9, p. 4807–4820. DOI: 10.1109/TWC.2014.2331287

[3] GOLDSMITH, A., JAFAR, S. A., MARIC, I., et al. Breaking spectrum gridlock with cognitive radios: An information theoretic perspective. *Proceedings of the IEEE*, 2009, vol. 97, no. 5, p. 894–914. DOI: 10.1109/JPROC.2009.2015717

[4] HOSSEINI, S. A., ABOLHASSANI, B., SADOUGH, S. M. S. A new protocal for cooperative spectrum sharing in mobile cognitive radio networks. *Radioengineering*, 2015, vol. 24, no. 3, p. 757–764. DOI: 10.13164/re.2015.0757

[5] DU, H., RATNARAJAH, T. Robust transceiver beamforming in MIMO cognitive radio via second-order cone programming. *IEEE Transactions on Signal Processing*, 2011, vol. 60, no. 2, p. 781–792. DOI: 10.1109/TSP.2011.2174790

[6] KIM, S. J., GIANNAKIS, G. B. Optimal resource allocation for MIMO ad hoc cognitive radio networks. *IEEE Transactions on Information Theory*, 2011, vol. 57, no. 5, p. 3117–3131. DOI: 10.1109/TIT.2011.2120270

[7] GUI, X., KANG, G. X., ZHANG, P. Sum-rate maximising in cognitive MIMO ad-hoc networks using weighted MMSE approach. *Electronics Letters*, 2012, vol. 48, no. 19, p. 1240–1242. DOI:10.1049/el.2012.1472

[8] ZHANG, Y., ANESE, E., GIANNAKIS, G. B. Distributed optimal beamformers for cognitive radios bobust to channel uncertainties. *IEEE Transactions on Signal Processing*, 2012, vol. 60, no. 12, p. 6495–6508. DOI: 10.1109/TSP.2012.2218240

[9] YAO, Y., LI, G., XU, J., et al. Space alignment based on regulaized inversion precoding in cognitive transmission. *Radioengineering*, 2015, vol. 24, no. 3, p. 824–829. DOI: 10.13164/re.2015.0824

[10] GOMADAM, K., CADAMBE, V. R., JAFAR, S. A. A distributed numerical approach to interference alignment and applications to wireless interference networks. *IEEE Transactions on Information Theory*, 2011, vol. 57, no. 6, p. 3309–3322. DOI: 10.1109/TIT.2011.2142270

[11] CADAMBE, V. R., JAFAR, S. A. Interference alignment and degrees of freedom of the K-user interference channel. *IEEE Transactions on Information Theory*, 2008, vol. 54, no. 8, p. 3425–3441. DOI: 10.1109/TIT.2008.926344

[12] PAPAILIOPOULOS, D., DIMAKIS, A. Interference alignment as a rank constrained rank minimization. *IEEE Transactions on Signal Processing*, 2012, vol. 60, no. 8, p. 4278–4288. DOI: 10.1109/TSP.2012.2197393

[13] ALEXANDROPOULOS, G. C., PAPADIAS, C. B. A reconfigurable iterative algorithm for the K-user MIMO interference channel. *Signal Processing*, 2013, vol. 93, no. 12, p. 353–3362. DOI: 10.1016/j.sigpro.2013.05.027

[14] MADDAH-ALI, M. A., MOTAHARI, A. S., KHANDANI, A. K. Communication over MIMO X channels: Interference alignment, decomposition and performance analysis. *IEEE Transactions on Information Theory*, 2008, vol. 54, no. 8, p. 3457–3470. DOI: 10.1109/TIT.2008.926460

[15] CASTANHEIRA, D., SILVA, A., GAMEIRO, A. Set optimization for efficient interference alignment in heterogeneous networks. *IEEE Transactions on Wireless Communications*, 2014, vol. 13, no. 10, p. 5648–5660. DOI: 10.1109/TWC.2014.2322855

[16] MEN, H., ZHAO, N., JIN, M., et al. Optimal transceiver design for interference alignment based cognitive radio networks. *IEEE Communications Letters*, 2015, vol. 19, no. 8, p. 1442–1445. DOI: 10.1109/LCOMM.2015.2442243

[17] REZAEI, F., TADAION, A. Interference alignment in cognitive radio networks. *IET Communications*, 2014, vol. 8, no. 10, p. 1769–1777. DOI: 10.1049/iet-com.2013.0731

[18] PERLAZA, S. M., FAWAZ, N., LASAULCE, S., et al. From spectrum pooling to space pooling: Opportunistic interference alignment in MIMO cognitive networks. *IEEE Transactions on Signal Processing*, 2010, vol. 58, no. 7, p. 3728–3741. DOI: 10.1109/TSP.2010.2046084

[19] SHARMA, S. K., CHARZINOTAS, S., OTTERSTEN, B. Interference alignment for spectral coexistence of heterogeneous networks. *EURASIP Journal on Wireless Communications and Networking*, 2013, vol. 2013, no. 1, p. 1–14. DOI: 10.1186/1687-1499-2013-46

[20] ALI, S. S., CASTANHEIRA, D., SILVA, A., et al. Transmission cooperative strategies for MIMO-OFDM heterogeneous networks. *Radioengineering* , 2015, vol. 24, no. 2, p. 431–441. DOI: 10.13164/re.2015.0431

[21] ZHANG, R., LIANG, Y. C. Exploiting multi-antennas for opportunistic spectrum sharing in cognitive radio networks. *IEEE Journal of Selected Topics in Signal Processing*, 2008, vol. 2, no. 1, p. 88–102. DOI: 10.1109/JSTSP.2007.914894

[22] MOSLEH, S., ABOUEI, J., AGHABOZORGI, M. R. Distributed opportunistic interference alignment using threshold-based beamforming in MIMO overlay cognitive radio. *IEEE Transactions on Vehicular Technology*, 2014, vol. 63, no. 8, p. 3783–3793. DOI: 10.1109/TVT.2014.2305849

[23] KRIKIDIS, I. Space alignment for cognitive transmission in MIMO uplink channels. *EURASIP Journal on Wireless Communications and Networking*, 2010, vol. 2010, no. 1, p. 1–6. DOI: 10.1155/2010/465157

[24] LU, E., MA, T., LU, I. T. Interference alignment-like behaviors of MMSE designs for general multiuser MIMO systems. In *Proceedings of the IEEE Global Telecommunications Conference (GLOBECOM 2011)*. 2011, p. 1–5. DOI: 10.1109/GLOCOM.2011.6134284

[25] SHEN, H., LI, B., TAO, M., et al. MSE-based transceiver designs for the MIMO interference channel. *IEEE Transaction on Wireless Communications*, 2010, vol. 9, no. 11, p. 3480–3489. DOI: 10.1109/TWC.2010.091510.091836

[26] HORN, R. A., JOHNSON, C. R. *Matrix Analysis*. Cambridge University Press: 1986. ISBN: 0-521-30586-1

[27] GRANT, M., BOYD, S. CVX: MATLAB software for disciplined convex programming, version 2.1. Available at: http://cvxr.com/cvx

[28] PEAUCELLE, D., HENRION, D., LABIT, Y., et al. Users guide for seumi interface 1.04, 2002. Available at: http://homepages.laas.fr/peaucell/software/sdmguide.pdf

[29] ISLAM, H., LIANG, Y. C., HOANG, A. T. Joint power control and beamforming for cognitive radio networks. *IEEE Transactions on Wireless Communications*, 2008, vol. 7, no. 7, p. 2415–2419. DOI: 10.1109/TWC.2008.061003

[30] KHANDAKER, M. R. A., RONG, Y. Transceiver optimization for multihop MIMO relay multicasting from multiple sources. *IEEE Transaction on Wireless Communications*, 2014, vol. 13, no. 9, p. 5162–5172. DOI: 10.1109/TWC.2014.2322361

About the Authors . . .

Ha Hoang KHA received the B.Eng. and M.Eng. degrees from Ho Chi Minh City University of Technology, in 2000 and 2003, respectively, and the Ph.D. degree from the University of New South Wales, Sydney, Australia, in 2009, all in Electrical Engineering and Telecommunications. From 2000 to 2004, he was a research and teaching assistant with the Department of Electrical and Electronics Engineering, Ho Chi Minh City University of Technology. He was a visiting research fellow at the School of Electrical Engineering and Telecommunications, the University of New South Wales, Australia, from 2009 to 2011. He was a postdoctoral research fellow at the Faculty of Engineering and Information Technology, University of Technology Sydney, Australia from 2011 to 2013. He is currently a lecturer at the Faculty of Electrical and Electronics Engineering, Ho Chi Minh City University of Technology, Vietnam. His research interests are in digital signal processing and wireless communications, with a recent emphasis on convex optimization techniques in signal processing for wireless communications.

Hung Quang TA received his B.Eng. and M.Eng. in Electronics and Telecommunications from the Hanoi University of Science and Technology (HUST), Hanoi, in 1996 and 2000, respectively. He received his PhD in Electrical Engineering & Telecommunications from the University of New South Wales (UNSW), Australia, in 2012. Currently, he is a lecturer at Faculty of Information Technology, Hanoi University, Vietnam. His research interests include optimization techniques, signal processing, image processing, embedded systems and cryptography.

Energy Harvesting-based Spectrum Access with Incremental Cooperation, Relay Selection and Hardware Noises

Tan N. NGUYEN [1], *Tran Trung DUY* [2], *Gia-Thien LUU* [2], *Phuong T. TRAN* [1], *Miroslav VOZNAK* [1,3]

[1] Wireless Communications Research Group, Faculty of Electrical and Electronics Engineering, Ton Duc Thang University, No. 19 Nguyen Huu Tho Street, Tan Phong Ward, District 7, Ho Chi Minh City, Vietnam
[2] Posts and Telecommunications Institute of Technology, 11 Nguyen Dinh Chieu St., Dist. 1, Ho Chi Minh City, Vietnam
[3] VSB Technical University of Ostrava, 17. listopadu 15/2172, 708 33 Ostrava - Poruba, Czech Republic

{nguyenhattan, tranthanhphuong}@tdt.edu.vn, {trantrungduy, lgthien}@ptithcm.edu.vn, miroslav.voznak@vsb.cz

Abstract. *In this paper, we propose an energy harvesting (EH)-based spectrum access model in cognitive radio (CR) network. In the proposed scheme, one of available secondary transmitters (STs) helps a primary transmitter (PT) forward primary signals to a primary receiver (PR). Via the cooperation, the selected ST finds opportunities to access licensed bands to transmit secondary signals to its intended secondary receiver (SR). Secondary users are assumed to be mobile, hence, optimization of energy consumption for these users is interested. The EH STs have to harvest energy from the PT's radio-frequency (RF) signals to serve the PT-PR communication as well as to transmit their signals. The proposed scheme employs incremental relaying technique in which the PR only requires the assistance from the STs when the transmission between PT and PR is not successful. Moreover, we also investigate impact of hardware impairments on performance of the primary and secondary networks. For performance evaluation, we derive exact and lower-bound expressions of outage probability (OP) over Rayleigh fading channel. Monte-Carlo simulations are performed to verify the theoretical results. The results present that the outage performance of both networks can be enhanced by increasing the number of the ST-SR pairs. In addition, the outage performance of both primary and secondary networks is severely degraded with the increasing of hardware impairment level. It is also shown that fraction of time used for EH and positions of the secondary users significantly impact on the system performance.*

Keywords

Cognitive radio, relay selection, energy harvesting, hardware impairments, outage probability

1. Introduction

Recently, energy harvesting (EH) has been gained much attention as a promising technique to prolong lifetime for energy-limited wireless networks without recharging batteries [1]. The EH systems allow wireless devices to collect energy from radio frequency (RF) and convert the harvested energy into direct current power by internal inverter circuits. To enhance performances for the EH networks, in terms of outage probability, error rate and diversity gain, cooperative relaying protocols [2] were considered as an efficient solution. The authors in [3] studied a dual-hop relaying protocol with EH and a greedy switching policy. In [4], the authors proposed two EH-based relaying protocols: time switching-based relaying (TSR) and power splitting-based relaying (PSA). In [5], the amplify-and-forward (AF) relay harvests the energy from the source, which is used to relay the source data to the destination. Moreover, the authors in [5] proposed optimization methods to maximize the end-to-end instantaneous channel capacity in both half-duplex and full-duplex relay modes. In [6], closed-form expressions of average channel capacity and throughput for EH-based decode-and-forward (DF) networks were derived. Cooperative relaying schemes with multiple source-destination pairs communicating with one EH relay were proposed in [7]. Furthermore, the authors in [7] proposed various power allocation strategies and evaluated the performances via both simulations and analyzes.

With the rapid increasing of wireless devices and systems, spectrum scarcity becomes a critical issue due to emergence of wireless services. To overcome this problem, Mitola [8] introduced cognitive radio (CR) concept, in which licensed users (primary users (PUs)) can share licensed bands to unlicensed users (secondary users (SUs)). The basic idea of the CR technique is that two wireless systems coexist and operate at the same spectrum resources. However, they have different priorities: PUs can use the licensed bands any time, while SUs can use the spectrum with lower priority [9]. In conventional CR method [10], SUs must detect the presence/absence of PUs. If there are vacant bands detected, SUs can access them to transmit the secondary data. Recently, researchers have proposed two spectrum sharing methods in which SUs can use the licensed bands without detecting PUs' operations. In the first method, named underlay CR [11], [12], PUs and SUs can use the licensed bands at the same time, provided that the co-channel interference from the secondary transmission must be lower than a maximum threshold required by PUs. In the second method, named overlay CR [13–15], SUs can use licensed bands but they must help PUs enhance the quality of service (QoS). In particular, the secondary transmitters (STs) play a role as relays for the primary network and via this assistance, they can find opportunities to access the licensed bands.

So far, most of the published papers have assumed that transceiver hardware is perfect. However, in practice, the transceiver hardware of wireless devices is imperfect because it is affected by impairments such as amplifier-amplitude non-linearity, I/Q imbalance and phase noise [16]. Hence, the hardware impairments (HI) need to be taken into account when evaluating performances of wireless relay networks. In [17], outage probability (OP) of two-way relay networks with the hardware noises at relay was investigated. The authors in [18] proposed and evaluated the outage performance of proactive relay selection protocols in co-channel interference networks. In [19], the authors investigated the joint impact of the imperfect hardware and the wireless power transfer on the outage performance of two-way underlay CR. The results in [16–18] have presented that the presence of HI degrades the system performances over fading channels.

In practical wireless networks, users are usually in motion, which requires extra energy in addition to energy used for signal transmission. Moreover, CR secondary users also consume energy for spectrum sensing process. Therefore, it is very imperative that energy efficiency must be considered for secondary users in CR networks. To the best of our knowledge, there are several reports related to cooperative CR models using the EH technique. In particular, in [20], the ST is deployed with a rechargeable battery which can harvest energy from the environment. The authors in [21] proposed an optimal spectrum access for EH-based CR networks, where the ST at the beginning of each time slot needs to determine whether to remain idle so as to conserve energy, or to execute spectrum sensing to acquire knowledge of the current spectrum occupancy state. In [22], [23], the authors studied the performance of the secondary networks operating on underlay mode. Published works [24], [25] evaluated the performances of both primary and secondary networks in overlay CR environment, where a single EH-based ST uses the AF or DF technique to forward the combined signals to both primary receiver (PR) and secondary receiver (SR). The authors in [26] proposed a cooperative spectrum access protocol in which the SU can harvest the energy from the primary signals and then assists the primary data transmission using Alamouti technique. Li et al. [27] also proposed a spectrum sharing method based on competitive price game model.

In this paper, we propose a new cooperative spectrum sharing relaying protocols, where the best EH-based ST is chosen to assist the data transmission between the nodes PT and PR. We also propose an incremental relaying cooperation [2] in which the PR only requires the help from STs when the communication between the PT and PR is not successful. Different with the schemes proposed in [24–26], the proposed scheme includes multiple ST-SR pairs and only the best ST is selected for the cooperation. Moreover, the impact of hardware impairments on the outage performance of the primary and secondary networks is also investigated. For performance evaluation, we derive exact and lower-bound closed-form expressions of outage probability for both networks over Rayleigh fading channel. We then perform Monte-Carlo simulations to verify the theoretical derivations.

The rest of this paper is organized as follows. The system model of the proposed protocol is described in Sec. 2. In Sec. 3, we evaluate the performance of the proposed scheme. The simulation results are shown in Sec. 4 and Sec. 5 concludes this paper.

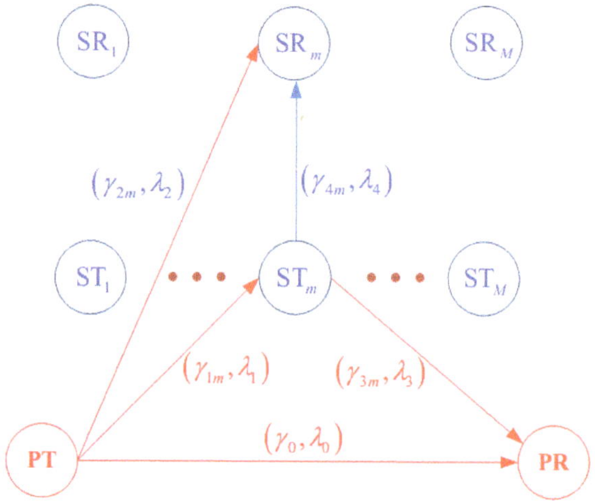

Fig. 1. System model of the proposed protocol.

2. System Model

In Fig. 1, we present the system model of the proposed scheme, where the primary network includes one PT-PR pair, while there are M ST-SR pairs in the secondary network. The PT attempts to transmit its data to the PR with the help of STs, i.e., $\text{ST}_m (m = 1, 2, ..., M)$. Via cooperation, the ST_m can access the licensed band to transmit its data to the SR_m.

Assume that all of the terminals are equipped with a single antenna and operate on half-duplex mode. We also assume that the STs (SRs) are close together and form a cluster, and hence, the distances from the PT to STs (SR) are assumed to be the same [11]. Let us denote d_0, d_1, d_2, d_3 and d_4 as the distances of the PT – PR, PT – ST_m, PT – SR_m,

ST_m–PR and ST_m–SR_m links, respectively. We also denote $h_{PT,PR}$, h_{PT,ST_m}, h_{PT,SR_m}, $h_{ST_m,PR}$ and h_{ST_m,SR_m} as channel coefficients of the PT – PR, PT – ST_m, PT – SR_m, ST_m–PR and ST_m–SR_m links, respectively. We assume that all of the links are modeled to be block and flat Rayleigh fading channels, which remain constant during an interval T and change independently over different intervals. As mentioned in [11], channel gains γ_0, γ_{1m}, γ_{2m}, γ_{3m} and γ_{4m} ($\gamma_0 = |h_{PT,PR}|^2$, $\gamma_{1m} = |h_{PT,ST_m}|^2$, $\gamma_{2m} = |h_{PT,SR_m}|^2$, $\gamma_{3m} = |h_{ST_m,PR}|^2$, $\gamma_{4m} = |h_{ST_m,SR_m}|^2$) are exponential random variables (RVs) with parameters λ_0, λ_1, λ_2, λ_3 and λ_4, respectively [11]. Moreover, to take path-loss into account, the parameters can be expressed as a function of the distance and the path-loss exponent by [11]: $\lambda_0 = d_0^{\chi}$, $\lambda_1 = d_1^{\chi}$, $\lambda_2 = d_2^{\chi}$, $\lambda_3 = d_3^{\chi}$ and $\lambda_4 = d_4^{\chi}$, respectively, where χ is path-loss coefficient.

We assume that the STs are limited-energy terminals which must harvest energy from the RF signals generated by the PT. It is also assumed that the nodes STs and SRs have enough energy for processing the control messages in set-up phases [23] as well as for decoding the received data.

The operation of the proposed protocol is split into three sub-blocks. Similar to the time switching scheme in [23], a duration of αT is used for the STs to harvest the energy from the PT, a duration of $(1-\alpha)\,\mathrm{T}/2$ for the STs and the PR to receive the data from the PT, and a duration of $(1-\alpha)\,\mathrm{T}/2$ is employed to forward the data from the selected ST to the PR and the intended SR. Then, the energy that the ST_m can harvest is given as [23, eq. (13)] [1]:

$$E_m = \eta\alpha \mathrm{T} P\gamma_{1m} \tag{1}$$

where η ($0 < \eta \leq 1$) is the energy conversion efficiency that depends on the internal inverter circuit in the STs, and P is the transmit power of the PT.

Hence, the transmit power of the ST_m over the time $(1-\alpha)\,\mathrm{T}/2$ can be obtained by [23, eq. (14)]:

$$P_m = \frac{E_m}{(1-\alpha)\,\mathrm{T}/2} = \frac{2\eta\alpha P\gamma_{1m}}{1-\alpha} = \mu P\gamma_{1m} \tag{2}$$

where $\mu = 2\eta\alpha/(1-\alpha)$.

At the next sub-block, the PT transmits its data to the PR, which is also received by the ST_m and SR_m. Under the imperfect hardware, the received signal at the node X, X $\in$ {ST_m, SR_m, PR}, can be given as

$$y_X = \sqrt{P}h_{PT,X}\left(x_P + \eta_{t,PT}\right) + \eta_{r,X} + n_X \tag{3}$$

where x_P is the primary signal transmitted by the PT, n_X is the additive white Gaussian noise (AWGN), $\eta_{t,PT}$ and $\eta_{r,X}$ are the noises caused by the hardware impairments at the transmitter PT and the receiver X, respectively. Similar to [18], n_X, $\eta_{t,PT}$ and $\eta_{r,X}$ are modeled as zero-mean Gaussian noises with variance of N_0, κ_{PT}^t and $\kappa_X^r P|h_{PT,X}|^2$, respectively, where κ_{PT}^t and κ_X^r indicate the level of hardware impairments at the nodes PT and X. From (3), the achievable data rate between the nodes PT and PR can be calculated by

$$\begin{aligned} C_0 &= \frac{(1-\alpha)\,\mathrm{T}}{2}\log_2\left(1 + \frac{P|h_{PT,PR}|^2}{\left(\kappa_{PT}^t + \kappa_{PR}^r\right)P|h_{PT,PR}|^2 + N_0}\right), \\ &= \frac{(1-\alpha)\,\mathrm{T}}{2}\log_2\left(1 + \frac{\Psi\gamma_0}{\kappa_{PT,PR}\Psi\gamma_0 + 1}\right) \end{aligned} \tag{4}$$

where $\Psi = P/N_0$ is the average transmit signal-to-noise ratio (SNR), $\kappa_{PT,PR} = \kappa_{PT}^t + \kappa_{PR}^r$ is total hardware impairment level.

Similarly, we can obtain the instantaneous channel capacity of the PT – ST_m and PT – SR_m links, respectively as

$$\begin{aligned} C_{1m} &= \frac{(1-\alpha)\,\mathrm{T}}{2}\log_2\left(1 + \frac{\Psi\gamma_{1m}}{\kappa_{PT,ST_m}\Psi\gamma_{1m} + 1}\right), \\ C_{2m} &= \frac{(1-\alpha)\,\mathrm{T}}{2}\log_2\left(1 + \frac{\Psi\gamma_{2m}}{\kappa_{PT,SR_m}\Psi\gamma_{2m} + 1}\right) \end{aligned} \tag{5}$$

where $\kappa_{PT,ST_m} = \kappa_{PT}^t + \kappa_{ST_m}^r$ and $\kappa_{PT,SR_m} = \kappa_{PT}^t + \kappa_{SR_m}^r$.

At the end of the second sub-block, the PR attempts to decode the received signal. If this node can decode the source signal successfully, it informs the decoding status by generating an ACK message. In this case, the STs and SRs remove the primary signal from their buffers and use the third sub-block to transmit the secondary data [2].

To optimize the performance of the secondary network, we propose a strategy to select the best ST-SR pair. At first, let us consider the signal received at the SR_m due to the transmission of the ST_m:

$$y_{SR_m} = \sqrt{P_m}h_{ST_m,SR_m}\left(z_m + \eta_{t,ST_m}\right) + \eta_{r,SR_m} + n_{R_m} \tag{6}$$

where z_m is the signal transmitted by the ST_m and η_{t,ST_m} is the noise caused by the hardware impairments at the ST_m which can be modeled as zero-mean Gaussian noise with variance of $\kappa_{ST_m}^t$.

From (2) and (6), the instantaneous channel capacity of the ST_m – SR_m link can be given as

$$C_{4m} = \frac{(1-\alpha)\,\mathrm{T}}{2}\log_2\left(1 + \frac{\mu\Psi\gamma_{1m}\gamma_{4m}}{\kappa_{ST_m,SR_m}\mu\Psi\gamma_{1m}\gamma_{4m} + 1}\right) \tag{7}$$

where $\kappa_{ST_m,SR_m} = \kappa_{ST_m}^t + \kappa_{SR_m}^r$.

From (7), the best ST-SR pair can be selected by the following method:

$$ST_a - SR_a : \gamma_{1a}\gamma_{4a} = \max_{m=1,2,\ldots,M}\left(\gamma_{1m}\gamma_{4m}\right). \tag{8}$$

[1] As mentioned in [19], hardware impairments are not taken into the harvested energy.

[2] Because the transmission between the PT and the PR is successful, the primary network allows the secondary users to use the third sub-block to transmit their signals.

Equation (8) implies that the ST-SR pair which provides the highest channel gain of the ST-SR links is selected for the communication at the third sub-block.

Next, let us consider the event that the decoding status at the PR is unsuccessful. In this case, it sends back a NACK message to request a retransmission from one of the STs. We denote $\mathcal{W}_{\mathrm{SR}}$ as a set of the SRs that can decode the primary signal successfully. Without loss of generality, we can assume that $\mathcal{W}_{\mathrm{SR}} = \{\mathrm{SR}_1, \mathrm{SR}_2, .., \mathrm{SR}_{N_\mathrm{R}}\}$, where N_R $(0 \leq N_\mathrm{R} \leq M)$ is the cardinality of $\mathcal{W}_{\mathrm{SR}}$. Similarly, each SR will feedback the ACK (or NACK) message to indicate the successful (or unsuccessful) decoding status [3].

If there is at least one SR decoding the primary signal correctly ($N_\mathrm{R} \geq 1$), from the successful STs, i.e, $\mathrm{ST}_1, \mathrm{ST}_2, .., \mathrm{ST}_{N_\mathrm{R}}$, we propose a method to select the ST for the cooperation at the next sub-block as follows:

$$\mathrm{ST}_b : P_b = \max_{j=1,2,\ldots,N_\mathrm{R}} \left(P_j\right) \text{ or } \gamma_{1b} = \max_{j=1,2,\ldots,N_\mathrm{R}} \left(\gamma_{1j}\right) \quad (9)$$

where the ST providing the maximum harvested energy (or the highest channel gain between the PT and STs) is selected as the best candidate.

If the node ST_b can decode the primary signal x_P successfully, it combines linearly x_P and its own signal z_b, follows the strategy given in [15] as

$$x_c = \sqrt{\beta P_b} x_\mathrm{P} + \sqrt{(1-\beta) P_b} z_b \quad (10)$$

where βP_b and $(1-\beta) P_b$ are the fractions of the total transmit power P_b, which are allocated to the signals x_P and z_b, respectively.

Then, the ST_b broadcasts the combined signal x_c, and the received signals at the PR and SR_b can be given, respectively by

$$\begin{aligned} y_\mathrm{PR} &= \sqrt{\beta P_b} h_{\mathrm{ST}_b,\mathrm{PR}} \left(x_\mathrm{P} + \eta_{t,\mathrm{ST}_b,1}\right) \\ &+ \sqrt{(1-\beta) P_b} h_{\mathrm{ST}_b,\mathrm{PR}} \left(z_b + \eta_{t,\mathrm{ST}_b,2}\right) + \eta_{r,\mathrm{PR}} + n_\mathrm{PR}, \\ y_{\mathrm{SR}_b} &= \sqrt{\beta P_b} h_{\mathrm{ST}_b,\mathrm{SR}_b} \left(x_\mathrm{P} + \eta_{t,\mathrm{ST}_b,3}\right) \\ &+ \sqrt{(1-\beta) P_b} h_{\mathrm{ST}_b,\mathrm{SR}_b} \left(z_b + \eta_{t,\mathrm{ST}_b,4}\right) + \eta_{r,\mathrm{SR}_b} + n_{\mathrm{SR}_b}. \end{aligned} \quad (11)$$

It is noted from (11) that the variances of the hardware impairments $\eta_{t,\mathrm{ST}_b,u}$, $\eta_{r,\mathrm{PR}}$ and η_{r,SR_b} are $\kappa^t_{\mathrm{ST}_b}$, $\kappa^r_{\mathrm{PR}} P_b |h_{\mathrm{ST}_b,\mathrm{PR}}|^2$ and $\kappa^r_{\mathrm{SR}_b} P_b |h_{\mathrm{ST}_b,\mathrm{SR}_b}|^2$, respectively, where $u = 1, 2, 3, 4$.

Moreover, because the SR_b obtained the signal x_P before, it can remove the interference component $\sqrt{\beta P_b} h_{\mathrm{ST}_b,\mathrm{SR}_b} x_\mathrm{P}$ from the received signal. After canceling the interference, the signal y_{SR_b} can be rewritten by

$$\begin{aligned} y^*_{\mathrm{SR}_b} &= \sqrt{(1-\beta) P_b} h_{\mathrm{ST}_b,\mathrm{SR}_b} \left(z_b + \eta_{t,\mathrm{ST}_b,4}\right) \\ &+ \sqrt{\beta P_b} h_{\mathrm{ST}_b,\mathrm{SR}_b} \eta_{t,\mathrm{ST}_b,3} + \eta_{r,\mathrm{SR}_b} + n_{\mathrm{SR}_b}. \end{aligned} \quad (12)$$

Combining (2), (11) and (12), we respectively obtain the achievable capacity for the $\mathrm{ST}_b - \mathrm{PR}$ and $\mathrm{ST}_b - \mathrm{SR}_b$ links as

$$\begin{aligned} C_{3b} &= \frac{(1-\alpha)\mathrm{T}}{2} \log_2 \left(1 + \frac{\beta \mu \Psi \gamma_{1b} \gamma_{3b}}{\left(1 - \beta + \kappa_{\mathrm{ST}_b,\mathrm{PR}}\right) \mu \Psi \gamma_{1b} \gamma_{3b} + 1}\right), \\ C_{4b} &= \frac{(1-\alpha)\mathrm{T}}{2} \log_2 \left(1 + \frac{(1-\beta) \mu \Psi \gamma_{1b} \gamma_{4b}}{\kappa_{\mathrm{ST}_b,\mathrm{SR}_b} \mu \Psi \gamma_{1b} \gamma_{4b} + 1}\right) \end{aligned} \quad (13)$$

where $\kappa_{\mathrm{ST}_b,\mathrm{PR}} = \kappa^t_{\mathrm{ST}_b} + \kappa^r_{\mathrm{PR}}$ and $\kappa_{\mathrm{ST}_b,\mathrm{SR}_b} = \kappa^t_{\mathrm{ST}_b} + \kappa^r_{\mathrm{SR}_b}$.

Next, we consider the case where there is no SR receiving the primary signal successfully, i.e., $N_\mathrm{R} = 0$. In this case, one of the STs have to use the total harvested energy to serve the PR. Let $\mathcal{W}_{\mathrm{ST}}$ as a set of STs that can decode the primary signal successfully. Without loss of generality, we can assume that $\mathcal{W}_{\mathrm{ST}} = \{\mathrm{ST}_1, \mathrm{ST}_2, .., \mathrm{ST}_{N_\mathrm{T}}\}$, where $N_\mathrm{T}(0 \leq N_\mathrm{T} \leq M)$ is the cardinality of $\mathcal{W}_{\mathrm{ST}}$. It is obvious that if $N_\mathrm{T} = 0$, the system cannot select any STs for the retransmission, and hence the primary signal is dropped [4]. Otherwise, the best ST is chosen by the following selection strategy:

$$\mathrm{ST}_c : \gamma_{3c} = \max_{j=1,2,\ldots,N_\mathrm{T}} \left(\gamma_{3j}\right) \quad (14)$$

where the successful ST having the highest channel gain between itself and the PR is selected as the best relay.

Then, the received signal at the PR can be given by

$$y_\mathrm{PR} = \sqrt{P_c} h_{\mathrm{ST}_c,\mathrm{PR}} \left(x_\mathrm{P} + \eta_{t,\mathrm{ST}_c}\right) + \eta_{r,\mathrm{PR}} + n_\mathrm{PR}. \quad (15)$$

Finally, the instantaneous data rate of the ST_c-PR link can be formulated by

$$C_{3c} = \frac{(1-\alpha)\mathrm{T}}{2} \log_2 \left(1 + \frac{\mu \Psi \gamma_{1c} \gamma_{3c}}{\kappa_{\mathrm{ST}_c,\mathrm{PR}} \mu \Psi \gamma_{1c} \gamma_{3c} + 1}\right) \quad (16)$$

where $\kappa_{\mathrm{ST}_c,\mathrm{PR}} = \kappa^t_{\mathrm{ST}_c} + \kappa^r_{\mathrm{PR}}$.

3. Performance Evaluation

For ease of analysis, we assume that the total hardware impairment levels are the same, i.e., $\kappa_{\mathrm{Y,Z}} = \kappa$, for all $\{\mathrm{Y, Z}\} \in \{\mathrm{PT, PR, ST}_m, \mathrm{SR}_m\}$. [5]

3.1 Mathematical Preliminaries

Firstly, it is well-known that cumulative density function (CDF) and probability density function (PDF) of an exponential RV Y with parameter λ_Y can be given, respectively as

$$F_\mathrm{Y}(y) = 1 - \mathrm{e}^{-\lambda_\mathrm{Y} y}, \; f_\mathrm{Y}(y) = \lambda_\mathrm{Y} \mathrm{e}^{-\lambda_\mathrm{Y} y}. \quad (17)$$

[3] When the SR decodes the primary signals correctly, it can remove the primary signal component from the signals received from the ST [14, 15].

[4] In this case, the PT would start a new transmission without sharing the licensed band to the secondary network because the STs cannot help the PR retransmit the data.

[5] When the hardware impairment levels are different, with the same manner we also obtain exact and asymptotic expressions of outage probability for both networks.

Next, let us consider a RV $Y_{\max}$, i.e., $Y_{\max} = \max_{i=1,2,\ldots,K}(Y_i)$, where K is a positive integer and Y_i is an exponential RV whose parameter is λ_Y. Hence, the CDF of $Y_{\max}$ can be given as (see in [28, eq. (7)])

$$F_{Y_{\max}}(y) = \sum_{m=0}^{K}(-1)^m C_K^m e^{-m\lambda_Y y} \tag{18}$$

where $C_K^m = [K!/m!/(K-m)!]$.

Then, the corresponding PDF can be obtained by

$$f_{Y_{\max}}(y) = \sum_{m=0}^{K-1}(-1)^m C_{K-1}^m K\lambda_Y e^{-(m+1)\lambda_Y y}. \tag{19}$$

We now consider a RV Z_* that is product of two exponential RVs Z_1 and Z_2 ($Z_* = Z_1 Z_2$), whose parameters are Ω_1 and Ω_2, respectively. The CDF of Y_* can be formulated by

$$F_{Z_*}(z) = \Pr[Z_1 Z_2 < z] = \int_0^{+\infty} f_{Z_1}(t) F_{Z_2}(z/t)\,dt. \tag{20}$$

Using the CDF and PDF obtained in (17) for (20), and then applying [29, eq. (3.324.1)] for the corresponding integral, we obtain

$$F_{Z_*}(z) = 1 - \sqrt{4\Omega_1\Omega_2 z}K_1\left(\sqrt{4\Omega_1\Omega_2 z}\right) \tag{21}$$

where $K_1(.)$ is modified Bessel function of the second kind [29].

3.2 Outage Probability Analysis

Outage probability is defined by the probability that the achievable rate at a receiver is below a target rate, i.e., R_{th}. Moreover, the receiver can be assumed to correctly decode received signals if the data rate is higher than R_{th}.

At first, notations used in this sub-section can be listed as follows:

$$\theta = 2^{\frac{2R_{\text{th}}}{(1-\alpha)T}} - 1,\ \rho_0 = \frac{\theta}{(1-\kappa\theta)\Psi},$$
$$\rho_1 = \frac{\theta}{[\beta-(1-\beta+\kappa)\theta]\Psi},\ \rho_2 = \frac{\theta}{(1-\beta-\kappa\theta)\Psi}. \tag{22}$$

Now, the outage probability of the primary network can be formulated by

$$\begin{aligned} P_{\text{PR}}^{\text{out}} &= \Pr[C_0 < R_{\text{th}}]\Pr[N_{\text{R}} = 0]\times \\ &\left(\Pr[N_{\text{T}} = 0] + \sum_{u=1}^{M} C_M^u \Pr\left[\left\{\begin{array}{l} N_T = u-1 \\ C_{1c} \ge R_{th} \\ C_{3c} < R_{th}\end{array}\right.\right]\right) + \\ &\Pr[C_0 < R_{\text{th}}]\sum_{m=1}^{M} C_M^m \Pr[N_{\text{R}} = m]\Pr[C_{1b} < R_{\text{th}}] + \\ &\Pr[C_0 < R_{\text{th}}]\sum_{m=1}^{M} C_M^m \Pr[N_{\text{R}} = m]\Pr\left[\left\{\begin{array}{l} C_{1b} \ge R_{th} \\ C_{3b} < R_{th}\end{array}\right.\right]. \end{aligned} \tag{23}$$

In (23), $\Pr[N_{\text{T}} = x]$ and $\Pr[N_{\text{R}} = y]$ are probabilities that the number of the successful SRs and STs equals x and y, respectively.

Proposition 1: The outage probability $P_{\text{PR}}^{\text{out}}$ can be calculated by

$$P_{\text{PR}}^{\text{out}} = \begin{cases} \text{OP}_{\text{PR}}^1, & \text{if } \theta < \beta/(1-\beta+\kappa) \\ \text{OP}_{\text{PR}}^2, & \text{if } \beta/(1-\beta+\kappa) \le \theta < 1/\kappa \\ 1, & \text{if } \theta \ge 1/\kappa \end{cases} \tag{24}$$

where OP_{PR}^1 and OP_{PR}^2 are given by (31) and (32). *Proof*: see Appendix A.

From (24)-(32), we can observe that the exact expressions of the outage probability are still in integral form, which is difficult to use to design and optimize the considered system. Hence, our next objective is to derive approximate closed-form expressions of the outage performance at high transmit SNR.

Proposition 2: At high SNR values, i.e., $\Psi = P/N_0 \to +\infty$, the outage probability $P_{\text{PR}}^{\text{out}}$ can be approximated by closed-form expressions as follows:

$$P_{\text{PR}}^{\text{out}} \overset{\Psi\to+\infty}{\approx} \begin{cases} \text{OP}_{\text{PR}}^{1,\infty}, & \text{if } \theta < \beta/(1-\beta+\kappa) \\ \text{OP}_{\text{PR}}^{1,\infty}, & \text{if} \beta/(1-\beta+\kappa) \le \theta < 1/\kappa \end{cases} \tag{25}$$

where, $\text{OP}_{\text{PR}}^{1,\infty}$ and $\text{OP}_{\text{PR}}^{2,\infty}$ are calculated as in (33) and (34). *Proof*: at high Ψ regimes, we obtain the following approximation:

$$\int_{\rho_0}^{+\infty} e^{-ax}e^{-\frac{b}{x}}dx \overset{\Psi\to+\infty}{\approx} \int_0^{+\infty} e^{-ax}e^{-\frac{b}{x}}dx \overset{\Psi\to+\infty}{\approx} \sqrt{\frac{4b}{a}}K_1\left(\sqrt{\frac{4b}{a}}\right) \tag{26}$$

where a and b are positive real numbers.

Then, using (26) for the corresponding integrals in (31) and (32), we respectively obtain (33) and (34).

Similarly, the outage probability of the secondary network can be formulated by the following formula:

$$\begin{aligned} P_{\text{SR}}^{\text{out}} &= \Pr[C_0 \ge R_{\text{th}}]\Pr[C_{4a} < R_{\text{th}}] \\ &+ \Pr[C_0 < R_{\text{th}}]\Pr[N_{\text{R}} = 0] \\ &+ \Pr[C_0 < R_{\text{th}}]\sum_{m=1}^{M} C_M^m \Pr[N_{\text{R}} = m]\Pr[C_{1b} < R_{\text{th}}] \\ &+ \Pr[C_0 < R_{\text{th}}]\sum_{m=1}^{M} C_M^m \Pr[N_{\text{R}} = m] \\ &\times \Pr[C_{1b} \ge R_{\text{th}}, C_{4b} < R_{\text{th}}]. \end{aligned} \tag{27}$$

Proposition 3: The exact outage probability of the secondary network can be computed by

$$P_{\text{SR}}^{\text{out}} = \begin{cases} \text{OP}_{\text{SR}}^1, & \text{if } \theta < (1-\beta)/\kappa \\ \text{OP}_{\text{PR}}^2, & \text{if } (1-\beta)/\kappa \le \theta < 1/\kappa \\ 1, & \text{if } \theta \ge 1/\kappa \end{cases} \tag{28}$$

where OP^1_{SR} and OP^2_{SR} can be found from (35) and (36). *Proof*: see Appendix B.

Also, the outage probability OP^1_{SR} is still in integral form. Hence, we attempt to find an approximate closed-form for OP^1_{SR} as below.

Proposition 4: The outage probability OP^1_{SR} can be approximated at high Ψ region as in (37). *Proof*: similar to the proof of Proposition 2.

For performance comparison, we introduce the direct transmission (DT) protocol, in which the PT communicates with the PR without the help of the STs. In this protocol, the data rate of the PT-PR link is given by

$$C^{DT}_{PT-PR} = \log_2\left(1 + \frac{\Psi\gamma_0}{\kappa\Psi\gamma_0 + 1}\right). \quad (29)$$

The outage probability of DT protocol can be expressed by

$$P^{out}_{DT} = \Pr\left[C^{DT}_{PT-PR} < R_{th}\right] = \begin{cases} 1, & \text{if } \vartheta \geq 1/\kappa \\ 1 - e^{-\frac{\lambda_0\vartheta}{1-\kappa\vartheta}}, & \text{if } \vartheta < 1/\kappa \end{cases} \quad (30)$$

where $\vartheta = 2^{R_{th}}$-1.

$$\begin{aligned} OP^1_{PR} &= \left(1 - e^{-\lambda_0\rho_0}\right)\left(1 - e^{-\lambda_2\rho_0}\right)^M\left(1 - e^{-\lambda_1\rho_0}\right)^M \\ &+ \left(1 - e^{-\lambda_0\rho_0}\right)\left(1 - e^{-\lambda_2\rho_0}\right)^M \sum_{u=1}^{M} C^u_M\left(1 - e^{-\lambda_1\rho_0}\right)^{M-u} e^{-(u-1)\lambda_1\rho_0}\sum_{v=0}^{u}(-1)^v C^v_u\int_{\rho_0}^{+\infty}\lambda_1 e^{-\lambda_1 x} e^{-\frac{v\lambda_3\rho_0}{\mu x}}dx \\ &+ \left(1 - e^{-\lambda_0\rho_0}\right)\sum_{m=1}^{M} C^m_M\left(1 - e^{-\lambda_2\rho_0}\right)^{M-m} e^{-m\lambda_2\rho_0}\left(1 - e^{-\lambda_1\rho_0}\right)^m \\ &+ \left(1 - e^{-\lambda_0\rho_0}\right)\sum_{m=1}^{M} C^m_M\left(1 - e^{-\lambda_2\rho_0}\right)^{M-m} e^{-m\lambda_2\rho_0}\sum_{t=0}^{m-1}(-1)^t C^t_{m-1} m\lambda_1\left[\frac{e^{-(t+1)\lambda_1\rho_0}}{(t+1)\lambda_1} - \int_{\rho_0}^{+\infty} e^{-(t+1)\lambda_1 x} e^{-\frac{\lambda_3\rho_1}{\mu x}}dx\right], \end{aligned} \quad (31)$$

$$\begin{aligned} OP^2_{PR} &= \left(1 - e^{-\lambda_0\rho_0}\right)\left(1 - e^{-\lambda_2\rho_0}\right)^M\left(1 - e^{-\lambda_1\rho_0}\right)^M \\ &+ \left(1 - e^{-\lambda_0\rho_0}\right)\left(1 - e^{-\lambda_2\rho_0}\right)^M \sum_{u=1}^{M} C^u_M\left(1 - e^{-\lambda_1\rho_0}\right)^{M-u} e^{-(u-1)\lambda_1\rho_0}\sum_{v=0}^{u}(-1)^v C^v_u\int_{\rho_0}^{+\infty}\lambda_1 e^{-\lambda_1 x} e^{-\frac{v\lambda_3\rho_0}{\mu x}}dx \\ &+ \left(1 - e^{-\lambda_0\rho_0}\right)\sum_{m=1}^{M} C^m_M\left(1 - e^{-\lambda_2\rho_0}\right)^{M-m} e^{-m\lambda_2\rho_0}. \end{aligned} \quad (32)$$

$$\begin{aligned} OP^{1,\infty}_{PR} &= \left(1 - e^{-\lambda_0\rho_0}\right)\left(1 - e^{-\lambda_2\rho_0}\right)^M\left(1 - e^{-\lambda_1\rho_0}\right)^M \\ &+ \left(1 - e^{-\lambda_0\rho_0}\right)\left(1 - e^{-\lambda_2\rho_0}\right)^M . \sum_{u=1}^{M} C^u_M\left(1 - e^{-\lambda_1\rho_0}\right)^{M-u} e^{-(u-1)\lambda_1\rho_0} . \left(e^{-\lambda_1\rho_0} + \sum_{v=1}^{u}(-1)^v C^v_u\sqrt{\frac{4v\lambda_1\lambda_3\rho_0}{\mu}}K_1\left(\sqrt{\frac{4v\lambda_1\lambda_3\rho_0}{\mu}}\right)\right) \\ &+ \left(1 - e^{-\lambda_0\rho_0}\right)\sum_{m=1}^{M} C^m_M\left(1 - e^{-\lambda_2\rho_0}\right)^{M-m} . e^{-m\lambda_2\rho_0}\left(1 - e^{-\lambda_1\rho_0}\right)^m \\ &+ \left(1 - e^{-\lambda_0\rho_0}\right)\sum_{m=1}^{M} C^m_M\left(1 - e^{-\lambda_2\rho_0}\right)^{M-m} . e^{-m\lambda_2\rho_0}\sum_{t=0}^{m-1}(-1)^t C^t_{m-1} m\left(\frac{e^{-(t+1)\lambda_1\rho_0}}{t+1} - \sqrt{\frac{4\lambda_1\lambda_3\rho_1}{\mu(1+t)}}K_1\left(\sqrt{\frac{4(1+t)\lambda_1\lambda_3\rho_1}{\mu}}\right)\right), \end{aligned} \quad (33)$$

$$\begin{aligned} OP^{2,\infty}_{PR} &= \left(1 - e^{-\lambda_0\rho_0}\right)\left(1 - e^{-\lambda_2\rho_0}\right)^M\left(1 - e^{-\lambda_1\rho_0}\right)^M \\ &+ \left(1 - e^{-\lambda_0\rho_0}\right)\left(1 - e^{-\lambda_2\rho_0}\right)^M \sum_{u=1}^{M} C^u_M\left(1 - e^{-\lambda_1\rho_0}\right)^{M-u} e^{-(u-1)\lambda_1\rho_0}\left(e^{-\lambda_1\rho_0} + \sum_{v=1}^{u}(-1)^v C^v_u\sqrt{\frac{4v\lambda_1\lambda_3\rho_0}{\mu}}K_1\left(\sqrt{\frac{4v\lambda_1\lambda_3\rho_0}{\mu}}\right)\right) \\ &+ \left(1 - e^{-\lambda_0\rho_0}\right)\sum_{m=1}^{M} C^m_M\left(1 - e^{-\lambda_2\rho_0}\right)^{M-m} e^{-m\lambda_2\rho_0}. \end{aligned} \quad (34)$$

$$\mathrm{OP}_{\mathrm{SR}}^{1}=\mathrm{e}^{-\lambda_0\rho_0}\left(1-\sqrt{\frac{4\lambda_1\lambda_4\rho_0}{\mu}}K_1\left(\sqrt{\frac{4\lambda_1\lambda_4\rho_0}{\mu}}\right)\right)^{M}+\left(1-\mathrm{e}^{-\lambda_0\rho_0}\right)\left(1-\mathrm{e}^{-\lambda_2\rho_0}\right)^{M}$$
$$+\left(1-\mathrm{e}^{-\lambda_0\rho_0}\right)\sum_{m=1}^{M}C_M^m\left(1-\mathrm{e}^{-\lambda_2\rho_0}\right)^{M-m}\mathrm{e}^{-m\lambda_2\rho_0}\left(1-\mathrm{e}^{-\lambda_1\rho_0}\right)^{m}$$
$$+\left(1-\mathrm{e}^{-\lambda_0\rho_0}\right)\sum_{m=1}^{M}C_M^m\left(1-\mathrm{e}^{-\lambda_2\rho_0}\right)^{M-m}\mathrm{e}^{-m\lambda_2\rho_0}\sum_{t=0}^{m-1}(-1)^tC_{m-1}^t m\lambda_1\left[\frac{\mathrm{e}^{-(t+1)\lambda_1\rho_0}}{(t+1)\lambda_1}-\int_{\rho_0}^{+\infty}\mathrm{e}^{-(t+1)\lambda_1x}\mathrm{e}^{-\frac{\lambda_4\rho_2}{\mu x}}\mathrm{d}x\right], \quad (35)$$

$$\mathrm{OP}_{\mathrm{SR}}^{2}=\mathrm{e}^{-\lambda_0\rho_0}\left(1-\sqrt{\frac{4\lambda_1\lambda_4\rho_0}{\mu}}K_1\left(\sqrt{\frac{4\lambda_1\lambda_4\rho_0}{\mu}}\right)\right)^{M}+1-\mathrm{e}^{-\lambda_0\rho_0}. \quad (36)$$

$$\mathrm{OP}_{\mathrm{SR}}^{1}\overset{\Psi\to+\infty}{\approx}\mathrm{e}^{-\lambda_0\rho_0}\left(1-\sqrt{\frac{4\lambda_1\lambda_4\rho_0}{\mu}}K_1\left(\sqrt{\frac{4\lambda_1\lambda_4\rho_0}{\mu}}\right)\right)^{M}+\left(1-\mathrm{e}^{-\lambda_0\rho_0}\right)\left(1-\mathrm{e}^{-\lambda_2\rho_0}\right)^{M}$$
$$+\left(1-\mathrm{e}^{-\lambda_0\rho_0}\right)\sum_{m=1}^{M}C_M^m\left(1-\mathrm{e}^{-\lambda_2\rho_0}\right)^{M-m}\mathrm{e}^{-m\lambda_2\rho_0}\left(1-\mathrm{e}^{-\lambda_1\rho_0}\right)^{m}$$
$$+\left(1-\mathrm{e}^{-\lambda_0\rho_0}\right)\sum_{m=1}^{M}C_M^m\left(1-\mathrm{e}^{-\lambda_2\rho_0}\right)^{M-m}\mathrm{e}^{-m\lambda_2\rho_0}\sum_{t=0}^{m-1}(-1)^tC_{m-1}^t m\lambda_1\left(\frac{\mathrm{e}^{-(t+1)\lambda_1\rho_0}}{t+1}-\sqrt{\frac{4\lambda_1\lambda_4\rho_2}{\mu(1+t)}}K_1\left(\sqrt{\frac{4(1+t)\lambda_1\lambda_4\rho_2}{\mu}}\right)\right). \quad (37)$$

4. Numerical Results and Discussion

In this section, we present Monte Carlo simulations to verify the derivations in Sec. 3. For the simulation environment, we consider a two-dimensional X-Y networks in which PT, PR, STs, SRs are respectively placed at $(0,0)$, $(1,0)$, $(x_{\mathrm{ST}},0)$ and $(x_{\mathrm{ST}},0.25)$, respectively, where $0<x_{\mathrm{ST}}<1$. In all of the simulations, the time block is normalized by 1 (T = 1) and the path-loss exponent is fixed by 4 ($\chi=4$).

In Figures 2 and 3, we respectively present the outage probability of the primary and secondary networks as a function of Ψ in dB. The parameters of these figures are fixed by $R_{\mathrm{th}}=1$, $x_{\mathrm{ST}}=0.5$, $\kappa=0.01$, $\alpha=0.1$, $\beta=0.95$, $\eta=0.5$ and $M\in\{1,2,3,6\}$. From Fig. 2, we can see that the outage performance of the primary network significantly enhances, as compared with the DT protocol. Moreover, it can be observed that the outage probability decreases with the increasing the number of the ST-SR pairs. As observed from Fig. 3, the outage performance of the secondary network is also better with high M values. It is worthy noting from Figures 2–3 that the simulation results match very well with the exact theoretical results and the approximate theoretical results rapidly converge to the exact ones.

Figure 4 illustrates the outage performance of both networks as a function of the co-ordinate x_{ST} when $R_{\mathrm{th}}\in\{1.5,2\}$, $\kappa=0$, $\alpha=0.1$, $\beta=0.95$, $\eta=0.5$, $M=2$ and $\Psi=0$ dB. We can observe from Fig. 4 that the outage probability rapidly increases with the increasing of R_{th}. It is also seen that the outage performance of the secondary network in the proposed protocol decreases when the value of x_{ST} increases. It is due to the fact that the link distances, i.e., PT-ST and PT-SR, increase when x_{ST} increases, which reduces the probability that the nodes ST and SR can decode the primary data successfully (or decreases the probability that STs can access the licensed bands as well as the probability that SRs can remove the interference component from the primary data). Moreover, the position of the nodes ST also impacts on the performance of the primary network in the proposed scheme. In particular, when $R_{\mathrm{th}}=1.5$, the outage probability increases when the value of x_{ST} changes from 0.05 to 0.95. More interesting, with $R_{\mathrm{th}}=2$, there exists an optimal value for x_{ST} at which the outage probability of the primary network is lowest. In almost of the values of x_{ST} and R_{th}, the primary network in our scheme outperforms that in the DT protocol. This figure also presents that by placing the nodes ST at appropriate positions, the proposed method will provide high performance gain, as compared with the DT one. Again, the simulation and analytical results are in good agreement, which validates the correction of our derivations.

In Fig. 5, we investigate the impact of the hardware impairments on the performance of both networks. In this simulation, we assign the values to the parameters as follows: $R_{\mathrm{th}}=1$, $x_{\mathrm{ST}}=0.15$, $\alpha=0.2$, $\beta=0.9$, $\eta=0.75$, $M=3$ and $\Psi=5$ dB. We can see that the outage probability of the considered protocols increases with the increasing of the value κ. Moreover, the outage performance of the DT

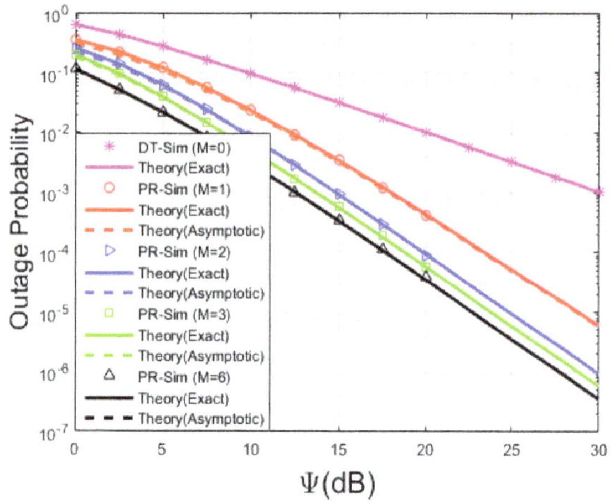

Fig. 2. Outage probability of the primary network as a function of the transmit SNR (Ψ) in dB when $R_{th} = 1$, $x_{ST} = 0.5$, $\kappa = 0.01$, $\alpha = 0.1$, $\beta = 0.95$, $\eta = 0.5$ and $M \in \{1, 2, 3, 6\}$.

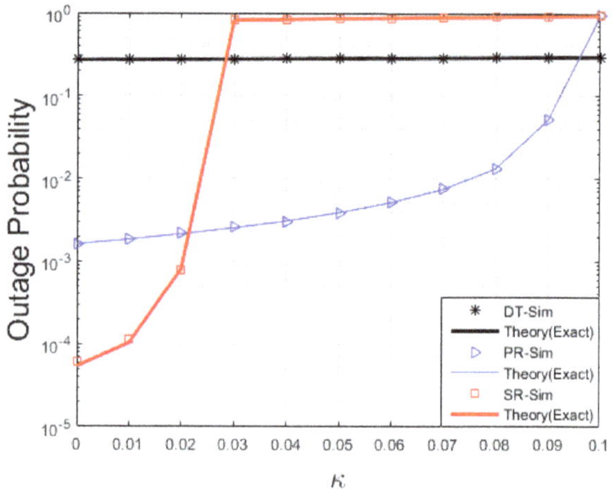

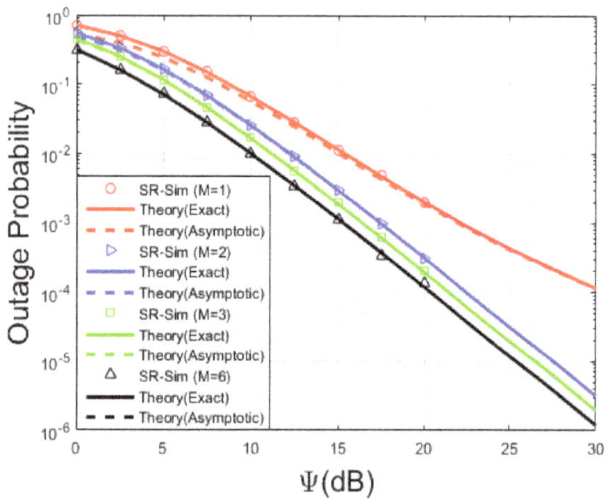

Fig. 3. Outage probability of the secondary network as a function of the transmit SNR (Ψ) in dB when $R_{th} = 1$, $x_{ST} = 0.5$, $\kappa = 0.01$, $\alpha = 0.1$, $\beta = 0.95$, $\eta = 0.5$ and $M \in \{1, 2, 3, 6\}$.

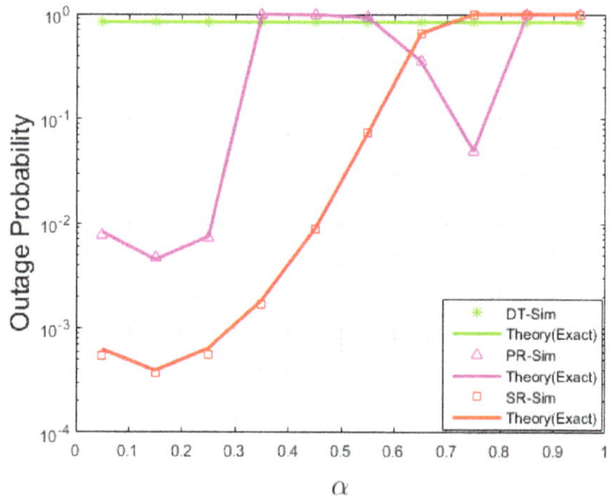

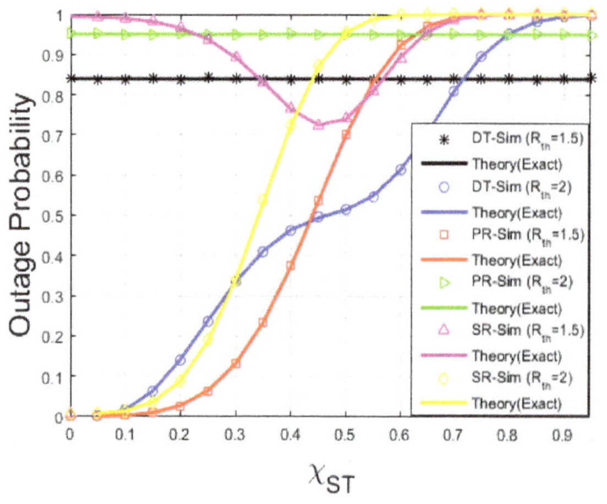

Fig. 4. Outage probability of the primary and secondary networks as a function of x_{ST} when $R_{th} \in \{1.5, 2\}$, $\kappa = 0$, $\alpha = 0.1$, $\beta = 0.95$, $\eta = 0.5$, $M = 2$ and $\Psi = 0$ dB.

Fig. 5. Outage probability of the primary and secondary networks as a function of κ when $R_{th} = 1$, $x_{ST} = 0.15$, $\alpha = 0.2$, $\beta = 0.9$, $\eta = 0.75$, $M = 3$ and $\Psi = 5$ dB.

Fig. 6. Outage probability of the primary and secondary networks as a function of α when $R_{th} = 1.5$, $x_{ST} = 0.1$, $\kappa = 0$, $\beta = 0.95$, $\eta = 1$, $M = 3$ and $\Psi = 0$ dB.

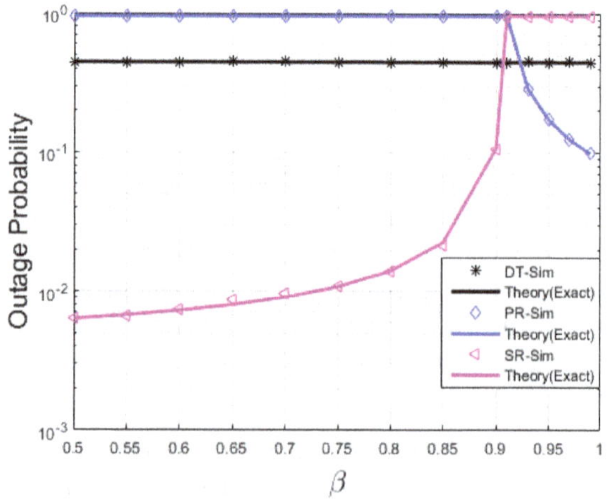

Fig. 7. Outage probability of the primary and secondary networks as a function of β when $R_{th} = 1.5$, $x_{ST} = 0.25$, $\kappa = 0.01$, $\alpha = 0.1$, $\eta = 0.25$, $M = 2$ and $\Psi = 5$ dB.

protocol only changes slightly, while that of the proposed scenario significantly degrades.

Figure 6 shows the impact of the fraction of time used for the energy harvesting time slot (α) on the outage performance with $R_{\text{th}} = 1.5$, $x_{\text{ST}} = 0.1$, $\kappa = 0$, $\beta = 0.95$, $\eta = 1$, $M = 3$ and $\Psi = 0$ dB. As seen from this figure, the performance of the primary and secondary networks varies with the change of the α. However, it can be observed that there exists the optimal value α^* so that the performance of the primary and secondary networks is best.

In Fig. 7, we investigate the impact of the fraction of the transmit power allocated to the primary signal (β) on the system performance. The simulation parameters of this figure are $R_{\text{th}} = 1.5$, $x_{\text{ST}} = 0.25$, $\kappa = 0.01$, $\alpha = 0.1$, $\eta = 0.25$, $M = 2$ and $\Psi = 5$ dB. We can see that the performance of the primary (secondary) network is better (worse) with high (low) β values. In this figure, the outage probability of the primary network (secondary) network almost equals 1 when β is less (higher) than 0.93 (0.91).

5. Conclusions

In this paper, we proposed an overlay spectrum access protocol to enhance the performance of the primary and secondary networks. The main contribution of this paper is to derive exact and lower-bound closed-form expressions of the outage probability, which were verified by computer simulations.

The results presented that by selecting appropriate parameters, the outage performance of both networks could be improved significantly. In particular, the proposed system can be optimized by appropriately designing the fraction of time block used for the energy harvesting process and the fraction of the transmit power allocated to the primary signal. In addition, increasing the number of the ST-SR pairs and selecting the STs with the optimal position could also enhance performance for both primary and secondary networks.

Acknowledgments

This research is funded by Vietnam National Foundation for Science and Technology Development (NAFOSTED) under grant number 102.01-2014.33.

References

[1] ZHOU, X.,ZHANG, R., HO, C. K. Wireless information and power transfer: Architecture design and rate-energy tradeoff. *IEEE Transactions on Communications*, Nov. 2013, vol. 61, no. 11, p. 4754–4767. DOI: 10.1109/TCOMM.2013.13.120855

[2] LANEMAN, J. N., TSE, D. N. C, WORNELL, G. W. Cooperative diversity in wireless networks: Efficient protocols and outage behavior. *IEEE Transactions on Information Theory*, Dec. 2004, vol. 50, no. 12, p. 3062–3080. DOI: 10.1109/TIT.2004.838089

[3] KRIKIDIS, I., TIMOTHEOU, S., SASAKI, S. RF energy transfer for cooperative networks: Data relaying or energy harvesting? *IEEE Communications Letter*, Nov. 2012, vol. 16, no. 11, p. 1772–1775. DOI: 10.1109/LCOMM.2012.091712.121395

[4] NASIR, A., DURRANI, S., KENNEDY, R. Relaying protocols for wireless energy harvesting and information processing. *IEEE Transactions on Wireless Communications*, Jul. 2013, vol. 12, no. 7, p. 3622–3636. DOI: 10.1109/TWC.2013.062413.122042

[5] KRIKIDIS, I., ZHENG, G., OTTERSTEN, B. Harvest-use cooperative networks with half/full-duplex relaying. In *Proceedings of the 2013 IEEE Wireless Communications and Networking Conference (WCNC2013)*. Shanghai (China), 2013, p. 4256–4260. DOI: 10.1109/WCNC.2013.6555261

[6] NASIR, A., DURRANI, S., KENNEDY, R. Throughput and ergodic capacity of wireless energy harvesting based DF relaying network. In *Proceedings of the 2014 IEEE International Conference on Communications (ICC2014)*. Sydney (Australia), June 2014, p. 4066–4071. DOI: 10.1109/ICC.2014.6883957

[7] DING, Z., PERLAZA, S., ESNAOLA, I., et al. Power allocation strategies in energy harvesting wireless cooperative networks. *IEEE Transactions on Wireless Communications*, Feb. 2014, vol. 13, no. 2, p. 846–860. DOI: 10.1109/TWC.2013.010213.130484

[8] MITOLA, J., MAQUIRE, G. Q. J. Cognitive radio: Making software radios more personal. *IEEE Personal Communications*, Aug. 1999, vol. 6, no. 4, p. 13–18. DOI: 10.1109/98.788210

[9] HAYKIN, S. Cognitive radio: Brain-empowered wireless communications. *IEEE Journal on Selected Areas in Communications*, Feb. 2005, vol. 23, no. 2, p. 201–220. DOI: 10.1109/JSAC.2004.839380

[10] HUANG, X. L., HU, F., WU, J., et al. Intelligent cooperative spectrum sensing via hierarchical Dirichlet process in cognitive radio networks. *IEEE Journal on Selected Areas in Communications*, Oct. 2014, vol. 33, no. 5, p. 771–787. DOI: 10.1109/JSAC.2014.2361075

[11] DUY, T. T., KONG, H. Y. Performance analysis of incremental amplify-and-forward relaying protocols with nth best partial relay selection under interference constraint. *Wireless Personal Communications*, Aug. 2013, vol. 71, no. 4, p. 2741–2757. DOI: 10.1007/s11277-012-0968-9

[12] DUY, T. T., KONG, H. Y. Adaptive cooperative decode-and-forward transmission with power allocation under interference constraint. *Wireless Personal Communications*, Jan. 2014, vol. 74, no. 2, p. 401–414. DOI: 10.1007/s11277-013-1292-8

[13] SIMEONE, O., STANOJEV, I., SAVAZZI, et al. Spectrum leasing to cooperating secondary ad hoc networks. *IEEE Journal on Selected Areas in Communications*, Jan. 2008, vol. 26, no. 1, p. 203–213. DOI: 10.1109/JSAC.2008.080118

[14] HAN, Y., PANDHARIPANDE, A., TING, S. H. Cooperative decode-and-forward relaying for secondary spectrum access. *IEEE Transactions on Wireless Communications*, Oct. 2009, vol. 8, no. 10, p. 4945-4950. DOI: 10.1109/TWC.2009.081484

[15] DUY, T. T., KONG, H. Y. Performance analysis of two-way hybrid decode-and-amplify relaying scheme with relay selection for secondary spectrum access. *Wireless Personal Communications*, Mar. 2013, vol. 69, no. 2, p. 857-878. DOI: 10.1007/s11277-012-0616-4

[16] BJORNSON, E., MATTHAIOU, M., DEBBAH, M. A new look at dual-hop relaying: Performance limits with hardware impairments. *IEEE Transactions on Communications*, Nov. 2013, vol. 61, no. 11, p. 4512–4525. DOI: 10.1109/TCOMM.2013.100913.130282

[17] MATTHAIOU, M., PAPADOGIANNIS, A. Two-way relaying under the presence of relay transceiver hardware impairments. *IEEE Communications Letters*, Jun. 2013, vol. 17, no. 6, p. 1136–1139. DOI: 10.1109/LCOMM.2013.042313.130191

[18] DUY, T. T., DUONG, T. Q., DA COSTA, D., et al. Proactive relay selection with joint impact of hardware impairment and co-channel interference. *IEEE Transactions on Communications*, May 2015, vol. 63, no. 5, p. 1594–1606. DOI: 10.1109/TCOMM.2015.2396517

[19] NGUYEN, D. K., MATTHAIOU, M., DUONG, T. Q., et al. RF energy harvesting two-way cognitive DF relaying with transceiver impairments. In *Proceedings of the 2015 IEEE Conference on Communication Workshop (ICCW2015)*. London (England), Jun. 2015, p. 1970–1975. DOI: 10.1109/ICCW.2015.7247469

[20] GAO, X., XU, W., LI, S., et al. An online energy allocation strategy for energy harvesting cognitive radio systems. In *Proceedings of the 8th International Conference on Wireless Communications and Signal Processing (WCSP2013)*. Hangzhou (China), Oct. 2013, p. 1–5. DOI: 10.1109/WCSP.2013.6677085

[21] PARK, S., HONG, D. Optimal spectrum access for energy harvesting cognitive radio networks. *IEEE Transactions on Wireless Communications*, Dec. 2013, vol. 12, no. 12, p. 6166–6179. DOI: 10.1109/TWC.2013.103113.130018

[22] MOUSAVIFAR, S., LIU, Y., LEUNG, C., et al. Wireless energy harvesting and spectrum sharing in cognitive radio. In *Proceedings of the 80th IEEE Vehicular Technology Conference (VTC Fall)*. Vancouver (Canada), Sep. 2014, p. 1–5. DOI: 10.1109/VTCFall.2014.6966232

[23] SON, P. N., KONG, H. Y. Exact outage analysis of energy harvesting underlay cooperative cognitive networks. *IEICE Transactions on Communications*, 2015, vol. E98-B, no. 4, p. 661–672. DOI: 10.1587/transcom.E98.B.661

[24] WANG, Z., CHEN, Z., LUO, L., et al. Outage analysis of cognitive relay networks with energy harvesting and information transfer. In *Proceedings of the 2014 IEEE International Conference on Communications (ICC2014)*. Sydney (Australia), Jun. 2014, p. 4348–4353. DOI: 10.1109/ICC.2014.6884004

[25] SON, P. N., HAR, D., KONG, H. Y. Joint power allocation for energy harvesting and power superposition coding in cooperative spectrum sharing. *Computing Research Repository*. [Online] Cited 2015. Available at: arxiv.org/abs/1510.02460

[26] ZHAI, C., LIU, J., ZHENG, L. Relay based spectrum sharing with secondary users powered by wireless energy harvesting. *IEEE Transactions on Communications*, May 2016, vol. 64, no. 5, p. 1875–1887. DOI: 10.1109/TCOMM.2016.2542822

[27] LI, Y. B., YANG, R., LIN, Y., et al. The spectrum sharing in cognitive radio networks based on competitive price game. *Radioengineering*, Sep. 2012, vol. 21, no. 3, p. 802–808. ISSN: 1805-9600

[28] DUY, T. T., KONG, H. Secrecy performance analysis of multihop transmission protocols in cluster networks. *Wireless Personal Communications*, Jun. 2015, vol. 82, no. 4, p. 2505–2518. DOI: 10.1007/s11277-015-2361-y

[29] GRADSHTEYN, I. S., RYZHIK, I. M. *Table of Integrals, Series, and Products*. 6th ed. San Diego, CA (USA): Academic Press, 2000. ISBN: 978-0-12-294757-5

About the Authors . . .

Tan N. NGUYEN was born in 1986 in Nha Trang City, Vietnam. He received B.S. and M.S. degrees in Electronics and Telecommunications Engineering from Ho Chi Minh University of Natural Sciences, Ho Chi Minh City, Vietnam in 2008 and 2012, respectively. In 2013, he joined the Faculty of Electrical and Electronics Engineering of Ton Duc Thang University, Vietnam as a lecturer. He is currently pursuing his Ph.D. degree in Electrical Engineering at VSB Technical University of Ostrava, Czech Republic. His major interests are cooperative communications, cognitive radio, and physical layer security.

Tran Trung DUY was born in Nha Trang city, Vietnam, in 1984. He received the B.E. degree in Electronics and Telecommunications Engineering from the French-Vietnamese training program for excellent engineers (PFIEV), Ho Chi Minh City University of Technology, Vietnam in 2007. In 2013, he received the Ph.D degree in electrical engineering from University of Ulsan, South Korea. In 2013, he joined the Department of Telecommunications, Posts and Telecommunications Institute of Technology (PTIT), as a lecturer. His major research interests are cooperative communications, cognitive radio, and physical layer security.

Gia-Thien LUU was born in 1981, in Quang Nam Province, Viet Nam. He graduated in Physics from Ho Chi Minh City University of Pedagogy, Viet Nam in 2003. In 2007, he obtained his MSc degree in Laser technology from Ho Chi Minh City University of Technology. He obtained Ph.D degree from Orleans University, France in 2013 with the thesis on the development of methods for time delay estimation of the electromyography signals. He is currently an Assistant Professor at Post and Telecommunication Institute of Technology. His main interests concern the development of methods in signal processing and their applications to the biomedical engineering and Telecommunication.

Phuong T. TRAN (corresponding author) was born in 1979 in Ho Chi Minh City, Vietnam. He received B.Eng. and M.Eng degrees in Electrical Engineering from Ho Chi Minh University of Technology, Ho Chi Minh City, Vietnam in 2002 and 2005, respectively. In 2007, he became a Vietnam Education Foundation Fellow at Purdue University, U.S.A., where he received his Ph.D. degree in Electrical and Computer Engineering in 2013. In 2013, he joined the Faculty of Electrical and Electronics Engineering of Ton Duc Thang University, Vietnam and served as the Vice Dean of Faculty since October 2014. His major interests are in the area of wireless communications and network information theory.

Miroslav VOZNAK born in 1971 is an associate professor with the Department of Telecommunications, Technical University of Ostrava, Czech Republic and foreign professor with Ton Duc Thang University in Ho Chi Minh City, Vietnam. He received his Ph.D. degree in telecommunications in 2002 at the Technical University of Ostrava. He is a senior researcher in the Supercomputing center IT4Innovations in Ostrava, Czech Republic, a member of editorial boards of several journals and boards of conferences. Topics of his research interests are IP telephony, wireless networks, speech quality and network security.

Appendix A: A Detailed Derivation of (24)

At first, we calculate $\mathrm{OP}_{\mathrm{PR}}^{1}$ in (24). Under the condition $\theta < \beta/(1-\beta+\kappa)$, the probability $\Pr[C_0 < R_{\mathrm{th}}]$, $\Pr[C_{1b} < R_{\mathrm{th}}]$, $\Pr[N_{\mathrm{R}} = 0]$, $\Pr[N_{\mathrm{T}} = 0]$ and $\Pr[N_{\mathrm{R}} = m]$ can be computed, respectively as

$$\begin{aligned}
\Pr[C_0 < R_{\mathrm{th}}] &= 1 - \mathrm{e}^{-\lambda_0\rho_0},\\
\Pr[C_{1b} < R_{\mathrm{th}}] &= \left(1 - \mathrm{e}^{-\lambda_1\rho_0}\right)^m,\\
\Pr[N_{\mathrm{R}} = 0] &= \left(1 - \mathrm{e}^{-\lambda_2\rho_0}\right)^M,\\
\Pr[N_{\mathrm{T}} = 0] &= \left(1 - \mathrm{e}^{-\lambda_1\rho_0}\right)^M,\\
\Pr[N_{\mathrm{R}} = m] &= \left(1 - \mathrm{e}^{-\lambda_2\rho_0}\right)^{M-m}\mathrm{e}^{-m\lambda_1\rho_0}. \qquad \text{(A.1)}
\end{aligned}$$

Next, considering the probability $\Pr\left[\left\{\begin{array}{l} N_T = u-1 \\ C_{1c} \geq R_{th} \\ C_{3c} < R_{th}\end{array}\right.\right]$ which can be given by

$$\begin{aligned}
&\Pr[N_{\mathrm{T}} = u-1, C_{1c} \geq R_{\mathrm{th}}, C_{3c} < R_{\mathrm{th}}]\\
&= \left(1 - \mathrm{e}^{-\lambda_1\rho_0}\right)^{M-u}\mathrm{e}^{-(u-1)\lambda_1\rho_0}\Pr\left[\gamma_{1c} \geq \rho_0, \mu\gamma_{1c}\gamma_{3c} < \rho_0\right]\\
&= \left(1 - \mathrm{e}^{-\lambda_1\rho_0}\right)^{M-u}\mathrm{e}^{-(u-1)\lambda_1\rho_0}\\
&\times \int_{\rho_0}^{+\infty} f_{\gamma_{1c}}(x)F_{\gamma_{3c}}\left(\frac{\rho_0}{\mu x}\right)\mathrm{d}x. \qquad \text{(A.2)}
\end{aligned}$$

By using (17) for the PDF $f_{\gamma_{1c}}(x)$ and (18) for the CDF $F_{\gamma_{3c}}(\rho_0/x\mu)$, we obtain (A.3) by

$$\begin{aligned}
&\Pr[N_{\mathrm{T}} = u-1, C_{1c} \geq R_{\mathrm{th}}, C_{3c} < R_{\mathrm{th}}]\\
&= \left(1 - \mathrm{e}^{-\lambda_1\rho_0}\right)^{M-u}\mathrm{e}^{-(u-1)\lambda_1\rho_0}\\
&\times \sum_{v=0}^{u}(-1)^v C_u^v \int_{\rho_0}^{+\infty}\lambda_1\mathrm{e}^{-\lambda_1 x}\mathrm{e}^{-\frac{v\lambda_3\rho_0}{\mu x}}\mathrm{d}x. \qquad \text{(A.3)}
\end{aligned}$$

Next, we can formulate the probability $\Pr\left[\left\{\begin{array}{l} C_{1b} \geq R_{\mathrm{th}} \\ C_{3b} < R_{\mathrm{th}}\end{array}\right.\right]$ as

$$\begin{aligned}
\Pr[C_{1b} \geq R_{\mathrm{th}}, C_{3b} < R_{\mathrm{th}}] &= \Pr\left[\gamma_{1b} \geq \rho_0, \mu\gamma_{1b}\gamma_{3b} < \rho_1\right]\\
&= \int_{\rho_0}^{+\infty} f_{\gamma_{1b}}(x)F_{\gamma_{3b}}\left(\frac{\rho_1}{\mu x}\right)\mathrm{d}x. \qquad \text{(A.4)}
\end{aligned}$$

Combining (17), (19) and (A.4), and after some manipulations, we arrive at

$$\begin{aligned}
&\Pr[C_{1b} \geq R_{\mathrm{th}}, C_{3b} < R_{\mathrm{th}}] = \sum_{t=0}^{m-1}(-1)^t C_{m-1}^t m\lambda_1\\
&\times\left[\frac{\mathrm{e}^{-(t+1)\lambda_1\rho_0}}{(t+1)\lambda_1} - \int_{\rho_0}^{+\infty}\mathrm{e}^{-(t+1)\lambda_1 x}\mathrm{e}^{-\frac{\lambda_3\rho_1}{\mu x}}\mathrm{d}x\right]. \qquad \text{(A.5)}
\end{aligned}$$

Then, substituting (A.1), (A.3) and (A.5) into (23), we obtain $\mathrm{OP}_{\mathrm{PR}}^{1}$ as expressed in (31).

Next, when $\beta/(1-\beta+\kappa) < \theta < 1/\kappa$, it is obvious that

$$\begin{aligned}
&\Pr[C_{1b} \geq R_{\mathrm{th}}, C_{3b} < R_{\mathrm{th}}]\\
&= \Pr[C_{1b} \geq R_{\mathrm{th}}] = 1 - \left(1 - \mathrm{e}^{-\lambda_1\rho_0}\right)^m. \qquad \text{(A.6)}
\end{aligned}$$

Combining (A.1), (A.3), (A.6) and (23), the probability $\mathrm{OP}_{\mathrm{PR}}^{2}$ can be obtained as in (32).

Finally, when $\theta \geq 1/\kappa$, we can observe that the primary network is always in outage, i.e., $P_{\mathrm{PR}}^{\mathrm{out}} = 1$.

Appendix B: A Detailed Derivation of (27)

At first, we consider the first case: $\theta < (1-\beta)/\kappa$. In this case, by using (21), we obtain

$$\begin{aligned}
\Pr[C_{4a} < R_{\mathrm{th}}] &= \Pr\left[\gamma_{1a}\gamma_{4a} \leq \frac{\rho_0}{\mu}\right]\\
&= \left(1 - \sqrt{\frac{4\lambda_1\lambda_4\rho_0}{\mu}}K_1\left(\sqrt{\frac{4\lambda_1\lambda_4\rho_0}{\mu}}\right)\right)^M. \qquad \text{(B.1)}
\end{aligned}$$

Similar to (A.5), we have

$$\begin{aligned}
&\Pr[C_{1b} \geq R_{\mathrm{th}}, C_{4b} < R_{\mathrm{th}}] = \sum_{t=0}^{m-1}(-1)^t C_{m-1}^t m\lambda_1\\
&\times\left[\frac{\mathrm{e}^{-(t+1)\lambda_1\rho_0}}{(t+1)\lambda_1} - \int_{\rho_0}^{+\infty}\mathrm{e}^{-(t+1)\lambda_1 x}\mathrm{e}^{-\frac{\lambda_4\rho_2}{\mu x}}\mathrm{d}x\right]. \qquad \text{(B.2)}
\end{aligned}$$

Substituting $\Pr[C_0 \geq R_{\mathrm{th}}] = \mathrm{e}^{-\lambda_0\rho_0}$, (A.1), (B.1) and (B.2) into (27), the outage probability $\mathrm{OP}_{\mathrm{SR}}^{1}$ can be obtained by (35).

Let us consider the second case where $(1-\beta)/\kappa \leq \theta < 1/\kappa$, similar to (A.6), we can obtain $\mathrm{OP}_{\mathrm{PR}}^{2}$, as given in (36).

Permissions

All chapters in this book were first published in RADIOENGINEERING, by Spolecnost pro radioelektronicke inzenyrstvi[CZ]; hereby published with permission under the Creative Commons Attribution License or equivalent. Every chapter published in this book has been scrutinized by our experts. Their significance has been extensively debated. The topics covered herein carry significant findings which will fuel the growth of the discipline. They may even be implemented as practical applications or may be referred to as a beginning point for another development.

The contributors of this book come from diverse backgrounds, making this book a truly international effort. This book will bring forth new frontiers with its revolutionizing research information and detailed analysis of the nascent developments around the world.

We would like to thank all the contributing authors for lending their expertise to make the book truly unique. They have played a crucial role in the development of this book. Without their invaluable contributions this book wouldn't have been possible. They have made vital efforts to compile up to date information on the varied aspects of this subject to make this book a valuable addition to the collection of many professionals and students.

This book was conceptualized with the vision of imparting up-to-date information and advanced data in this field. To ensure the same, a matchless editorial board was set up. Every individual on the board went through rigorous rounds of assessment to prove their worth. After which they invested a large part of their time researching and compiling the most relevant data for our readers.

The editorial board has been involved in producing this book since its inception. They have spent rigorous hours researching and exploring the diverse topics which have resulted in the successful publishing of this book. They have passed on their knowledge of decades through this book. To expedite this challenging task, the publisher supported the team at every step. A small team of assistant editors was also appointed to further simplify the editing procedure and attain best results for the readers.

Apart from the editorial board, the designing team has also invested a significant amount of their time in understanding the subject and creating the most relevant covers. They scrutinized every image to scout for the most suitable representation of the subject and create an appropriate cover for the book.

The publishing team has been an ardent support to the editorial, designing and production team. Their endless efforts to recruit the best for this project, has resulted in the accomplishment of this book. They are a veteran in the field of academics and their pool of knowledge is as vast as their experience in printing. Their expertise and guidance has proved useful at every step. Their uncompromising quality standards have made this book an exceptional effort. Their encouragement from time to time has been an inspiration for everyone.

The publisher and the editorial board hope that this book will prove to be a valuable piece of knowledge for researchers, students, practitioners and scholars across the globe.

List of Contributors

Hulya Ozturk
Dept. of Mathematics, Gebze Technical University, Gebze, 41400 Kocaeli, Turkey

Korkut Yegin
Dept. of Electrical and Electronics Engineering, Ege University, Bornova, 35100 Izmir, Turkey

Richa Barsainya and Tarun K. Rawat
Division of Electronics and Communication, Netaji Subhas Institute of Technology, Sector-3, Dwarka, 110078 New Delhi, India

Kavishaur Dwarika
School of Engineering, University of KwaZulu-Natal, King George V Avenue, 4041 Durban, South Africa

Hongjun Xu
School of Engineering, University of KwaZulu-Natal, King George V Avenue, 4041 Durban, South Africa
School of Information and Electronic Engineering, Zhejiang Gongshang University, P.R. China

Huijun Hou
School of Electronics and Information Engineering, Harbin Institute of Technology, Harbin 150001, P. R. China

Xingpeng Mao
School of Electronics and Information Engineering, Harbin Institute of Technology, Harbin 150001, P. R. China
Collaborative Innovation Center of Information Sensing and Understanding, Harbin Institute of Technology, Harbin 150001, P. R. China

Bakhtiar Ali, Nida Zamir and Muhammad Fasih Uddin Butt
Department of Electrical Engineering, COMSATS Institute of Information Technology Islamabad, Pakistan

Soon Xin Ng
Department of Electronics and Computer Science, University of Southampton, Southampton SO17 1BJ, United Kingdom

Lukas Sekanina, Zdenek Vasicek, Vojtech Mrazek
Faculty of Information Technology, IT4Innovations Centre of Excellence, Brno University of Technology, Czech Republic

Satyendra Kumar and Hariom Gupta
Dept. of Electronics and Communication Engineering, Jaypee Institute of Information Technology - Noida, India

Kaushik Saha
Samsung R&D Institute India - Delhi, India

Jan Marek, Jiri Hospodka and Ondrej Subrt
Dept. of Circuit Theory, Czech Technical University in Prague, Technická 2, 166 27 Praha, Czech Republic

Ranko Vojinovic
Faculty of Information Technology, Bulevar Sv. Petra Cetinjskog 22, 81000, Podgorica, Montenegro

Milos Dakovic
University of Montenegro, Faculty of Electrical Engineering, Dz. Vasingtona bb, 81000, Podgorica, Montenegro

Ning Tai, Chao Wang, Weiwei Wu and Naichang Yuan
College of Electronic Science and Engineering, National University of Defense Technology, Changsha, Hunan, China

Liguo Liu
College of Electronic Engineering, Naval University of Engineering, Wuhan, Hubei, China

Muhammad Imran and Bruce A. Harvey
Department of Electrical & Computer Engineering, College of Engineering Florida State University, 2525 Pottsdamer St, Tallahassee, FL 32310, USA

Tan N. Nguyen and Phuong T. Tran
Wireless Communications Research Group, Faculty of Electrical and Electronics Engineering, Ton Duc Thang University, No. 19 Nguyen Huu Tho Street, Tan Phong Ward, District 7, Ho Chi Minh City, Vietnam

Hoang-Sy Nguyen and Miroslav Voznak
Wireless Communications Research Group, Faculty of Electrical and Electronics Engineering, Ton Duc Thang University, No. 19 Nguyen Huu Tho Street, Tan Phong Ward, District 7, Ho Chi Minh City, Vietnam
VSB Technical University of Ostrava, 17. listopadu 15/2172, 708 33 Ostrava - Poruba, Czech Republic

Dinh-Thuan Do
Faculty of Electronics Technology, Industrial University of Ho Chi Minh City, Vietnam

Dinh-Thuan Do
Faculty of Electronics Technology, Industrial University of Ho Chi Minh City, 12 Nguyen Van Bao St., Go Vap Dist., Ho Chi Minh City, Vietnam

Hoang-Sy Nguyen
Wireless Communications Research Group, Faculty of Electrical and Electronics Engineering, Ton Duc Thang University, 19 Nguyen Huu Tho St., Tan Phong Ward, Dist. 7, Ho Chi Minh City, Vietnam

Miroslav Voznak
Wireless Communications Research Group, Faculty of Electrical and Electronics Engineering, Ton Duc Thang University, 19 Nguyen Huu Tho St., Tan Phong Ward, Dist. 7, Ho Chi Minh City, Vietnam
Faculty of Electrical Engineering and Computer Science, Technical University of Ostrava, 17. listopadu 2172/15, 708 33 Ostrava-Poruba, Czech Republic

Thanh-Sang Nguyen
Binh Duong University, Binh Duong Province, Vietnam

Shourong Hou, Yujie Zhou, Hongming Liu and Nianhao Zhu
Dept. of Electronic Engineering, Shanghai Jiao Tong University, 800 Dongchuan Road, Shanghai, China

Jan Kubak and Pavel Sovka
Dept. of Circuit Theory, Faculty of Electrical Engineering, Czech Technical University in Prague

Miroslav Vlcek
Dept. of Applied Mathematics, Faculty of Transportation Sciences, Czech Technical University in Prague

Kriti Garg and Dharmendra K. Upadhyay
Dept. of Electronics and Communication Engineering, Netaji Subhas Institute of Technology, Sector-3, Dwarka, 110078 New Delhi, India

Jiri Milos, Ladislav Polak, Stanislav Hanus and Tomas Kratochvil
Dept. of Radio Engineering, Brno University of Technology, Technická 3082/12, 616 00 Brno, Czech Republic

Gnane Swarnadh Satapathi and Srihari Pathipati
Dept. of Electronics and Communication Engineering, National Institute of Technology Karnataka, Mangalore, 575 025 Karnataka, India

Ha Hoang Kha
Faculty of Electrical and Electronics Engineering, Ho Chi Minh City University of Technology, VNU-HCM, Vietnam

Hung Quang Ta
Faculty of Information Technology, Hanoi University, Vietnam

Tran Trung Duy and Gia-Thien Luu
Posts and Telecommunications Institute of Technology, 11 Nguyen Dinh Chieu St., Dist. 1, Ho Chi Minh City, Vietnam

Miroslav Voznak
Wireless Communications Research Group, Faculty of Electrical and Electronics Engineering, Ton Duc Thang University, No. 19 Nguyen Huu Tho Street, Tan Phong Ward, District 7, Ho Chi Minh City, Vietnam
VSB Technical University of Ostrava, 17. listopadu 15/2172, 708 33 Ostrava - Poruba, Czech Republic

Index

www.ingramcontent.com/pod-product-compliance
Lightning Source LLC
Chambersburg PA
CBHW080656200326
41458CB00013B/4878
9781632386656